PROPERTY OF

SIGMA-ALDRICH COMPANY LTD

NMR Shift Reagents

Author
Thomas J. Wenzel, Ph.D.
Assistant Professor of Chemistry
Bates College
Lewiston, Maine

CRC Press, Inc.
Boca Raton, Florida

Library of Congress Cataloging-in-Publication Data

Wenzel, Thomas J.
  NMR shift reagents.

  Bibliography: p.
  Includes index.
  1. Lanthanide shift reagents. 2. Nuclear magnetic
resonance spectroscopy. 3. Stereochemistry.
I. Title.
QD77.W49 1987   543'.0877   87-6559
ISBN 0-8493-5298-3

This book represents information obtained from authentic and highly regarded sources. Reprinted material is quoted with permission, and sources are indicated. A wide variety of references are listed. Every reasonable effort has been made to give reliable data and information, but the author and the publisher cannot assume responsibility for the validity of all materials or for the consequences of their use.

All rights reserved. This book, or any parts thereof, may not be reproduced in any form without written consent from the publisher.

Direct all inquiries to CRC Press, Inc., 2000 Corporate Blvd., N.W., Boca Raton, Florida, 33431.

© 1987 by CRC Press, Inc.

International Standard Book Number 0-8493-5298-3

Library of Congress Card Number 87-6559
Printed in the United States

FOREWORD

This book is intended to serve as a practical guide and source book for users of lanthanide NMR shift reagents. The book purposefully focuses on lanthanide shift reagents because of their overwhelming popularity and utility. Four distinct classes of lanthanide shift reagents are covered: achiral lanthanide tris β-diketonates, chiral lanthanide tris β-diketonates, binuclear lanthanide(III)-silver(I) complexes, and water-soluble lanthanide complexes.

Chapter 2 provides an extensive description of the use of lanthanide shift reagents for the study of achiral substrates. Most of the work described in this chapter has been performed using achiral lanthanide tris β-diketonates. The selection process for application and the experimental techniques necessary to purify and employ the lanthanide tris β-diketonates are described in detail. A thorough review of the range of achiral compounds that have been studied with the aid of lanthanide shift reagents is provided. The compounds are categorized according to the functional group responsible for bonding with the shift reagent. They are described in order of increasing structural complexity (i.e., aliphatic acyclic, olefinic acyclic, monocyclic, bicyclic, etc.). Achiral polyfunctional substrates are described separately from monofunctional substrates and are categorized under the class of compound to which the bonding functional group belongs.

Chapter 3 describes the application of lanthanide shift reagents to the study of chiral substrates. A discussion of the experimental techniques for handling and using chiral lanthanide tris β-diketonates is included. Procedures that successfully employ lanthanide shift reagents in assigning absolute configurations of compounds are described. A review of the range of compounds for which chiral lanthanide shift reagents have been used in the determination of enantiomeric excess is also provided. Keeping with the format of the previous chapter, the compounds are categorized according to the functional group responsible for bonding; however, mono- and polyfunctional substrates have not been discussed separately.

Binuclear lanthanide(III)-silver(I) complexes that are effective shift reagents for olefins, aromatics, certain halogenated compounds, phosphines, and organic salts are described in Chapter 4. The criteria for the selection and use of achiral and chiral analogs are discussed. Specific compounds that have been studied using these reagents are reviewed. Water-soluble lanthanide shift reagents and their applications are described in Chapter 6. The use of these reagents with biologically important compounds including amino acids and proteins, nucleotides, carbohydrates, and membranes are described. The development and utility of water-soluble shift reagents for cations, and of chiral water-soluble reagents, are also covered.

Since the theory of NMR shift reagents is described in several excellent review articles, the discussion of this topic (Chapter 5) has been kept to a minimum. This chapter's principal focus is on the validity of structural information obtained by fitting lanthanide shift data using the simplified dipolar shift equation. The assumptions necessary to perform such a structural analysis, and the likelihood of their being met with lanthanide shift reagents, are discussed. The methodology for assessing certain variables in the fitting process has been refined and is presented as such. In cases where conflicting opinions remain, I have opted to report all sides of the issue, rather than to defend one view over the other.

I am grateful to many people who provided assistance and support while I worked on this project. I would like to thank Bates College for providing me with funds to offset the costs incurred in researching and writing this work. I also want to thank Joan for her unending patience, and Dave for his unending encouragement. Finally, I am deeply indebted to Bob Sievers for having the confidence in me to successfully undertake such a project.

Thomas J. Wenzel

THE AUTHOR

Thomas J. Wenzel, Ph.D. is Assistant Professor of Chemistry at Bates College at Lewiston, Maine.

Dr. Wenzel received his B.S. degree from Northeastern University, Boston, Massachusetts in 1976. He obtained his Ph.D. in 1981 from the University of Colorado, Boulder, Colorado. He was appointed Assistant Professor of Chemistry at Bates College in 1981.

Dr. Wenzel is a member of the American Chemical Society and Phi Kappa Phi. He serves on the board of directors of the Natural Resources Council of Maine. He received a National American Chemical Society Analytical Chemistry Summer Fellowship (1980) and the University of Colorado Award for Creative Research (1981).

Dr. Wenzel has been the recipient of research grants from the National Science Foundation, Research Corporation, and the Petroleum Research Fund. He has received grants for educational endeavors from the National Science Foundation and the Pittsburgh Conference and Exposition. He currently carries out research with the aid of undergraduate students in the areas of NMR shift reagents, gas chromatography, and liquid chromatography.

To Joan and Erica

TABLE OF CONTENTS

Chapter 1
Introduction
I. Structural Studies 2
II. Future Prospects 3

Chapter 2
Achiral Lanthanide Chelates
I. Shift Reagent Selection 5
   A. Metal 5
   B. Ligand 7
II. Experimental Techniques 9
   A. Solvent 9
   B. Shift Reagent Preparation and Purification 10
   C. Scavengers 11
   D. Procedures for Use of LSR 12
   E. Recovery of the Substrate 13
   F. High Field NMR 13
III. Applications 14
   A. Monofunctional Substrates 14
      1. Alcohols and Phenols 14
      2. Ketones 23
      3. Esters 31
      4. Lactones 35
      5. Aldehydes 36
      6. Carboxylic Acids 37
      7. Anhydrides 38
      8. Ethers 38
      9. Peroxides 41
      10. Amines 41
      11. Nitrogen Heterocycles 44
      12. Nitriles 46
      13. Amides 47
      14. Lactams and Imides 49
      15. Carbamates 50
      16. Nitrogen Oxides 50
      17. Nitrosos 51
      18. Azoxys 52
      19. Nitros 52
      20. Isocyanates 52
      21. Oximes 53
      22. Azos and Azides 53
      23. Imines 54
      24. Substrates with a Sulfur Atom 54
         a. Sulfoxides 55
         b. Sulfones and Sulfanate Esters 56
         c. Sulfites, Sultones, and Sulfate Diesters 56
         d. Sulfines 57
         e. Thiolsulfinates, Sulfinamates, and Thiacarbamates 58
         f. Sulfides and Thiols 58

			g.	Thiocyanates and Isothiocyanates............................. 59
			h.	Thioamides, Thioketones, Thioacetates, and Thiocarbamates... 59
		25.		Substrates with a Phosphorus Atom............................. 60
			a.	Phosphates 60
			b.	Phosphonates, Phosphorinanes, and Oxazophospholines 60
			c.	Phosphine Oxides 61
			d.	Phosphoranes.. 63
			e.	Phosphites... 63
			f.	Phosphines 64
			g.	Phosphoramides 64
		26.		Organohalides... 64
	B.			Organic Salts... 64
	C.			Metal Complexes and Substrates with a Silicon Atom............... 66
		1.		Metal β-Diketonates 66
		2.		Organometallics.. 68
		3.		Silicon and Germanium-Containing Compounds 70
	D.			Polyfunctional Substrates.. 71
		1.		Alcohols 72
		2.		Ketones 79
		3.		Esters 87
		4.		Lactones 92
		5.		Aldehydes 94
		6.		Acid Chlorides... 95
		7.		Ethers.. 95
		8.		Peroxides 99
		9.		Amines 99
			a.	Primary... 99
			b.	Secondary 100
			c.	Tertiary.. 101
		10.		Nitrogen Heterocycles 102
		11.		Nitriles... 108
		12.		Amides 109
		13.		Lactams and Imides.. 112
		14.		Nitrogen Oxides, Nitrones, and Azoxys 115
		15.		Nitros 116
		16.		Oximes 116
		17.		Azos, Azines, and Azides 117
		18.		Imines 118
		19.		Substrates with a Sulfur Atom 118
		20.		Substrates with a Phosphorus Atom......................... 120
	E.			Polymers 121
IV.				Influences of LSR on Coupling Constants 123
V.				Separation of Diastereotopic Protons.................................. 124
VI.				Secondary Deuterium Isotope Effects 124
VII.				Chemically Induced Dynamic Nuclear Polarization 125

Chapter 3
Studies of Chiral Substrates with Lanthanide Tris Chelates
I.		Shift Reagent Selection 127
	A.	Metal... 127

	B.	Ligands 127
II.	Experimental Techniques 129
	A.	Solvent 129
	B.	Preparation and Purification 129
	C.	Procedures for Using Chiral LSR.. 130
III.	Assignment of Absolute Configuration... 131
	A.	Circular Dichroism Spectra 131
	B.	MTPA Esters... 131
	C.	Camphanate Esters 133
	D.	Preparation of Amides... 134
	E.	Other Derivatives... 134
IV.	Determination of Optical Purity... 135
	A.	Enantiomeric Resolution with Achiral LSR.................................... 135
	B.	Application of Chiral LSR 136
		1. Alcohols 136
		2. Ketones 140
		3. Esters 142
		4. Lactones 145
		5. Aldehydes 145
		6. Carboxylic Acids 145
		7. Ethers 146
		8. Amines 147
		9. Nitrogen Heterocycles 148
		10. Amides 149
		11. Lactams 149
		12. Nitrosoamines.. 150
		13. Organoborons 150
		14. Substrates with a Sulfur Atom 150
		15. Substrates with a Phosphorus Atom................................... 152
		16. Metal Complexes 153
		17. Organic Salts 154

Chapter 4
Binuclear Lanthanide(III)-Silver(I) Shift Reagents

I.	Shift Reagent Selection 155
	A.	Metal... 155
	B.	Ligand... 156
II.	Experimental Techniques 157
	A.	Solvent 157
	B.	Shift Reagent Preparation and Purification 157
	C.	Use of Binuclear Reagents 158
III.	Structure and Theory... 158
IV.	Applications 159
	A.	Olefins 159
	B.	Alkynes 160
	C.	Aromatics 160
	D.	Halogenated Compounds 161
	E.	Phosphines 162
	F.	Polyfunctional Substrates 162
	G.	Multifunctional Substrates 163
	H.	Organic Salts 165
	I.	Chiral Binuclear Reagents 167

Chapter 5
Theory of Lanthanide Shift Reagents

I. Shift Mechanism ... 169
    A. Pseudocontact and Contact Shifts ... 169
    B. Pseudocontact Shift Equation ... 169
II. Separation of Dipolar and Contact Shifts ... 171
III. Stoichiometry and Symmetry ... 172
    A. X-Ray Crystallography ... 173
    B. Low-Temperature NMR Studies ... 174
    C. Vapor Phase Osmometry ... 174
    D. Infrared Spectroscopy ... 174
    E. Optical Spectroscopy ... 175
    F. Luminescence Spectroscopy ... 175
    G. Circular Dichroism ... 175
    H. Mass Spectrometry ... 175
    I. Graphical or Iterative Procedures ... 176
    J. Conclusions and "Effective Axial Symmetry" ... 176
IV. Application of the Pseudocontact Shift Equation ... 177
    A. Complexation Shifts ... 178
    B. Bound Shifts ... 178
    C. Location of the Principal Magnetic Axis ... 180
    D. Atomic Coordinates and L-S Bond Length ... 180
    E. Significance Testing ... 181
    F. Limitations of Configurational and Conformational Assignment ... 181
V. Relaxation Phenomena with LSR ... 183

Chapter 6
Water-Soluble Lanthanide Shift Reagents

I. Shift Reagent Selection ... 185
II. Applications to Biological Systems ... 186
    A. Amino Acids, Peptides, and Proteins ... 186
    B. Nucleotides and Nucleic Acids ... 190
    C. Carbohydrates ... 191
    D. Membranes ... 193
    E. Cations ... 194
III. Other Applications ... 197
    A. Carboxylates ... 197
    B. Phenols ... 200
    C. Oxides ... 200
    D. Amines ... 201
    E. Nitrogen Heterocycles ... 201
    F. Substrates with a Sulfur Atom ... 202
    G. Organometallics ... 202
IV. Chiral Shift Reagents ... 202

References ... 205

Index ... 269

Chapter 1

INTRODUCTION

Lanthanide NMR shift reagents were first introduced by Hinckley in 1969.[1] In his original report, the dipyridine adduct of tris(2,2,6,6-tetramethyl-3,5-heptanedionato) europium, $Eu(dpm)_3py_2$, was employed as the NMR shift reagent. It was later shown that the anhydrous lanthanide complexes of dpm, $Ln(dpm)_3$, were more effective as NMR shift reagents than the pyridine adduct.[2] The lanthanide tris β-diketonates have three properties that make them especially useful as NMR shift reagents.

The first is that they are soluble in most organic solvents including chloroform, carbon tetrachloride, carbon disulfide, and benzene. The second is that the lanthanide ion in a tris β-diketonate complex can expand its coordination number by bonding to suitable electron pair donors. The lanthanide ion is a hard Lewis acid and therefore forms donor-acceptor complexes with hard Lewis bases. Most oxygen- and nitrogen-containing compounds are examples of hard Lewis bases. The exchange of donor between its complexed and uncomplexed form is usually rapid on the NMR time scale. The spectrum of the donor is therefore a time average of its complexed and uncomplexed form.

The final property that distinguishes lanthanide shift reagents (LSR) from transition metal analogs is that the shifts observed in the NMR spectrum of a donor in the presence of a LSR are highly predictable. This is a result of the magnetic properties of the lanthanide ions. There are two possible mechanisms that cause shifts in the NMR spectrum of a donor compound bonded to a paramagnetic metal ion. The contact mechanism involves a situation in which there is a finite probability of finding the unpaired electron of the metal at the nucleus whose NMR spectrum is being recorded. A pseudocontact, or through-space, mechanism involves a situation in which the unpaired electrons exert a magnetic field on a nucleus by a dipole-dipole interaction. Shifts that result from a pseudocontact mechanism, as is largely the case with the lanthanide ions, are highly predictable. Shifts that result from a contact mechanism, as observed with transition metal NMR shift reagents such as $Ni(acac)_2$ and $Co(acac)_2$, are difficult to predict.

One other factor that has certainly led to the widespread application of lanthanide shift reagents is their ease of use. The compounds are air stable and require little in the way of specialized handling.

Since Hinckley's first report, a number of notable studies extending the general use and number of lanthanide shift reagents have been reported. Sanders and Williams observed that the shifts in the NMR spectrum of a donor in the presence of $Ln(dpm)_3$ were considerably larger than those in the presence of $Ln(dpm)_3 py_2$.[2] The smaller shifts with the pyridine adduct result because of the competitive effects of pyridine for coordination sites on the shift reagent. In 1971, Rondeau and Sievers reported that the lanthanide tris chelates of 6,6,7,7,8,8,8-heptafluoro-2,2-dimethyl-3,5-octanedione H(fod) were effective NMR shift reagents for a wide range of donor compounds.[3] Their effectiveness was especially pronounced for classes of compounds such as ethers that are rather weak donors. The complexes with fod are also considerably more soluble in organic solvents than those with dipivaloylmethane (dpm), a factor which limits the utility of the dpm complexes in some instances.

The first report of a chiral lanthanide NMR shift reagent was by Whitesides and Lewis in 1970.[4] In this study, lanthanide tris chelates with the ligand 3-(*tert*-butylhydroxymethylene)-*d*-camphor were employed. Chiral lanthanide tris chelates with the ligands 3-(trifluoroacetyl)-*d*-camphor[5] and 3-(heptafluorobutyryl)-*d*-camphor[5,6] have been studied and are generally more effective than the *t*-butyl derivative. In 1974, Whitesides and co-workers described lanthanide complexes with the ligand *d,d*-dicampholylmethane.[7] The complexes with *d,d*-dicampholylmethane are the most effective chiral lanthanide NMR shift reagents for the study of enantiomers. Applications of chiral shift reagents are described in Chapter 3.

The lanthanide tris β-diketonates are not suitable shift reagents for soft Lewis bases such as olefins, aromatics, halogenated compounds, and phosphines. Lanthanide shift reagents for olefins were first reported by Evans and co-workers in 1975.[8] In their study, silver(I) heptafluorobutyrate was used with a lanthanide tris β-diketonate to form a binuclear shift reagent complex. In 1980 it was reported that superior binuclear reagents resulted when silver β-diketonates with the ligands 6,6,7,7,8,8,8-heptafluoro-2,2-dimethyl-3,5-octanedione and 1,1,1-trifluoro-2,4-pentanedione were mixed with lanthanide chelates of fod.[9] A simplified set of equilibria for this process is shown in Equations 1 and 2. The binuclear

$$Ag(\beta-dik) + D \rightleftharpoons Ag(\beta-dik)D \quad (1)$$

$$Ln(\beta-dik)_3 + Ag(\beta-dik)D \rightleftharpoons [Ln(\beta-dik)_4]AgD \quad (2)$$

complexes utilize the silver as a bridge between the soft Lewis base and the lanthanide ion. Chiral binuclear derivatives can also be employed to study enantiomers. Applications of binuclear reagents are discussed in Chapter 4.

The interest in lanthanide NMR shift reagents for organic solvents has led to more research into the development of lanthanide NMR shift reagents for use in aqueous solutions. The aquated ions are suitable for many applications in which an aqueous shift reagent is needed. A few examples of chiral lanthanide NMR shift reagents for use in aqueous solutions have been reported. Studies employing lanthanide shift reagents in aqueous solutions are described in Chapter 6.

I. STRUCTURAL STUDIES

In 1958, McConnell and Robertson derived an equation that described the effects of a shift reagent that operates by a pseudocontact mechanism.[10] One form of the pseudocontact shift equation is shown in Equation 3.

$$\frac{\Delta \nu}{\nu} = K[(g_z^2 - \tfrac{1}{2}g_x^2 - \tfrac{1}{2}g_y^2)((3\cos^2\theta - 1)/r^3) - \frac{3}{2}(g_x^2 - g_y^2)(\sin^2\phi \cos 2\Omega/r^3)]$$

In the event that the shift reagent-donor complex has axial (C_3 or greater) symmetry, the g_x and g_y terms are equal, and the second term is zero. The simplified form of the pseudocontact shift equation then results. In this equation, r is the distance between the lanthanide ion and the nucleus of interest; and Θ is the angle between the principal magnetic axis and the line drawn from the lanthanide ion to the nucleus. The constraint of axial symmetry in the shift reagent-donor complex would seem to necessitate C_3 or greater symmetry of the donor molecule. If the donor bonds equally at three equivalent rotamers, however, the symmetry requirement for application of the simplified form of the pseudocontact shift equation is met.[11]

Tris β-diketonate complexes with either symmetrical or unsymmetrical ligands have the potential to exhibit C_3 symmetry. In addition, the shift mechanism with lanthanide shift reagents is largely pseudocontact in nature. This is observed because the unpaired electrons of the lanthanide ions reside in 4f orbitals. These orbitals are effectively shielded by filled 5s and 5p orbitals and do not participate to any significant extent in covalent bonding with the donor. These two factors make it tempting to rigorously apply the simplified form of the pseudocontact shift equation to lanthanide shift reagent studies. Optimization of the r and Θ values of the ^1H and ^{13}C nuclei of a donor by iterative computer programs theoretically provides a powerful solution structural technique.

The validity of applying the simplified pseudocontact equation to lanthanide shift reagent data has been debated in the literature. The issues surrounding the use of the simplified pseudocontact equation are discussed in Chapter 5.

II. FUTURE PROSPECTS

The field of NMR shift reagents experienced an explosive growth after Hinckley's first paper. Effective shift reagents are now available for most classes of compounds. The theoretical concepts are largely understood. The validity of using the simplified pseudocontact shift equation has been discussed in detail in the literature. A number of reviews of varying breadth have been published.[12-53] Most of the recent papers on NMR shift reagents consist either of refinements of existing knowledge or application of shift reagents to new compounds.

The rather recent development of effective binuclear shift reagents may spark more development and improvement in this area. Water-soluble shift reagents will be used more frequently as high-field instruments make NMR spectroscopy a practical technique for many biochemical studies. Chiral aqueous NMR shift reagents may therefore see more growth. Finally, the tempting nature of the application of the simplified pseudocontact shift equation will continue to spur work dealing with aspects of this problem.

Chapter 2

ACHIRAL LANTHANIDE CHELATES

Lanthanide tris β-diketonates are effective NMR shift reagents for the study of organic compounds that contain a hard Lewis base functional group. Among the classes of compounds in this category are most of those that contain an oxygen or nitrogen atom.

The term polyfunctional substrate is used in this chapter to denote compounds that contain more than one potential donor group. The donor groups in polyfunctional compounds compete for coordination sites on the shift reagent. The association of a donor group with the shift reagent is dependent on electronic effects and steric hindrances. In a compound with two donor groups, there are four possible modes of bonding. The first two(I,II) involve situations in which complexation of only one of the two donor groups is observed. The third(III) is when both groups bond independently to the shift reagent. In this case, the NMR spectrum of the substrate reflects a weighted contribution from the two bound forms. The final situation(IV) is one in which the substrate bonds in a chelate manner. Low lanthanide to substrate (L:S) ratios quite often favor situation one or two, whereas high L:S values often favor the third case. Applications of LSR to specific classes of monofunctional and polyfunctional substrates are described in this chapter.

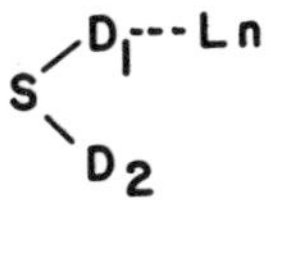

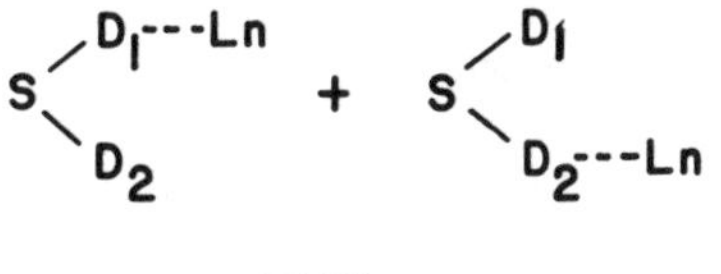

I. SHIFT REAGENT SELECTION

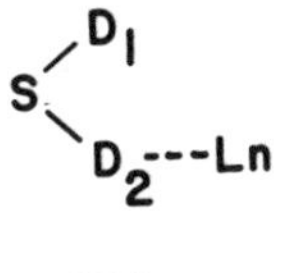

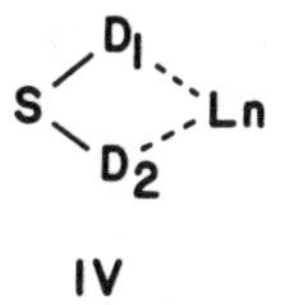

There are two variables that enter into the selection of a lanthanide shift reagent, the choice of the metal and the choice of a ligand. While this may seem to provide a rather endless variety of shift reagents to choose from, the selection is not that involved a process, and the choices are not generally as great as they may at first seem.

A. Metal

The lanthanide metals that can be employed in shift reagents include the 3⁺ ions of praseodymium, neodymium, samarium, europium, dysprosium, terbium, holmium, erbium, thulium, and ytterbium. Lanthanum(III) and lutetium(III) are not suitable for use because they are diamagnetic. Cerium(III) is unstable and tends to oxidize to the diamagnetic Ce(IV) ion in β-diketonate complexes. Promethium is radioactive and gadolinium(III) has an f^7 configuration. The f^7 configuration is isotropic and will not produce pseudocontact shifts in the NMR spectrum of a substrate.

Europium(III) has a $7F_0$ ground state that is nondegenerate and has no Zeeman splitting.[54] The Eu(III) ion in the ground state cannot cause contact or pseudocontact shifts. The first excited state of Eu(III) is a $7F_1$ state and is significantly populated at room temperature. This energy state accounts for the shift reagent properties of Eu(III).[54] The broadening with the paramagnetic lanthanide ions is not that severe due to short electron spin relaxation times. This results in lanthanide ions from the presence of closely spaced energy levels and strong spin-orbit coupling.[55] At lower temperatures, the electron spin relaxation is slower and broadening is more pronounced.[56]

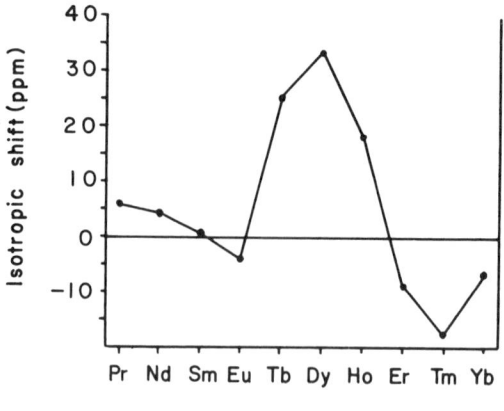

FIGURE 1. Isotropic shifts at 298 K for the 4-CH$_3$ resonance of 4-picoline with Ln(dpm)$_3$. Negative values correspond to downfield shifts. (From Horrocks, W. D., Jr. and Sipe, J. P., III, *Science*, 177, 994, 1972. Copyright 1972 by the AAAS and reprinted with permission.)

The constant in the simplified form of the dipolar shift equation contains terms that reflect the magnetic susceptibility anisotropy of the particular lanthanide ion.[57] This constant is determined by subtracting one half of the x and one half of the y components of the magnetic susceptibility values from the z component. Since the unpaired

$$\Delta H/H = K[\chi_z - \tfrac{1}{2}(\chi_x + \chi_y)](3\cos^2\theta - 1/r^3)$$

electrons in the lanthanide series reside in shielded 4f orbitals, these terms tend to be constant for a particular ion from complex to complex. As a result, if x_z is less than $\tfrac{1}{2}(x_x + x_y)$ for a metal, complexes with that metal will shift the resonances in the NMR spectrum of a substrate upfield. If x_z is greater than $\tfrac{1}{2}(x_x + x_y)$, the resonances will be shifted downfield. The magnitude of the absolute value of this term determines whether a particular metal produces relatively large or small shifts in the NMR spectrum of a substrate.

In Figure 1, the shifts recorded for the methyl resonance of 4-picoline as the metal of the shift reagent is varied are shown.[57] From this figure, it can be seen that complexes of Pr, Nd, Sm, Tb, Dy, and Ho are upfield shift reagents with a relative ordering of the shifts of Dy > Tb > Ho > Pr > Nd > Sm. Complexes with Eu, Er, Yb, and Tm are downfield shift reagents with a relative ordering Tm > Er > Yb > Eu.

The observation of so-called "wrong-way" shifts has been noted on a number of occasions using LSR. "Wrong way" shifts are only observed for certain nuclei of a substrate rather than all of the nuclei of a substrate. Two reasons for "wrong way" shifts are either a change in the sign of the geometric term of the simplified dipolar shift equation,[58-69] or the presence of a large contact shift.[70] The geometric term of the dipolar equation changes its sign at 54.736 and 125.264°. Angles between the values of 54.7 and 125.3° are not typically noted in studies with LSR.[61] Cases in which angles between these values are observed usually involve large and rigid substrates.

In viewing Figure 1, one might be tempted to select dysprosium and thulium as the best metals to employ in shift reagents. In fact these are seldom used in studies with LSR for two reasons. The first is that any paramagnetic substance introduces broadening into a spectrum and the broadening with a shift reagent is proportional to the square of the shift.[71] While dysprosium and thulium are powerful shift reagent metals, the broadening of the spectrum of a substrate in their presence tends to obliterate the fine structure that results from coupling. This information is usually of critical importance in structure determinations.

The second reason is that many of the less powerful metals still induce shifts that are large enough to produce first-order spectra. In a sense, the dysprosium and thulium analogs often bring about excessively large shifts that can make finding and assigning all of the resonances difficult. As a result, the most preferred metals are europium as a downfield reagent and praseodymium as an upfield reagent. Ytterbium chelates are also useful for certain studies in which a downfield shift reagent is needed.

Downfield reagents are usually preferred over the upfield analogs. The resonances furthest downfield in the unshifted spectrum are usually closest to the electron-withdrawing group that bonds to the shift reagent. These resonances often exhibit the largest shift and remain the furthest downfield in the shifted spectrum. A first-order spectrum with an upfield shift reagent requires a complete inversion of the spectrum. The resonances of nuclei closest to the electron-withdrawing group are now furthest upfield. Larger shifts are necessary to achieve a first-order spectrum with an upfield shift reagent than with a downfield reagent. The inversion of spectra with upfield shift reagents in some respects contradicts our sense of NMR spectra.

Europium tris β-diketonates are the most popular shift reagents for the study of achiral substrates. The europium chelates induce shifts large enough for most applications and the broadening is so slight as to leave the fine structure due to coupling intact.

B. Ligand

Lanthanide complexes with a wide variety of achiral β-diketonate ligands have been tested as NMR shift reagents; however, only two have found any widespread acceptance. These are complexes with the ligands 2,2,6,6-tetramethyl-3,5-heptanedione, H(dpm), and 6,6,7,7,8,8,8-heptafluoro-2,2-dimethyl-3,5-octanedione, H(fod). The ligand 2,2,6,6-tetramethyl-3,5-heptanedione is sometimes given the shorthand notation of H(thd). The dpm notation, which comes from the common name dipivaloylmethane, is used in this text because of its more frequent use in the literature. The structure of these two and certain other useful achiral β-diketone ligands that have been evaluated in NMR shift reagents are shown in Table 1. Lanthanide chelates with the dpm and fod ligands are available from a number of commercial sources.

Lanthanide complexes with ligands more fully fluorinated than fod such as 1,1,1,5,5,6,6,7,7,7-decafluoro-2,4-heptanedione,[72,73] 1,1,1,2,2,6,6,7,7,7-decafluoro-3,5-heptanedione,[74-76] and 1,1,1,2,2,3,3,7,7,8,8,9,9,9-tetradecafluoro-4,6-nonanedione[76-79] have been tested as NMR shift reagents. These complexes reportedly produce larger shifts than the chelates of fod and dpm in the NMR spectra of certain substrates. The shifts are not significantly larger, and these complexes tend to be more difficult to prepare and retain in anhydrous crystalline forms. As a result, their use has been limited.

One potential benefit of the highly fluorinated ligands is the elimination of the t-butyl resonance observed with both the fod and dpm complexes. With the fod complexes, this resonance can obliterate much of the spectrum between 1 to 4 ppm. Deuterated fod ligands have sometimes been used to eliminate such an interference. The t-butyl resonance of the complex of Eu(III) with dpm is usually upfield of TMS; however, the chemical shift of this resonance does depend on the particular substrate under study.[80-82] Lanthanide complexes with the ligands 4,4,5,5,6,6,6-heptafluoro-1-phenyl-1,3-hexanedione, 4,4,5,5,6,6,6-heptafluoro-1-(2-naphthyl)-1,3-hexanedione,

Table 1
LIGANDS

	Name	Abbreviation	Structure
I	2,2,6,6-tetramethyl-3,5-heptanedione	H(dpm)	$(CH_3)_3CCCH_2CC(CH_3)_3$ with two C=O
II	6,6,7,7,8,8,8-heptafluoro-2,2-dimethyl-3,5-octanedione	H(fod)	$(CH_3)_3CCCH_2CCF_2CF_2CF_3$ with two C=O
III	1,1,1,5,5,6,6,7,7,7-decafluoro-2,4-heptanedione	H(dfhd)	$CF_3CCH_2CCF_2CF_2CF_3$ with two C=O
IV	1,1,1,2,2,3,3,7,7,8,8,9,9,9-tetradecafluoro-4,6-nonanedione	H(tfn)	$CF_3CF_2CF_2CCH_2CCF_2CF_2CF_3$ with two C=O
V	—	—	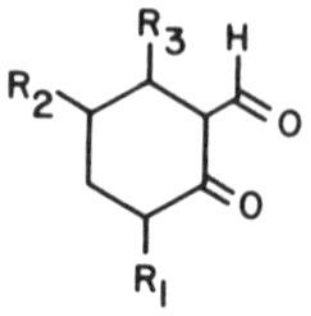 (cyclohexane-diketone with R_1, R_2, R_3 substituents)
VI	1-(2-heptafluorofuran)-1,1-difluoro-4-(2-thienyl)-2,4-butanedione	—	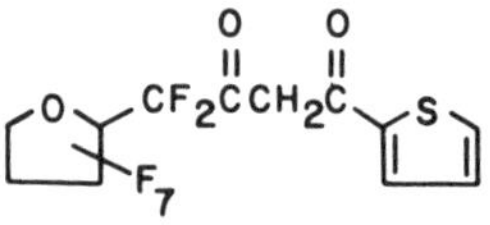 (furan)-CF_2CCH_2C-(thienyl) with two C=O, F_7 on furan

4,4,5,5,6,6,6-heptafluoro-1-(2-thienyl)-1,3-hexanedione, and 4,4,5,5,6,6,6-heptafluoro-1-(2-furyl)-1,3-hexanedione have been found to function as effectively as chelates with fod.[83] The complexes with these ligands are rather easy to prepare and retain in a crystalline state. They each have resonances in the aromatic region of the spectrum and are, therefore, complementary to complexes with the fod ligands.

Other ligands that have been employed in lanthanide shift reagents include dibenzoylmethane;[84-87] 1-phenyl-1,3-butanedione;[85,86,88] 1,1,1-trifluoro-5,5-dimethyl-2,4-hexanedione;[68,89-91] 1,1,1,2,2-pentafluoro-6,6-dimethyl-3,5-heptanedione;[90] 4,4,4-trifluoro-1-(2-thienyl)-1,3-butanedione;[92-95] 1-phenyl-4,4,4-trifluoro-1,3-butanedione;[96] 1-(4-fluorophenyl)-4,4,5,5,5-pentafluoro-1,3-pentanedione;[97,98] 1-(4-fluorophenyl)-4,4,5,5,6,6,6-heptafluoro-1,3-hexanedione;[97,98] 1-trifluoromethoxy-1,1,2,2-tetrafluoro-5-phenyl-3,5-pentanedione;[99] 1-trifluoromethoxy-1,1-difluoro-5,5-dimethyl-2,4-hexanedione;[100,101] and 1-trifluoromethoxy-1,1,2,2-tetrafluoro-6,6-dimethyl-3,5-hexanedione.[100,101] Complexes with these ligands generally have been found to be less effective than complexes with either fod or dpm. Goryushko employed complexes with the ligands 1-(2-heptafluorofuran)-1,1-difluoro-4-phenyl-2,4-butanedione and 1-(2-heptafluorofuran)-1,1-difluoro-4-(2-thienyl)-2,4-butanedione and reported that the shifts were larger than those with chelates with fod.[101] Potapov et al.[102,103] has studied a series of shift reagents with ligands with structure V in Table 1, but did not compare the results to those with fod or dpm. The derivative in which $R_1 = R_2 = H$ and $R_3 =$ t-butyl was reported to be the best example. Li evaluated complexes with the ligand 1-phenyl-3-methyl-4-heptafluorobutyryl-5-pyrazolone as NMR shift reagents.[104] Briggs et al.[105] observed only small shifts in the spectra of alcohols and ketones in $CDCl_3$ with $Pr(NO_3)_3$ $(OPPh_3)_3$, $Pr(ClO_4)_3[OP(NMe_3)_3]_4$, $Eu(NO_3)_3(OAsPh_3)_4$, $Ce(NCS)_3$ $(OPPh_3)_4$, and lanthanide perchlorates. A variety of adducts of tris chelates with fod[106,107] and dpm[107] have been evaluated as NMR shift reagents. None of these adducts seems to offer any advantages over the anhydrous complexes. Horrocks and Wong[108] have employed lanthanide prophyrin complexes of the general formula Ln(III)TAP(β-dik) (TAP = tetraaryl porphine) as shift reagents for heterocyclic amines and alcohols. These were found to be less effective than the β-diketonate analogs.

Lanthanide complexes with the ligands 2,4-pentanedione[84,87,109,110] and 1,1,1-trifluoro-2,4-pentanedione[106] have also been tested as shift reagents. The complexes with these ligands, in addition to exhibiting reduced solubility compared to those with fod and dpm in relatively nonpolar solvents, are ineffective as shift reagents.

The reasons why R groups of the ligand alter shift reagent effectiveness are not understood. One general trend is that ligands with rather bulky R groups such as t-butyl and -C_3F_7 are more effective in shift reagent applications. It has been postulated that bulky R groups may lead to a more specific orientation of the donor in the shift reagent-donor complex.[110,111] In complexes with ligands with small R groups, a large variety of orientations would be observed, and the specificity necessary to produce large shifts in the spectrum of the substrate would not be achieved.

It has also been postulated that the -C_3F_7 group of the fod ligand withdraws electron density from the metal and results in a more positive metal than observed in the dpm complexes.[3] The metal in the complex with fod would then be a harder Lewis acid and would bond more strongly to Lewis bases. Such reasoning has been used to explain the effectiveness of chelates of fod compared to those of dpm.[3,73,76,112] A similar reasoning has also been used to explain the effectiveness of complexes with dfhd (-CF_3 replacing t-butyl) compared to those of fod.[72] Other evidence indicates, however, that the Lewis acidity of the metal is not the only mechanism for explaining the differences between complexes of fod and dpm.[80] With certain substrates, larger shifts are observed with chelates of dpm.[80,113] One definite advantage of the complexes with fod over those with dpm is their increased solubility in nonpolar solvents. The solubility of complexes with dpm is improved in the presence of a substrate,[114] but is still less than the solubility of complexes with fod.

The complexes with fod and dpm are especially effective as NMR shift reagents and are relatively easy to store and use. Complexes with either of these ligands are preferred for the study of achiral substrates. The development of improved ligands (larger shifting) for NMR shift reagents seems unlikely since we cannot explain the influence of the ligands, and most of the obvious ligands have already been tested.

II. EXPERIMENTAL TECHNIQUES:

In this section, the experimental techniques commonly employed in the storage and use of achiral lanthanide tris β-diketonates will be described.

A. Solvent

The lanthanide tris chelates of fod and dpm are soluble in a wide range of organic solvents including CCl_4, $CDCl_3$, CH_2Cl_2,[115,116] tetrachloroethane,[56] CS_2,[113,117-119] benzene,[113,115,119] toluene,[120] cyclohexane,[116,121] n-heptane,[115] and acetone.[113,122] Chelates of fod have been used in studies in which $AsCl_3$ was the solvent.[123] It should be kept in mind that a solvent such as acetone that can coordinate with the shift reagent will significantly reduce the shifts in the spectrum of a substrate.[113]

A number of workers have evaluated the effects of various solvents on the magnitude of lanthanide-induced shifts.[113,115-119,124-129] Walters[117,118] has recommended the use of carbon disulfide with complexes of H(dpm) because of increased solubility in this solvent. The solubility of Eu(dpm)$_3$ in 0.5 ml of CS_2, CCl_4, and $CDCl_3$ was reported to be 600 mg, 130 mg, and 130 mg, respectively.[117] The greater solubility permitted larger shifts to be obtained in the spectra of substrates.

The ordering of the magnitude of lanthanide shifts as a function of solvent varies somewhat from report to report.[113,115,119,126,127,129] To summarize these results, the largest shifts are observed in nonpolar solvents such as n-heptane and cyclohexane, intermediate shifts are observed in CS_2 and CCl_4, and smaller shifts occur in benzene and chloroform. The solvent dependence seems to manifest itself in the value of the association constant between the chelate and the substrate. Larger association constants are noted in the most nonpolar solvents.[115,116,125,126] Whether or not a solvent such as $CDCl_3$ lowers the association constant through competition with the substrate is not fully known. The spectrum of chloroform is not shifted in the presence of the tris chelates to any significant degree, which would seem to rule out any competitive effect.[115] The geometry of the shift reagent-substrate complex also appears constant with changes of the solvent.[115,126,129] Whatever the explanation for the observed solvent effects, the differences in lanthanide shifts between n-heptane and $CDCl_3$ are not so significant as to rule out $CDCl_3$ as a solvent for studies with LSR. In fact, CCl_4 and $CDCl_3$ are by far the most commonly used solvents. It has been recommended that CCl_4 be used for quantitative studies and that solvents such as $CDCl_3$ and benzene only be employed for qualitative studies in which spectral-dispersion is sought.[128]

The purity of a solvent is an important consideration and will influence the magnitude of the lanthanide shifts.[130] Water is of most concern, but other impurities have been noted. For carbon tetrachloride, it is recommended that it be stored over sodium hydroxide pellets prior to use.[82,131,132] In addition to water, phosgene and deuterium chloride may be found as impurities in chloroform. Storage over sodium bicarbonate[132,133] or magnesium foil,[133] or passage through basic alumina,[134] will remove DCl and phosgene. The chloroform should then be stored over molecular sieves until use.[133,134]

The most common internal reference compound is tetramethylsilane (TMS), which is presumed to exhibit no shift in the presence of the lanthanide tris chelates. The TMS resonance often interferes with upfield shift reagents, and benzene has been recommended as an internal reference[135] in these cases. TMS frequently contains small amounts of tetrahydrofuran as an impurity that can be removed by washing with sulfuric acid and potassium bicarbonate.[136] The TMS should then be distilled and stored over molecular sieves.[136]

B. Shift Reagent Preparation and Purification

With the widespread commercial availability of the lanthanide chelates with fod and dpm, most people probably purchase their NMR shift reagents. The H(fod) and H(dpm) ligands are also commercially available from a number of sources and the lanthanide chelates can be synthesized by literature methods.[137,138] No matter what the source, the purity of the lanthanide tris β-diketonate is a concern. Impurities that may ultimately affect the resulting shifts in the NMR spectrum of a substrate can take a number of forms.

The most common is a situation in which the complex is hydrated. The ineffectiveness of hydrated lanthanide tris chelates relative to the anhydrous complexes has been demonstrated.[73,127] The chelates with fod can be effectively dried for most applications by placing them *in vacuo* (1 torr) over P_4O_{10} for 24 hr. For Eu(fod)$_3$, a noticeable color change from pale yellow to bright yellow occurs as it is converted to its anhydrous form. It has been reported, however, that vacuum pumping at room temperature is only suitable for rigorous drying of the later lanthanide chelates of fod.[139] The earlier complexes probably exist as hemi-hydrates, and the bridging water in the hemi-hydrate crystal can only be removed by pumping at elevated temperatures.[139] The chelates with dpm are best dried *in vacuo* (1 torr) in an Abderholden at 100 C-115 (water or toluene) over P_4O_{10} for 24 hr. For volatile complexes such as those of fod or dfhd, sublimation may occur as drying is attempted at elevated temperatures. Hexane has been reported to be a suitable solvent for drying chelates of fod in an Abderholden.[140] All commercially available complexes should be dried prior to use. All LSR should be stored in a dessicator over P_4O_{10} after they have been dried. Some workers go so far as to recommend storage and handling of the LSR under an atmosphere of dry nitrogen in a glove bag.[128] This procedure would only be necessary under the most stringent attempts to apply LSR in quantitative fitting of the dipolar shift equation.

There are other impurities besides waters of hydration that have been observed with commercially available or laboratory-synthesized NMR shift reagents. Occasionally, impurities that are insoluble in organic solvents are obtained. These impurities are most likely lanthanide oxides or hydroxides that can result in the synthesis of the complexes if too much base is added to the reaction. Recrystallization or sublimation of the complex can eliminate these impurities.

A second possible impurity that presents more problems is the presence of a binuclear complex of the form $M[Ln(\beta\text{-dik})_4]$.[141] Complexes such as these have been observed in some commercial sources of lanthanide shift reagents with the fod ligand. The binuclear complex is soluble in organic solvents, and certain examples of these can be sublimed.[142] It has been reported that the binuclear complex can be removed by dissolving the shift reagent in CCl_4 and shaking with 0.1 M HCl or 0.2 M $LnCl_3$. The $Ln(fod)_3$ complex must then be dried and purified before use.[141]

For studies in which exceedingly accurate concentrations of the shift reagent are necessary, such as detailed analyses employing the pseudocontact shift equation, purification by either recrystallization or sublimation must be performed. Recrystallization should always be carried out in a noncoordinating solvent such as cyclohexane[143] or methylcyclohexane[144] to insure against the possibility of isolating a lanthanide chelate-donor complex.[145] Sublimation is often preferred to recrystallization as a purification method, but will sometimes give particles that are "magnetic" in nature and are difficult to handle when weighing small quantities.

The presence of only one symmetrical t-butyl resonance in the spectrum of $Eu(fod)_3$ has been reported as a reliable criterion for shift reagent purity.[146-148] The observation of other peaks, at one time believed to represent dimerization of the chelates in solution,[149] has been ascribed to impurities of the shift reagent.[141,146,148] We usually check a new batch of a LSR by running the spectrum of a substrate and comparing it to previous results. The small expenditure of time and money is worth the effort.

C. Scavengers

No matter how much effort is put into cleaning and drying the glassware, and purifying the solvent, substrate, and shift reagent, small amounts of impurities or scavengers will be present in shift reagent experiments. In many instances, this represents no particular problem to the experimenter. If the lanthanide shift data are to be employed in either a detailed structural analysis involving fitting of the dipolar shift equation, or to extrapolate the values for chemical shifts of resonances in the unshifted spectrum (δo), the presence of scavengers is a problem that must be taken into account.[127,150,151]

The presence of scavengers is apparent in a plot of the lanthanide-induced shifts (LIS) vs. [L]/[S]. The effect of scavengers on such a plot is shown in Figure 2.[151] At low [L]/[S] ratios, the scavenger competes with the substrate, and the measured LIS are lower than they would be in the absence of scavengers. At higher ratios, the effects of the scavenger are minimized, and a linear plot is obtained. Methods to correct for the influence of scavengers have been described by Shapiro et al.[151] and Armitage and Hall.[127] At least one well-characterized proton that is resolved in the unshifted spec-

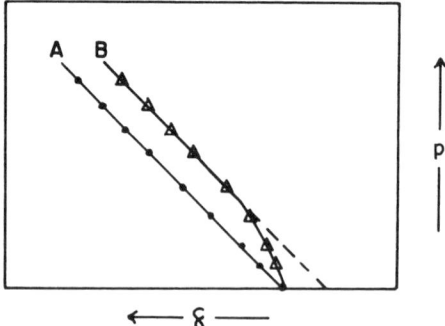

FIGURE 2. Plot of ratio of shift reagent concentration to substrate concentration(p) at constant S_o against the magnitude of the lanthanide-induced shift(S). (A) scavenger absent; (B) scavenger present. (From Shapiro, B. L., Shapiro, M. J., Godwin, A. D., and Johnston, M. D., *J. Magn. Reson.*, 8, 402, 1972. With permission.)

trum is necessary. Data for this resonance are collected, and a linear fit to the data, ignoring values at low [L]/[S] ratios, is obtained. The difference between the extrapolated and actual δo value is used to calculate a corrected set of [L]/[S] ratios. These ratios are then applied to data for the other resonances of the substrate, and extrapolation now results in correct values for δo.

D. Procedures for Use of LSR

Usually it is best to record a series of spectra with variable [L]/[S] ratios when performing studies with LSR. If the principal interest is in obtaining spectral dispersion, it is often adequate to obtain a series of spectra as small amounts of the shift reagent are added to a solution of the substrate under study. The shift reagent can be added as either a solid or as small aliquots of a concentrated stock solution. The latter procedure is more useful with the more soluble complexes of fod. Under these conditions, a gradual change from a second-order to first-order spectrum is observed. Such a gradual change often makes assignment of the resonances an easier task. A second advantage of obtaining a series of spectra involves a frequent situation in which a complete first-order spectrum of the substrate is never attained. Unique resonances for many or all of the nuclei may be obtained, however, at different steps of the series. Evaluation, including decoupling data, of each spectrum of the series permits a more complete assignment.

For analyses in which highly accurate values of the concentrations of the shift reagent and substrate are necessary, the above procedure is not adequate. For these cases, the method of incremental dilution described by Shapiro and Johnston is a better procedure.[136] In this method, the first spectrum of the series is recorded at a high [L]/[S] ratio. The initial solution is prepared by weighing the shift reagent into the NMR tube. An aliquot of a known concentrated solution of the substrate is added and the amount of substrate is determined gravimetrically. Additional solvent and TMS is then added to achieve the desired concentration of substrate. Subsequent spectra are obtained after addition of aliquots of a stock solution of substrate with a concentration identical to that of the sample in the NMR tube. The amount of each aliquot is measured gravimetrically. The concentration of substrate remains constant throughout the procedure, and the concentration of shift reagent, which decreases during the analysis, is determined gravimetrically. After each addition, the sample should be allowed to warm in the probe for a period of 5 min. Errors of 10 to 20 Hz (at 100 MHz) have been reported if the sample is not allowed to reach thermal equilibrium.[147]

The incremental dilution method offers a number of advantages for analyses requiring high accuracy.[136] The weight of each additional aliquot of substrate is higher than the weight of solid shift reagent that would be added, thereby reducing relative errors in the weight. The reduced handling of the shift reagent in the method of incremental dilution minimizes the possibility of hydration when in contact with the air. Errors that might result from changes in the volume on mixing were reported to be less than 1%.[136]

Addition of the shift reagent to a solution will change the bulk diamagnetic susceptibility of the solution. This has the effect of moving unshifted resonances such as the TMS or chloroform peaks upfield relative to their previous setting on the instrument. The movement of these resonances is more pronounced for lanthanide ions that cause larger shifts in the spectrum of a substrate. With a shift reagent containing dysprosium, the change in location of these resonances can be as much as 10 to 20 ppm. The distance between the TMS and chloroform resonances remains constant, however, and the TMS resonance can be reset to zero using the field adjustment on the instrument.

The chemical shift of the resonances of the shift reagent may also vary as the nature and concentration of the substrate is varied. As an example, in the NMR spectrum of 1-octanol (0.2 M) with Yb(fod)$_3$ (0.05 M) in CDCl$_3$, the t-butyl resonance of the fod ligands is observed in the region from 2.6 to 3.5 ppm. In the NMR spectrum of 1-aminooctane (0.2 M) with Yb(fod)$_3$ (0.05 M) in CDCl$_3$, the t-butyl resonance of the ligands is observed in the region from −2.2 to −2.9 ppm.[152]

E. Recovery of the Substrate

In some instances, it may be necessary or desirable to recover the substrate used in a study with a LSR for future work. Procedures employing silica gel in column[153,154] or thin-layer chromatography have been described.[2] Elution of the chelate is achieved with a solvent such as CH$_2$Cl$_2$ or CHCl$_3$. The lanthanide chelate is collected in the early fractions. If Eu(III) is employed as the shift reagent, it is convenient to monitor its presence by its bright red luminescence under short wave UV light.[153] The substrate is then eluted from the column with a more polar solvent.

Alumina can also be used as the column material.[155,156] Desreux et al.[156] have reported that the slightly acidic silica liberates some of the ligand from the shift reagent, which may then be present as an impurity in the substrate, and recommends the use of alumina. Alcohols can be eluted from a column of alumina with benzene while the LSR is retained on the column.[155] It is not recommended that the shift reagent be reused after any of these purification procedures.

F. High Field NMR

Employing LSR with high field NMR instrumentation may present problems in the form of excessive signal broadening. This broadening is not due to shortening of T_1 values, but results from the slower tumbling rate of the shift reagent-donor complex compared to the free donor.[157] Broadening from such chemical exchange phenomena will be more pronounced at higher fields. Bulsing et al. have described a method using the Carr-Purcell-Meiboom-Gill spin-lock sequence to remove severely broadened resonances, but not affect the remaining portion of the spectrum.[157] An example is shown in Figure 3. The severely broadened peak from the shift reagent in Figure 3a interferes with the methylene resonance of benzyl alcohol. A half-echo experiment removes this peak from the spectrum as seen in Figure 3b.

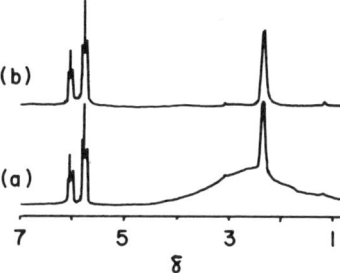

FIGURE 3. 270 MHz NMR spectrum of benzyl alcohol (PhCH$_2$OH) and Pr(dpm)$_3$ in CDCl$_3$ solution: (a) normal spectrum; (b) effect of simple half-echo experiment with t = 5 ms. (From Bulsing, J. M., Sanders, J. K. M., and Hall, L. D., *J. Chem. Soc. Chem. Commun.*, 1201, 1981. With permission.)

III. APPLICATIONS

A. Monofunctional Substrates
1. Alcohols and Phenols

Compounds containing the hydroxyl functional group represent the largest class of monofunctional compounds that have been studied by NMR shift reagents. Hinckley's first report on LSR involved the study of cholesterol.[1] Hydroxyl-containing compounds studied with the aid of LSR include simple aliphatic alcohols, olefinic alcohols, monocyclic derivatives such as cyclopentanols and cyclohexanols, a variety of rigid multicyclic derivatives, steroids, and phenols. The scope and variety of alcohols studied by LSR are too numerous to permit detailed descriptions of each. Instead, brief descriptions of the types of information that can be gained through the use of LSR with hydroxyl-containing compounds will be given.

LSR have been employed to convert the proton NMR spectra of straight chain aliphatic alcohols from second- to first-order.[2,105,129,158-164] The europium shift reagents are capable of resolving the proton resonances up to eight carbons removed from the hydroxyl group in a straight chain alcohol.[158] The position of the olefin group in 11-eicosenol could not be located using Eu(fod)$_3$.[158] The spectrum of 1-octanol was not first-order in the presence of Pr(dpm)$_3$; however, addition of Ho(dpm)$_3$, a more powerful upfield shift reagent, did result in a first-order spectrum.[159] It is possible to analyze mixtures of straight chain alcohols with LSR. The NMR spectrum of a mixture containing C-1 through C-7 straight chain alcohols in the presence of Eu(dpm)$_3$ exhibits at least one unique unobstructed resonance for each component.[161]

Branched chain aliphatic alcohols and alcohols containing aromatic substituent groups have been studied with LSR.[2,84,116,125,129,163-183] The ^{1}H and ^{13}C shifts for all of the 6-carbon alcohols in the presence of Eu(dpm)$_3$ (^{1}H) and Yb(dpm)$_3$(^{13}C) have been reported.[167] This information was used to determine the weighted average conformation of the substrates in their complexed form. In certain instances, the coupling constants changed as the concentration of shift reagent was increased. This indicates that the average conformation of the substrate varies on complexation with the shift reagent. The shifts in the ^{13}C NMR spectrum of other simple branched chain alcohols with LSR have been reported in the literature.[165,174] The presence of a small contact shift was reported for the resonance of the carbon atoms close to the site of the hydroxyl group.[165]

The separation of the resonances of diastereotopic protons in a variety of branched chain aliphatic alcohols has been noted.[170,173] These separations are most pronounced for protons close to the location of the hydroxyl group. This observation is quite common, and caution must always be used when interpreting the spectra of substrates with diastereotopic protons in the presence of LSR.

LSR have been used to study a number of acyclic carbinols with one or more phenyl substituent groups.[175-177,179-183] These studies have been concerned with either assigning the relative configurations of two diastereomers[175,180-183] or determining the conformation of the substrate.[176,177,179] Several workers have reported that complexation of the LSR did not alter the conformation of these substrates.[179,180] The conformation of 2-phenyl-1-propanol was determined with the aid of shift data with Eu(fod)$_3$ and Pr(fod)$_3$.[176,177] Isobutanol and 2,2-diphenylethanol were employed as model substrates in these studies. The authors cautioned against using only shift data with LSR for drawing conclusions about the conformation of flexible molecules.[177]

LSR have been used to simplify the ^{1}H NMR spectrum of benzyl alcohol,[2] to locate the position of deuterium substitution on 1-methylbenzylalcohol,[168] and to distinguish a series of phenyl-t-butyl carbinols.[172] LSR have been used to quantify mixtures of the three methylbenzylalcohols.[178]

The importance of steric and electronic effects in the value of the association constant between an alcohol and a LSR has been noted.[116,125,169,171,184,185] A relationship between the paramagnetic shift and steric strain energies and inductive effects was observed in a study of aliphatic alcohols.[116] Eu(fod)$_3$ in squalane has been employed as a stationary phase in a gas chromatographic column. The interaction of alcohols with Eu(fod)$_3$ was evaluated by comparing the retention times between a column with only squalane and one with squalane and Eu(fod)$_3$.[184] In this study, it was found that lengthening the chain, which increases the inductive effect, increases the association constant. Branching of the chain, which serves to increase the steric hindrance of the donor, decreased the association constant. The increased steric effects proceeding from primary to secondary to tertiary alcohols led to a reduction of the association constant. The ^{19}F NMR spectra of certain polyfluorinated alcohols show no shifts in the presence of chelates of dpm.[186] The hydroxyl group in these compounds is too weak a base because of the electron-withdrawing properties of the fluorine atoms.

Some fluorinated alcohols do associate with the fod complexes, and the shifts in the ^{1}H and ^{19}F spectra of compounds such as 2,2,3,3-tetrafluoropropanol and 4-chloro-3,3,4-trifluoro-2-methyl-2-butanol have been reported.[187]

Methanol has been studied with Eu(fod)$_3$ and Pr(fod)$_3$ in the nematic solvent p-ethoxybenzylidene-p-n-butyl aniline.[188] Coupling between the hydroxyl and methyl group was observed. It was found, however, that the shift reagents altered the value of the coupling constant.

The ^{17}O NMR spectrum of methanol has been recorded in the presence of lanthanide chelates of dpm.[189] Nitromethane was used as the internal oxygen reference because it bonds weakly, if at all, to the dpm chelates. From this study, the dpm chelates of Dy(III) and Tb(III) were determined to be the best for ^{17}O studies. This conclusion was based on a consideration of shift magnitude and line broadening. Measurement of the ^{17}O NMR spectrum of a polyfunctional substrate in the presence of a LSR was recommended as a method to determine the site of complexation.

LSR have been used to study olefinic compounds with hydroxyl functional groups.[58,190-195] Shift data with LSR have been used to distinguish *trans,trans*-2,4-hexadienol from the *trans,cis* isomer.[190] Eu(dpm)$_3$ has been used to determine the configuration about the double bond in a variety of alkenol compounds.[191,192] The compounds nerol(I) and geraniol(II) have been distinguished in the presence of Eu(dpm)$_3$.[193] The carotenoid rhodopin has been studied through the aid of lanthanide shift data.[58] The shifts in the spectrum of *trans*-retinol with Eu(fod)₃ have been reported.[194]

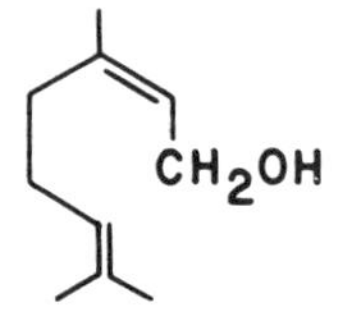

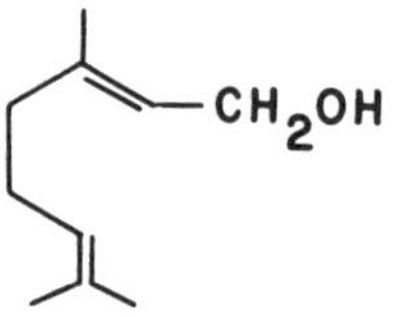

Cyclic alcohols or cyclic compounds with hydroxyl-containing substituent groups have been the focus of a large number of studies. These include derivatives of cyclopropane,[159,196-202] cyclobutane,[159,203-205] cyclopentane,[203,204,206-210] cyclohexane,[61,63,65,114,147,162,185,203,211-225] and cycloheptane.[226-228] The spectra of these compounds are quite complicated, and the use of LSR can provide valuable information for configurational and conformational analysis.

Eu(fod)₃ has been used successfully to distinguish the *cis* (III) and *trans* isomers (IV) of chrysanthemyl alcohols.[197] A study of cyclopropylmethanols with LSR has been reported.[200] The *cis trans* stereochemistry of substituent groups on the ring of cyclopropyl methanols has been determined with the aid of Eu(fod)₃.[201] The Z(V) or E(VI) configuration of 1-hydroxymethyl-2-alkyl-3-ethylidenecyclopropanes was assigned on the basis of studies with Eu(dpm)₃.[201]

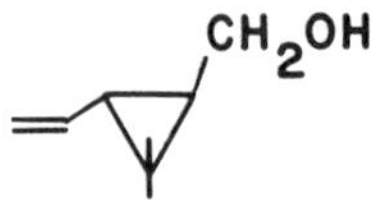

The configuration about the cyclopropane ring[199] and the first olefin bond in the long chain carbon substituent group[198] of presqualene alcohols was assigned through the use of Eu(dpm)₃. Eu(fod)₃ has been employed to distinguish the *cis* and *trans* isomers of presqualene alcohols.[197]

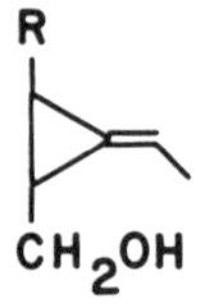

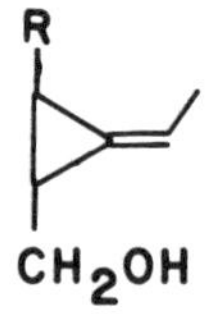

Shift data with Eu(fod)₃ have been used to assign the configuration of a highly substituted cyclobutane ring in which one of the substituents was a hydroxymethyl group.[205] A series of benzocycloalkenols in which the cycloalkene contained four (VII), five, or six carbon atoms has been studied with the aid of LSR.[203,204] In the presence of Eu(dpm)₃, the aromatic coupling constants were obtained for these compounds. By comparison with the 220 and 300 MHz spectra of the unshifted substrate, it was found that the aromatic coupling

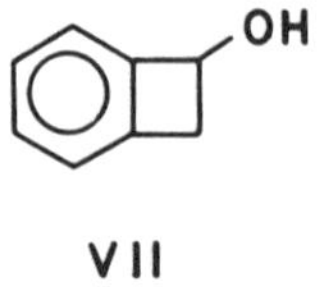

constants varied as the concentration of shift reagent was increased. This probably results from a change in the conformation of the cycloalkene ring on complexation with the shift reagent. It was also reported in this study that the smaller the ring, the larger the association constant between the substrate and shift reagent. It was reasoned that as the ring is made smaller, the hydroxyl group is less coplanar with the ring, and steric effects are reduced.[203]

The spectrum of cyclopentanol is first-order in the presence of Eu(dpm)$_3$[209] or Eu(fod)$_3$,[207] and the position of deuterium substitution in cyclopentanol can be determined using LSR.[207] Lanthanide shift data for the ^{13}C resonances have been used to distinguish the *cis* and *trans* configurations of 3-methylcyclopentanol and 1,3-dimethylcyclopentanol.[210]

The shifts in the ^{19}F and ^{1}H spectra of trifluoroindanols(VIII) in the presence of Eu(fod)$_3$ have been used to determine if the OH and F substituent groups on the cyclopentene ring are *cis* or *trans* to each other.[206] It was also

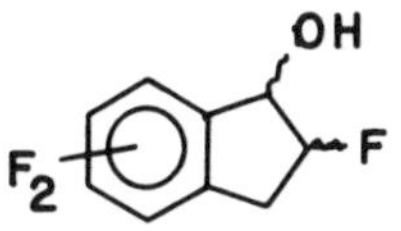

VIII

possible to determine the substitution pattern of the two fluorine atoms on the aromatic ring.[206] A force field model has been used to predict the lanthanide position and relative shifts for indanol, *cis*- and *trans*-2-methylindanol, and 7-methylindanol.[229] The position of substituent groups on fluoren-9-ols has been determined with the aid of LSR.[208]

Shapiro and co-workers[61,63,147,211] have studied highly substituted cyclohexanols including *cis*- and *trans*-1,5,5-trimethyl-3-(1-naphthyl)cyclohexanol with LSR. Using the shift data, it was possible to determine both the conformation of the ring and the axial or equatorial preference of the substituent groups. The superior shifting ability of chelates of Pr(III) vs. those of Eu(III) for aromatic groups can be evidenced by the results for *cis*-3-(1-naphthyl)-1,3,5,5-tetramethylcyclohexan-1-ol.[63] The use of Pr(dpm)$_3$ resulted in a first-order spectrum. Another interesting observation about this substrate is that the H-3′ and H-4′ resonances of the naphthyl moiety shift in the opposite direction to that of the other resonances.[61,63] These protons have a Θ value that causes a change in sign of the angle term of the pseudocontact shift equation. The *cis* and *trans* isomers of 3-(1-naphthyl)-1,3,5,5-tetramethylcyclohexan-1-ol were readily distinguished with LSR.[225]

For *cis*-4-*tert*-butyl-1-phenylcyclohexanol the *m*- and *p*-hydrogen atoms in the phenyl ring lie at an angle of approximately 55° in the lanthanide complex and have negligible lanthanide-induced shifts.[65] In the *trans* compound, all of the resonances shift downfield.

Tert-butylcyclohexanols have been the subject of a number of studies with LSR.[114,213,218,219] These are often used as model substrates because of the strong equatorial preference of the *t*-butyl group. *Cis*- and *trans*-4-*t*-butylcyclohexanol are readily distinguished using LSR.[114,213,218] The shifts resulting from the addition of Pr(dpm)$_3$ have been used to determine the stereochemistry of *cis*- and *trans*-4-*tert*-butyl-1,2,2-trimethylcyclohexanol.[219] LSR have also been used to determine the position of deuterium substitution in *trans*-4-*t*-butylcyclohexanol.[213] The association constants for *cis*- and *trans*-4-*t*-butylcyclohexanol and *cis*- and *trans*-4-*t*-butyl-1-methyl-1-cyclohexanol with Eu(dpm)$_3$ and Eu(fod)$_3$[185] have been compared. Significantly larger association constants were observed with Eu(fod)$_3$. Conformational analysis of 1-methylcyclohexanol[220] and 1-alkylcyclohexanols[223] have been performed by fitting of lanthanide shift data. Both studies found that complexation of the shift reagent altered the conformation of the ring in favor of the less sterically hindered conformer. An equatorial arrangement of the hydroxy group was favored in the presence of the shift reagent.[223]

Cis- and *trans*-alphabenzyl cyclohexanols can be distinguished through the use of LSR.[217] The axial and equatorial arrangement of substituent groups of some highly substituted cyclohexanes with a hydroxymethyl group has been determined using shift data obtained with Eu(fod)$_3$.[216] The stereochemistry of IX was assigned on the basis of lanthanide shift data.[221]

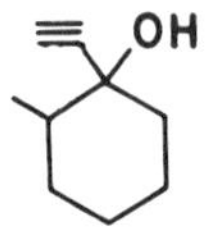

IX 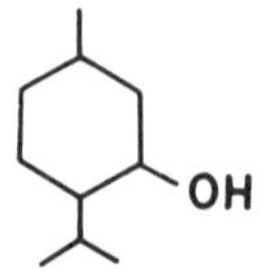   X

The effects of LSR on the rotation of the isopropyl group in menthol(X) has been studied.[212] This study was performed using tetrachloroethylene as the solvent. The barrier to rotation of the complexed substrate was determined by elevating the sample temperature. The ^{19}F NMR spectrum of the menthol derivative with a perfluorinated isopropyl group has been recorded in the presence of a LSR.[214] Two -CF$_3$ resonances were observed without the shift reagent and the shifts of the two resonances were of different magnitudes in the presence of the shift reagent. A variety of monocyclic terpene isomers including menthols, carveols, and pulegols have been studied with Eu(dpm)$_3$.[215] The *cis* and *trans* isomers of these compounds were readily distinguished through the use of the shift reagent. It was also possible to assess the effect that the location of the hydroxyl group had on the rotation of the isopropyl group. The shifts in the ^{13}C spectra of menthol and other monoterpenes with Eu(fod)$_3$ have been reported.[224]

The *Z* and *E* isomers of 1,4-dimethyl-4-nitrocyclohexa-2,5-dienols were assigned with the aid of ^{13}C shift data with Eu(fod)$_3$.[222]

Eu(fod)$_3$ has been employed to determine the most probable conformation of 4-cycloheptene carbinol-8-d$_2$.[227] Shifts in ^{13}C NMR spectra of several methyl-substituted cycloheptanols in the presence of europium shift reagents have been reported.[226] These data were useful in assigning the ^{13}C spectra and assessing certain configurational and conformational aspects of many of the compounds. Shift data with Eu(fod)$_3$ has been used to facilitate assignment of the resonances and determination of the conformation of 5*H*-dibenzo[a,d]cyclohepten-5-ol.[228]

A wide range of multicyclic hydroxyl-containing compounds have been studied with the aid of LSR. A large number of these have involved compounds with the norbornane skeleton(XI).[77,114,229-246] The compounds *exo*- and *endo*-2-norbornanol and

XI

exo- and *endo-*dehydronorbornanol have been distinguished in the presence of Eu(dpm)$_3$.[230,233,235] Substituted derivatives can also be successfully studied with the aid of LSR. Both the ^{1}H[114,241] and ^{13}C[240] NMR spectra of borneol are converted into essentially first-order spectra and can be assigned in the presence of LSR. Shifts in the ^{1}H NMR spectrum of borneol with Pr(dpm)$_3$ fit well with the pseudocontact shift equation.[241] Schneider and Wiegand have discussed some of the problems in using the dipolar shift equation to fit lanthanide shift data for isoborneol, isofenchols, and fenchols.[243] The rotamer preference of isoborneol and *endo-*5-norbornen-2-ol has been determined by fitting of lanthanide shift data.[244] A force field model has been used to predict the position of the lanthanide ion and relative shift values for borneol and isoborneol.[229] Evidence of a contact shift mechanism in the ^{1}H NMR spectrum of borneol and isoborneol with Eu(dpm)$_3$ has been reported.[114] LSR have been used to aid the assignment of the ^{13}C NMR spectrum of isoborneol.[239]

Exo- and *endo-*2-norbornanol and 18 methyl-substituted derivatives have been studied with LSR.[233,235] Partially deuterated derivatives were analyzed to aid in the assignment of some of the spectra. The *exo* and *endo* derivatives were distinguished in the presence of the LSR. The coupling constants of the substrate were not affected by the presence of the shift reagent.[235] Eu(dpm)$_3$ has been used to determine the position of deuterium substitution in the alcohol of *exo-*norbornyltrifluoroacetate.[77] Shift data with Eu(dpm)$_3$ has been used to assign the position of deuterium substitution in borneol,[245] isoborneol,[245] and *exo-*2-norbornenol.[246]

Shift data with Eu(dpm)$_3$ has been used to distinguish the different configurations of methyl derivatives of 5-norbornene-2-ols.[232] Eu(dpm)$_3$ has been employed to provide more interpretative ^{1}H NMR spectra for 2-*exo-*methyl-*endo-*fenchol and 2-*endo-*methyl-*exo-*fenchol.[237]

The *syn* or *anti* position of the hydroxy group in 7-hydroxynorbornene has been assigned using LSR.[236] LSR were also used in this study to determine the positions of deuterium substitution in the deuteration products of *syn-* and *anti-*7-hydroxynorbornene and 7-hydroxynorbornadiene. The data with the shift-reagent permitted the distinction of *exo* and *endo* isomers.[236] A comparison of the Eu(III) chelates of fod and dpm proved inconsistent in that no one chelate resulted in larger shifts for all of the substrates.[236] Eu(dpm)$_3$ has been used to aid in the structural assignment of 1,2,3,4,5-*endo-*7-*syn-*hexamethyl-7-*anti-*hydroxybicyclo[2.2.1]hept-2-ene.[238]

The ^{1}H NMR spectra of 2-*exo-*hydroxymethyl-3-*endo-*methylnorbornene and the corresponding 2-*endo-*hydroxymethyl-3-*exo-*methyl isomer were assigned using Eu(dpm)$_3$.[242] The shift data obtained in this report were used to assess the rotational preference of the hydroxymethyl group. No significant changes of the coupling constants were noted on addition of the LSR. Changes have been observed for the coupling constants alpha to the coordination site in 2-*endo-*hydroxymethyl-5-norbornene; however, the coupling constants did not vary in a systematic fashion as the concentration of shift reagent was increased.[234]

The configurations of 3-nortricyclanol and the *cis-* and *trans-*methyl isomers of 3-nortricyclanol were determined through the use of shift data with Eu(dpm)$_3$.[231]

A number of studies of adamantanols with LSR have been reported.[64,113,129,165,185,209,218,229,244,247-252] Unique resonances are observed for all the protons of 1-adamantanol and 2-adamantanol in a series of spectra obtained with increasing amounts of shift reagent.[250] The assignment of the ^{13}C NMR spectrum of 2-adamantanol has been confirmed with the aid of Eu(dpm)$_3$.[247] Contact contributions to the shifts in the ^{13}C NMR spectrum of certain resonances of adamantanol in the presence of LSR have been reported.[165] Contact contributions were only noted for carbons close to the site of bonding. Eu(dpm)$_3$, Eu(fod)$_3$, and Yb(fod)$_3$ have been employed to study a series of 2-alkyl-2-adamantanols.[64,249] Well-resolved spectra were obtained for a variety of 2-alkyl derivatives, but the shifts could only be used to obtain qualitative information about the preferred conformation of the alkyl substituent group. The shifts in some hydroxyalkyl adamantanes have been correlated with the pseudocontact shift equation.[209]

A force field model has been used to predict the position of the lanthanide ion and relative shifts for 1- and 2-adamantanol.[229] Greater steric effects in 2-adamantanol compared to 1-adamantanol reportedly caused a longer lanthanide-donor bond length.[218] This study also attempted to determine if steric factors caused a rotational preference of the hydroxyl group on bonding with the shift reagent.[218] This was difficult to assess, but with some compounds a rotational preference was indicated. The rotational preference of the hydroxy group in 2-adamantanol has been studied by fitting shift data with Eu(dpm)$_3$.[244] The position of deuterium substitution in 1-methyl-2-adamantanols has been determined through an analysis of the ^{2}H spectrum in the presence of Pr(fod)$_3$.[251]

LSR have been used to aid in the determination of the conformation of thujane(bicyclo[3.1.0]hexane) derivatives.[253] No change in the coupling constants was observed on addition of the shift reagent. This indicates that the shift reagent does not alter the conformation of the substrate. The effects of Eu(fod)$_3$ on the ^{13}C spectra of certain thujane derivatives have been reported.[224] LSR have been used to assign the *exo* and *endo* configurations of the hydroxyl group in substituted bicyclo[3.1.0]hexenes.[254] Shift data with Eu(fod)$_3$ have been used to determine the *exo* and *endo* stereochemistries of 6,6-diarylbicyclo[3.1.0]hexan-3-exo-ols.[255]

Eu(dpm)$_3$ has been used to differentiate the *exo* and *endo* position of the hydroxyl group in 4,7,7-trimethyltricyclo[2.2.1.0^{2,6}]heptan-3-ol,[256] and to study the configuration of 6-hydroxybicyclo[3.2.0]hept-2-ene.[248] Lanthanide shift data facilitated a determination of the conformation of *cis*- and *trans*-4,4,6-trimethylbicyclo[4.1.0]heptan-2-ol.[257]

Substrates such as verbanols,[258-260] verbenols,[258,260] nopinols,[258-260] myrtanols,[224,260] myrtenols,[224] and pinocarveols,[261,262] which have the same basic ring skeleton(XII), have been the focus of studies with LSR.

XII

Lanthanide shift data have been used to assign ^{1}H[262] and ^{13}C[224,260] spectra. The configurations of the hydroxy and methyl groups in methylverbenols, -verbanols, and -nopinols have been determined with the aid of Eu(dpm)$_3$.[258] It was also possible to assign the conformation of these substrates through an evaluation of the coupling constants obtained in the shifted spectra. Using *cis*- and *trans*-pinocarveol as model substrates, Hinckley and Brumley described the errors that may result in using lanthanide shift data to assign conformations.[261]

A method for determining the configurations of multicyclic ring compounds that is based on a comparison with a compound of known configuration has been described.[263,264] The shifts with a LSR are measured for the known compound. Ratios of the shifts of each proton to a specific reference proton are calculated. The same procedure is performed for the compound of unknown configuration. A poor correlation of the two sets of ratios is obtained if the configurations are different. The method was reported to be suitable for flexible compounds.

LSR have been used to aid in the structural assignment of bicyclooctenes.[265-267] XIII was determined to exist as a pair of diastereomers based on an analysis of the spectrum with Eu(dpm)$_3$.[267] This conclusion was reached

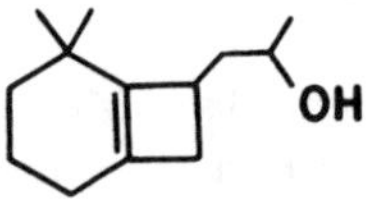

XIII

since the methyl group attached to the carbinol carbon appeared as two doublets in the shifted spectrum.[267]

The ^{1}H and ^{13}C NMR spectra of 9-hydroxybicyclo[3.3.1]nonane in the presence of LSR have been used in an attempt to determine the preferred conformation.[268] The shift data only allowed the elimination of one of three possible conformations. A conformational study of exo- and endo-bicyclo[3.3.1]nonan-3-ol has been performed with the aid of Eu(dpm)$_3$.[269] The preferred conformations of bicyclo[6.1.0]nona-2,4,6-trienes have been determined with the aid of LSR.[270] Configurational assignments of a bicyclononene[271] and tricyclononadiene[272] have been performed with the aid of LSR.

LSR have been employed in the determination of the stereochemistries of 7-hydroxybicyclo[4.3.1]decatriene,[273] 8-exo-vinyl-8-endo-hydroxytricyclo[5.3.0.0^{2,10}]deca-3,5-diene,[274] and pentacyclo[5.3.0.0^{2,5}.0^{3,9}.0^{4,8}]decan-6-ol and derivatives.[275] Cubenol(XIV) and epicubenol, which differ in the configuration of the carbinol carbon, have been distinguished through the use of Eu(dpm)$_3$.[276]

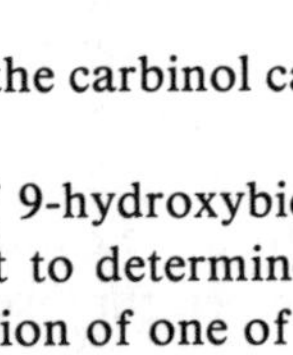

XIV

Both the configuration and conformation of bicycloundecanols have been studied with the aid of LSR.[277] The orientation of the hydroxyl group in two tricyclic dodecatrienes has been determined using Eu(dpm)$_3$.[278] Shift data with a LSR have been used to assign the structure and stereochemistry of isomeric perhydrophenalenols.[279] The configuration and conformation of tricyclo[4.4.1.1^{2,5}]dodecan-11-ols have been assigned on the basis of shift data with Eu(fod)$_3$.[280] The exo and endo isomers of 4-hydroxy-4,5-dihydroaldrin (XV) were distinguished using Eu(dpm)$_3$.[281]

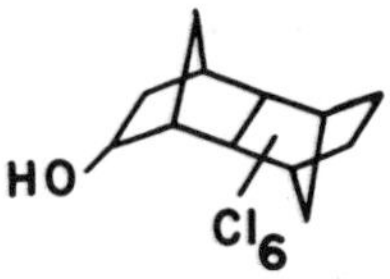

XV

A variety of hydroxy steroids have been studied with the aid of LSR. In these instances, the substrates are so large and their NMR spectra so complex that a completely first-order ^{1}H NMR spectra is not obtained with the shift reagent. LSR have been applied with success to the assignment of ^{13}C NMR spectra,[79,282-289] ring conformations close to the site of the hydroxyl group,[62,288,290-294] and methyl resonances of steroids.[2,143,292,295-302]

Vitamin D analogs have been the topic of a number of studies with LSR.[288,290,291,294] Shift data with LSR have been used to aid in the assignment of the A and seco B ring of vitamin D$_3$ and dihydrotachysterol$_3$.[294] Shifts in the ^{13}C NMR spectra with Eu(fod)$_3$ have been used to assign the conformation of the A-ring of vitamin D analogs.[288] The shifts in the ^{1}H spectra were evaluated in an attempt to determine the epimer assignment of the A ring in vitamin D analogs. It was found that the two epimers could only be distinguished when samples of both isomers were studied.[290] The A-ring conformation of calciferol has been assigned using lanthanide shift data.[291] It was reported that complexation of the LSR did not perturb the conformation of calciferol. The changes in conformational preference with temperature were also determined through the use of shift data with a LSR.

The ^{1}H[1,2,303,304] and ^{13}C[282,285,286,304] spectra of cholesterol have been studied with LSR. Shift data in the ^{1}H and ^{13}C spectra, and relaxation data with Gd(dpm)$_3$ have been employed to fit the structure of cholesterol.[304] LSR have been used to distinguish steroid stereoisomers that differ in the geometry of the functional group and/or the A/B ring junction of the steroid skeleton.[62] All of the ^{13}C resonances of 5α-cholestan-3β-ol have been assigned based on a study with Yb(dpm)$_3$.[287] The importance of accounting for the complexation shifts when assigning ^{13}C resonances was stressed in this study. Complexation shifts are measured using a diamagnetic La(III) or Lu(III) complex.

A further caution when assigning ^{13}C spectra in the presence of LSR is the observation of contact shifts for carbons close to the site of complexation.[282,286] Yb(III) shift reagents should be used to minimize contact shifts.[286]

Resolution of the A-ring resonances and some of the B- and C- ring resonances of androstan-2β-ol were observed with Eu(dpm)$_3$.[292] For D-ring hydroxy steroids, the shifts with a LSR usually permit assignment of all of the resonances in the ^{1}H spectrum of both the C and D ring.[292] The configuration of the A and B ring of trachyloban-19-ol has been determined by comparing shift data with LSR to compounds of known configuration.[293] LSR have been used to distinguish isomeric pairs of sterols.[300-302] With the aid of Yb(fod)$_3$, it was possible to assign the absolute configuration at position C-24 of the side chain and to determine the geometry of double bonds in the side chain of XVI.[302]

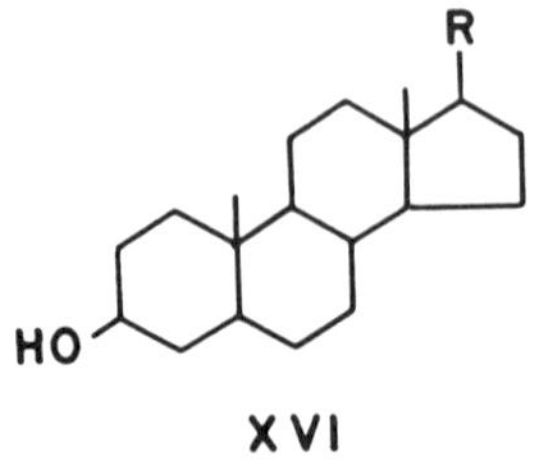

XVI

Compounds with five rings have been studied with the aid of LSR.[2,143,292,295,297,298,305] These studies have focused on an evaluation of the methyl resonances because of the complexity of the remainder of the spectrum. Complete resolution of the eight methyl resonances of friedelan-3β-ol has been achieved with Eu(dpm)₃.[292] Lanthanide shift data have been used to determine the position of methyl substitution[298] or differences in configurations in a series of similar compounds.[297] The number and position of attachment of methyl groups in 20 triterpenoids has been determined using shift data with Eu(dpm)₃.[143] An algorithm for the automatic sorting of signals in LSR experiments on substrates with complex spectra has been described.[305] The procedure was reported to be especially useful for assigning ¹³C spectra, and its use was demonstrated by assigning the eight methyl resonances of β-amyrin.

Pr(dpm)₃ has been used to aid in the assignment of the ¹³C spectra of a series of derivatives of pimarol.[283] Shift data with Eu(fod)₃ has been used to successfully determine the location of deuterium substitution in certain steroids.[296]

Phenols can be studied with the chelates of fod,[306-312] but the chelates of dpm decompose in solutions of phenolic compounds.[2,309,313] The magnitude of the shifts observed in the NMR spectra of phenols with LSR is less than those for aliphatic alcohols.[310] The shifts in the spectra of a variety of alkyl- and nitro-substituted phenols have been recorded with Eu(fod)₃, and it was found that both electronic and steric effects were important in determining the magnitude of the shifts.[313] This has also been noted in a ¹³C NMR study of 16 phenolic compounds in the presence of Eu(fod)₃.[306] In these studies, a pronounced reduction in shifts was observed when the proton *ortho* to the hydroxy group was replaced with a *t*-butyl group.[306] The chelates of fod can be used to determine the substitution pattern of alkyl phenols.[308,312] Mixtures of phenols[308] such as *o*-, *m*-, and *p*-cresol can be quantitated in the presence of Eu(fod)₃.[309,310] The shifts observed in the NMR spectra of phenols in the presence of Eu(fod)₃ are larger than those in the presence of a shift reagent consisting of a germanium complex with a macrocyclic ligand.[313]

2. Ketones

Monofunctional ketones have been widely studied with the aid of LSR. There has been considerable discussion in the literature as to the exact nature of the bonding between a ketone and a LSR. The most common representation of the bonding involves two-site complexation at each of the lone pairs of electrons on the carbonyl group. A representation of this type of bonding is shown

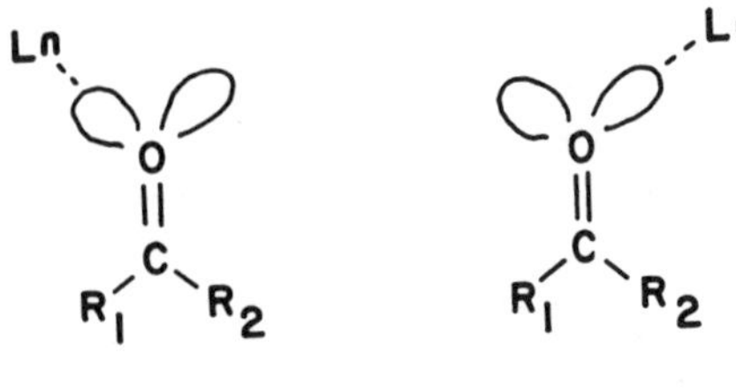

A number of investigators have described data with LSR that indicate two-site, rather than straight-on, bonding of the lanthanide species.[78,299,314-320] If the R groups are equivalent, the shift reagent will bond to the substrate equally at each lone pair. If the R groups are different, it has been reported that a higher proportion of the bonding occurs at the least sterically hindered side of the carbonyl group.[78,299,315,317,318] In a study of straight chain ketones, it was noted that the complex Eu(tfn)₃ was especially sensitive to steric effects and exhibited a pronounced preference for the less hindered position.[78]

Raber et al., in a study of adamantanone and methyladamantanones, reported that consistently better fits of the data were obtained if the bonding of the LSR and ketone was described by a one-site model with a linear array of the Ln—O=C atoms.[321] A rational for the orbitals involved in one-site bonding was presented in this report. If the oxygen atom of the carbonyl group is sp, rather than sp_2 hybridized, the oxygen has an unused sp and p-orbital. The

lanthanide ion can then act as a sigma(sp) and pi(p) acceptor. The pi bonding would involve the unfilled 5d orbitals of the lanthanide ion. Unequal steric effects of the R group of the ketone would result in distortions from linearity.

Hofer performed a study on 2- and 3-alkyl-1-indanones and reported that both the one- and two-site model were equally suited for fitting of the data.[322,323] Maps of the magnetic field in the region of the substrate were not significantly different for the two models.[323] The only appreciable differences were noted for nuclei close to the carbonyl group. A somewhat similar finding was noted by Filippova et al. in a study of olefinic ketones.[324] For many substrates, no distinction could be drawn between the fits with the one- or two-site model. In certain instances, however, better fits were obtained with the two-site model. These results are in direct contrast to the findings of Abraham et al. who reported that the one-site model was unsatisfactory in the analysis of bicyclo[3.1.0]hexan-3-one,[325] cyclohexanone, and 4-tert-butylcyclohexanone.[326] Abraham et al. recommend either a two-site model,[325-327] or more recently, have come out in favor of a four-site model.[325,328,329]

Brooks and Sievers, in a report employing Eu(fod)$_3$ in squalane as a gas chromatographic stationary phase, systematically studied the influences of steric and electronic effects on the strength of the interaction between ketones and LSR.[184] Stronger complexation was observed as one R group of the ketone was made longer. This was believed to result from the increased inductive effect of the larger R group. Branching of one of the R groups led to variable results that were determined by a combination of both an increased inductive effect and increased steric hindrance. Moving the position of the carbonyl group toward the center of the chain increased the steric effects and decreased the association constant. Cyclization of the ketone led to pronounced increases in the association constant due to a large decrease in steric effects. A pronounced steric effect, leading to a decrease in the association constant, has been observed for the series adamantanone(I), norcamphor(II), and camphor(III).[218] The association constants of 1-acetyladamantane and 1-bromoacetyladamantane with Eu(fod)$_3$ have been determined and are smaller by a factor of ten for the bromo derivative.[185] This difference was ascribed to electron-withdrawing effects of the bromine atom.

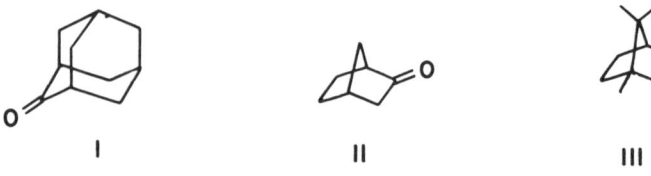

Contact shifts in the ^{13}C NMR spectra of ketones with LSR have been reported.[134,165,282,330,331] The contact contribution in ketones is often larger than in alcohols because of the pi orbitals.[165] Contact shifts are experienced by carbons more removed from the site of complexation in extended pi systems. As an example, more carbon resonances exhibited the presence of a contact shift mechanism in cyclohexenone than in cyclohexanone or 4-t-butylcyclohexanone.[165] In a study of the ^{13}C NMR of keto steroids with Eu(fod)$_3$, contact shifts were noted for the carbonyl carbon and the carbons α and β to the carbonyl.[282] With Pr(fod)$_3$, contact shifts were observed for the carbonyl carbon and carbons α to the carbonyl. Chelates with Yb(III) produce the smallest contact shifts.[134,330,331] Since contact shifts can result in "wrong way" shifts for certain resonances,[282] it is important to consider their presence when studying the ^{13}C NMR spectra of ketones with LSR. It has been advised that the complexation shift be measured when studying the ^{13}C NMR of ketones with LSR.[165,330]

LSR have been employed with acyclic[58,59,78,165,190,315,320,324,330,332-340] and aromatic[190,332,334,335,341-345] ketones, monocyclic ketones,[59,147,150,165,196,211,212,215,226,314,316,330,346-358] multicyclic ketones,[134,218,254,265,268,269,275,316-319,353,359-373] and ketosteroids.[62,79,282,295,298,374]

Chelates of fod have been employed in the study of the preferred conformation of 2-butanone.[59,320] The conformational analysis of the *cis* and *gauche* form of 2-butanone could be performed successfully using ^{13}C and ^1H shift data, but a similar analysis of 3-methyl-3-butanone was not sensitive enough to provide meaningful results.[320] The influence of the R group on the conformation of IV has been investigated through the use of LSR.[375] Eu(fod)$_3$ was used in low concentrations to resolve the ^2H resonances of acetone-d$_6$ and acetonitrile-d$_3$.[376] The relaxation times of the deuterium were also measured in this study.

IV

Shift data obtained with Eu(dpm)$_3$ have been used to study conformational properties of substituted acetophenones and benzophenones.[335] The shift data confirmed the presence of the equilibrium shown below. An analysis of this process

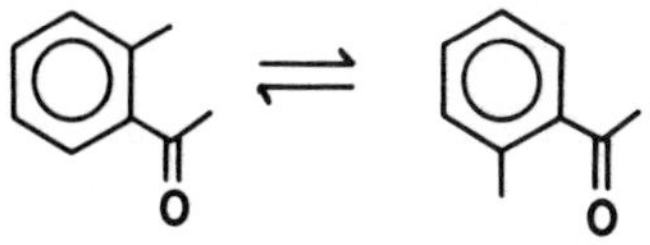

attempted by low temperature NMR studies was unsuccessful.[335] Shift data with Yb(fod)$_3$ in the ^1H and ^{13}C spectra of acetophenone[344] and 2,4,6-trimethylacetophenone[345] have been applied in a similar conformational study. It was reported that the complexation shifts in the ^{13}C spectra of these conjugated substrates were significant,[343] and corrections should be made for their presence.[344,345] The conformational properties of symmetrically substituted benzophenones have also been studied with the aid of LSR.[335] The conformational preference of 1-p-tolyl-2-phenyl-1-propanone was determined on the basis of shift data with Eu(fod)$_3$ and Yb(fod)$_3$.[179] It was reported in this study that the LSR did not alter the conformation of the substrate. The influence of electronic and steric effects on the association constant of acetophenone derivatives with Eu(dpm)$_3$ has been discussed.[341,342]

LSR have been employed in the study of the conformations of substituted α,β-unsaturated ketones(V).[190,315,333,334] Factors that enhance the probability that complexation of the LSR would alter the conformational preference of the substrate were discussed for examples in which R$_0$ was a methyl group.[315] The shifts of rigid α,β-unsaturated ketones with a known configuration have been used as a model by which the conformation of flexible substrates could be assigned.[333] Shift data with Eu(dpm)$_3$ were used to assess the planarity of substrates in which R$_0$ was a phenyl group and R$_1$, R$_2$, and R$_3$ were varied.[334] Both the configuration and conformation of dienones(VI) have been determined with the aid of LSR.[324,337,339]

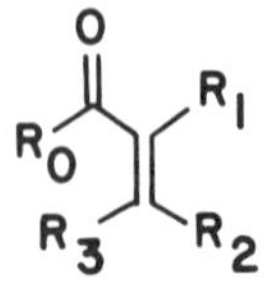

Cyclopropyl ketones have been the focus of several studies with LSR.[196,202,354,377,378] Conformational analysis of the cyclopropanoyl ring in a series of cyclopropyl ketones was performed by comparing the shifts with Eu(dpm)$_3$ in the spectra of unknowns to those in the spectra of compounds with cyclopropane rings with known conformations.[354] The conformations of 17 different methyl cyclopropyl ketones have been studied with the aid of LSR.[377] The Z and E isomers of 1-acetyl-2-alkyl-3-ethylidenecyclopropane were distinguished in the presence of Eu(dpm)$_3$.[202]

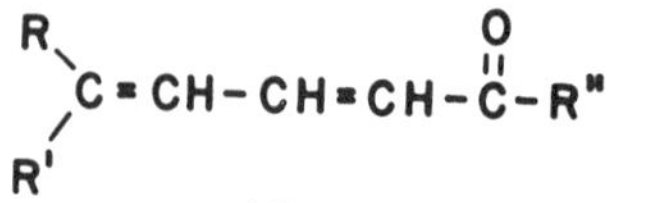

The configurations of substituted acetyl cyclopentanes and acetyl cyclohexanes have been determined through the application of LSR.[356] The shifts in the ^{1}H NMR spectra of cyclic ketones with 5 to 15 carbon atoms in the presence of Eu(dpm)$_3$ and Pr(dpm)$_3$ have been reported.[350,351] Complexation shifts measured with La(fod)$_3$ have been reported for the ^{13}C spectra of cyclopentenone and cyclohexenone and benzo derivatives of cyclopentanone and cyclohexanone.[343] The magnitude of the complexation shift in the spectrum of these substrates can be related to the extent of pi conjugation. The effects of conjugation and steric factors on the association constants of C$_5$, C$_6$, and C$_7$ cyclic ketones have been assessed.[379]

The most widely studied cyclic ketones are cyclohexanones.[59,147,150,165,211,212,215, 224,260,314,316,326,327,329,330,333,343,346,348,352,353,355,357,358,379-383] A number of reports have considered the validity of applying LSR in the study of conformationally mobile systems. Many of these studies have employed substituted cyclohexanones that exist in one conformation as model substrates. One popular model compound is 4-t-butylcyclohexanone, or a derivative, since the t-butyl group has a strong preference for the axial position.[59,150,314,316,333,346,352,358] In the presence of LSR, the axial and equatorial protons bonded to the same carbon atoms in 4-t-butylcyclohexanone exhibit different shifts,[333] Abraham et al. have reported, however, that the tert-butyl group of 4-tert-butylcyclohexanone causes a ring puckering,[381] and that 4-phenylcyclohexanone,[381,383] or especially trans-2-decalone,[329,383] are better conformationally rigid models of the cyclohexane ring.

A procedure for determining the mole fractions of the equatorial and axial conformers of 2-alkylcyclohexanones has been described.[59,352] The method is based on a comparison of the measured shifts of resonances of the unknown substrate to those observed for the axial and equatorial positions in 4-t-butylcyclohexanone. A similar procedure was used to analyze the rotamer position of substituent groups in 2-alkyl-4-tert-butylcyclohexanones.[358] The geometry of the shift reagent-substrate complex was set using the rigid ring, and analysis of the conformationally mobile side chain was then performed. Resolution of diastereotopic protons of the alkyl group was observed in some cases.[358] It was not known whether complexation of the shift reagent caused alterations of the rotamer populations.

Shapiro and co-workers have carried out studies on highly substituted cyclohexanones with LSR.[147,150,211,353] Much of their work has involved a detailed analysis of the stoichiometry of shift reagent-substrate complexes[147,150,211] and is discussed in Chapter 5. The information obtained with the aid of LSR permitted the distinction of the cis,cis and cis,trans isomers of 3-(1-naphthyl)-5,5-dimethylcyclohexanone.[211] The configurational assignment was not based on fitting the shift data to the pseudocontact shift equation. Instead, it involved an analysis of the multiplet pattern of the resonances that could be obtained from the spectrum with the shift reagent.[211] The distance dependence of the shifts can be used to assign qualitatively the conformation of highly substituted cyclohexanones.[150] In this report, a variety of 3-(aryl)-3,5,5-trimethylcyclohexanones were studied. In some instances, the rotations of the aryl group were sufficiently slow so that unique resonances were observed for symmetrically disposed protons.[150] One other observation in these studies was that complexation of the LSR sometimes changed the value of coupling constants.[353] Changes in the geminal coupling constants in 3,5,5-trimethyl-3-(p-chlorophenyl)cyclohexanone were observed as the concentration of LSR was increased. Since similar results were observed with camphor, a rigid compound, the change of coupling constants were the result of electron-withdrawing effects of the LSR,[353] rather than conformational changes. It has been reported that the conformations of 2-, 3-, and 4-methylcyclohexanone and 2-chlorocyclohexanone were not altered upon complexation with Yb(fod)$_3$.[327]

Monocyclic ketoterpenes have been the focus of studies with LSR.[212,215,224,260,333,355] The rotational properties of the isopropyl group in menthone and some of its epimers have been studied.[212,215,355] Shift data with LSR have been used to distinguish epimeric monocyclic ketoterpenes as well as resolve and quantity mixtures of epimers.[355] The shifts in the ^{13}C spectra of several monoterpenes with Eu(fod)$_3$ have been reported.[224,260]

The position of deuterium substitution in 3,4-dimethylcyclohexanone has been determined using LSR.[357] The methyl doublets of the cis and trans isomers of VII were completely resolved in the presence of Eu(dpm)$_3$.[348] The methyl resonance of the cis isomer exhibited the larger shift since it is closer to the LSR. The structures of polymethylated cyclohexadienones have been assigned on the basis of shift data with Eu(fod)$_3$.[380]

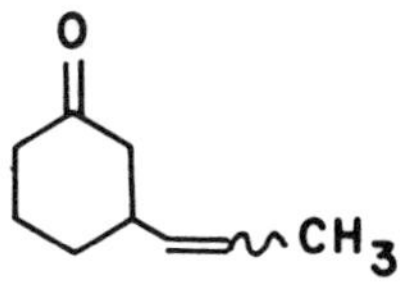

VII 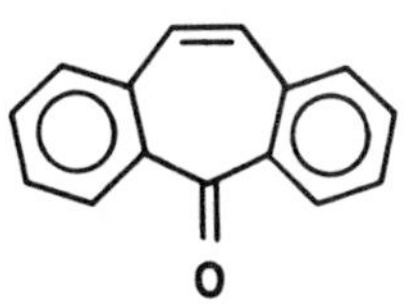   VIII

The shifts in the ^{13}C NMR spectrum of cycloheptanone with Eu(dpm)$_3$ and Eu(fod)$_3$ have been used to assign the resonances in the unshifted spectrum.[226] The ^{1}H and ^{13}C spectra and conformation of 5H-dibenzo[a,d]cyclohepten-5-one(VIII) were assigned on the basis of a study with a LSR.[228] A study of the ring inversion of 4,4,7,7-tetramethylcyclononanone in the presence of Eu(dpm)$_3$ in carbon disulfide has been performed.[347] Two experimental conditions were employed. One was to hold the temperature constant at 5°C and vary the concentration of LSR. The second was to hold the concentration of LSR constant and vary the temperature. It was possible to determine the low temperature symmetry of the substrate. It was also found that the coalescence temperature was higher in the presence of the LSR.

The shifts in the ^{13}C and ^{1}H NMR spectra of bicyclo[3.1.0]hexan-3-one in the presence of Yb(fod)$_3$ have been used to assign the conformation.[319,325,384] The configurations of highly substituted bicyclo[3.1.0]hexenones have been determined with the aid of LSR.[254]

A number of studies with LSR have utilized camphor as a substrate.[218,243,245,316,317,353,361,364,373] Hinckley's second paper on lanthanide shift reagents employed Eu(dpm)$_3$.py$_2$ to assign the methyl resonances of camphor.[361] Camphor has been used as a substrate in the assessment of steric effects on complexation,[218] and in a detailed analysis of the nature of shift reagent coordination with the carbonyl group.[317] The shifts in the spectrum of camphor with Eu(dpm)$_3$, Pr(dpm)$_3$, and Yb(dpm)$_3$ have been used to assess the validity of the simplified pseudocontact shift equation.[316] The effects of LSR on the coupling constants in ketones have been evaluated for camphor.[353] Camphor and other bicycloheptanones with known configurations have been used as models to assign the configuration of unknown methyl substituted bicycloheptanones.[364] Data with Eu(dpm)$_3$ have been used to determine the position of deuterium substitution in camphor.[245]

LSR have been employed in the analysis of α,α-dichlorocyclobutanones that result from the adducts of olefins and dichloroketene.[349,366] In one study, model compounds were used as a basis to assign the configuration of reaction products IX and X.[366]

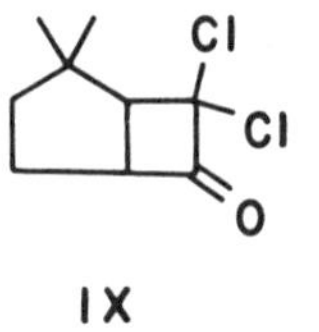

IX 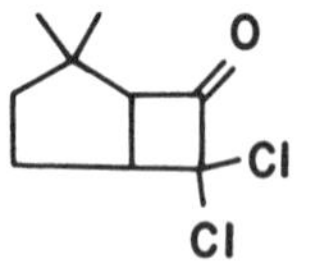   X

The protons α and β to the carbonyl group in the cyclobutane ring were distinguished in the presence of LSR.[349]

Data with Eu(dpm)$_3$ were used to assign the methyl resonances of XI so that the position of deuterium substitution in a reaction scheme could be determined.[359] The ^{13}C spectra of 5,5-dimethyl- and 6,6-dimethyl-2-norbornanone were assigned by studying deuterated analogs in the presence of LSR.[385] The structure of XII was determined with the aid of shift data with Eu(dpm)$_3$.[360]

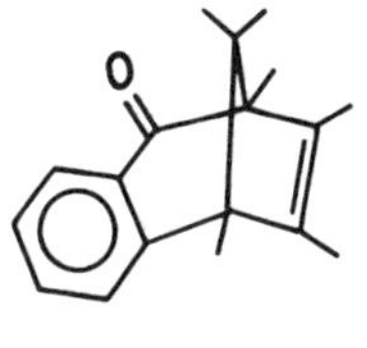

XI 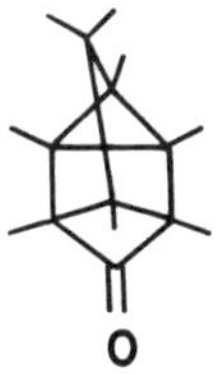   XII

The *syn* and *anti* isomers of spiro[3.4]octan-1-one were distinguished in the presence of Eu(dpm)$_3$, and it was possible to determine the position of deuterium substitution in a partially deuterated derivative.[372] The stereochemistry of the spirodecanone XIII was assigned on the basis of data from a study with LSR.[386]

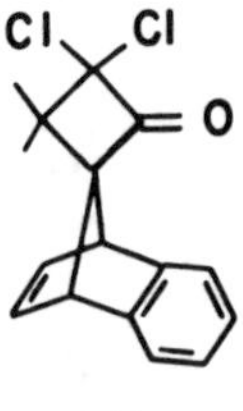

XIII

The shifts in the ^{13}C spectra of a series of bicyclo[4.2.0]oct-7-en-3-ones with Eu(dpm)$_3$ have been reported.[387]

The conformation of bicyclo[3.3.1]nonan-9-one(XIV) has been the focus of three reports involving LSR.[268,269,371] Schneider et al. studied the ^{1}H and ^{13}C spectra with Eu(fod)$_3$ and Yb(fod)$_3$, but was unable to generate only one conformation or conformational average that fit the data.[268]

XIV

This study did rule out the boat-boat conformer as a contributing form. Raber investigated the shifts in the ^{1}H spectrum with Eu(fod)$_3$ and reported that the chair-chair conformer alone could not explain the data.[371] A contribution of the chair-boat conformation resulted in the best fit. Vegar and Wells have reported that LSR do not alter the conformation of bicyclo[3.3.1]nonan-3-one.[269]

LSR have been used to determine the ring conformation of nootkatone(XV).[363]

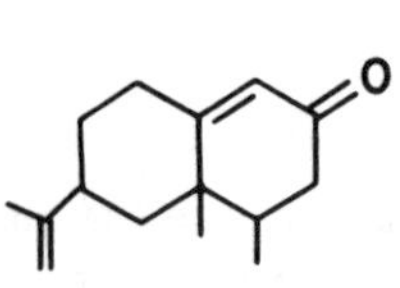

XV

By studying the shifts in the spectrum of nootkatone with LSR, the structure and spectra of one of its isomers were assigned.[362] The ring structure of valeranone has been determined with the aid of LSR.[318] Model cyclohexanones were used in this study to help in the assignment of resonances and structure. LSR have been used to assign the proton resonances of pentacyclo[5.3.0.0.^{2,5}.0^{3,9}.0^{4,8}]decan-6-one.[275] The conformation and relative energies of the conformers of *cis*- and *trans*-2-decalone have been determined with the aid of ^{1}H and ^{13}C shift data with Yb(fod)$_3$.[329] The conformation of *trans*-9-alkyl and *trans*-10-alkyl-2-decalones have been studied in the presence of Eu(fod)$_3$.[382]

Shift data and Eu(fod)$_3$ were used to assign the bridgehead protons of XVI.[367] The relative stereochemistries of 4-oxo-4,5-dihydroaldrin(XVII) and δ-ketoendrin(XVIII) were determined through studies with Eu(dpm)$_3$.[281]

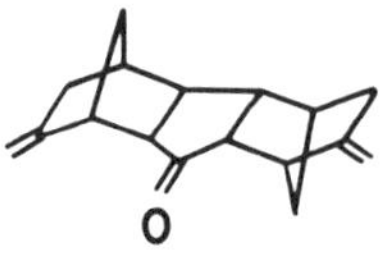

XVI 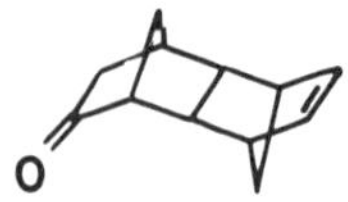   XVII    XVIII

The isomers XIX and XX have been distinguished through the application of Eu(fod)$_3$.[370] The olefin resonances of XIX and XX exhibited shifts of different magnitudes because of their respective distances from the carbonyl group.

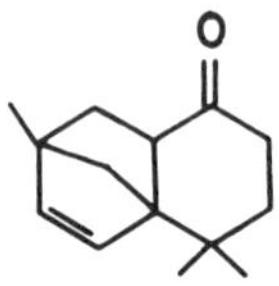

XIX 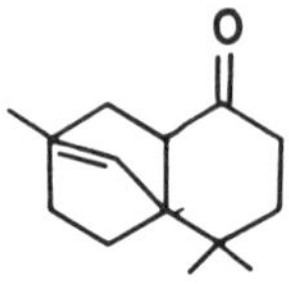   XX

The extent of deuteration of a benzocyclohexanone derivative has been determined using shift data with Eu(fod)$_3$.[368] The configuration of the olefin substituent of XXI was determined by the use of Eu(fod)$_3$.[369]

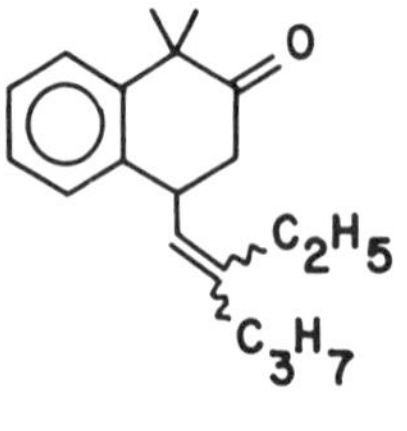

XXI    XXII 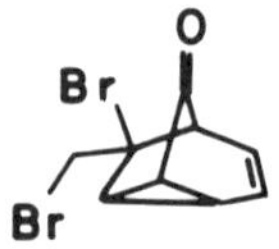   XXIII

Shift data with LSR have been used to assign the position of deuterium substitution in XXII[246] and the stereochemistry of the bromomethyl group in XXIII and XXIV.[388] The relative shifts of the vinylic protons with Pr(fod)$_3$ have been used to distinguish XXV from XXVI.[389]

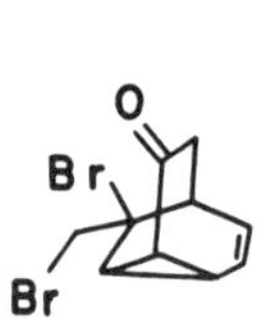

XXIV 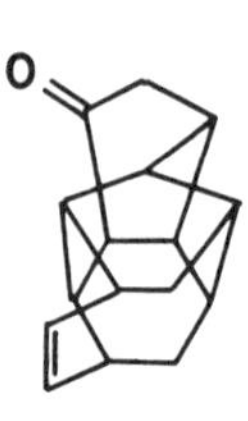   XXV    XXVI

The shifts in the ^{1}H and ^{13}C NMR spectra of indanone,[134,333] fluorenone,[134,343,365,390] and phenalenone[134] with LSR have been reported. A significant contact shift for some of the carbons of phenalenone was noted in one study.[134] Shift data with Eu(dpm)$_3$ helped in the structural assignment of 9,9-dimethyl-1,2,3,4-tetrahydroanthrone.[265]

Adamantanone has been employed as a substrate in studies designed to assess the validity of the simplified pseudocontact shift equation,[316] the nature of the bonding between ketones and LSR,[325,328] and the influence of steric effects on the association of ketones and LSR.[218,379]

A number of keto steroids have been examined with the aid of LSR. As with the hydroxy steroids, studies of keto steroids have focused on the assignment of certain ring conformations,[62,366,374] methyl resonances,[295,298-300] or ^{13}C resonances.[79,264,282,289] The conformations of the A rings of 4,4-dimethyl-3-keto steroids have been assigned by comparing the shifts with Eu(fod)$_3$ to those of steroids with known ring conformations.[374] The eight methyl resonances of 12-ursen-11-one have been assigned by studying the spectrum with Eu(fod)$_3$.[295,298]

Eu(fod)$_3$ and Pr(fod)$_3$ have been utilized in assigning the ^{13}C resonances of the 3-, 4-, 15-, 16-, and 17-keto androstanes.[282] The shifts in the ^{13}C NMR spectra of bromo-substituted keto steroids with Yb(tfn)$_3$ have been reported.[79] The dibromo compound XXVII did not exhibit any shifts in the presence of LSR because of either electronic or steric effects.[79] Shift data with Eu(fod)$_3$ enabled the determination of positions of deuterium substitution in steroids with the 13(17)-en-16-one unit.[391] Mixtures of testosterone and its 17-methyl- and 7β,17α-dimethyl-derivatives were analyzed and quantitated in the presence of Eu(fod)$_3$.[392]

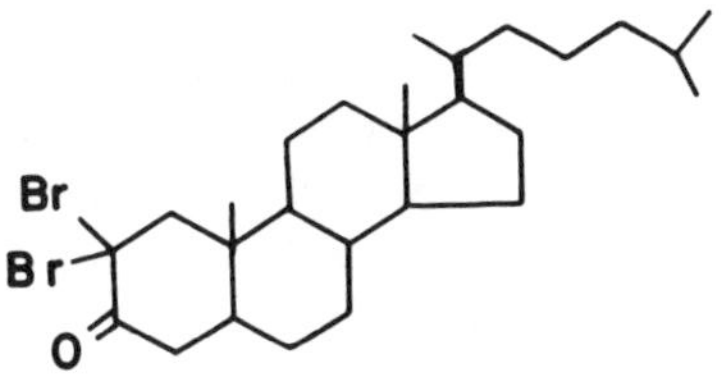

XXVII

3. Esters

The carboxy group of esters will be considered as a monofunctional donor. It has been recognized that the ester group bonds to LSR through the carbonyl, rather than the ether, oxygen.[393,394] This was confirmed by recording the shifts in the ^{17}O NMR spectrum of ethyl acetate in the presence of Dy(dpm)$_3$.[189] The shift in the oxygen resonance of the carbonyl oxygen was considerably larger than that of the methoxy oxygen. The bonding of a LSR to the carbonyl group of an ester is similar in nature to that previously described for ketones. Contact shifts in the carbonyl carbon resonance of esters have been noted.[331] A study of the influence of electronic effects on the association constant of 1-substituted adamantanes with ester functional groups with Eu(fod)$_3$ has been reported.[185] Replacement of the hydrogen atoms of the substituent group with fluorine atoms resulted in a significant reduction in the association constant. A correlation of the association constant with the gas phase basicity of the substrate was noted in this report.[185]

A number of studies have described the influence of LSR on the spectra of methyl esters of long chain fatty acids.[117,158,166,395-402] The resonances for the hydrogen atoms on the first eight carbons of methyl petroselinate (I) were resolved with Eu(fod)$_3$.[158,395] This facilitated a determination of the location of the olefin bond. Neither report, however, was able to resolve enough of the chain to determine the position of the olefin group in methyl oleate (II).[158,395] One study noted a cross-over of some

$$CH_3OC(CH_2)_4CH=CH(CH_2)_{10}CH_3 \quad\quad CH_3OC(CH_2)_7CH=CH(CH_2)_7CH_3$$
(with C=O above each OC)

I   II

resonances and recommended recording a series of spectra in which each contains an additional increment of LSR.[395] The spectra of methyl esters of fatty acids with Eu(dpm)$_3$ have been recorded in CS$_2$.[117] This report involved the study of methyl butyrate through methyl nonanoate, and first-order spectra were obtained for all substrates except the methyl nonanoate. The shifts in the ^{13}C spectra of methyl hexanoate through methyl dodecanoate with Eu(dpm)$_3$, Eu(fod)$_3$, and Yb(fod)$_3$ have been reported.[400] Complete resolution of all the carbon resonances of methyl dodecanoate with Yb(fod)$_3$ was noted in this study. Eu(dpm)$_3$ was also used to quantify mixtures of methyl butyrate, methyl hexanoate, and methyl heptanoate.[117] In these mixtures, the three methoxy singlets were resolved in the presence of the shift reagent. Using this same technique, mixtures of methyl oleate and methyl elaidate isomers, which differ only in the configuration of the double bond, were quantified with Eu(fod)$_3$.[396] Methyl oleate and methyl elaidate can also be distinguished by their ^{13}C spectra in the presence of Yb(fod)$_3$.[399] LSR have been employed to determine the amount of various unsaturated fatty acids in fatty acid mixtures.[401]

LSR have been used to determine the position of deuterium substitution in methyl esters of aliphatic acids.[398] This determination was performed by comparison of the ^{1}H spectrum of the fully protonated and partially deuterated analogs. The extent of deuterium substitution of 2,2-dideuterio fatty acid methyl esters has been determined through the use of LSR.[402] Deuterium NMR has been used to analyze the position and amount of deuterium substitution in methyl nonanoate and methyl 2-methyloctanoate.[397] This study found that deuterium NMR was preferable to ^{1}H NMR since lower levels of ^{2}H substitution could be detected. The shifts in the ^{19}F NMR spectra of trifluoroethyl esters with a variety of shift reagents have been reported.[403]

The preferred conformation of the methyl ester of 3-phenylbutyric acid has been determined through the analysis of shift data with Eu(fod)$_3$.[404] In this study, the shifts of the substrate were compared to the shifts of esters with well-characterized conformations.[404] Shift data with LSR have also been used to confirm that simple esters occupy an *s-trans* conformation.[394] This conformation was expected because of steric and dipolar effects.

The configuration of olefin bonds in substrates with an ester moiety have been determined through analyses with LSR.[190,405,406] The shifts with Pr(fod)$_3$ enabled the distinction of the isomers of III.[405]

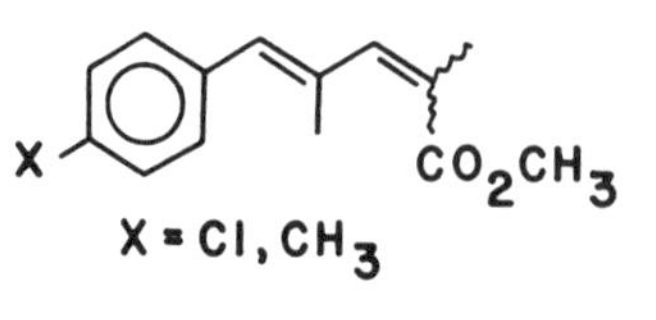

III (X-phenyl-CH=CH-CO$_2$CH$_3$, X = Cl, CH$_3$)   IV (R$_1$R$_2$C=CR$_3$-O$_2$CCH$_3$)

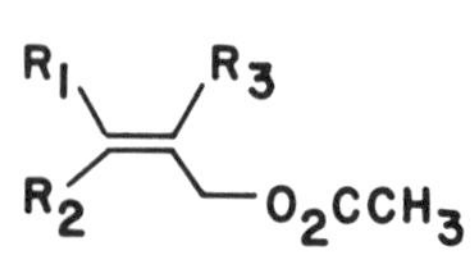

Shift data with Eu(fod)$_3$ has been used to distinguish the *trans,trans* from the 2-*trans*-4-*cis* isomer of the methyl ester of sorbic acid.[190] It was also possible to determine the configuration of the olefin bond in methyl cinnamate using LSR.[190]

Esters of structure IV were studied with the aid of LSR.[406] These compounds were prepared by the esterification of the corresponding alcohol. The stereochemistry about the double bond was assigned on the basis of shift reagent experiments. It was also possible to quantify mixtures of the *cis* and *trans* isomers in the presence of the shift reagent by monitoring the acetoxyl methyl resonance. Data obtained with LSR have been used to assign the *cis* and *trans* configuration of neryl acetate(V) and geranyl acetate(VI).[193]

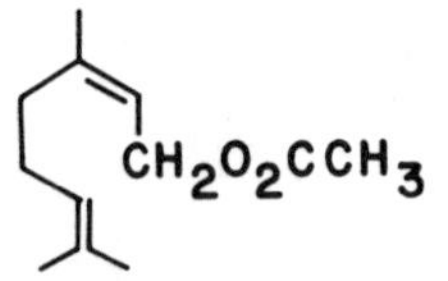

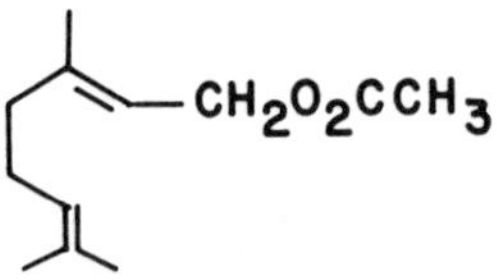

LSR have been used to study more complex olefinic esters,[407-409] including tocopherols and derivatives of vitamin A. Isomers of vitamin A that differed in the configuration of double bonds were distinguished using LSR.[408] It was also possible to distinguish and quantify mixtures of tocopherols and retinols by analyzing the acetate signals in the presence of LSR.[407] A similar procedure can be used to analyze mixtures of esterified cresols.[407] Mixtures of *o*-, *m*-, and *p*-toluic esters can be resolved and quantified in the presence of LSR.[178,410] The ratios of the association constants of methyl benzoate and phenyl acetate were determined by measuring the shifts of the methyl resonances of mixtures of these substrates in the presence of Eu(dpm)$_3$.[411]

The shifts in the esterification products of phenols with LSR have been measured and used as a basis for investigating polymers with similar functional groups.[412] A systematic correlation of the lanthanide-induced shifts with the Hammett sigma constants has been noted in the spectra of *m*- and *p*-substituted aromatic methyl esters.[410] Esters of 5,6,7,8-tetrahydro-α-naphthoic acids have been identified in petroleum concentrates with the aid of Eu(fod)$_3$.[413]

The configurations of *cis* and *trans*-methylchrysanthemate(VII) have been determined from lanthanide shift data, and mixtures of the two can be quantified in the presence of a LSR.[197,393] The *Z* and *E* isomers of methyl 2-alkyl-3-ethylidene-1-carboxylatecyclopropane(VIII) have been distinguished on the basis of shift data with Eu(dpm)$_3$.[202]

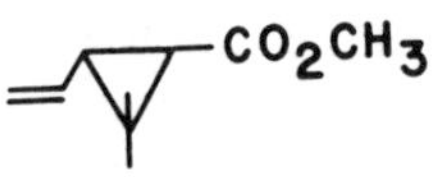

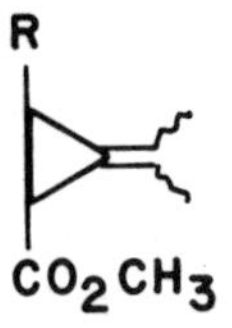

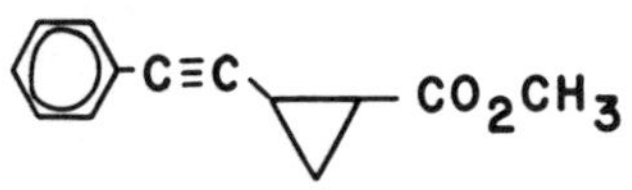

The spectrum of IX is first-order in the presence of a lanthanide shift reagent.[196] LSR have been used to assign all of the resonances and distinguish the *cis* and *trans* isomers of methyl and ethyl 2-phenylcyclopropane carboxylate.[378,414]

Shift data with Eu(fod)$_3$ enabled the distinction of 1- and 2-acetoxycyclopentadiene and 1- and 2-[(ethoxycarbonyl)methyl]cyclopentadiene.[416] The *cis* or *trans* configuration of the esters of certain 1-alkyl-4-methyl-4-nitrocyclohexa-2,5-dienols was assigned on the basis of shifts in the ^{13}C spectra with Eu(fod)$_3$.[222]

A procedure to assign multiple peaks of a mixture to a particular isomer has been described and its use demonstrated on syn- and anti-2-methoxycarbonyl spiro[cyclopropane-1,1'-indene](X).[417]

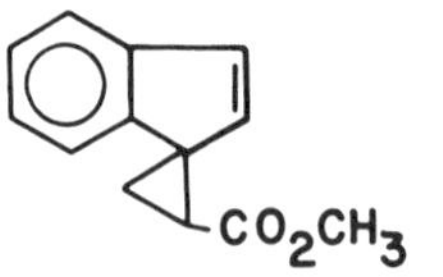

X

One resonance from one of the isomers is selected, and the ratio of the shifts of each proton to the reference signal is calculated. If the signals are from the same compound as the reference signal, a plotting procedure yields a linear plot. Nonlinear results are obtained for resonances from the other isomer.[417]

Eu(fod)$_3$ has been employed in the determination of the axial and equatorial positions of substituent groups in highly substituted cyclohexane derivatives.[216] The substrates were prepared by acetylating the corresponding alcohol. The shifts for the esters were smaller than those for the alcohols. The configurations of methyl esters of p-menthan-7-oic, p-menth-2-en-7-oic, and p-mentha-2,5-dien-7-oic acid have been assigned on the basis of studies with LSR.[418] The configurations and conformations of 2- and 4-methyl-1-(ethoxycarbonyl)cyclohexane were determined through the use of Eu(fod)$_3$.[419] LSR have been used to analyze and assign the methyl resonances for a series of eight monocyclic monoterpene derivatives.[215]

The spectra of multicyclic ring compounds with seven [77,224,420-422] and nine [270,423] carbons and an ester functional group have been analyzed in the presence of LSR. The equatorial and axial forms of substituted bicyclo[4.1.0]heptanes(XI) were distinguished on the basis of shift data with LSR.[421] The structure and configuration of a bicyclo[4.1.0]heptene (XII) was assigned with the aid of Eu(fod)$_3$.[422] The shifts in the ^{13}C spectrum of XIII with Eu(fod)$_3$ have been reported.[224]

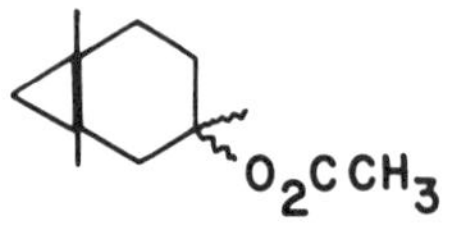

XI 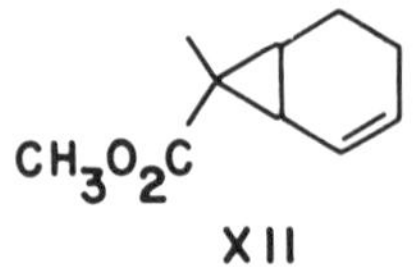   XII 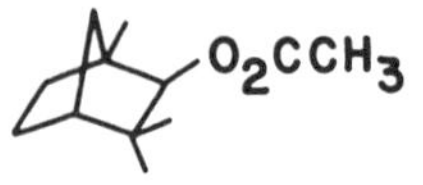   XIII

The spectra of trifluoroacetate esters of exo-norbornanol have been recorded in the presence of a number of LSR.[77] The shifts with Eu(fod)$_3$ and Eu(tfn)$_3$, however, were not satisfactory. The electron-withdrawing effects of the —CF$_3$ group lowers the basicity of the carbonyl group to the point that only weak complexation occurs. Large enough shifts were observed with Yb(fod)$_3$ to enable the determination of the distribution of deuterium in the substrates.[77] LSR have been employed in assigning the configuration of the tricycloheptanes XIV.[420]

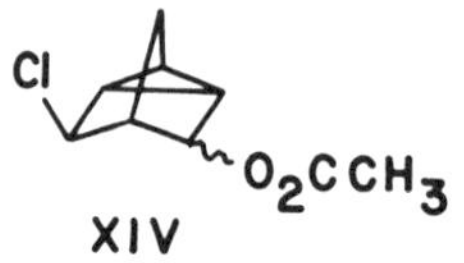

XIV

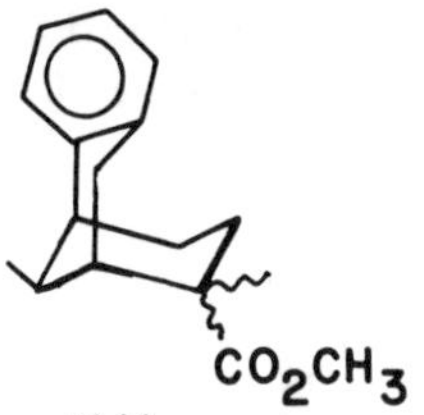

XV 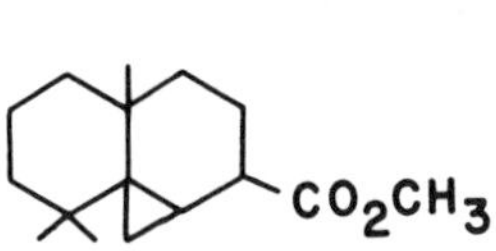   XVI

Shift data with Eu(dpm)₃ were utilized in the structural and stereochemical assignments of bicyclo[3.3.1]nonene derivatives(XV).[423] Shift data with a LSR have been used to assign the configuration and conformation of bicyclo[6.1.0]nona-2,4,6-trienes substituted with an ester functional group.[270] The conformation of hinokiic acid was assigned on the basis of lanthanide shift data of the corresponding methyl ester(XVI).[424]

A number of investigations have described the application of LSR to steroids with ester functional groups.[62,143,289,295,298-300,307,425,426] LSR have been used to assign the methyl resonances in the ¹H NMR spectra of 12-ursen-28-oate,[295,298] methyl betulinate, methyl oleanolate,[143] acetyl sterols,[426] and other acetylated steroids.[299,300] With the acetyl sterols it was possible to differentiate the 24 α and 24 β methyl groups through the use of Ho(fod)₃.[426] An inequivalence of the isopropyl methyl resonances at C-26 and C-27 was also noted in this study. LSR have been used to aid in the assignment of the ¹³C spectra of acetylsteroids.[289,307] The shifts in the presence of a LSR for a series of epimeric esters prepared from their corresponding alcohols have been reported.[62] The shifts were larger for the original alcohols; however, the epimeric distinction in the shifted spectra was more pronounced for the ester derivatives. Mixtures of vitamin D₂ isomers have been analyzed through the use of LSR.[425] The corresponding alcohols were acetylated and the methyl resonances of the acetyl group were resolved on addition of a shift reagent. Such a procedure was suitable for the analysis of mixtures containing up to eight different isomers.[425]

4. *Lactones*

The shifts in the ¹⁷O NMR spectrum of γ-butyrolactone in the presence of Dy(dpm)₃ confirmed that lactones bond to LSR through the carbonyl oxygen.[189] The shifts in the spectrum of β-angelicalactone(I) with Eu(fod)₃ and Pr(fod)₃ were best fit to the structure if the LSR was situated at the lone pair *anti* to the ring oxygen.[427]

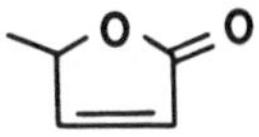

I 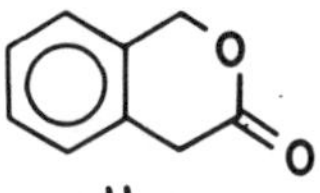   II

The compounds γ-butyrolactone and 3-isochromanone(II) have been employed as model substrates in a study of the conformation of esters.[394] These two substrates were chosen because they are locked into an *s-cis* conformation. The shifts in the ¹³C spectrum of 3-isochromanone with Yb(fod)₃ have been reported.[307] The magnitude of the shifts in the compounds γ-butyrolactone and 32aB,4,5,6,7,7-a8-hexahydro-2(³H)-benzofuranone have been measured and used to determine the site of complexation of LSR with polyfunctional substrates with a lactone group.[428]

The shifts in the spectrum of 3-phenyl-7-methylcoumarin with Pr(fod)₃ have been reported.[429] LSR have been employed to determine the configuration and conformation of cis- and trans-3,5-dimethylvalerolactone(III).[430] Conformational analysis

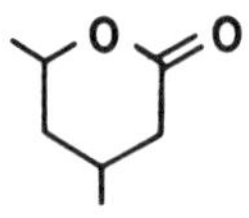

III 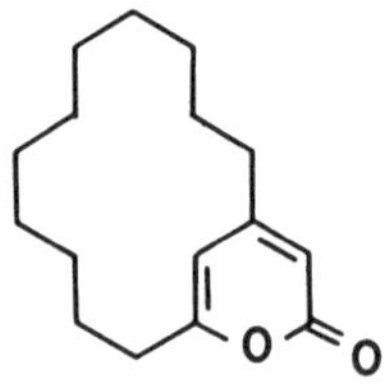 IV

of this compound was achieved by an evaluation of the coupling constants and dihedral angles present in the rigid ring system. Data obtained with Eu(fod)$_3$ have been used to aid in the structural assignment of IV.[431] The stereochemistry of V was assigned on the basis of shift data with Eu(fod)$_3$.[201]

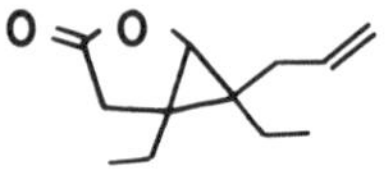

V

Tori et al.[432] have employed a novel method of conformational analysis on two ten-member ring-sesquiterpenes. The LSR was Eu(fod)$_3$, and the substrates were costunolide(VI) and dihydrocostunolide. The method

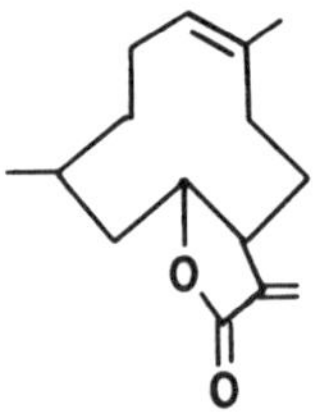

VI

involves measuring changes in the nuclear Overhauser effect in the presence of a LSR. The enhanced relaxation brought about by the paramagnetic lanthanide ion causes a decrease in the NOE leading to a decrease in the intensity of a resonance. The decrease in NOE is related to a $1/r_6$ dependency and can be used in conformational analysis.

5. Aldehydes

A number of investigations of aldehydes with LSR have been reported.[224,315,335,342-345,375,433-436] With aldehydes, there is a strong preference for the LSR to bond at the less hindered lone pair of the carbonyl group.[315,345,435] LSR have been applied in the study of vinyl aldehydes with structure I.[315,433]

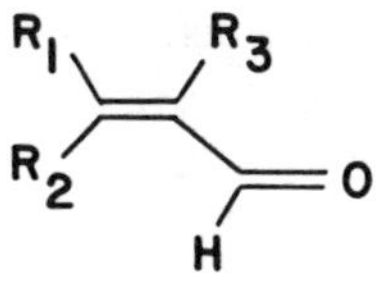

Using shift data, it was possible to determine both the configuration about the double bond and the conformation of the compound. LSR have been used to study the interconversion of the conformers of methyl-2-benzaldehyde, methyl-3-benzaldehyde, and chloro-3-benzaldehyde.[335] Abraham has published several reports describing conformational analyses of benzaldehydes by fitting of lanthanide shift data.[343-345,434-436] Both ^{1}H and ^{13}C data were

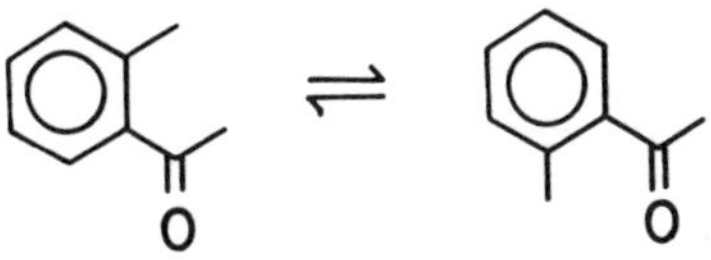

used in an effort to improve the fit. The complexation shifts for benzaldehydes were significant for some carbon resonances and provided a means to study the extent of pi type interactions.[343] It was reported that the shifts in the ^{13}C spectra of benzaldehydes in the presence of a LSR could be explained by a pure dipolar mechanism after accounting for the presence of complexation shifts.[434,435]

The conformation of 2-phenylpropionaldehyde was determined by an analysis of data recorded with a shift reagent.[375] It was reported that complexation of the shift reagent did not perturb the conformation of the substrate. The shifts in the ^{13}C spectrum of II with Eu(fod)$_3$ have been reported.[224]

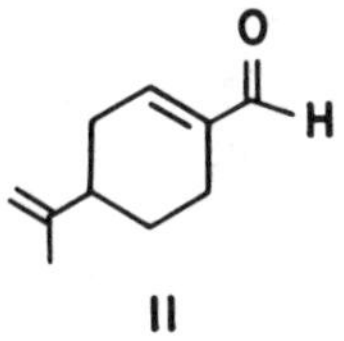

6. Carboxylic Acids

Lanthanide chelates of dpm are, for the most part, unstable in solutions of carboxylic acids, and a precipitate forms.[437] Chelates of fod, however, do not decompose in the presence of carboxylic acids and can be employed with these substrates.[73,218,311,437-440] Solutions of Eu(fod)$_3$ and butyric acid were kept for up to 7 days without any observable precipitation.[73]

The shifts in the NMR spectrum of benzoic acid, adamantane-1-carboxylic acid, crotonic acid, and undecylic acid with Eu(dpm)$_3$ and Eu(fod)$_3$ have been reported.[437] Decomposition of the Eu(dpm)$_3$ was observed with the first two substrates. For the other compounds, the shifts were slightly less in magnitude than those for the corresponding esters. The NMR spectrum of n-hexanoic acid is converted to a first-order spectrum in the presence of Eu(fod)$_3$.[311] No decomposition of the shift reagent was observed, and the magnitude of the shifts remained stable over a period of several days.

Shift data with Eu(fod)₃ have been used to distinguish *cis*- and *trans*-3-alkenoic acids.[438] The conformations of dipropyl-, diethyl-, and dimethylacetic acid were studied with the aid of LSR and compared to the conformations of the corresponding amides.[439] LSR are suitable for the study of amino acid derivatives as demonstrated by the shifts reported in the spectrum of *N*-trifluoroacetyl-*d*-alanine.[73]

7. Anhydrides

Eu(fod)₃ has been employed in the determination of the solution structure of I.[441] The shifts in the spectrum were indicative of coordination of the LSR at the carbonyl groups. The results of the solution structure were compared to those found in the solid state X-ray crystal structure of the same compound.[441]

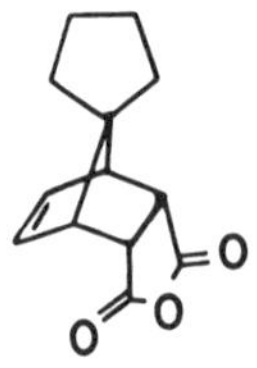

I

8. Ethers

The ether functional group is a rather weak Lewis base. As a result, ethers bond more weakly to LSR than many other classes of compounds with oxygen atoms.[185] In their initial report on NMR shift reagents with the fod ligand, Rondeau and Sievers compared the shifts in the NMR spectrum of di-*n*-butyl ether with Eu(fod)₃, Pr(fod)₃, Eu(dpm)₃, and Pr(dpm)₃.[3] The larger shifts with the complexes with fod were attributed to the greater Lewis acidity of the lanthanide metal.

An extended pi system tends to further weaken the interaction between ethers and LSR. The shifts in the spectra of methoxy-*n*-butane and anisole (methoxybenzene) were recorded with the chelates of fod, and a smaller association constant was indicated for the aromatic ether.[129,135] The ¹³C and ¹H shifts in the spectra of 1-methoxy- and 2-methoxynaphthalene with Eu(fod)₃ and Yb(fod)₃ were much smaller than those for similar ketone derivatives.[134] No usable shifts were obtained in the spectrum of methoxyfluorene with Yb(fod)₃.[134] Employing Yb(dpm)₃, large shifts were noted in the ¹H NMR spectrum of tetrahydrofuran, while essentially no shifts were observed in the spectrum of furan.[112,442] Sievers et al. have reported that good shifts are observed in the ¹H NMR spectrum of anisole when lanthanide chelates with dfhd are used.[72]

Substantial shifts in the NMR spectra of ethers can be obtained if steric hindrance about the ether oxygen is reduced by holding the R groups back in a rigid configuration.[129,379] As a result, the two most common types of ether substrates to which LSR have been applied are cyclic ethers[443-446] and epoxides.[244,248,273,307,379,428,447-456]

The shifts in the spectrum of 1,2-epoxydodecane with Eu(fod)₃ have been reported.[447] Resolution of the hydrogen resonances of the first seven methylene groups were observed. Resolution of the two diastereotopic hydrogen atoms on the methylene group alpha to the epoxide was also noted. The shifts in the ¹H NMR spectra of 1,2-epoxyoctane and cyclohexene oxide with Eu(dpm)₃ have been reported.[428] The conformation of I has been assigned on the basis of shift data with Yb(dpm)₃ and the coupling constants that could be obtained in the shifted spectrum.[452]

I   **II**

The structures of *cis*- and *trans*-4-*tert*-butylcyclopentane-1,2-epoxide were assigned on the basis of shift data with a LSR.[456]

The shifts in the spectra of 1,4-dihydronaphthalene-1,4-*endo*-oxide(II), 1,2,3,4-tetrahydronaphthalene-*endo*-oxide, and benzonorbornadiene-*exo*-oxide(III) have been fit to structures using the dipolar shift equation.[454,455] Significantly better fits of the data were obtained if the angle term was incorporated into the procedure. Shift data with Eu(dpm)$_3$ have been used to confirm the stereochemistry of the pesticides eldrin and dieldrin, and to determine the stereochemistry of photodieldrin, a breakdown product of these pesticides.[450] The *syn* or *anti* position of the epoxide ring relative to carbon-2 in IV has been determined with the aid of LSR.[273] A subsequent report by Wing,[244] however, questioned some of the assignments in the earlier report.[273]

III   **IV**

LSR have been used to confirm the configuration of di-spiro(bicyclo[4.2.1]non-3-ene-9,2'-oxirane-3',9"-bicyclo-[4.2.1]non-3"-ene(V).[451] The shifts in the ^{13}C spectrum of VI and VII with Yb(fod)$_3$ have been reported.[307]

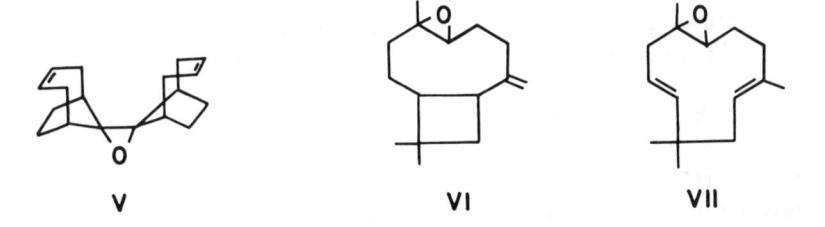

V   **VI**   **VII**

The configurations of the epoxide groups in *trans*-epoxy-2,3,3-bicyclo(3.3.0)octane and *trans*-epoxy-2,3-bicyclo(4.3.0)nonane have been determined through the use of data recorded with a LSR.[448] The investigators reported that the application of LSR to the determination of these configurations was superior to any other method available.

The shifts with Eu(dpm)$_3$ have been used to distinguish VIII from IX.[248] The shift data also provided information about the conformation of the two compounds.

VIII   **IX**

Shift data with Eu(dpm)$_3$ were successfully used to determine the conformation of 2-methyltetrahydropyran, but did not result in a reliable conformation for cis-4-methyl-2-(2′-methyl-1′-propenyl)tetrahydropyran.[446] Spectra recorded with Eu(fod)$_3$ permitted the distinction of 7,8-cis-endo-diphenyl-2-oxabicyclo[4.2.0]octane(X) from 7,7-diphenyl-2-oxabicyclo[4.2.0]octane(XI).[443]

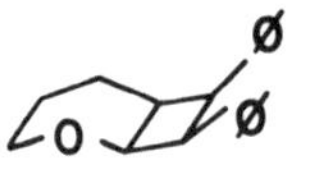

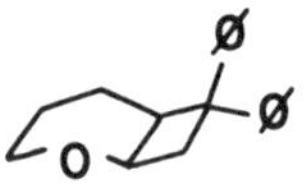

The distinction of the two isomers was based on an analysis of the coupling constants that were obtained in the shifted spectra. Shift data with these substrates indicated that the shift reagent bonded preferentially to the oxygen lone pair pointing below the plane of the ring. This lone pair is removed from the steric hindrance of the phenyl groups.

A determination of the conformations of a series of substituted 3-oxabicyclo[3.3.1]nonanes(XII) has been facilitated through the use of LSR.[444] The shift data for the

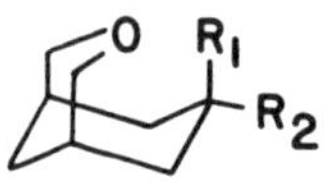

compounds under study were compared to data collected on 4-methyltetrahydropyran and 2-oxadamantone. Using both ^{13}C and ^{1}H shift data and coupling constants that could be determined in the shifted spectra, the investigators were able to determine the conformations and degree of ring flattening of the substrates. Similar to the results with X and XI,[443] a preferential bonding of the shift reagent to the exo lone pair on the oxygen, which is removed from any steric hindrance created by R$_1$, was implied.[444] The assignment of the aromatic resonances of XIII was facilitated by recording the spectrum with Pr(fod)$_3$.[445]

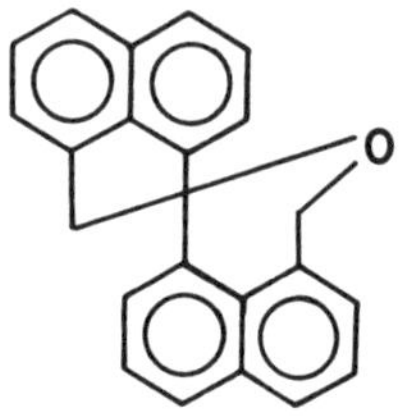

LSR have been employed in the study of methyl ethers.[168,177,222,243,273,307,379,403,407,456-460] Only small shifts were recorded in the ^{19}F spectra of trifluoroethyl ethers with Eu(fod)$_3$.[403] Shift data with Eu(fod)$_3$ have been used to aid in a determination of the conformation of 2-phenylpropyl methyl ether.[177] The structure of 1-methoxy-4-bromo-2-butyne was assigned on the basis of a shift reagent experiment.[460] The methyl resonances of isoanhydrovitamin A were resolved in the presence of Eu(dpm)$_3$.[407] The isomers of XIV have been distinguished with the aid of Eu(fod)$_3$.[168] The methyl ethers of the Z and E isomers of 1,4-dimethyl-4-nitrocyclohexa-2,5-dienols were distinguished using ^{13}C lanthanide shift data in conjunction with other available data.[222]

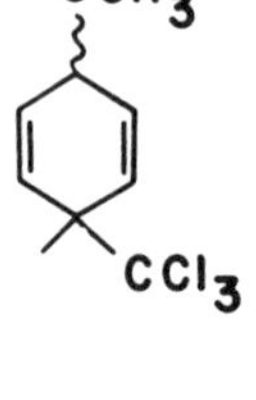

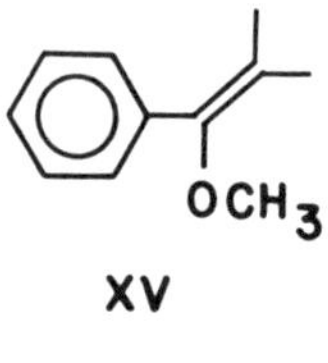

Enol ethers of the methoxy styrene family(XV) have been studied with the help of Eu(dpm)$_3$.[459] It was possible to determine both the geometry and conformation of these substrates through the use of the shift reagent. It was also found that the degree of conjugation between the ring and double bond influenced the basicity of the ether. The basicity of the ether influenced the magnitude of the shifts in the presence of the shift reagent.[459] Shifts in the ^{1}H [Eu(dpm)$_3$] and ^{13}C [Yb(fod)$_3$] spectra of cis-4-tert-butylcyclohexyl methyl ether have been reported.[243,307] Both the stereochemistry and conformation of 7-methoxybicyclo[4.3.1]decatriene have been determined through the application of a LSR.[273] Coupling constants obtained in the shifted spectrum permitted an assignment of the exo or endo position of the methoxy group and whether the ring was in a half-chair or half-boat conformation. LSR have been used to aid in the determination of the configuration of the cyclopropane ring in XVI and XVII.[457]

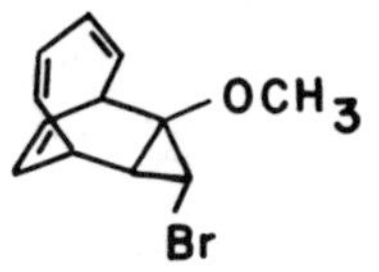

XVI 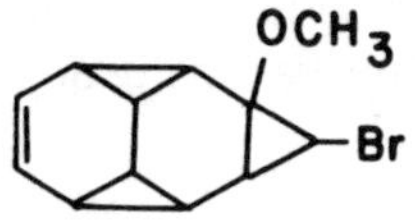   XVII

9. Peroxides

The spectra of t-butylhydroperoxide(I) and cumene hydroperoxide(II) have been recorded in the presence of Eu(fod)$_3$.[166] Shifts were

$(CH_3)_3C-O-O-H$

I

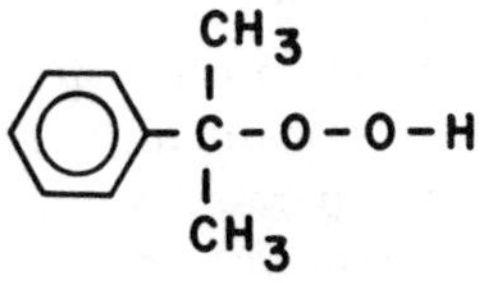

II

observed in the spectra on addition of the shift reagent; however, extra peaks were noted. The intensity of the two sets of resonances varied systematically with the concentration of the shift reagent. The total area of the two sets of resonances remained constant. One possible explanation offered by the investigators is that two discrete complexes, with the structures shown below, form with peroxide substrates.[166]

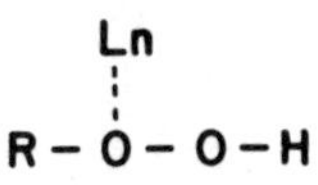

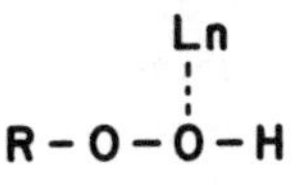

$$\begin{array}{cc} \text{Ln} & \text{Ln} \\ \vdots & \vdots \\ R-O-O-H & R-O-O-H \end{array}$$

The NMR spectra of di-tert-butyl peroxide and dicumene peroxide did not exhibit shifts in the presence of Eu(fod)$_3$.[166] This is most likely the result of excessive steric hindrance.

10. Amines

The interaction between amines and LSR is highly dependent on steric[169,442,461-468] and electronic[169,186,442,461,466,469,470] effects. The most pronounced steric effect is that observed with increased substitution at the nitrogen atom. The association constants of amines decrease significantly from primary to N-substituted to N,N-disubstituted derivatives.[442,461,462,464,467,468] This has been reported for mono-, di-, and trialkyl amines[169,442,467] and for a series such as aniline, N-alkyl-, and N,N-dialkylaniline.[461,464]

In addition to the degree of N-substitution, the steric nature of the substituent groups is important in determining the association constant of amines with LSR. The shifts in the spectrum of t-butylamine in the presence of a shift reagent were considerably less than those for n-alkylamines.[169,465,467] A correlation between the steric strain energies and lanthanide shifts has been observed for a series of aliphatic amines.[169,467]

Ring compounds such as anilines or piperidines exhibit pronounced steric effects when substituted at positions ortho or alpha to the nitrogen atom.[461,463,464,468] Shifts with LSR are observed to decrease for the series aniline, o-toluidine, and 2,6-xylidine.[464] The effect of such substitution on the association constant is less pronounced, however, than substitution at the nitrogen atom.[461,464]

Electronic effects have been reported to be less important than steric effects in determining the association constants of amines with LSR.[442,468] Nevertheless, electronic effects influence the degree of interaction. Large shifts were observed in the spectrum of pyrrolidine (I), while essentially no shifts were noted by pyrrolle (II), in the presence of LSR.[442] The ^{1}H and

I II

^{19}F shifts decrease significantly for the series aniline, 4-fluoroaniline, trifluoroaniline, and pentafluoroaniline.[469] Essentially no shifts are observed in the ^{19}F spectrum of n-perfluorobutylamine.[186] In each of these cases, the electron-withdrawing fluorine atoms lower the basicity of the nitrogen atom and reduce the association constant. The shifts in the spectra of para-substituted anilines, which have similar steric effects, have been shown to correlate with the basicity of the nitrogen atom.[461,470] A similar correlation has been observed for aliphatic amines.[169]

The presence of contact shifts in the spectra of amines has been noted in several reports.[89,124,165,469,471-476] Many of these reports have involved an analysis of the shifts in the ^{13}C spectrum of the substrate.[89,165,471-474] With some compounds, a contact shift mechanism was also reported to occur in the ^{1}H[124,469,474,475] and ^{19}F[89,469] spectra.

Contact shifts are typically noted for carbons alpha and beta to the amine nitrogen.[471-473] With europium chelates, this often results in an upfield shift for the resonance of the carbon atom beta to the amine.[471-473] This observation is more pronounced if the carbon atom at the beta position is substituted.[471,472] Contact contributions tend to correlate with the association constant and are less for N-substituted derivatives.[471] In extended pi systems, the contact contribution is larger and extends over more bonds.[89,475]

The spectra of a number of acyclic amines have been studied in the presence of LSR.[129,165,169,176,177,186,442,462,465-467,471,473,477-479] Many of these reports have focused on an assessment of steric effects, electronic effects, and the presence of a contact shift mechanism and have already been described. The shifts in the ^{14}N spectra of n-propylamine and t-butylamine with Dy(dpm)$_3$ and Dy(fod)$_3$,[465] and n-propylamine with Eu(dpm)$_3$ and Yb(dpm)$_3$, have been reported.[466] A conformational analysis of 2-phenylpropylamine was facilitated through the use of shift data with Eu(fod)$_3$ and Pr(fod)$_3$.[176,177] The compounds isobutylamine and 2,2-diphenylethylamine were used as model substrates in this study.[176] The relative configuration of the two diastereomers of N-(2-phenyl-1-methyl)propyl-N-phenylamine were assigned from an experiment with LSR.[479] The conformational preferences of butylamines were determined through the analysis of shift data with Eu(dpm)$_3$[129]

Aromatic amines have been the focus of a number of studies with LSR.[89,178,186,307,442,461,464,469,470,474,475,480,481] Most of these have employed derivatives of aniline and assessed factors such as steric effects, electronic effects, and the presence of a contact shift mechanism. Shift data with Eu(fod)$_3$ have been used to assign the resonances and determine the position of deuterium substitution in III.[481]

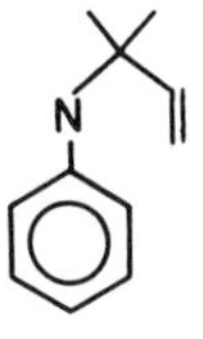

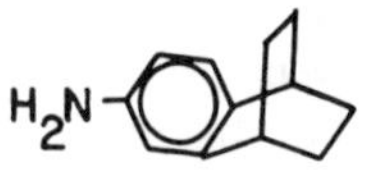

The assignment of the bridgehead resonances of IV and V was facilitated through the use of Pr(fod)$_3$.[480] Mixtures of o-, m-, and p-methylbenzylamine have been quantified by analysis with Eu(fod)$_3$.[178] The spectrum of amphetamine has been studied in the presence of Eu(fod)$_3$.[478] The diastereotopic protons of the methylene group were resolved in the presence of the LSR. The resonances were assigned on the basis of distance and angle considerations. This information also permitted a calculation of the relative population of each rotamer of amphetamine in the complexed form.

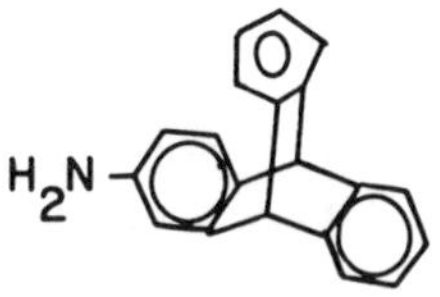

Monocyclic amines such as pyrrolidine,[129,165,442,465,466] piperidine,[129,307,442,462,463,482] and cyclohexyl amines[442,483] have been studied with the aid of LSR. The shifts in the ^{1}H,[165,442] ^{13}C,[165] and ^{14}N[465,466] spectra of pyrrolidine with various LSR have been reported. Shift data with Eu(dpm)$_3$ have been used to determine the configuration of 2-alkylpiperidines and 2,6-dialkylpiperidines.[463] Resolution of the resonances for axial and equatorial protons on the ring was observed in the presence of the LSR. It was also reported in this study that complexation of the LSR did not influence the conformation of these substrates. The shifts in the ^{13}C spectra of VI and VII with Yb(fod)$_3$ were measured and found to be larger for VII.[307] This is probably a result of improved steric and electronic effects in VII.

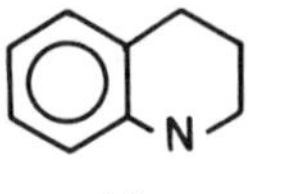

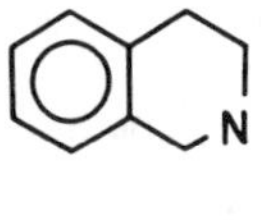

LSR have been utilized in the assignment of the configuration of the three isomers of 1,3-dimethylcyclohexylamine.[483] The association constant and bound shifts for trans-2-phenylcyclopropylamine with Eu(fod)$_3$ have been reported.[378] The N-methyl resonances of the cis and trans isomers of 3-chlorocyclobenzaprine(VIII) were more fully resolved in the presence of Eu(dpm)$_3$.[484]

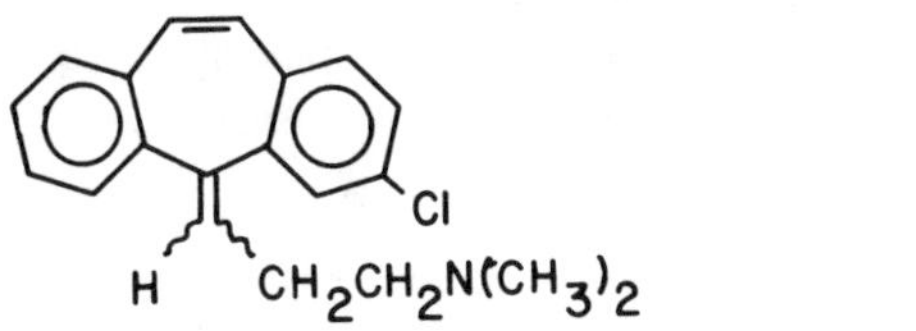

44 NMR Shift Reagents

Multicyclic amines have been substrates in studies with LSR.[124,165,307,442,471,472,485-487] Contact shifts have been noted in the $^1H^{124}$ and $^{13}C^{165,471}$ spectra of 1-aminoadamantane. The shifts in the $^1H^{442}$ and $^{13}C^{307}$ spectra of quinuclidine(IX) have been reported.

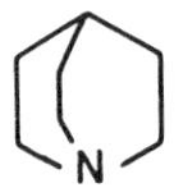

IX 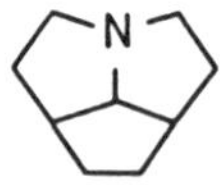   X 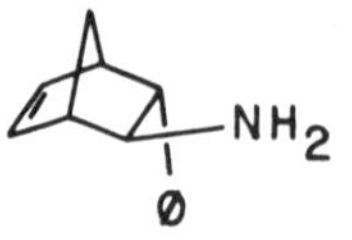   XI

The structural assignment of X was facilitated by the use of shift data with Eu(dpm)$_3$.[487]

The compounds 5-*exo*-amine-6-*endo*-phenyl- and 5-*endo*-amine-6-*exo*-phenylnorborn-2-ene(XI) have been distinguished on the basis of spectra obtained in the presence of Eu(dpm)$_3$.[485] Contact shifts have been reported in the ^{13}C NMR spectrum of *exo*-norbornylamine in the presence of Eu(dpm)$_3$ and Pr(dpm)$_3$.[472] The four isomeric 2-methyl-3-aminopinanes have been distinguished using Eu(dpm)$_3$.[486] Complete assignment of the resonances for each pure isomer was accomplished with the aid of the shift reagent.

11. Nitrogen Heterocycles

Compounds of the nitrogen heterocyclic class coordinate quite effectively to lanthanide tris β-diketonates. The first organic soluble LSR was the dipyridine adduct of Eu(dpm)$_3$.[1] Since nitrogen heterocycles bond strongly to LSR and have extended pi systems, it is not surprising that the presence of a contact shift mechanism has been reported in a number of instances.[71,469,488-493]

Downfield contact shifts for the carbon atoms alpha and gamma to the nitrogen atom, and upfield contact shifts for carbon atoms beta to the nitrogen were observed for nitrogen heterocycles with europium shift reagents.[491] It has been reported that the shifts in the ^{1}H NMR spectrum of pyridine,[71,492] quinoline,[490,491] isoquinoline,[491] and acridine[491] can be explained by a purely pseudocontact mechanism. Chalmers, however, reported the presence of a contact shift in the protons closest to the coordination site in quinoline.[488] Anomalous shifts in the ^{1}H spectra of a series of substituted pyridines have been reported and were believed to be due to the presence of contact shifts.[489]

Pyridine and its derivatives have been widely studied with the aid of LSR.[71,110,163,442,465,466,468,469,482,489,492,493,494-503] The impact that steric[442,468,495,500,502] and electronic effects[163,468,495,500,502] have on the association of pyridine derivatives with LSR have been evaluated. The most pronounced steric effects are observed when the pyidine ring is substituted ortho to the nitrogen atom.[442,495,500] Association constants of the order pyridine > 2-methylpyridine > 2,6-dimethylpyridine were obtained by performing competition experiments with two monofunctional substrates in solution with a LSR.[500] The position of the lanthanide ion in the complex with 2-methylpyridine is reportedly displaced away from the methyl substituent group. Substitution at the 3-, 4-, or 5-position in pyridine did not influence the location of the lanthanide ion, and linear bonding was observed.[501] The steric effects in quinoline were such that the lanthanide ion was positioned in a nonlinear arrangement away from the second ring(I). In 2,4-dimethylquinoline, the steric encumbrance of the methyl group was more pronounced than the other ring, and the lanthanide ion was positioned toward the side of the second ring(II).[501]

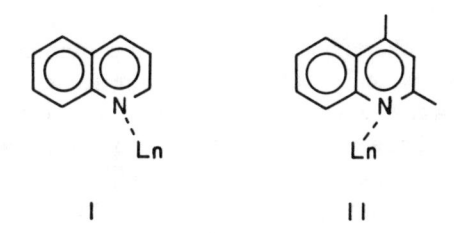

Electronic effects, while operative with pyridine derivatives, are less pronounced than steric effects.[468,500] The magnitude of lanthanide shifts for nitrogen heterocycles bears no simple relationship to the basicity as determined by pKa values.[495] For a series of compounds with similar steric properties such as 4-methylpyridine, pyridine, and 4-chloropyridine, electronic effects do influence the shifts.[500] For these three compounds, the shifts were largest for 4-methylpyridine and least for 4-chloropyridine. These relative values agree with what would be expected on the basis of electron-donating and -withdrawing properties of the substituent groups.

The shifts in the ^{14}N spectrum of pyridine in the presence of a variety of LSR have been reported.[465,466,498] From these studies, it was determined that chelates of Dy(III) and Yb(III) were the best upfield and downfield shift reagents for ^{14}N NMR, respectively.[498]

LSR have been used to distinguish the cis and trans isomers of 4[β(1-naphthyl)vinyl]pyridine.[499] The spectrum of pyridine has been recorded in the nematic phase N-(p-ethoxybenzylidene)-p'-n-butylaniline with Eu(dpm)$_3$.[494] From this study, it was concluded that LSR were not of much practical use in the study of the NMR spectra of oriented molecules. The shifts in the spectra of pyridine, quinoline, and benzo(f)quinoline with LSR have been compared to those of the corresponding N-oxides.[496] The shifts were smaller for the N-oxides, presumably because of the larger distance between the lanthanide ion and hydrogen atoms.

The spectra of quinolines[71,307,442,488,490,491,495,496,501,502,504,505] and isoquinolines[491,495,500,504,506] have been recorded in the presence of LSR. A number of these reports have investigated and reported the presence of contact shifts with these substrates.[71,488,490,491,504] Isoquinoline has a higher association constant than quinoline with LSR because of reduced steric effects.[495] Much the same as pyridines, the association constants of quinolines and isoquinolines with LSR are subject to pronounced steric effects when substituted at positions close to the nitrogen atom. The relative coordination abilities of a series of substituted quinolines has been reported as 6-chloro->6-chlor-2-methyl->6-chloro-8-methyl->2,8-dimethyl-.[500]

LSR have been used to determine the position of substituent groups in substituted isoquinolines.[506] It was possible to distinguish and confirm the structures of 5,6-dihydroquinoline and 7,8-dihydroquinoline with the aid of LSR.[505]

The shifts in the three-ring nitrogen heterocyclic compounds acridine(III),[491,495,500-502] phenanthridine(IV),[495] and benzo(f)quinoline (V)[496,500,504] with LSR have been reported.

12. Nitriles

In comparison to many other oxygen and nitrogen-containing functional groups, the cyano group coordinates rather weakly to LSR. The fod chelates, which are usually more effective LSR for weak Lewis bases, have been reported to produce shifts in the spectra of organonitriles that are twice as large as those with the dpm chelates.[507]

The cyano group bonds to LSR in a linear fashion,[508-510] and it has been noted that the association of organonitriles with LSR is not significantly influenced by steric effects.[510] The association constants for 32 organonitriles with Eu(fod)$_3$ in CCl$_4$ have been reported.[185,511] The differences in association constants were more dependent on electronic than steric effects. Contact shifts should be small for organonitriles because of the weak complexation with LSR. It has been reported, however, that the ^{13}C resonance of the cyano carbon in n-butylisocyanide exhibits a large contact shift.[471] The contact shift was large enough with Eu(fod)$_3$ and Pr(fod)$_3$ to shift this resonance in the wrong direction from what would be expected if only a pseudocontact mechanism was operative.

The application of LSR in the study of acyclic,[376,465,466,471,507-509,512,513] aromatic,[512,514,515] monocyclic,[452,508,510] and bicyclic[270,516-519] organonitriles has been described. The spectrum of heptanenitrile was first-order in the presence of Eu(fod)$_3$.[507] Shifts with Eu(fod)$_3$ were used to determine the molecular structure and conformer populations of the substrates methyl-, ethyl-, isopropyl-, and tertiary-butylnitrile.[509] The shifts in the ^{14}N NMR spectrum of acetonitrile with Dy(dpm)$_3$, Dy(fod)$_3$, Eu(dpm)$_3$, and Yb(dpm)$_3$ have been reported.[465,466]

LSR have also been used to study unsaturated acyclic organonitriles. The spectra of allyl cyanide[507,512] and acrylonitrile[507] have been recorded in the presence of LSR. The three methyl acrylonitrile isomers(I-III) have been distinguished with the aid of LSR.[508]

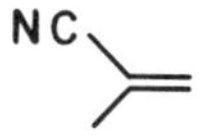

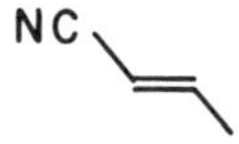

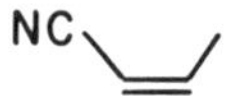

LSR have been used to assign the *cis* or *trans* configuration of olefins in which one of the substituent groups was a cyano group.[513]

Shifts in the spectra of benzylnitrile[514] and phenylnitrile[512] with LSR have been reported. Shift data with Eu(fod)$_3$ have been used to fit the structures of alkyl-substituted benzonitriles.[515] The best fits for unsymmetrical derivatives were obtained if the Ln–N≡C bond deviated slightly from linearity.[515] Various isomeric alkyl-substituted benzonitriles were distinguished on the basis of shift data with a shift reagent. LSR have been used to distinguish the isomers of IV.[508] Data with a LSR were used to determine the ring conformation of *cis*- and *trans*-1-cyano-2-vinylcyclobutane.[452]

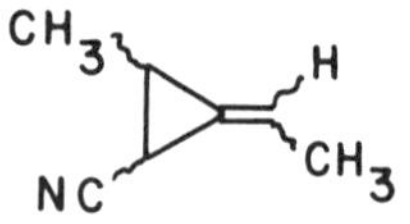

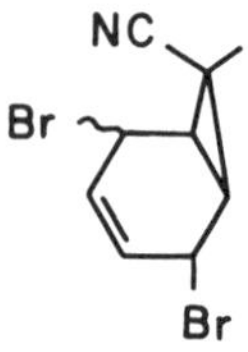

Shift data with Eu(fod)₃ has been used to determine the preferred conformation of cyclohexane carbonitrile.[510] The coupling constants did not change in the presence of the shift reagent, indicating that complexation of the shift reagent did not alter the conformation of the substrate. This result is not surprising in view of the small steric effects with organonitriles.[510] The isomers of V were distinguished with the aid of Eu(fod)₃.[517] LSR have been utilized in assigning the conformation of 9-cyanobicyclo[6.1.0]nona-2,4,6-triene(VI).[270] The two isomers of 9-cyano-4,9-dimethylbicyclo[6.1.0]nona-2,4,6-triene were distinguished in the presence of Eu(dpm)₃.[518] The position of trimethylsilyl groups in 2-cyanobicyclo[2.2.1]heptane derivatives have been confirmed through the use of Eu(fod)₃.[519] A fit of calculated to observed shift data with a LSR has been carried out for 1- and 2-adamantane carbonitriles.[516]

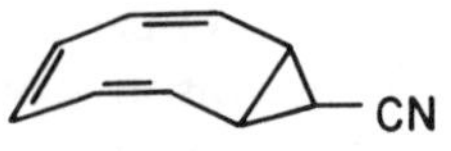

VI

13. Amides

The complexation between a LSR and a compound with the amide functional group occurs at the oxygen atom of the carbonyl.[315,520,521] It has been noted that coordination of the LSR occurs preferentially at the lone pair with the least amount of steric hindrance.[315,522-524] As the substituent groups of the substrate get larger, smaller shifts are observed due to the increased steric effect.[520] Electronic effects have also been shown to influence the association constants of amides with LSR.[185,525] The association constants of 1-substituted amides with substituent groups I and II with Eu(fod)₃ have been determined.[185] A considerably larger association constant was observed for I.

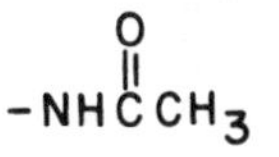

$$-NHCCH_3 \qquad -NHCCF_3$$
$$\text{I} \qquad\qquad \text{II}$$

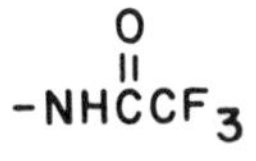

The presence of contact shifts in the ¹³C spectrum of N,N-dimethylformamides with Eu(dpm)₃ have been reported.[526] Contact shifts have also been noted in the carbonyl carbon resonance of amides and urethanes.[331] In some instances, the carbonyl carbon exhibited a "wrong way" shift because of the contact shift mechanism.

Most of the investigations of amides with LSR have studied the rotational properties of these substrates.[56,67,140,315,512,520-524,527-535]

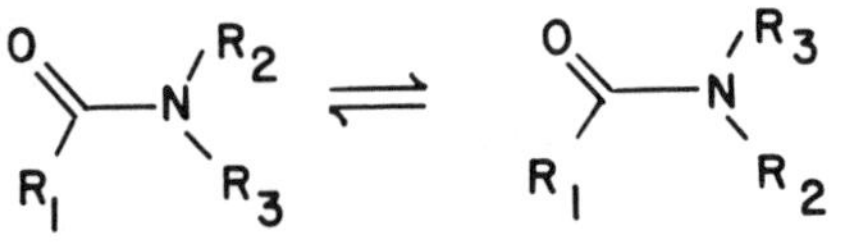

Rotation about the carbonyl-nitrogen bond is hindered, and LSR are useful for resolving the spectra of the two rotamers. The shift reagent causes larger shifts for the group in the *syn* position.[523,527,529,530] The compound 1-methylpyrrolidine-2-one(III), which has the *cis* configuration, has been utilized as a model for assigning the resonances of the rotational isomers of amides.[530] The compounds N-methylacetamide and N-pivaloylamide were employed as models in a study of the conformation of N-methylpropionamide and N-methylisobutyramide.[533]

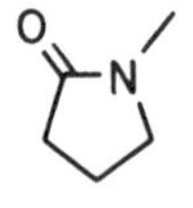

III

An important factor that must be considered in such studies is whether complexation of the shift reagent has any effect on the conformational preference of the substrate. Cohen-Addad and Cohen-Addad employed Eu(fod)$_3$ and Yb(fod)$_3$ in a study of the conformations of diethylacetamide,[531] dipropylacetamide,[531] and isobutyramide[532] and reported that complexation of the shift reagent did not alter the average solution conformation.[531] This finding is in contrast, however, to most other reports describing studies of amides with LSR. Finocchiaro has reported that for substituted amides the LSR preferentially coordinated to the conformer with the least steric hindrance.[315] If this confomer was not the only one present in solution, conformational changes in favor of this rotamer were observed in the presence of the shift reagent. This finding implies that the impact of a LSR on conformational alteration of an amide depends on the nature of the substituent groups. In a study of N,N-dimethylamides[523] and N-methylformamide,[523,534] Graham noted a similar finding to Finocchiaro in that stronger binding of the shift reagent occurred with the less sterically hindered *cis* conformer. As a result, more of the *cis* conformer was observed with the shift reagent. A method to account for the change in conformer population in the presence of a shift reagent has been described.[534] The conformation of α-heterosubstituted N,N-diethylacetamides has been reported to be strongly perturbed in the presence of Eu(dpm)$_3$ and Eu(fod)$_3$.[524]

The temperature at which coalescence of the methyl resonances of dimethylformamide and dimethylacetamide occurs is higher in the presence of Yb(dpm)$_3$.[512] The rotational barrier in dimethylformamide, dimethylacetamide, and dimethylpropionamide has also been shown to increase with increasing concentrations of Eu(fod)$_3$ or Pr(fod)$_3$.[56] The conformational preference of N-methylacetamide, acetanilide, and phenacetin varied with the concentration of LSR.[527] The conformational population of N,N-diisopropylamides with Eu(fod)$_3$ was different than that with Pr(fod)$_3$.[140] Montaudo et al. employed low shift reagent concentrations in a study of the influence of steric and conjugative effects on the barrier to rotation in amides.[528] The concentrations of LSR were kept low (L:S ratio of approximately 0.1 to 0.2) to minimize perturbing effects of the shift reagent. At the concentrations employed, they reported that the shift reagent did not alter the energy barrier to rotation. Cheng and Gutowsky have summarized practical approaches to the application of LSR to dynamic systems such as amides.[535]

Slow rotation about both the carbonyl-nitrogen and ring-nitrogen bond of 2,6-di-2-propylacetanilide(IV) was observed in a study with Eu(dpm)$_3$.[67] The isopropyl groups exhibited different shifts, and a "wrong-way" shift, due to an unusual angle term, was noted for one of the methyl resonances. Shift data with LSR were useful for studying geometrical isomerism in substrates with structure V.[536] Mixtures

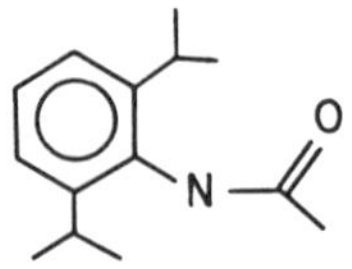

IV 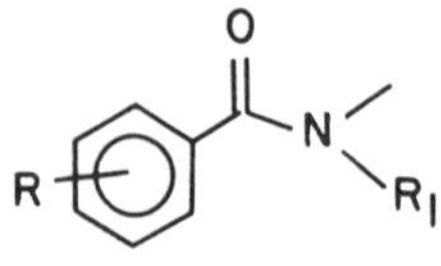   V

of o-, m-, and p-toluamide have been quantified in the presence of Eu(fod)₃.[178] All of the resonances of cis- and trans-N,N-dimethyl-2-phenylcyclopropane-1-carboxamide were assigned on the basis of shift data with Eu(fod)₃.[378,414]

The shifts in the ¹⁴N NMR spectrum of dimethylformamide with Dy(fod)₃, Dy(dpm)₃,[465] Eu(dpm)₃, and Yb(dpm)₃[466] have been reported. The shifts were much smaller than those for nitrogen-containing substrates in which the nigrogen atom bonded to the LSR.

14. Lactams and Imides

LSR have been applied in investigations of lactams.[189,307,512,525,530,537-543] The shifts in the ¹H and ¹³C NMR spectra of N-vinylpyrrolidone and N-methylpyrrolidone were recorded with Eu(fod)₃, and it was determined that the resonances of carbon atoms close to the site of complexation exhibited a contact shift.[537] The contact mechanism was reportedly larger than observed with ketones. The larger contact shift with lactams may result from conjugative interaction between the carbonyl and nitrogen atom.[537]

Shift data with LSR have been used to determine the preferred conformation of a wide variety of lactams.[539] This study included compounds with rings containing from 4 to 13 members. The stereochemistries of 2-acetidones(I) have been analyzed with the use of the shift reagents Eu(fod)₃, Pr(fod)₃, and TiCl₄.[541] The shifts

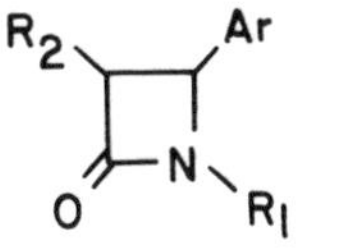

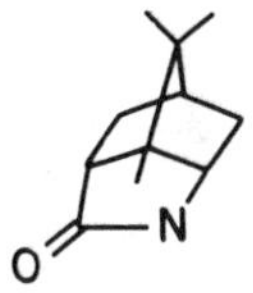

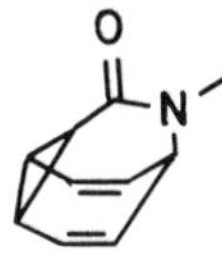

with the lanthanide reagents were significantly larger than those with TiCl₄. Assignment of the resonances and determination of the ring stereochemistry were possible from information gained through the use of LSR. The stereochemistry of 1-substituted-3-benzylidene-2-pyrollidones and -piperidones has been determined on the basis of shift data with Eu(fod)₃.[543] The shifts in the ¹⁷O spectrum of N-methylpyrrolidone with lanthanide chelates of dpm have been reported.[189] The shifts in the ¹³C spectrum of N-methylpiperidone with Yb(fod)₃ have been reported.[307] Shift data with a LSR have been used to confirm the structure of II[538] and III.[540]

The structure of IV was assigned on the basis of shifts in the ¹³C spectrum with Eu(fod)₃.[542]

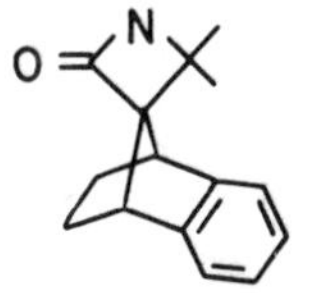

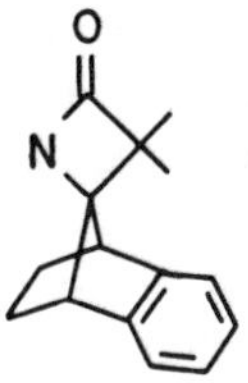

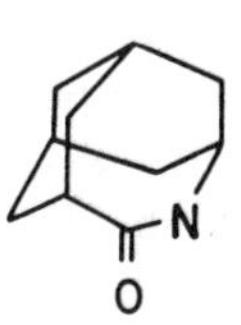

The data from the shift reagent experiment enabled a distinction between the correct structure(IV) and its isomer(V). The binding constant between Eu(dpm)₃ and VI has been reported.[525]

The assignment of the methyl resonances of camphorimides(VII) was facilitated through the use of Eu(dpm)₃.[544]

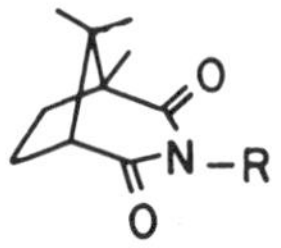

VII

15. Carbamates

LSR have been applied in the study of rotational properties of esters of carbamic acid.[139,521,545-548] As with amides and lactams, carbamates coordinate to the LSR through the carbonyl oxygen atom.[521,545,546] Supporting evidence for this conclusion is that the thia derivative

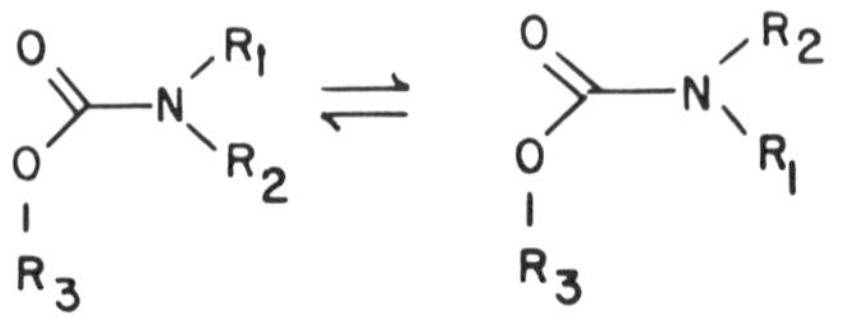

of trimethylcarbamate(I) exhibits no shifts in the presence of Eu(dpm)$_3$.[545]

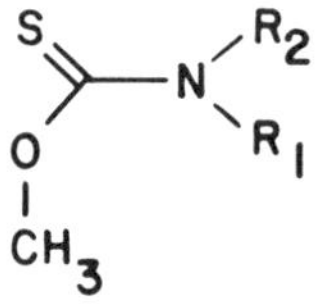

I

The shifts in the presence of a shift reagent are larger for the substituent group in the *cis* position,[546] and this distinction can be used to determine the conformer populations.[521] The methyl groups *cis* and *trans* to the carbonyl group in trimethylcarbamate were resolved in the presence of LSR.[139,546] The resonances were broad at low concentrations of LSR and sharpened as the concentration of shift reagent was increased. The LSR retards the isomerization of trimethylcarbamate, but by extrapolation to zero concentration of LSR, it was possible to determine values for the substrate in its uncomplexed form.[139]

16. Nitrogen Oxides

Nitrogen oxides coordinate to LSR through the oxygen atom.[66,89,111,496,549-551] The binding constants of nitrogen oxides with LSR were larger than those for the corresponding nitrogen heterocycles.[551] The shifts in the ^{1}H NMR spectra of pyridine N-oxide, quinoline N-oxide, and benzo(f)quinoline N-oxide in the presence of LSR, however, were smaller than those for the corresponding nitrogen heterocycles.[496] The smaller shifts in the spectra of the oxides result from the longer distance between the lanthanide ion and the hydrogen atoms. Contact shifts in the ^{13}C NMR spectra of 8-picoline N-oxide and pyridine N-oxide in the presence of LSR have been reported.[89] The contact shifts were more pronounced in pyridine N-oxide because of the extended pi system. The presence of a contact shift mechanism has also been reported in the ^{1}H spectra of pyridine N-oxide and alkyl derivatives with LSR.[549] In some instances, the contact shifts were large enough to result in "wrong way" shifts for some of the resonances.[549]

Shift data with Eu(dpm)$_3$ have been used to determine the configurations of α-phenyl-α-N-dimethylnitrones(I).[66] Pyrimidine 1-oxides(II) and 3-oxides(III) have been distinguished in the presence of LSR.[550] The distinction results because complexation occurs at the N-oxide oxygen rather than the ring nitrogen.

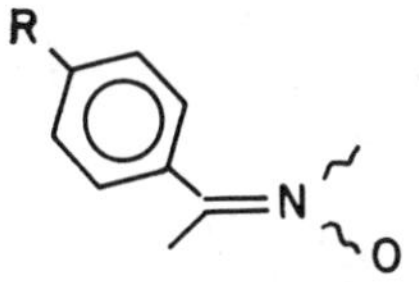

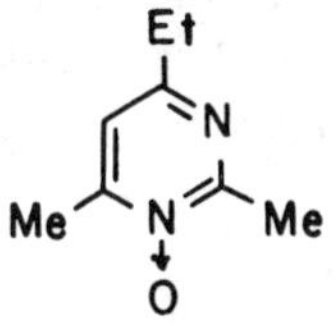

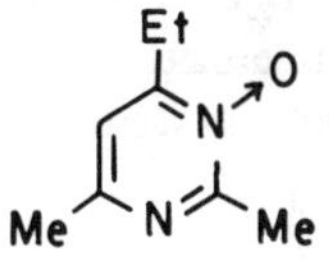

17. Nitrosos

Compounds with the nitroso group have been studied with the aid of LSR.[552-555] These reports agree that complexation of nitroso substrates with LSR occurs through the oxygen atom.[552-555] The conformational preference of ten dialkyl nitrosamines was determined using data with LSR.[555] Rotation about the nitrogen-nitrogen bond is slow, and larger shifts are observed for the group occupying the *syn* position.

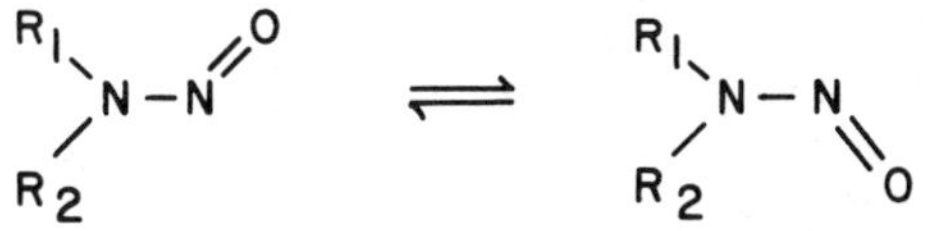

This report also stated that the conformational preference of the substrates was not altered on complexation with the LSR.

The population of the axial and equatorial conformers of 2-alkyl N-nitroso derivatives of piperidines were determined in a study with Eu(fod)$_3$.[554] In this report, N-nitroso-4-*tert*-butylpiperidine was employed as a model substrate because of its fixed conformation. Complexation of the LSR did not seem to alter the conformation of the substrates.

Shift data with Eu(dpm)$_3$ was used to aid in the assignment of the methyl resonances of N-nitroso camphidine(I).[552]

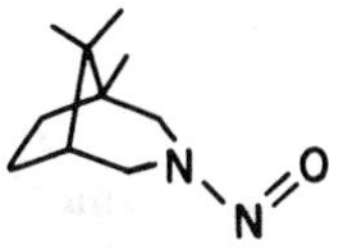

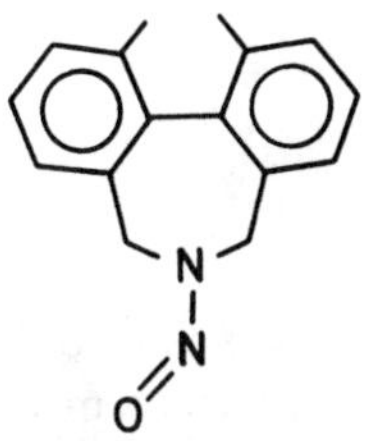

Six methyl resonances were observed in the presence of the shift reagent because of the hindered rotation about the nitrogen-nitrogen bond. The resonances of the four protons alpha to the nitrogen atom in N-nitroso-6,7-dihydro-1,11-dimethyl-5H-dibenz[c,e]azepine(II) were assigned on the basis of data with Eu(dpm)$_3$.[553]

18. Azoxys

The use of LSR with substrates containing an azoxy moiety has been described.[556-558] The exact mode of binding between the azoxy unit and the LSR has not been delineated. Snyder reported that the LSR probably coordinated with the oxygen of the azoxy unit.[557] Three bonding modes(I-III) were evaluated in a study of cis-azoxyalkanes by Bearden, and the best fits of the data were obtained if bonding took place at the nitrogen atom(III).[558] In each of these studies, the substrates were rigid bicyclic or multicyclic systems in which the azoxy unit was constrained to the cis configuration.

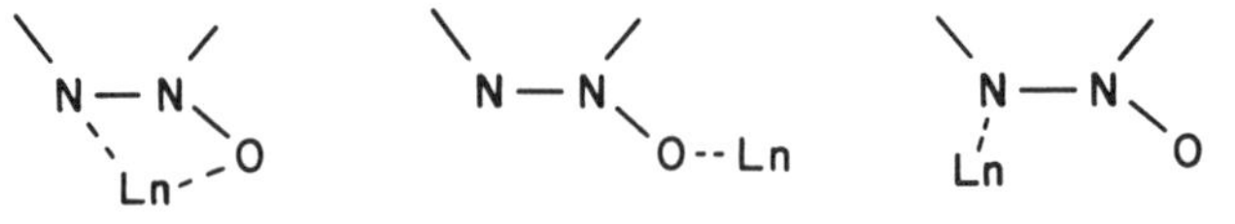

Azoxy compounds with two different substituent groups have four possible isomers(IV-VII). Eu(fod)$_3$ was utilized

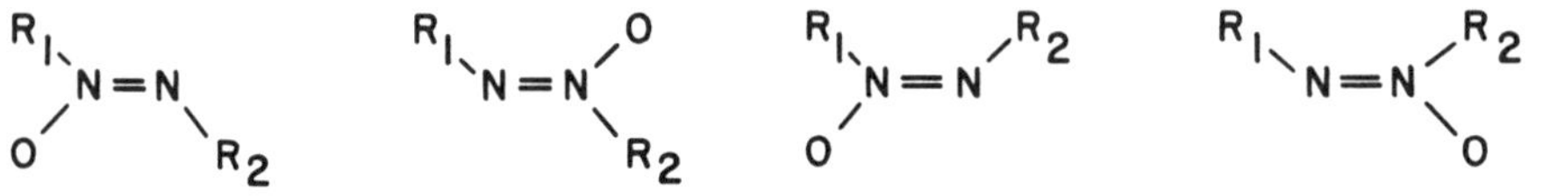

in an attempt to distinguish mixtures of the two cis isomers.[556] The shifted spectrum was complicated, however, and assignment of the resonances could only be obtained by analyzing one of the pure isomers with Eu(fod)$_3$.

The unsymmetrical nature of the azoxy unit in cis-azoxyalkanes has been confirmed by the analysis of spectra in the presence of Eu(dpm)$_3$.[557] The substrates in this report were rigid multicyclic compounds(VIII-X) that were constrained to the cis configuration. In each case, two bridgehead proton resonances were detected in the presence of the LSR.

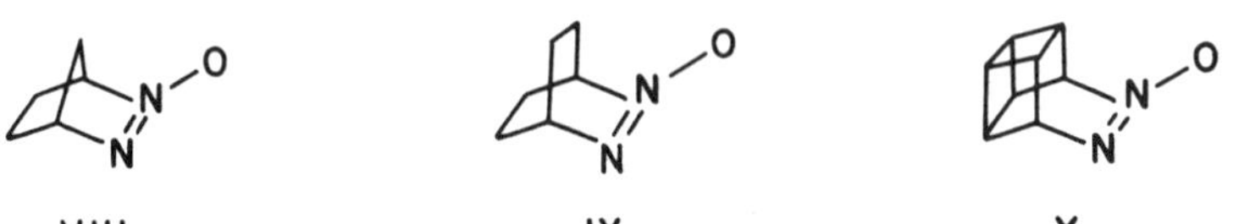

19. Nitros

Compounds with a nitro group bond weakly to lanthanide tris chelates.[72,512] Small shifts are observed in the spectrum of nitromethane in the presence of Yb(dpm)$_3$,[512] while essentially no shifts are observed for nitrobenzene. Sievers has reported that lanthanide chelates with the dfhd ligand more effectively bond to the nitro group than chelates with fod and dpm.[72] Shifts were observed in the spectra of nitromethane, p-nitrotoluene, and p-chloronitrobenzene in the presence of Eu(dfhd)$_3$.

20. Isocyanates

The geometry of 2-fluorophenylisocyanate(I) in solution was studied by an analysis of shift data with Eu(fod)$_3$.[559] It was possible to determine whether the N=C=O group was positioned cis or trans to the fluorine substituent. Bonding of the shift reagent reportedly occurred at the oxygen atom of the substrate.

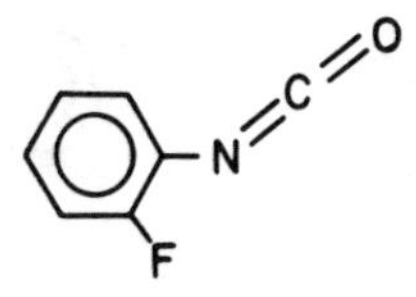

21. Oximes

Oximes coordinate effectively with LSR, but it is not known whether complexation occurs at the lone pair on the nitrogen or oxygen atom.[560-564] Shift data with Eu(dpm)$_3$ was used to determine the *syn*(I):*anti*(II) ratios for compounds in which R$_1$ = CH$_3$ and R$_2$ = ethyl, *n*-propyl, *n*-pentyl, *n*-hexyl, *i*-propyl, and *i*-butyl.[560] In this study larger shifts were observed for R$_2$ in the *anti* position.

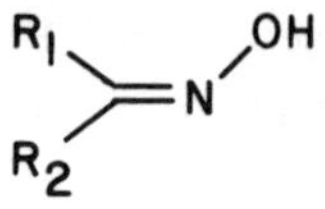

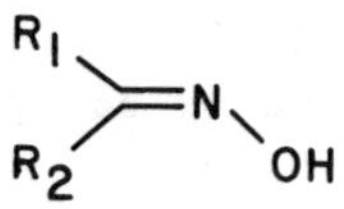

Wolkowski has also noted larger shifts for the substituent group *syn* to the −OH group in oximes.[562] It was concluded that the LSR bonded to the oxygen atom since this would result in a shorter distance to the *syn* group. The shift data were also used to distinguish the *syn* and *anti* isomers when R$_1$ = R$_2$. Fraser has reported, however, that shift data in the ¹H spectrum with LSR is unreliable for the assignment of oxime stereochemistry and recommended studying the shifts in the ¹³C spectrum.[564] The alpha carbons at the *syn* and *anti* position were readily distinguished in the presence of a LSR. The configurations of compounds III, IV, and V have been assigned with the aid of LSR.[562] The stereochemistry of camphoroxime(VI) has been assigned on the basis of shift data with Eu(dpm)$_3$.[563]

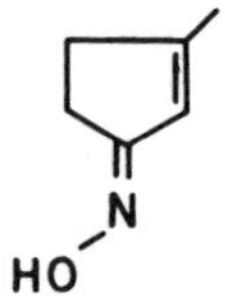

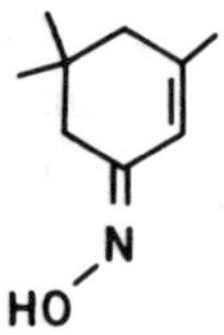

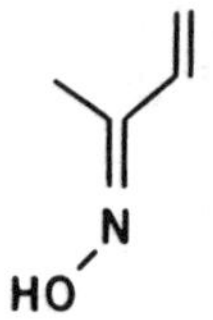

22. Azos and Azides

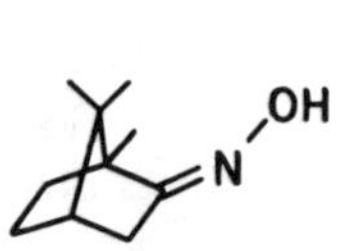

Compounds with the nitrogen linkages I and II can be studied with the aid of LSR. The magnitude of the shifts

$$R_1-N=N-R_2 \qquad\qquad R_1=N-N=R_2$$

I   II

is highly dependent, however, on the nature of the substrate. Large shifts were observed in the spectrum of *cis*-azobenzene(III) with Eu(fod)$_3$, while essentially no shifts were observed for the *trans* isomer(IV).[565] The spectra of azobenzenes

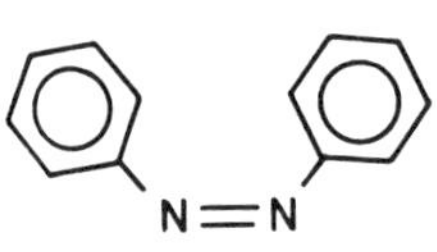

III 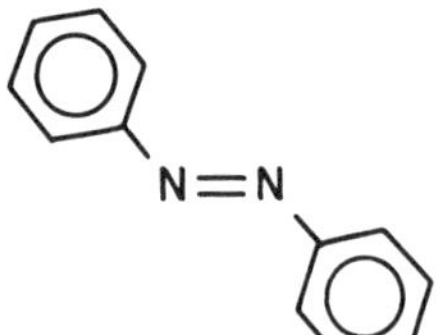   IV

were not shifted in the presence of Yb(dpm)$_3$.[512] The spectrum of azo-2-methylpropane was not shifted in the presence of Eu(fod)$_3$.[565]

Cyclic compounds with these linkages bond effectively to LSR.[565-568] Eu(dpm)$_3$ was used to aid in the structural assignment of 3,8-diphenyl-1,2-diaza-1-cyclooctene.[566] Eu(fod)$_3$ has been employed in the structural and conformational assignment of 2,3-diaza-1,3,5,7-cyclooctatetraene.[567] Large shifts in the spectrum of 1,4-dimethyl-2,3-diazabicyclo[2.2.2]oct-2-ene(V) have been observed with Eu(fod)$_3$.[565]

LSR have been used to determine the stereochemistry of VI and VII.[568]

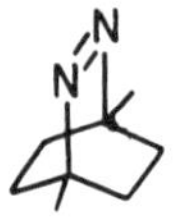

V 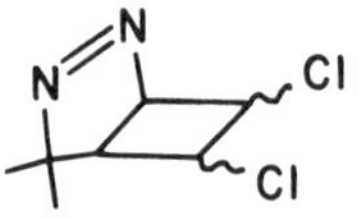   VI 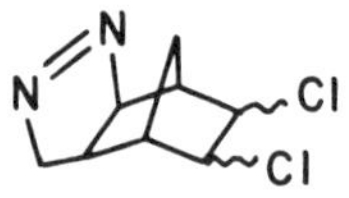   VII

The substitution about the diaza units in VI and VII is unsymmetrical, and preferential complexation of the LSR occurred at the nitrogen atom furthest removed from the methyl groups.[568]

The shifts in the ^{14}N NMR spectrum of CH$_3$N$_3$ with Dy(dpm)$_3$ and Dy(fod)$_3$ have been reported.[465] Only the first two nitrogen atoms exhibit shifts in the presence of the chelates.[465]

23. Imines

No shifts were observed in the spectra of certain mono- and diimines with Yb(dpm)$_3$.[512] Addition of Eu(fod)$_3$ to benzylideneanilines(I) caused an isomerization of the more stable *trans* isomer to the *cis* isomer.[569] Increasing the concentration of shift reagent resulted in a corresponding increase in the concentration of the *cis* isomer.

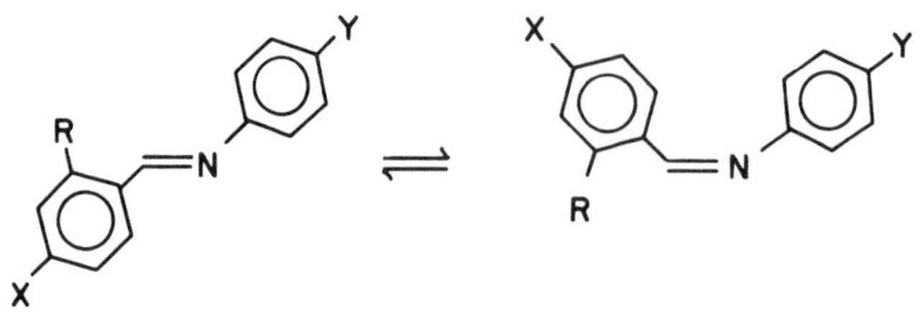

I

24. Substrates with a Sulfur Atom

Compounds containing a variety of functional groups with a sulfur atom have been studied with LSR. These include sulfoxides,[60,129,244,551,570-577] sulfones,[60,129,445,570,574,578] sulfites,[579-583] sulfines,[584,585] thiosulfinates,[586] sultones,[587] isothiocyanates,[514] thiocyanates,[514] sulfonate esters,[588,589] sulfate diesters,[590] thiacarbamates,[547] sulphinamates,[591] thioacetates,[592] thioketones,[593] thioamides,[570] sulfides,[60,76,129,551,570,593] and thiols.[119,247]

Most of these have an oxygen atom in the functional group that preferentially complexes to the LSR.

a. Sulfoxides

The shifts in the spectra of aliphatic sulfoxides such as dimethylsulfoxide,[551,570,573] and di-*n*-butylsulfoxide[573] have been reported. Sulfoxide substrates coordinate through the oxygen atom. The diastereotopic protons alpha and beta to the sulfoxide group in di-*n*-butylsulfoxide were resolved in the presence of a LSR.[573] Diastereotopic resolution of the methylene protons of I was observed in the presence of Eu(fod)$_3$.[577]

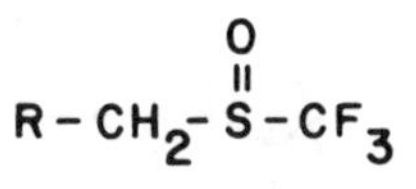

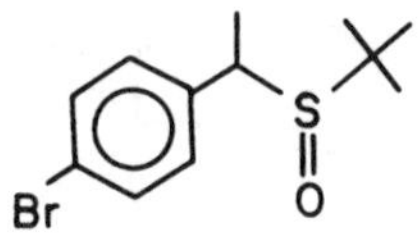

The diastereotopic methyl groups of 2-propyl phenyl sulfoxide were resolved in the presence of Eu(dpm)$_3$.[573] Eu(fod)$_3$ has been employed in the study of the solution conformation of *threo* and *erythro* 1-(*p*-bromophenyl)ethyl-*t*-butyl sulfoxide(II).[571] The conformations of alkyl *threo*- and *erythro*-1-phenylethyl sulfoxides were determined with the aid of shift data with a LSR.[576] The solution conformation of methyl phenyl sulfoxide has been determined in the presence of a LSR.[574]

Eu(dpm)$_3$ was employed to aid in the assignment of the configuration of III, IV, and V.[572] Selectively deuterated substrates were studied to facilitate assignment of resonances in the shifted spectra. The configuration of V was also determined in an experiment by Wing.[244] This same report noted the importance of considering the angle term of the dipolar shift equation when assigning resonances in the presence of a shift reagent. Using distances only, the resonances of 3,3-dimethylthietane-1-oxide(VI) were incorrectly assigned.[244]

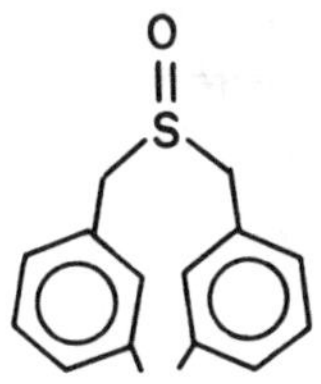

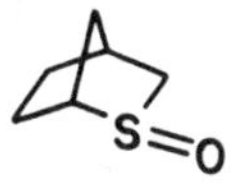

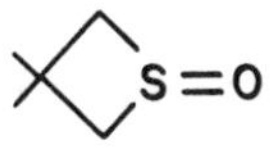

The reagents Eu(dpm)$_3$ and Yb(fod)$_3$ have been used to study the conformation of 3-substituted thietane-1-oxides.[575] The 3-*cis*, 3-*trans*, and 3,3-disubstituted derivatives were analyzed. Conformational assignment was facilitated by studying model compounds VII and VIII, which have locked conformations.

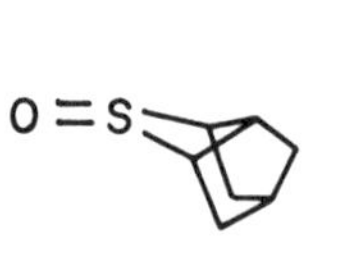

VII 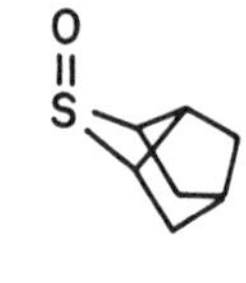   VIII

Shift data with Eu(dpm)₃ have been used to assign the configuration of the sulfinyl group in 5α-cholestan-2α,5-episulfoxides(IX,X).[60] By studying

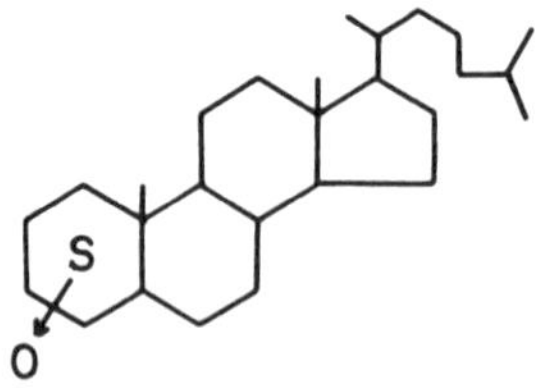

IX 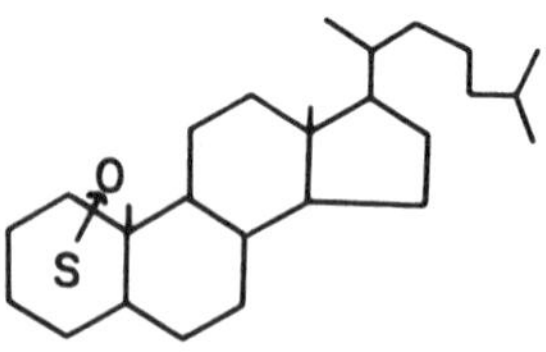   X

the sulfone derivative of IX, it was possible to distinguish the two sulfoxide isomers on the basis of their shift data with the LSR. It was reported in this work that the sulfinyl derivative exhibited a higher association constant with the LSR than the sulfone derivative.

b. Sulfones and Sulfanate Esters

The relative abilities of sulfoxides and sulfones to coordinate with a LSR appear to be substrate dependent. The shifts in the NMR spectrum of dimethyl sulfone with Eu(fod)₃[570] were larger than those for dimethylsulfoxide. A study of tetramethylene-sulfide, -sulfoxide, and -sulfone, however, exhibited an opposite finding as the order of shifts was sulfoxide > sulfone ≫ sulfide.[129] Shift data with Pr(fod)₃ was used to aid the assignment of the aromatic resonances of XI.[445]

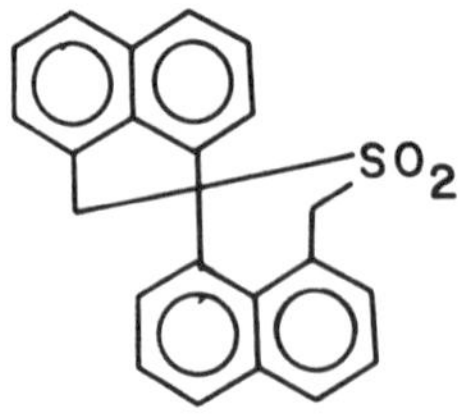

XI

The conformations of methylphenylsulfones and diphenylsulfones have been studied in the presence of LSR.[574] The structures of 2-thiolene and 3-thiolene 1,1-dioxides have been assigned on the basis of shift data with Eu(fod)₃.[578]

Mixtures of methyl esters of alkane and alkene sulfonates have been quantified through the use of Eu(dpm)₃ and Eu(fod)₃.[588,589] The methoxy signal of the sulfonate ester was a convenient signal to monitor.

c. Sulfites, Sultones, and Sulfate Diesters

A number of investigations of six-member cyclic sulfites(XII) with LSR have been reported.[579-583] These substrates complex to a LSR through the exocyclic oxygen atom.[580]

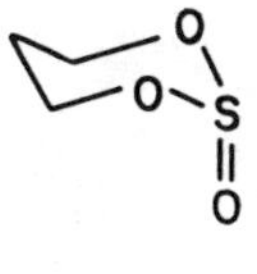

XII 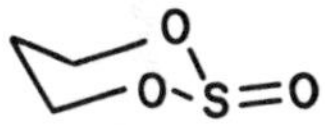   XIII 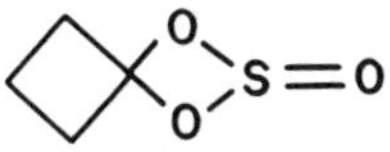   XIV

Highly substituted derivatives have been employed in studies designed to assess the validity of applying LSR to conformational analysis. These substrates can occupy a combination of the axial(XII), equatorial(XIII), and twist(XIV) conformations in solution.[581] Analysis of the shift data and coupling constants that were obtained in the presence of the LSR permitted a determination of the conformation of these substrates.[579,581-583] Complexation of the shift reagent, however, caused an alteration of the conformation in favor of the axial form.[579-583] If the thionyl group is fixed in an axial position in its complex with the LSR, the configuration of substituent groups on the substrate can be assigned.[580]

The shifts in the spectrum of 6-hexadecyl-1,2-oxathiane-2,2-dioxide(XV) with Eu(fod)$_3$-d$_{27}$ have been used for conformational analysis.[587]

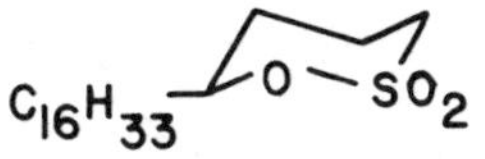

XV 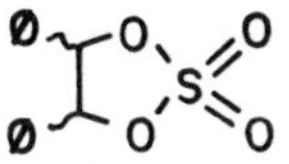   XVI

The configurations of 2,2-dioxo-1,3,2-dioxathiolanes(XVI) have been determined through the application of LSR.[590] The configuration of the sulfur atom was assessed before and after a reaction by recording the ^{17}O NMR spectrum. It was reported that the shift reagent preferentially bonded at the sulfinyl group removed from steric hindrances on the ring.[590]

d. Sulfines

Eu(dpm)$_3$ has been employed in a determination of both the configuration and conformation of substituted diphenyl sulfines(XVII).[584,585] The shift reagent coordinates with the sulfinyl oxygen atom and is located closer to the ring in the *syn* position. As a result, the resonances for the protons on the *syn* ring exhibit larger shifts. This difference can be used to distinguish the *Z* and *E* isomers of nonsymmetrical sulfines such as XVIII and XIX.[585]

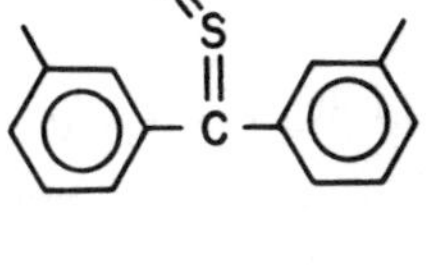

XVII 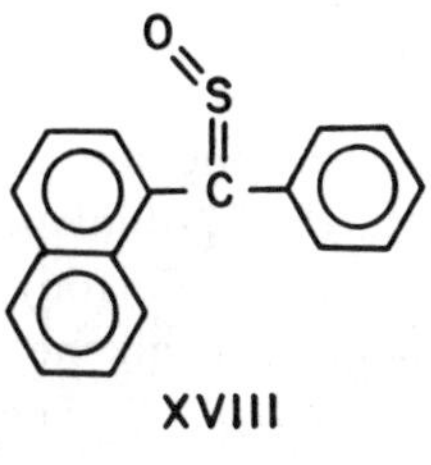   XVIII 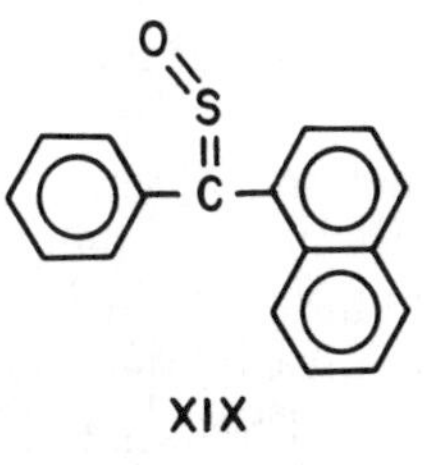   XIX

It is also possible to determine the conformational preference of sulfines with the aid of shift data with a LSR.[584] Such an evaluation only provides the conformer populations for the complexed substrate. Determination of the conformer populations by other means demonstrated that the shift reagent altered the conformational preference of the substrate.[584] The perturbation of the conformational equilibria was reported to favor the more polar conformer as well as the conformer that minimized steric interactions.

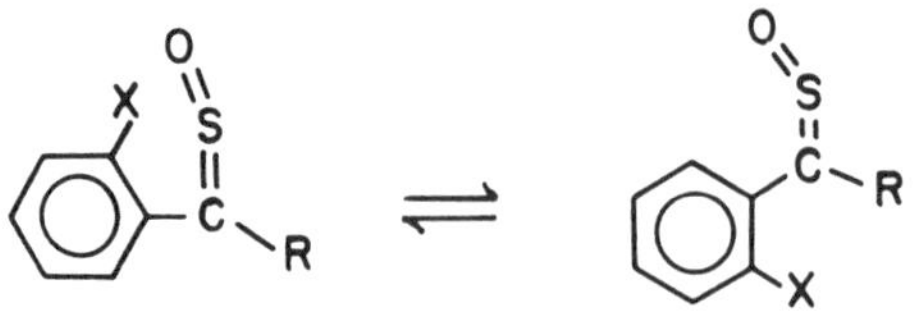

e. *Thiolsulfinates, Sulfinamates, and Thiacarbamates*

LSR have been used to study the compounds ethyl ethane-, isopropyl isopropane-(XX), and benzyl toluene thiolsulfinate.[586] No shifts

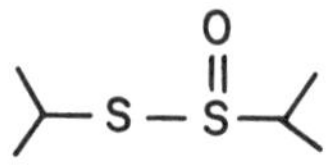

were observed in the NMR spectrum of di-isopropyldisulfide with Eu(fod)$_3$, confirming that the shift reagent coordinates at the oxygen atom of thiolsulfinates.[586] Four methyl doublets were observed in the shifted spectrum of isopropyl isopropane thiosulfinate. Only partial resolution of the two side groups was noted in the ethyl and benzyl derivatives.

The conformation of cyclic sulfinamates such as 2-oxo-1,2,3-oxathiazans(XXI) and 5,6-benzo-3,4-dihydro-2-oxo-1,2,3-oxathiazans(XXII) have been studied in the presence of Eu(fod)$_3$.[591]

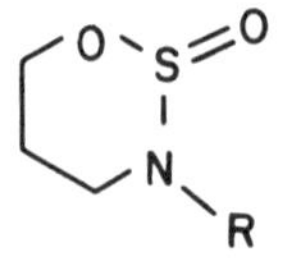

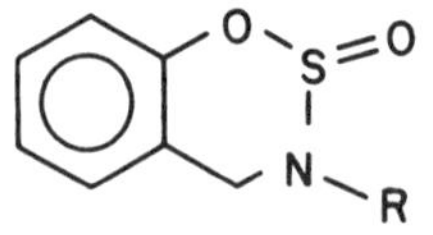

LSR have been used to resolve the methyl resonances of XXIII to facilitate study of the rotation about the N—C=O bond.[547]

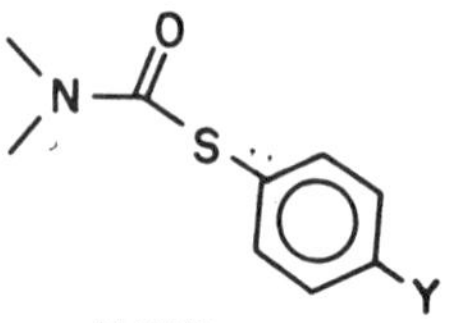

f. *Sulfides and Thiols*

Sulfides such as methylsulfide bond weakly,[570] if at all,[60,514] to LSR. No shifts were observed in the spectrum of diphenylsulfide in the presence of Eu(dpm)$_3$.[551] Only small shifts occurred in the spectra of sulfides in the presence of Eu(fod)$_3$.[593] The shifts were found to be somewhat larger if Eu(tfn)$_3$ was used as the shift reagent.[76]

Thiols bond weakly through the sulfur atom to the lanthanide ion in tris chelates and can be studied by LSR.[119,247] Larger shifts are observed for aliphatic thiols such as ethanethiol and propanethiol compared to aromatic thiols.[119] For ethanethiol, larger shifts were reported with Eu(dpm)$_3$ vs. Eu(fod)$_3$.[119] It was reported that Yb(dpm)$_3$ complexed only weakly with 2-adamantanethiol, but the shift data were adequate for confirming the assignment of the ^{13}C spectrum.[247]

g. Thiocyanates and Isothiocyanates

No shifts were observed in the spectra of methyl- and benzyl isothiocyanate(XXIV) in the presence of Eu(fod)$_3$.[514]

Ph—CH$_2$NCS   Ph—CH$_2$SCN

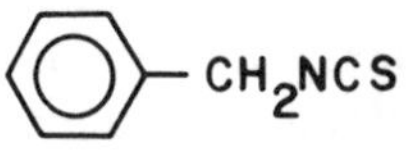

XXIV 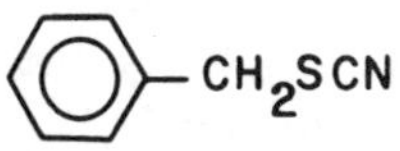   XXV

The spectra of the corresponding thiocyanates(XXV), however, exhibited substantial shifts in the presence of Eu(fod)$_3$.[514] These differences reflect the comparative donor properties of nitrogen and sulfur atoms.

h. Thioamides, Thioketones, Thioacetates, and Thiocarbamates

Thioamides(XXVI) reportedly bond to LSR through the sulfur atom.[427,521,529,594,595] The thiocarbonyl unit bonds only weakly to LSR, however, as evidenced by the small shifts in the spectrum of 2-thioadamantanone with Eu(fod)$_3$.[593] If

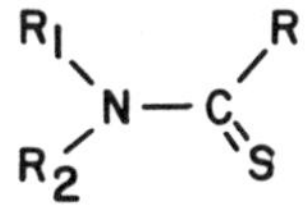

R$_1$\\
   N—C(R)=S\\
R$_2$/

XXVI

R$_1$ and R$_2$ are different, two rotational isomers occur. It is possible to distinguish and quantify these based on shift data with LSR.[529,594,595] If R$_1$ = R$_2$, as in diethyl thioformamide, two sets of signals are observed for the ethyl groups in the ^{13}C NMR spectrum in the presence of the shift reagent.[595] In this report, assignment of the ^{13}C resonances was achieved by using the LSR to obtain a first-order ^{1}H NMR spectrum. The proton resonances were then selectively decoupled facilitating assignment of the ^{13}C resonances.[595]

Esters of thiocarbamic acid(XXVII) reportedly bond to a LSR through the sulfur atom.[521,592]

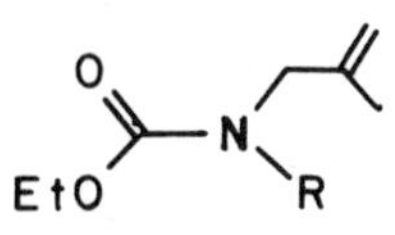

EtO—C(=O)—N(R)—CH=CH$_2$

XXVII

The *syn* and *anti* isomers are distinguishable with the aid of LSR, and the percent of each rotamer in a mixture can be determined. No shifts were observed, however, in the spectrum of O-ethylthioacetate and N,N-dimethylthiocarbamate esters with Eu(dpm)$_3$.[592]

The shifts in the spectra of thioamides with Eu(fod)$_3$ have been compared to corresponding seleno derivatives.[427] The NMR spectra of the seleno compounds were shifted in the presence of Eu(fod)$_3$, but the magnitude of the shifts was much less than those for the thioamides. It was also reported that the shifts in substituted thioureas with LSR were comparable to those for thioamides.[427]

25. Substrates with a Phosphorus Atom

A variety of functional groups containing a phosphorus atom bond to LSR. These include phosphates,[520,551,596-599] phosphorinanes,[582,583,600-602] phosphonates,[596-599,603-607] phosphine oxides,[597-599,603,608-612] phosphites,[520,551,596,597,613] phosphoranes,[613] oxazophospholines,[614,615] phosphoramides,[577,616-619] and phosphines.[415,596,597] For each class of compounds except phosphines, the functional group has an oxygen atom that preferentially complexes with the LSR. Many of those contain the phosphoryl group, which complexes quite strongly with LSR. As will be discussed, phosphines bond weakly, if at all, to the lanthanide tris chelates and are more appropriately studied with the binuclear shift reagents described in Chapter 4.

a. Phosphates

Phosphates(I) bond to the lanthanide ion through the phosphoryl oxygen atom.[520,551,596,597,599] Replacing the phosphoryl oxygen with sulfur, as in triethylthiophosphate, results in essentially no shifts in the presence of chelates of fod or dpm.[520,596] Large shifts are observed in the spectra of trimethyl- and triethylphosphate in the presence of $Eu(dpm)_3$.[520,596] Contact shifts have been reported to occur in the ^{31}P NMR spectrum of phosphates.[596,597] Contact contributions were noted with chelates of both fod and dpm, but were larger with the fod chelates.[596] Shift data with $Eu(dpm)_3$ and $Pr(dpm)_3$ were used to assign the stereochemistry of substituted vinyl phosphates(II).[598] The assignment

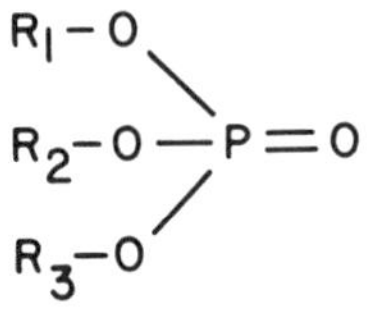

$$R_1-O\diagdown$$
$$R_2-O-P=O$$
$$R_3-O\diagup$$
$$\text{I}$$

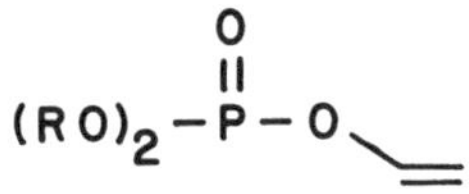

$$(RO)_2-\overset{\overset{O}{\|}}{P}-O\diagdown_=$$
$$\text{II}$$

was based on a comparison of the shifts of unknowns to those observed in compounds with known stereochemistry.

b. Phosphonates, Phosphorinanes, and Oxazophospholines

Phosphonates(III) bond to LSR in a manner similar to that for phosphates.[596,597,599,604-607] As with phosphates,

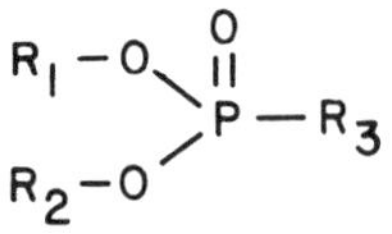

$$R_1-O\diagdown\overset{\overset{O}{\|}}{P}-R_3$$
$$R_2-O\diagup$$
$$\text{III}$$

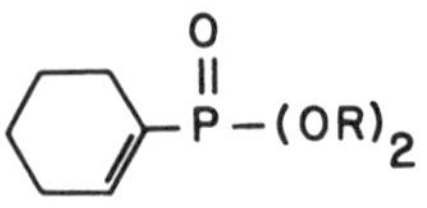

$$\underset{\text{IV}}{\bigcirc\!\!=\!\!\overset{\overset{O}{\|}}{P}-(OR)_2}$$

contact shifts have been reported in the ^{31}P NMR spectra of phosphonates.[597,607] The shifts in the NMR spectra of trimethyl phosphonate,[596] diethylphenyl phosphonate,[603] and triethyl phosphonate[597] with LSR have been reported. LSR have been employed to determine the stereochemistry of substituted vinyl phosphonates(IV).[598,605] The configuration of the olefin groups in 1,3-alkadienyl phosphonates have been determined through the use of LSR.[604,606]

It has been shown that substrates of the phosphorinane class(V) bond to LSR through the phosphoryl group.[600]

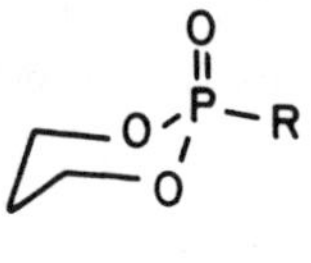

V 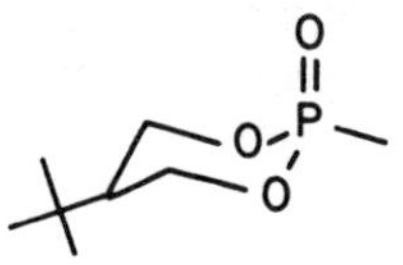   VI

The spectrum of trans-2-methyl-5-t-butyl-2-oxo-1,3,2-dioxaphosphorinane(VI) is first-order in the presence of Eu(dpm)$_3$.[600] No shifts were observed for derivatives in which the phosphoryl oxygen was replaced with a sulfur atom or a methoxy group.[600] The spectrum recorded with the shift reagent permitted assignment of the trans configuration of VI. LSR have been used by other workers to assign the cis or trans configuration of phosphorinanes.[602]

Conformational analysis of phosphorinanes with the aid of LSR is more difficult. A number of investigators have reported that the conformational preference of phosphorinanes is altered on complexation with a shift reagent.[582,583,601,602] The alteration reportedly occurs in favor of the conformer with the more basic phosphoryl group.[583]

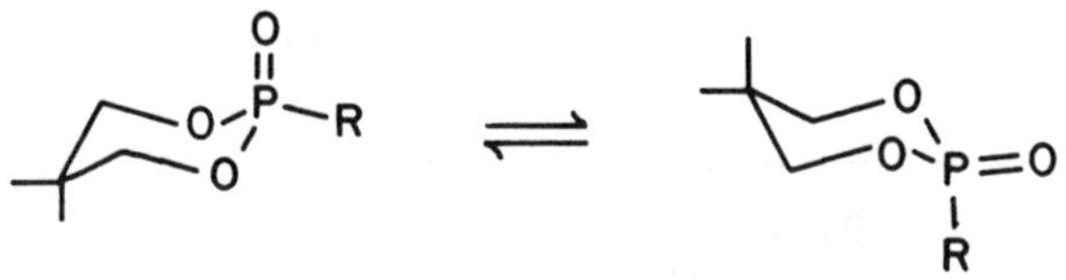

Through the use of coupling constants, which can be obtained in the shifted spectra, and shift data with the shift reagent, it is possible to determine the conformation of the complexed substrate. Caution must be exercised, however, if conclusions are to be made about the conformation of the uncomplexed form.

LSR are suitable for the study of oxazaphospholine-2-oxides(VII).[614,615] Once again, bonding occurs through the phosphoryl group.

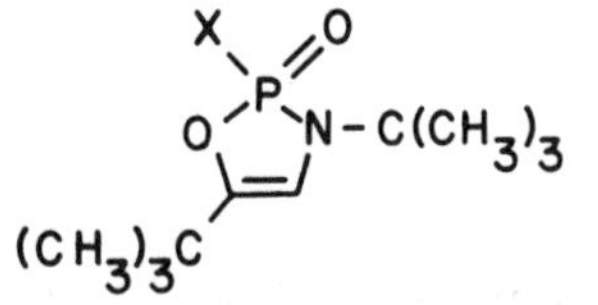

VII

c. Phosphine Oxides

Phosphine oxides(VIII) bond to LSR through the phosphoryl group, and contact contributions have been noted in the ^{31}P NMR spectra.[597] The shifts in simple phosphine oxides such as the trimethyl,[608] triethyl, and tributyl[610] derivatives have been reported. The conformation and dynamic behavior of phenyl and alkyl-substituted phosphine oxides were studied with the aid of ^{13}C and ^{31}P lanthanide shift data and relaxation data with Gd(dpm)$_3$.[612] Both the configuration and conformation of 1-substituted phospholane 1-oxides(IX) have been studied with the aid of LSR.[603,608-610]

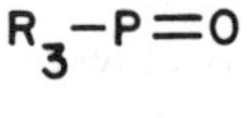

R$_3$-P=O

VIII 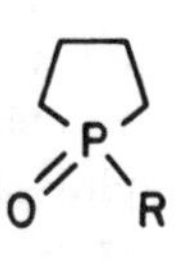   IX

The ¹H spectra of 1,2,5-triphenylphosphole oxide(X) 1-phenylphospholene oxide (XI), and 1-phenyl-3,4-dimethylpholene oxide were more interpretative in the presence of a LSR.⁶⁰³

X   XI

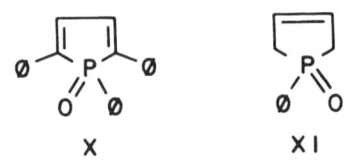

The two different protons alpha to the phosphorus atom in X and XI were resolved in the shifted spectra. The shifts in XII and XIII with Eu(dpm)₃ were reported and were in agreement with proposed structures for the compounds.⁶⁰⁹

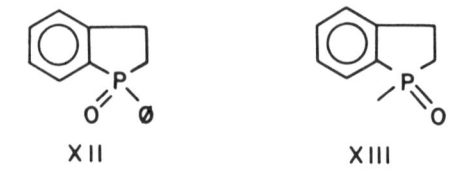

XII   XIII

The *cis* or *trans* configuration of compounds analogous to XIII, in which the methyl group was replaced by an ethyl or butyl group, was assigned on the basis of shift data with Eu(dpm)₃.⁶¹⁰ The two protons alpha to the phosphoryl group were resolved in the spectrum with the shift reagent, but those at the beta position were not.

The NMR spectra of phosphetan oxides(XIV) in the presence of LSR provide valuable information. A series of

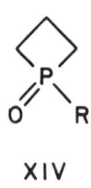

XIV

2,2,3,4,4-pentamethyl phosphetan oxides in which R was a variety of alkyl and aryl groups have been studied with Eu(dpm)₃.⁶¹¹ By first studying complexes with a known *cis* or *trans* configuration, it was possible to assign the configuration of an unknown derivative.

The shifts in the spectra of six-membered cyclic phosphine oxides such as XV, XVI, and XVII were reported and were in agreement with proposed structures.⁶⁰⁹

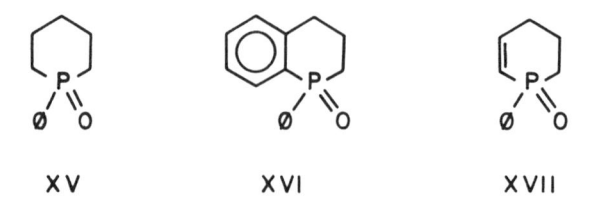

XV   XVI   XVII

Shift data for XV and other well-characterized phosphine oxides have been used to determine the structures of 1-phosphabicyclo[2.2.1]heptane 1-oxide(XVIII) and 1-phosphabicyclo[2.2.2]octane-1-oxide.⁶⁰⁸

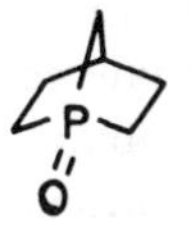

Bicyclic phosphine oxides(XIX,XX) have been studied with the aid of LSR.[603]

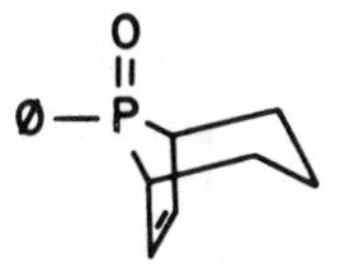

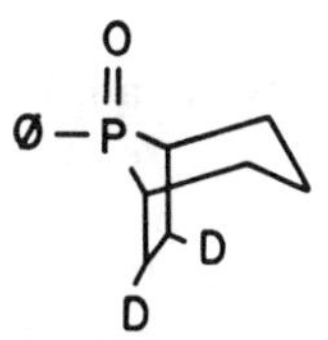

d. Phosphoranes

The shifts in the spectra of a series of spirophosphoranes(XXI, XXII, XXIII) have been recorded with Eu(fod)$_3$ and Eu(dpm)$_3$.[613]

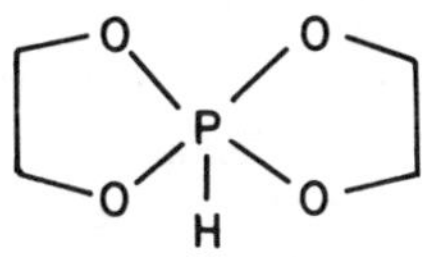

The equilibrium of some derivatives of XXII was displaced toward the phosphite on addition of the shift reagent. The polyfunctional phosphite(XXIV)

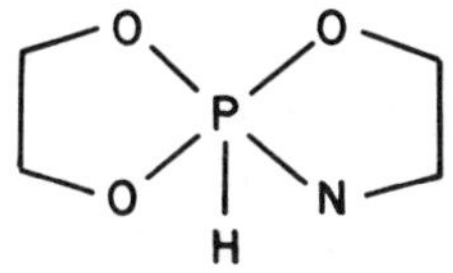

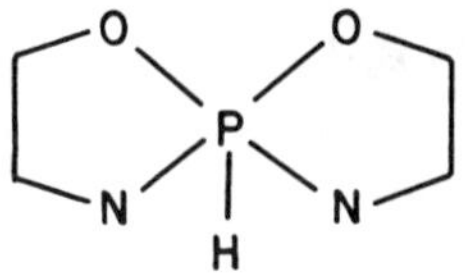

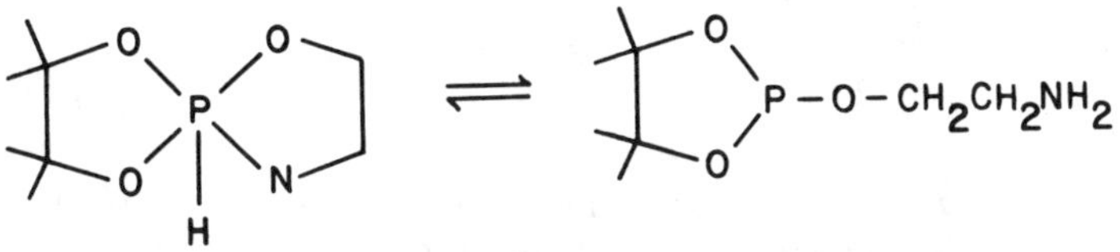

would be expected to complex to the shift reagent through the amine group.

e. Phosphites

Phosphites bond to LSR through the oxygen atoms,[597] and small contact shifts are observed in the ^{31}P NMR spectrum.[596,597] Significant shifts are observed in the ^{1}H NMR spectrum of triethyl phosphite in the presence of chelates of fod and dpm.[520,596] The interaction of phosphites with LSR is considerably weaker, however, than that for substrates with a phosphoryl group.[596]

The shifts in the ^{1}H NMR spectra of dimethylphenylphosphonite(XXV) and ethyldiphenylphosphonite(XXVI) are smaller than those for triethyl phosphite.[596] It has been reported that no binding occurs between triphenylphosphite and Eu(dpm)$_3$.[551]

$C_6H_5P(OCH_3)_2$ $(C_6H_5)_2POCH_2CH_3$

XXV   XXVI

f. Phosphines

The phosphorus atom of the phosphine group is a soft Lewis base and is not expected to coordinate effectively with a lanthanide tris β-diketonate complex. As a result, only small shifts, if any, are observed in the spectra of phosphines with lanthanide tris chelates. The ^{31}P NMR spectrum of triphenyl phosphine exhibits a small shift in the presence of Eu(dpm)$_3$.[415] No observable shifts were noted in the ^{1}H spectrum in this report.[415] Small shifts were noted in both the ^{1}H and ^{31}P spectra of ethyl diphenyl phosphine with Pr (fod)$_3$ and Yb (fod)$_3$.[596] Small shifts were also observed in the ^{1}H, ^{13}C, and ^{31}P spectra of triethyl phosphine with chelates of fod and dpm.[597] Since the phosphorus atom is a soft Lewis base, it effectively bonds to the silver atom of the binuclear shift reagents discussed in Chapter 4. As a result, the shifts for a phosphine substrate in the presence of a binuclear reagent are larger than those with a lanthanide tris β-diketonate.

g. Phosphoramides

The stoichiometry of the complexes of Eu(fod)$_3$[616] and Pr(fod)$_3$[617] with hexamethylphosphoramide have been reported. The kinetics of the exchange of hexamethylphosphoramide between its free and coordinated form have been studied.[618,619] The diastereotopic methylene protons of XXVII were resolved in the presence of Eu(fod)$_3$.[577]

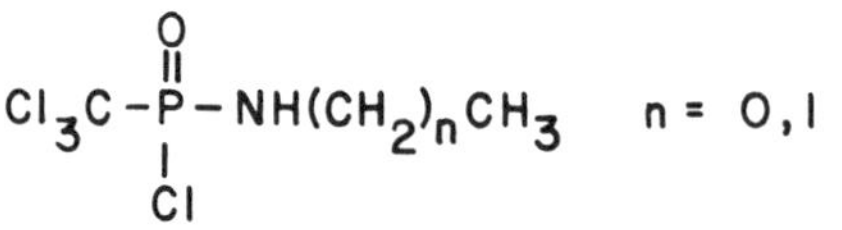

$$Cl_3C-\underset{\underset{Cl}{|}}{\overset{\overset{O}{\|}}{P}}-NH(CH_2)_nCH_3 \quad n = 0,1$$

XXVII

26. Organohalides

As a group, the organohalides constitute weak Lewis bases that bond rather poorly to lanthanide tris β-diketonates. Essentially no shifts are observed in the NMR spectra of organochlorides, -bromides, and -iodides in the presence of a lanthanide tris β-diketonate.[620] If one wishes to study compounds of these classes with shift reagents, the binuclear lanthanide(III)-silver(I) shift reagents described in Chapter 4 can be employed for certain examples. The lanthanide tris chelates can be used to shift the spectra of organofluorides.[120,121,620]

The shifts in the NMR spectra of 1-propylfluoride, 1-octylfluoride, and *sec*-octylfluoride with Eu(fod)$_3$ and Yb(fod)$_3$ have been measured.[620] Complete resolution of the eight proton resonances of *n*-octylfluoride was obtained with Yb(fod)$_3$. Substantially larger shifts were observed in the spectrum of *n*-octylfluoride with Eu(dfhd)$_3$ compared to Eu(fod)$_3$. It was postulated that the interaction between an organofluoride and lanthanide tris chelate could involve either an association of the electronegative C—F bond with the electropositive lanthanide ion, or a donor-acceptor bond utilizing lone pairs on the fluorine atom.[620]

The compounds ethylfluorogermane[121] and (h^5-C$_5$H$_5$)Mo(CO)$_3$GeF(C$_6$H$_5$)$_2$[120] exhibit shifts with Eu(fod)$_3$ that indicate that the compounds bond to the shift reagent through the fluorine atom.

B. Organic Salts

The somewhat surprising observation has been made that shifts are observed in the NMR spectra of the cation of organic salts in the presence of lanthanide tris β-diketonates. Substrates include those of the ammonium,[525,621-626] sulfonium,[622,627,628] phosphonium,[525,622] iminium,[629] and oxonium[625] classes. The most probable interaction involves one in which the anion of the organic salt complexes with the shift reagent to form an anionic lanthanide species. An electrostatic attraction between the cation and lanthanide anion species causes the shifts.

$$Ln(fod)_3 + R^+X^- = R[Ln(fod)_3X]$$

The importance of the anion of the organic salt on the magnitude of the shifts in the spectrum of the cation has been noted.[621,622,628,629] The general ordering of shifts is such that $Cl^- \sim SO_4^{2-} > Br^- > I^-$. Two reasons may explain this observation. The larger shifts with salts with a chloride counterion may result because it is a harder base than bromide or iodide and associates more strongly with the lanthanide tris chelate. This would shift the equilibria to the right.[621] An alternative explanation is that the association constants are independent of the anion and the smaller diameter of the Cl^- relative to Br^- and I^- reduces the distance between the lanthanide and the cation.[622]

Graves and Rose were the first to report shifts in the spectra of organic salts with lanthanide tris chelates.[621] They noted shifts in the spectra of cyanine dyes, N-ethylquinolinium chloride, and tetraethylammonium chloride in the presence of chelates of fod.[621] Shifts have been reported for a series of N-benzylpyridinium chlorides(I) with LSR.[624]

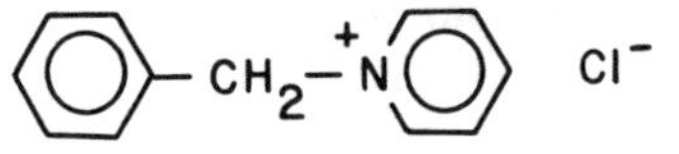

Ortho substitution on the benzyl group of I decreased the coordination between the substrate and the shift reagent. Substitution at three positions, two on the benzyl ring and one on the pyridine ring, led to a further decrease in coordination of the substrate.[624] Shifts in the spectra of N-phenylpyridinium perchlorates were observed in the presence of Eu(fod)$_3$.[625]

Polyfunctional substrates containing a tertiary amine and ammonium group have been investigated with LSR.[623] In the NMR spectrum of N-methylnicotinium iodide(II) with Eu(fod)$_3$, the shifts of the hydrogen resonances of the pyridine ring were larger than those for the pyrrolidine ring.

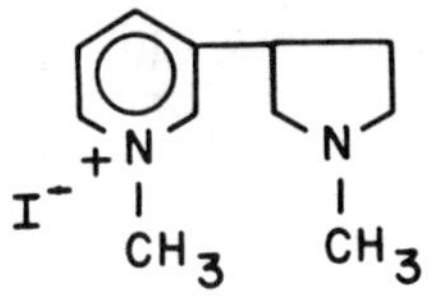

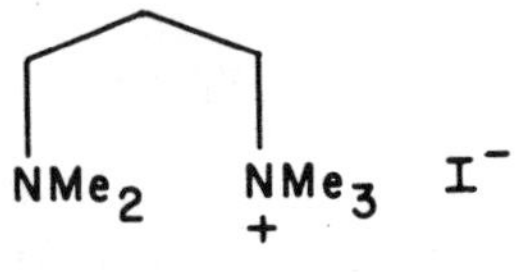

A similar result was obtained for [3-(dimethylamino)propyl]trimethylammonium iodide(III). The magnitude of the shifts indicated preferred complexation of the shift reagent at the ammonium group. In a mixture of N,N-dimethyldodecylamine and N,N,N-trimethyldodecylammonium chloride in the presence of Eu(fod)$_3$, larger shifts were obtained for the methyl groups of the salt. These results imply that complexation of the iodide ion with Eu(fod)$_3$ occurs more favorably than complexation of a tertiary substituted nitrogen atom. A study of the solution geometry of N-methylnicotinium

66   NMR Shift Reagents

iodide through fitting of shift data with Ln(fod)$_3$ chelates and relaxation data with Gd(fod)$_3$ has been carried out.[626]

The shifts for a sulfonium salt(IV) in the presence of Eu(fod)$_3$ were larger than those for the corresponding sulfide.[622] The spectra of trialkylsulfonium and dialkylarylsulfonium salts with Pr(fod)$_3$ have been reported.[627,628] These substrates reportedly coordinated to the LSR through the anion rather than the lone pair of the sulfur atom. A similar result was reported for iminium salts(V).[629]

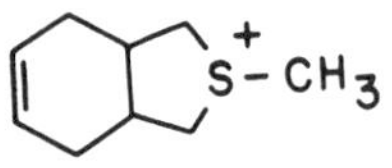

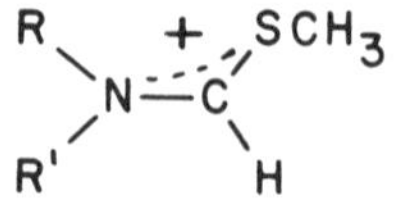

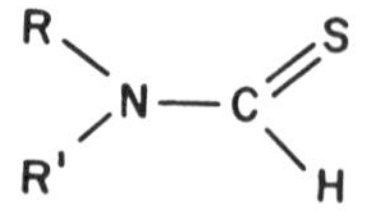

Complexation with the anion was indicated because of the absence of shifts in the spectrum of the corresponding sulfur compound(VI) in the presence of Eu(fod)$_3$.[629] Shifts have been observed in the spectra of phosphonium chloride and bromide, but were not noted in the iodide derivative.[525,622] The spectra of oxonium salts exhibit shifts with chelates of fod.[625]

It has recently been reported that larger shifts can be obtained in the spectra of ammonium salts using binuclear lanthanide(III)-silver(I) complexes.[630] A similar improvement has been noted for sulfonium salts and is expected for phosphonium, iminium, and oxonium salts. This work is described in Chapter 4.

C. Metal Complexes and Substrates with a Silicon Atom

A variety of metal complexes have been studied with the aid of LSR. These complexes interact with the shift reagent by bonding through a suitable functional group on one of the ligands. Substrates studied with LSR encompass metal β-diketonates,[70,133,631-640] organometallic complexes,[89,120,342,436,514,641-648] and compounds containing a silicon atom.[131,645,649-663]

1. Metal β-Diketonates

In a series of reports, Lindoy and co-workers have studied the effects of LSR on the spectra of metal β-diketonate and metal β-ketoimine complexes.[70,133,631-633,640] The β-ketoimine complexes contained ligands prepared from a Schiff's base condensation of ethylenediamine and a β-diketone and have structure I. The oxygen atoms of the ligand appear to bond in a chelate arrangement with the lanthanide ion.[70,631] Complexes with ligands in which the two oxygen atoms have been replaced by sulfur atoms exhibit no shifts in the presence of a LSR.[631]

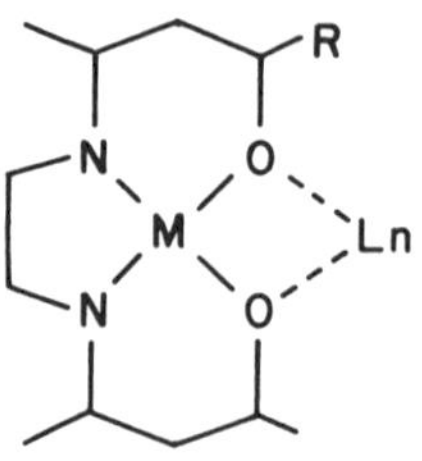

Varying the nature of R has pronounced effects on the bonding of the metal complex to the shift reagent. As the R group is made bulkier, steric hindrance increases, and the association constants decrease.[70,631,639] When R = CF_3, the basicity of the oxygen atoms is lowered to the point that no shifts are observed in the spectrum with a LSR.[70,631]

The nature of the shift reagent affects the magnitude of the shifts in the spectra of I.[70,636] The shifts recorded with Eu(dpm)$_3$ were much smaller than those with Eu(fod)$_3$.[70,636] The metal atom of the substrate influences the magnitude of the shifts as well. Larger shifts were observed in the spectra of complexes with Pd(II) and Pt(II) compared to Ni(II).[70]

Conformational analyses of the bridging group of I in which the amine used in the preparation was 1-methyl-1,2-diaminoethane have been performed.[70] Derivatives of II reportedly bonded to LSR in a chelate manner involving the two oxygen atoms and the azide unit.[640] Evidence for azide complexation with the shift reagent came from studying the IR spectrum. The assignment of both the ^{1}H and ^{13}C spectra of II was facilitated through the use of LSR. Contact shifts, apparent in the form of upfield shifts with Eu(fod)$_3$, were observed for certain of the carbon resonances.[640]

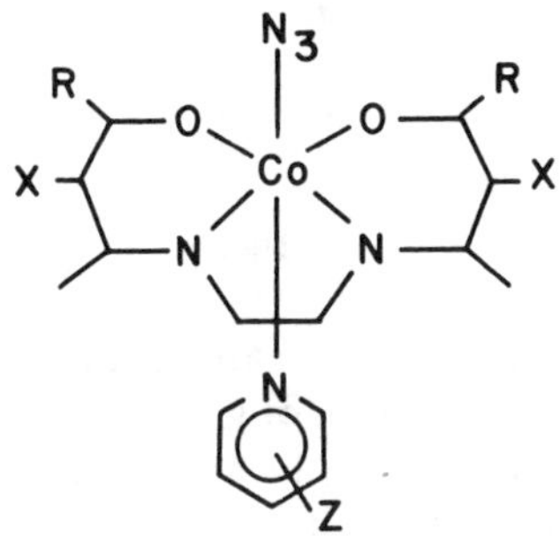

II

Tris β-diketonates such as Co(acac)$_3$ form adducts with Eu(fod)$_3$.[632,633,635-637,639] For some of these adducts, the exchange between the free and coordinated form at ambient probe temperatures is slow on the NMR time scale.[632,636,637] The LSR-Co(acac)$_3$ complex also undergoes slow intramolecular rearrangement, and as a result, two methyl resonances, corresponding to two chemically different sets of methyl groups, are observed. An X-ray structure of the Co(acac)$_3$-Eu(fod)$_3$ adduct has been carried out.[632] The substrate bonds to the shift reagent through three bridging oxygen atoms of the β-diketonate ligands on cobalt.

Steric effects influence the bonding of the Co(III) chelates with LSR. Bulky R groups on the β-diketone ligands reduce the association constants.[639] A decrease in the association constant is noted for the series Co(acac)$_3$, Co(bzac)$_3$(R$_1$ = CH$_3$, R$_2$ = phenyl), and Co(dbm)$_3$(R$_1$ = R$_2$ = phenyl) with Eu(fod)$_3$ in benzene.[639] Complexes of Co(III) with the dbm and dpm ligands do not interact with Eu(fod)$_3$.[633,639]

A number of reports have measured shifts in the NMR spectrum of the metal atom of the metal chelates with LSR.[635-638] The contact and pseudocontact contribution of the shifts in the ^{59}Co spectra of Co(acac)$_3$, Co(bzac)$_3$, and Co(mbzac)$_3$ with chelates of fod have been determined.[636,637] Large shifts were noted in the ^{59}Co spectra with the diamagnetic La(fod)$_3$ complex. For certain chelates, two cobalt resonances were noted in the presence of the shift reagent.[636] It was postulated that these are the resonances of the facial and meridial isomers. The shifts in the ^{27}Al spectrum of Al(acac)$_3$ with chelates of fod have been reported.[635] The exchange was slow at ambient probe temperatures, and signals were observed for the free and complexed chelate. The shifts in the ^{195}Pt spectrum of Pt(acac)$_2$ with Pr(fod)$_3$ have been reported.[638] A review article describing much of the work on the study of metal β-diketonates and β-ketoimines with LSR has been published.[634]

2. Organometallics

The shifts in the spectra of a series of tetraalkyl tin complexes with LSR have been reported.[664] For each substrate, one of the alkyl groups had a functional group, usually a hydroxyl group, that coordinated with the shift reagent. The shift data were suitable for determining the structure of many of the compounds. It was noted in certain instances that changes to the Sn-^{1}H coupling constants occurred in the presence of the shift reagent.

The spectra of (n^4-cyclopentadienyl) (n-cyclopentadienyl) cobalt complexes(III) have been studied with Eu(fod)$_3$ and Pr(fod)$_3$.[643] The carbonyl

III

group is the site of complexation since no shifts are observed for analogous complexes without the carbonyl group. Substitution of groups alpha to the carbonyl group caused a decrease in the association constant of the donor because of steric effects. Only small shifts were observed in the complex with *tert*-butyl groups at both positions alpha to the carbonyl. The metal complexes with a cyclopentadieneone ligand reportedly associated more strongly with the shift reagent than free cyclopentadieneone.[643] The cobalt reduces the ring conjugation, thereby making the carbonyl oxygen a stronger base.[643] The binding constants of IV and V with Eu(dpm)$_3$ have been determined.[342]

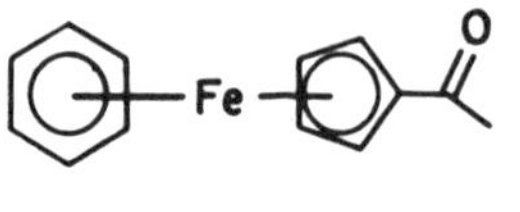

IV 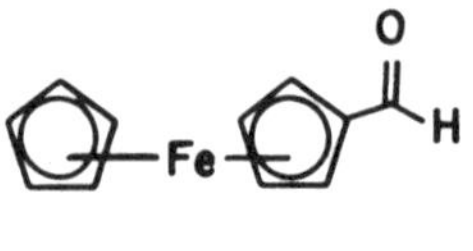   V

The spectra of metal carbonyl complexes with organic ligands have been studied with the aid of LSR.[89,641,642,644,648] The preferred site of coordination was usually a functional group of the organic ligand.[641,648] The shifts in the spectra of iron, manganese, and cobalt-containing metallocene derivatives with Eu(dpm)$_3$ have been measured.[641] VI-IX are examples of the ligands included in these complexes.

VI   VII 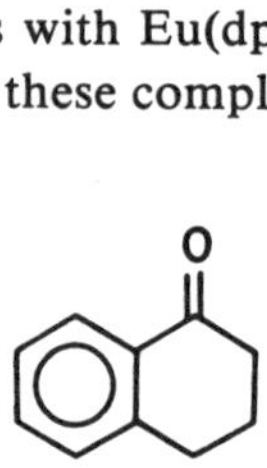   VIII 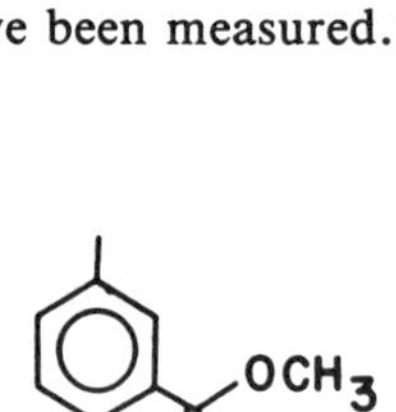   IX

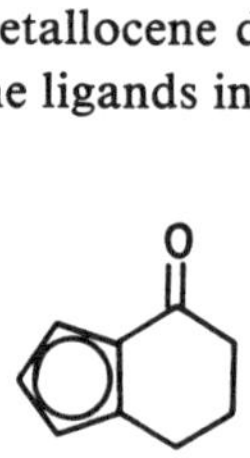

It was possible, with the aid of the lanthanide shift data, to determine the structure and stereoisomers of the complexes. The conformations of the cyclohexane ring of ligands VII and VIII were determined and compared for the complexed and free form.

Shifts in the ^{13}C and ^1H spectra of (tropone)Fe(CO)$_3$(X) with a LSR have been reported.[89,642] Assignments of resonances in the spectra of XI and XII were confirmed through a study with Eu(dpm)$_3$.[642]

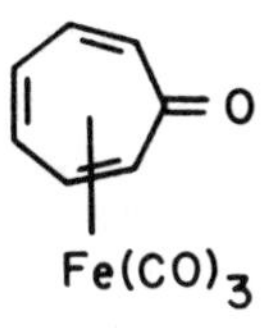

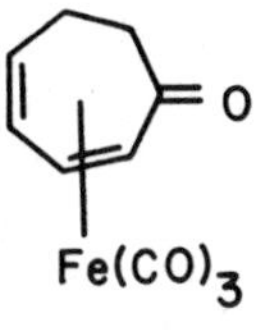

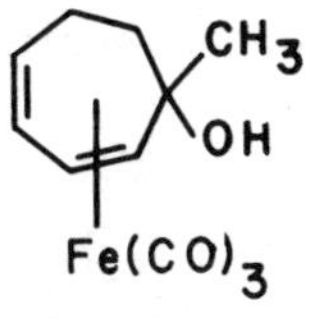

Willcott et al.,[644] in a study of tricarbonyl(*trans,trans*-3,5-heptadien-2-ol)iron (XIII) with Yb(dpm)$_3$,

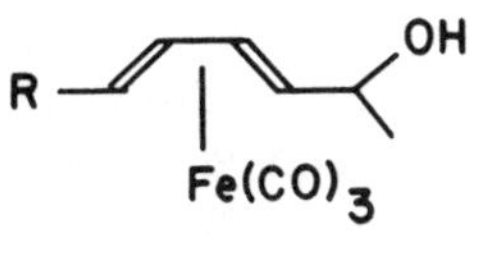

disputed earlier conclusions reported in the literature.[642] Willcott et al. reported that the results of studies with LSR were consistent with other data on the conformation of the compounds, but could not be used alone to assign either the configuration or conformation of the iron complexes.

Shift data with LSR were used to assign the stereochemistry about the double bond of XIV and XV.[89,645] Both XIV and XV

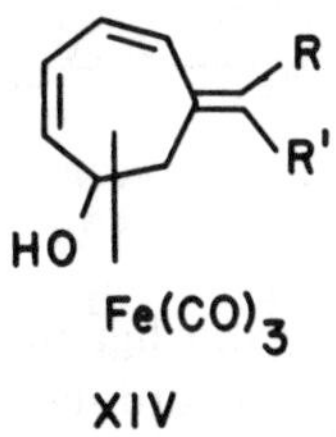

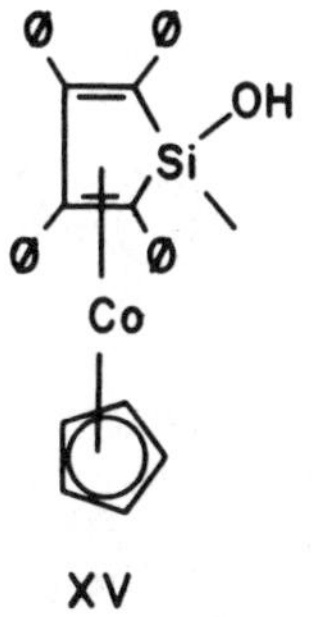

complexed to Eu(dpm)$_3$ through the hydroxyl group.[89,645] The lanthanide shift data allowed the stereochemistry of the 1-silacyclopentadienyl ring to be assigned. The conformations of *mono*- and *bis*-tricarbonylchromium complexes of dimethyl-1,1'-biphenyl-2,2'-dicarboxylate were determined through an experiment with LSR.[648] Substantial shifts have been noted in the spectra of cobalt complexes with α,β-unsaturated thioaldehydes(XVI) with chelates of fod.[766] The absence of shifts in the spectrum of XVII with LSR demonstrated that complexation of XVI involved the sulfur atom.

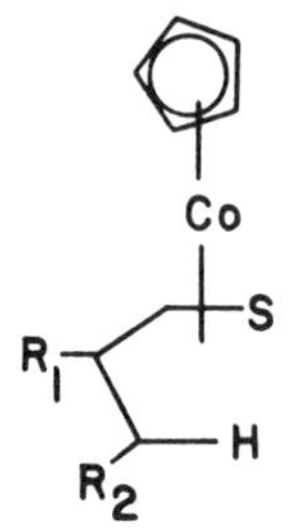

XVI 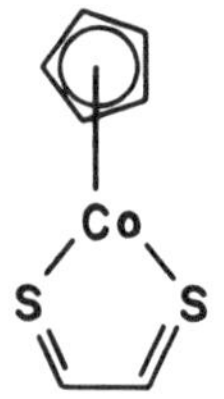   XVII

Metal complexes that do not contain a donor group in one of the organic ligands have been observed to coordinate to LSR through carbonyl,[120,646] cyano,[120,646] thiocyanate,[514] fluorine,[120] and even hydride or cyclopentadienyl[120] ligands. Eu(fod)$_3$ has been employed in the determination of the conformation and rotational barriers in a series of complexes with the formulas h^{5}C$_5$H$_5$Fe(CO)(CN)L (L = various phosphine groups) and h^{5}C$_5$H$_5$Ni(CN)PPh$_3$.[646] Complexation reportedly occurred at the cyano group in these complexes.

Marks has recorded the shifts in the spectra of a wide range of organometallic complexes with LSR in which bonding occurred through F, Cl, N$_3$, CO, and CN ligands.[120] Derivatives with Br, I, and -NCS ligands did not appear to complex with the shift reagent.[120,647] The complex (h^5-C$_5$H$_5$)Fe(CO)$_2$CN coordinated to the LSR through the CN group, (h^5-C$_5$H$_5$)Mo(CO)$_3$GeF(C$_6$H$_5$)$_2$ through the fluorine atom, and (h^5-C$_5$H$_5$)Fe(CO)$_2$CH$_3$ exhibited no shifts in the presence of Eu(fod)$_3$.[120] The spectrum of the dimer [(h^5-C$_5$H$_5$)Fe(CO)$_2$]$_2$, which contains bridging carbonyl groups, exhibited shifts in the presence of Eu(fod)$_3$. This finding is in accord with the observation that bridging carbonyl groups in organometallic complexes are more basic in nature than terminal carbonyl groups.[120] Complexes with more basic terminal CO ligands, such as (phen)[(C$_6$H$_5$)$_3$P]$_2$Mo(CO)$_2$, complex with Eu(fod)$_3$, and shifts are observed in the NMR spectrum. Shifts have also been recorded in the spectra of (h^5-C$_5$H$_5$)$_2$WH$_2$ and (h^5-C$_5$H$_5$)$_2$Sn in the presence of Eu(fod)$_3$.[120] Neither of these complexes has functional groups that would be expected to bond to a LSR. Tetrahydrofuran, which bonds effectively to lanthanide shift reagents, was added to solutions of the metal complexes and LSR. The unshifted spectra of the metal complexes resulted, thereby demonstrating that the complexes coordinated to the shift reagent. In the case of the tin complex, bonding must involve the cyclopentadienyl ligand. The tungsten complex could bond through a cyclopentadienyl ligand or the hydride group.[120]

Shifts have been observed in the spectra of metal complexes with a thiocyanate linkage in the presence of Eu(fod)$_3$.[514] No shifts were noted in similar complexes with isothiocyanate ligands. LSR therefore provide an easy method to distinguish these different forms of linkage isomerism.

Two reports have described shift reagents suitable for use with Lewis acids.[1532,1533] The shift reagent consists of a Co(II) complex with a clathro chelate ligand. The ligand has a phosphorus atom at the cap, and the lone pair of the phosphorus atom can coordinate with transition metal ions. The paramagnetic Co(II) ion causes shifts in the NMR spectra of the transition metal complex. The use of these reagents was demonstrated in resolving and assigning the *syn* and *anti* positions of the allyl group of a set of (allyl)PdCl$_2$ complexes.[1532,1533]

3. Silicon- and Germanium-Containing Compounds

The spectra of tertiary silanols with methyl and phenyl substituent groups are shifted in the presence of chelates of dpm.[650,658] The spectra of polymethylpolysiloxanes and

1,1,1,3,3,3-hexaalkylsiloxanes are not shifted in the presence of chelates of dpm.[650] This distinction has been utilized in the determination of methyldiphenylsilanol in tetramethyldiphenyldisiloxane.[658]

Pr(dpm)₃ has been employed in the study of the ¹H and ²⁹Si NMR spectra of trimethylsilylated sugars(XVIII).[131]

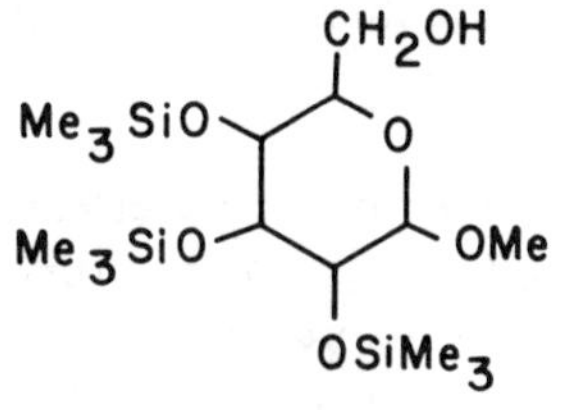

XVIII

The preferred site of complexation of these substrates was the underivatized hydroxyl group, although some complexation may occur at other functional groups. The bonding was selective enough so that the ¹H resonances could be assigned in the shifted spectra. Selective decoupling of the shifted proton resonances permitted assignment of the ²⁹Si resonances. The assignments were confirmed by carrying out double resonance experiments on partially deuterated derivatives.

Alkoxy-,[660,662,663] dialkoxy-,[651,659] and trialkoxysilanes[652,653,656] bond to LSR through the ether oxygen and shifts are observed in the NMR spectra. Similar observations were noted for germane compounds.[653,654,662] Cis- and trans-β-chlorovinyltrimethoxysilanes were distinguished on the basis of shift data with chelates of fod.[656] The conformation of chloromethyl dimethyl methoxysilane has been assigned using shift data with Eu(fod)₃ and Yb(fod)₃.[663] Shifts were also observed in the spectra of diacetoxysilanes in the presence of LSR.[652] Bonding of the methoxy oxygen to the shift reagent resulted in shifts in the spectrum of p-methoxyphenyltrimethylsilane.[657]

Silatranes can be studied with LSR.[649,655,661] The conformation of 1-methylsilatrane was determined on the basis of shift data with a LSR.[661] Complexation reportedly involved the oxygen, rather than nitrogen, atom.[661]

D. Polyfunctional Substrates

As described in the introduction to this chapter, polyfunctional substrates can bond in a variety of manners with a LSR. These range from a strong preference for one of the donor sites to a situation in which two or more donor sites have comparable binding ability, but bond independently to the shift reagent. Chelate bonding can occur if two or more donor atoms have an appropriate arrangement. Chelate bonding is often characterized by significant broadening in the spectrum of the substrate in the presence of the shift reagent.

The distinction between the first two situations presented above is not always clear. The ability of a particular donor group to bond to a LSR is dependent on both electronic and steric effects within the substrate. Differences in the bonding interaction of polyfunctional substrates with complexes of fod and dpm have also been observed on a number of occasions. A further complication is that the degree of preference of one donor site over another will often vary as the ratio of shift reagent to substrate is varied. At low L:S ratios, the shifts usually reflect bonding at either the most basic or least hindered site. As the concentration of LSR is raised, and competition among donor sites for the shift reagent is reduced, bonding of the substrate at a less preferential site is more likely to occur. Such a situation is usually indicated by nonlinearity in plots of the lanthanide shifts vs. L:S. An increase in the slope for nuclei close to the second binding site is observed at higher shift reagent concentrations. Appreciable binding of a polyfunctional substrate at two independent sites may either aid or hinder spectral dispersion and interpretation depending on the particular substrate under study. Fitting of the dipolar equation to the observed shifts under the condition of two independent binding sites is a more difficult, and often intractable, problem.

1. Alcohols

By virtue of favorable electronic and steric effects, the hydroxyl group effectively competes with most other donor groups for coordination to a LSR. Over 100 reports have described applications of LSR to multifunctional substrates containing a hydroxyl group, and in only a limited number of cases has preferred binding at another site been observed. Many of these have involved substrates such as steroids or carbohydrates in which the hydroxyl group is sterically hindered and complexation occurred at a carbonyl[79,388,665-667] or cyclic ether[668-670] group. Every one of these substrates had either secondary or tertiary hydroxyl groups. The phosphoryl group is a particularly effective donor and is the preferred site of complexation in I.[603]

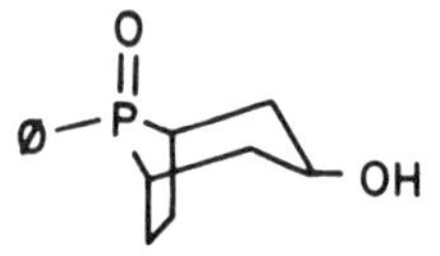

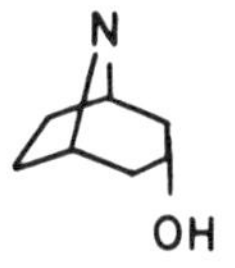

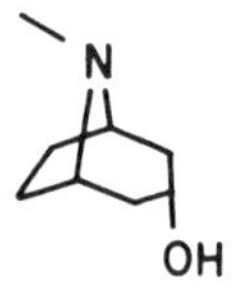

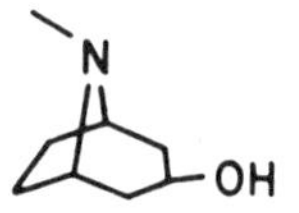

Secondary amines reportedly exhibit a higher affinity for bonding with LSR than do secondary alcohols.[671] Nortropine(II) complexed to LSR through the nitrogen atom. It has been reported that bonding of pseudotropine(III) with Eu(dpm)$_3$ occurs at the hydroxyl group.[671] The interaction between Yb(fod)$_3$ and IV has been reported to involve both donor groups with a preference for the tertiary amine.[307] Two methoxy groups ortho to each other on an aromatic ring form a particularly strong bidentate complex with LSR and compete with secondary hydroxyl groups for complexation.[672]

Chelate bonding in which one of the donors is a hydroxyl group has been implicated for a number of substrates. Proper location of the two functional groups, so that a stable chelate ring is formed, is necessary. Five-member chelate rings have been observed with 2-methoxyphenol(V),[306] 2-ethoxyethanol(VI),[129] and have been postulated for VII.[673]

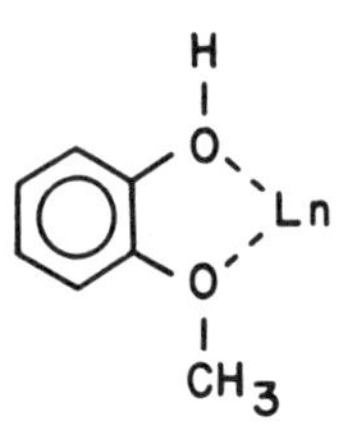

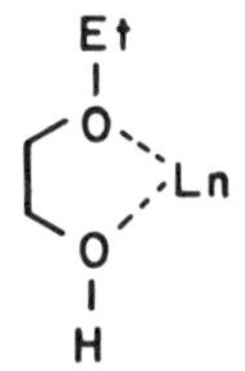

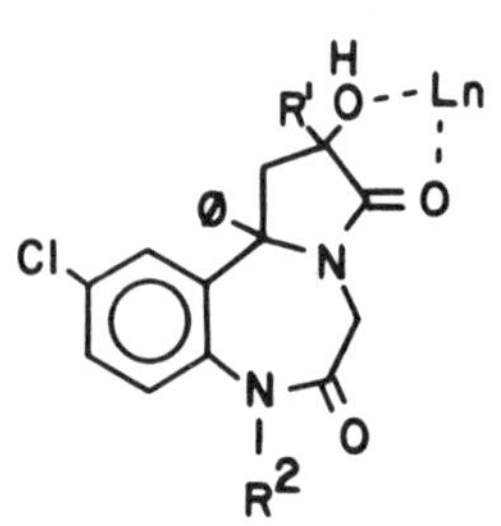

Six-member chelate rings are formed with the hydroxyl and thionyl group of VIII,[674] the hydroxy and carbonyl group of esters of 2,3-diphenyl-3-hydroxy propionic acid(IX),[675] and X.[676]

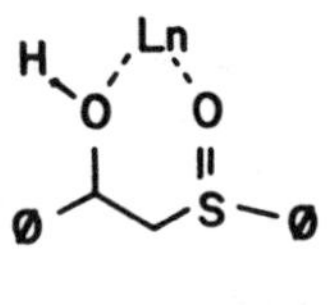

VIII

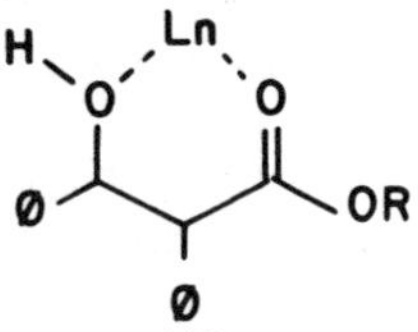

IX

XI reportedly bonds in a chelate manner through the hydroxyl group and ring oxygen.[677] This arrangement would result in an eight-member chelate ring.

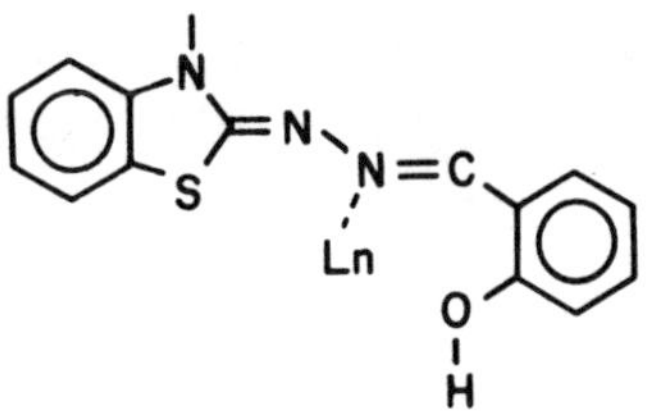

X

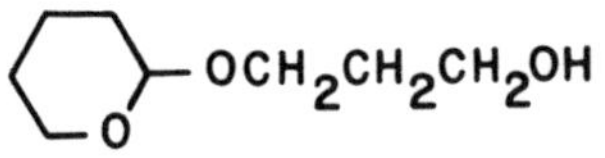

XI

Derivatized carbohydrates often contain two or more functional groups with an arrangement suitable for bidentate binding with a LSR.[678,679] This is especially pronounced in *trans*-fused benzylidene acetals having a *cis* arrangement of potential donor groups.[679]

LSR have been used to aid the structural assignments of methoxymethylphenylethylalcohols,[680] 3-(4-*tert*-butyl-2,6-dinitroanilino)-2-butanol,[681] and 1,1,3,3-tetramethylbutylphenoxyethanols.[155] The conformation of monoheptylglycol ether has been studied with Eu(dpm)$_3$ and Eu(fod)$_3$.[682] The assignment of the ^{13}C spectrum of *n*-dodecyloctaethyleneglycol monoether was facilitated through the use of Yb(dpm)$_3$.[683]

Eu(fod)$_3$ has been applied in the study of methyl esters of long-chain fatty acids that also contain a hydroxyl group.[447,684] At low concentrations of shift reagent, the shifts were indicative of complexation at the hydroxyl functional group of methyl ricinoleate, methyl-12-hydroxystearate, and methyl *erythro*- and *threo*-9,10 dihydroxystearate. Appreciable complexation at the ester group was noted at higher concentrations of shift reagent. The spectra of the polyfunctional fatty acid methyl esters in the presence of the shift reagent were considerably more interpretative than shifted spectra obtained for monofunctional fatty acid methyl esters.

The configurations about certain of the double bonds of prostaglandins have been determined with the aid of shift data with Eu(dpm)$_3$.[685] Studies of symmetrical carotenoids with Eu(dpm)$_3$ have been reported.[58] Shift data with LSR have been used to aid in assigning the structure of peridinin(XII).[686] Lanthanide shift data have been used to confirm the structure of *bis*[3-phenyl-5-oxoisoxazol-4-yl]pentamethineoxonol-(XIII).[687]

XII

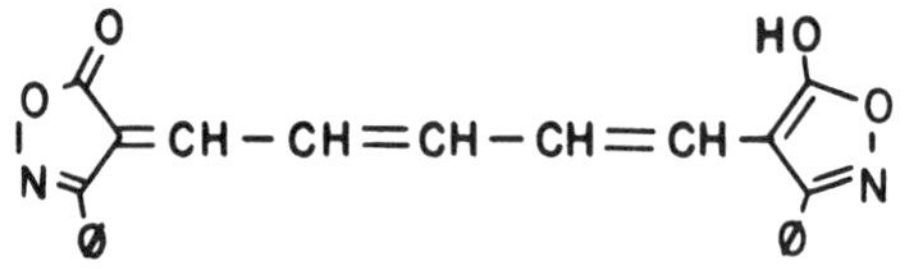

XIII

Shifts in the ^{13}C spectra of methoxyphenols with Eu(fod)$_3$ have been reported.[306] The intramolecular hydrogen bonding of 2-hydroxy-1-benzaldehyde and 3-hydroxy-2-naphthaldehyde have been studied in the presence of Eu(fod)$_3$ and Pr(fod)$_3$.[688] LSR have been used in the analysis of the conformation of substituted biphenyls(XIV).[689] Bidentate bonding was observed in symmetrical derivatives, and the strong bidentate complexation altered the conformation of the substrate.

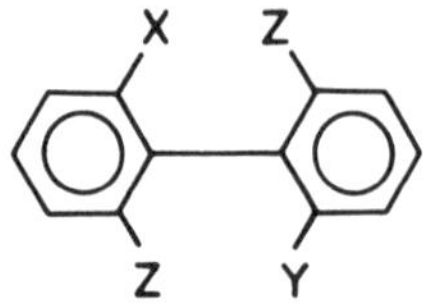

XIV

A variety of polyfunctional monocyclic substrates have been studied with the aid of LSR.[677,690-698] Data with Eu(fod)$_3$ were used to assign the stereochemistry of substituted cyclopropanols(XV).[692]

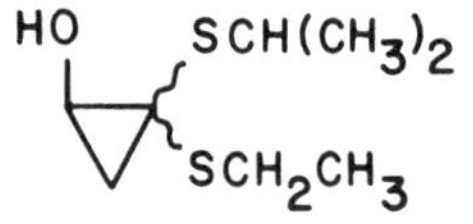

XV 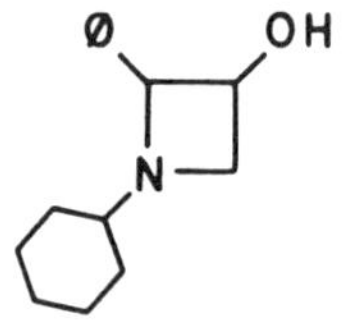   XVI

The diastereomers of 1-cyclohexyl-2-phenylazetidin-3-ol(XVI) were distinguished and assigned on the basis of shift data with a LSR.[695] The structure of the three isomers of 2-methyl-2-n-propylcyclopentane-1,3-diol were assigned through an analysis of the shifts in the spectra in the presence of Eu(dpm)$_3$.[691]

The complexation of Eu(dpm)$_3$ with a series of 2-(hydroxyphenylmethyl)-cyclohexanols has been evaluated.[690] It was reported that the two hydroxy groups exhibited equivalent, yet independent, binding with the shift reagent. The configurations of 2,6-dialkyl-3-piperidinols were assigned on the basis of shift data with Eu(dpm)$_3$.[693] The diastereomers of XVII were distinguished through the use of a LSR.[697]

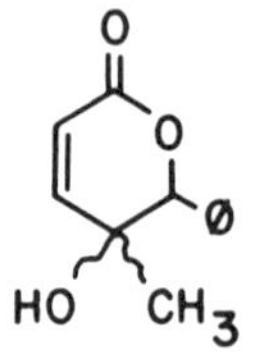

XVII

Data collected with Eu(dpm)$_3$ have been used to determine aspects of the structure and configuration of 3,3-dimethyl-4-(3-hydroxybutyl)-5-hydroxymethylcyclohexan-1-one.[696] The conformations of ethyl cis- and trans-2-hydroxy-1-cycloalkyl carboxylates (alkyl = C$_5$, C$_6$, C$_7$, and C$_8$) have been assigned on the basis of shift data with Eu(fod)$_3$.[698]

A wide range of multicyclic polyfunctional compounds with one or more hydroxyl groups have been studied with the aid of LSR. The conformation of the methoxy group in methoxy indanols was assigned using lanthanide shift data.[699] The stereochemistry of XVIII was assigned on the basis of studies with a LSR.[700] The two diastereomers of XIX were distinguished in the presence of Eu(dpm)$_3$.[672] The stereochemistries and conformations of XX and XXI were assigned on the basis of shift data with Eu(fod)$_3$.[255]

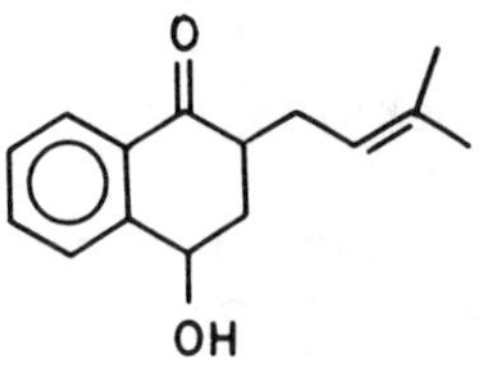

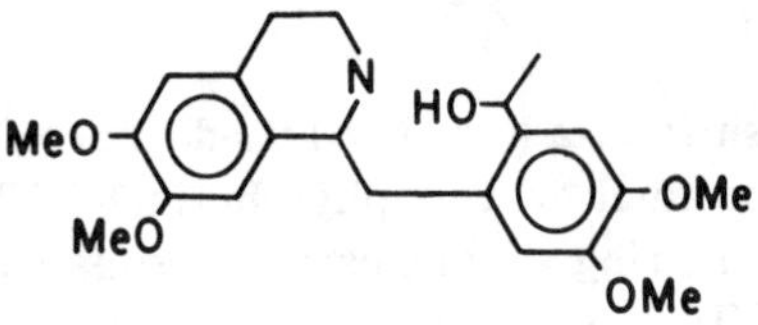

The structure of XXII was determined through the use of Eu(fod)$_3$.[701]

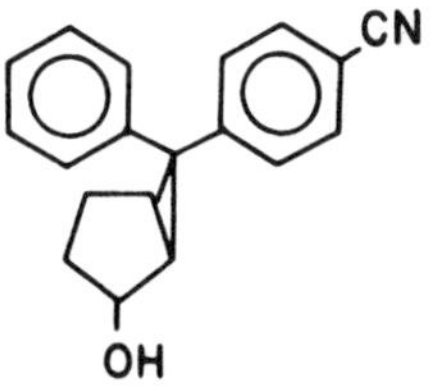

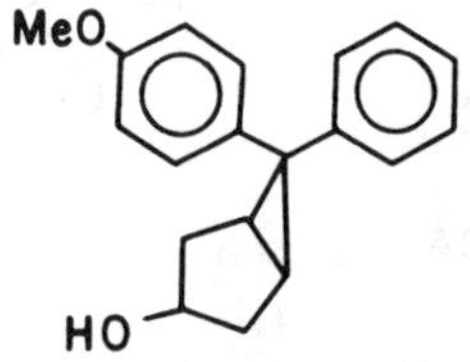

The structures and stereochemistries of bicyclo[3.2.0]heptanes[366] and -heptenes[702] have been assigned with the aid of Eu(fod)$_3$. Shift data with Eu(dpm)$_3$ have been used to determine the exo- or endo- position of deuterium substitution in 10-hydroxyisoborneol and 10-hydroxycamphor.[245]

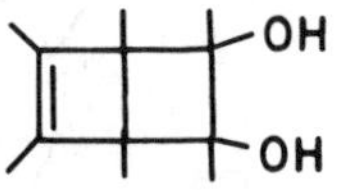

The shifts in the ^{13}C spectrum of quinine with Yb(fod)$_3$ have been reported.[307] The conformations of cis- and trans-9-hydroxybicyclo[3.3.1]nonane-endo-3-carboxylates have been determined with the aid of Eu(dpm)$_3$.[703] Complexation of the hydroxyl group was noted at low concentrations of shift reagent. Complexation at the ester group was observed at higher shift reagent concentrations. The coupling constants did not vary over the range of shift reagent concentrations employed. LSR have also been used in assigning the conformer ratios of 2,6-disubstituted bicyclo[3.3.1]nonanes[704] and a 3,7-disubstituted bicyclo[3.3.1]nonane.[705]

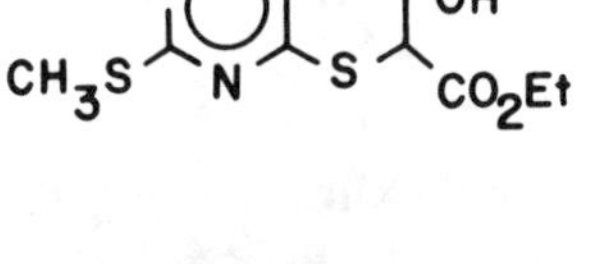

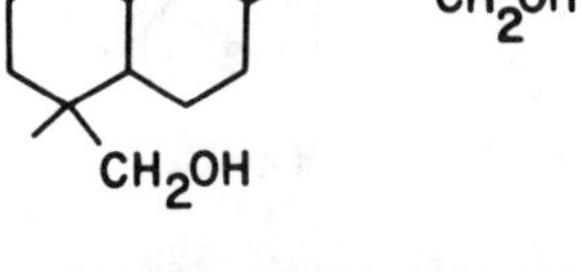

The shifts in the spectrum of XXIII in the presence of Eu(fod)$_3$ have been used to aid in its structural assignment.[706] LSR have been used to assign the configuration about the olefin bond in dihydropinifolic acid diol(XXIV) and to distinguish an isomer that differed in the configuration of the ring hydroxymethyl group.[707] The conformation of cannivonine b(XXV) has been studied with the aid of Eu(fod)$_3$.[708,709] Complexation at the hydroxyl group was favored at low concentrations of shift reagent. Complexation of the amine reportedly occurred at higher shift reagent concentrations. Partially deuterated derivatives of XXV were studied to facilitate assignment of the resonances.

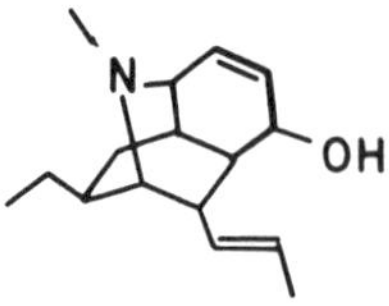

XXV

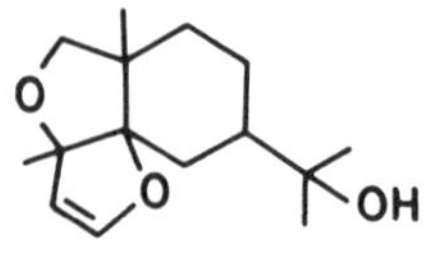

XXVI

Lanthanide shift data have been used to obtain the structure of the carbon skeleton of the natural product deacetylphytuberin(XXVI) and to determine the conformation of the cyclohexane ring.[710] The carbon resonances of the cyclopropane ring of illudin M-acetate (XXVII) have been assigned on the basis of shift data with Pr(fod)$_3$.[711]

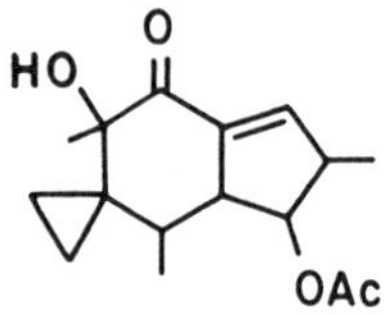

XXVII

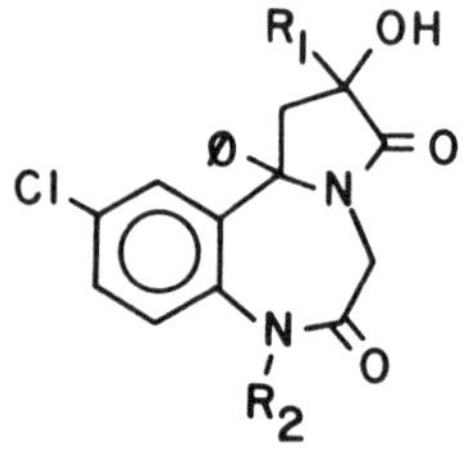

XXVIII

LSR have been used to determine certain aspects of the stereochemistry of tricyclic compounds such as 1,2,7,11b-tetrahydropyrrolo[1,2-d][1,4]benzodiazepine-3,6(5H)diones (XXVIII),[673] (XXIX),[388] and *exo/endo* isomeric dihydro- and tetrahydrodicyclopentadiene-9,10-diols(XXX,XXXI).[712]

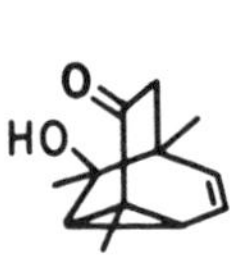

XXIX

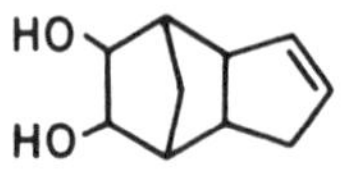

XXX

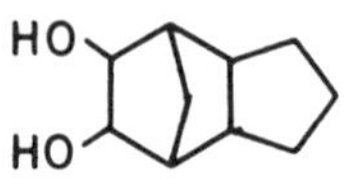

XXXI

Structural assignment of the hemiketal dimer XXXII prepared from 2-hydroxy-2-methylcyclobutanone was confirmed using shift data with a LSR.[713] The structure of XXXIII was assigned on the basis of shift data with Eu(fod)$_3$.[714] In particular, the stereochemistry of the methyl-substituted carbon of the lactone ring was assigned. The configuration of the

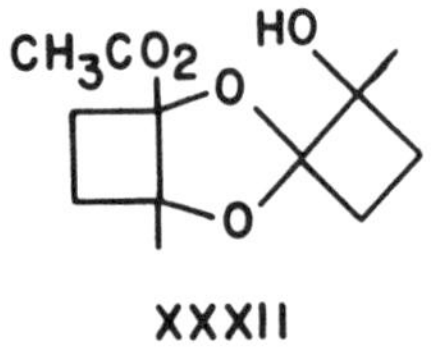

XXXII

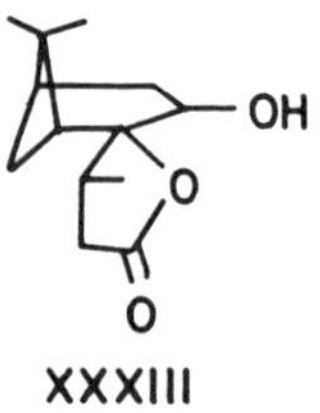

XXXIII

hydroxy group in XXXIV was determined using data collected with Eu(fod)$_3$.[715] The principal site of complexation appeared to be the hydroxyl group; however, the shifts were indicative of some complexation at the two epoxide rings. Substituted coumarins have been studied with the aid of Eu(fod)$_3$.[716] The shifts with Eu(dpm)$_3$ were used in conjunction with other data to assign the structure of XXXV.[717] Changing the

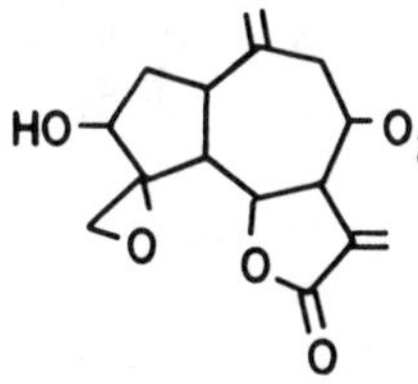

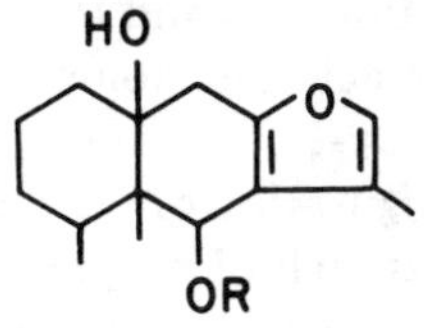

R group from −H to −CH$_3$ significantly reduced the lanthanide shifts for protons close to the −OR portion of XXXV.

Polyfunctional adamantane derivatives with hydroxyl substituent groups have been the focus of several studies with LSR.[718-722] The shifts in the spectrum of 2,4-adamantanediol in the presence of LSR were indicative of independent, but equal, complexation at both hydroxy groups.[719,721] The total shift for each resonance of 2,4-adamantanediol could be obtained by accounting for each of the two individual components.[719] LSR have been used to determine the configuration about substituted carbon atoms in polyfunctional adamantane derivatives.[718,721] A separation of the two ^{13}C methyl resonances of 4-hydroxy-1-isopropyl-2-oxaadamantane was observed in the presence of Yb(fod)$_3$.[722] The shifts and binding site of 2-hydroxy-1-(2-hydroxyethyl)adamantane with Eu(dpm)$_3$ have been reported.[720] Independent association at both hydroxyl groups was noted.

Determination of the structure of the natural product conferol (XXXVI) was facilitated with Eu(dpm)$_3$.[723] Shift data with LSR have been used to assign the configuration of hydroxyl groups in trans-4,5-dihydroxy-4,5-dihydroaldrin and cis-exo-4,5-dihydroxy-4,5-dihydroisodrin.[281] Eu(fod)$_3$ has been utilized in the determination of the stereochemistry of 9-hydroxy-7-oxatetracyclo[6,3,0,0^{2,6},0^{3,10}]undecane (XXXVII) and XXXVIII.[724]

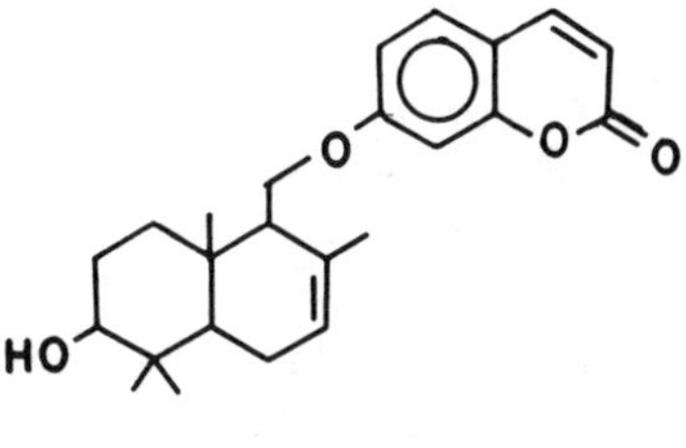

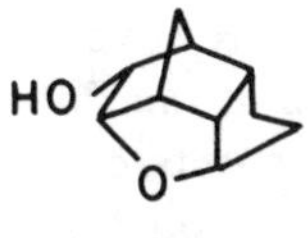

Hydroxyethyl groups of porphyrin compounds bond effectively to Eu(fod)$_3$.[725] The structural assignment of naphthomycin, which has hydroxy, ketone, and lactam functional groups, was facilitated through the use of Eu(fod)$_3$.[726] There was no discussion of the site of complexation of naphthomycin in this report.

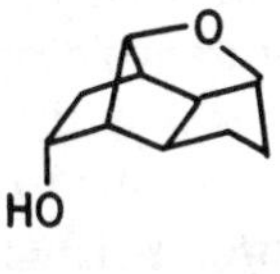

Polyfunctional steroids containing one or more hydroxyl groups have been the focus of several studies with LSR. As with monofunctional steroids, the shifts in the presence of a LSR are not large enough to provide first-order spectra, but are adequate for

probing certain aspects of the substrate. The structures of the A ring of 1α,19-diol derivatives of cholestane have been assigned on the basis of shift data with Eu(fod)$_3$.[727] The conformation of the B ring of 3β-hydroxy-β-homocholestan-7,8a-lactone was assigned on the basis of shift data with Eu(fod)$_3$.[667] LSR have been used to assign the methyl resonances of polyfunctional steroid derivatives,[295,728,729] and to determine the configuration about carbon atoms substituted with a hydroxyl group.[298,730-733]

Data collected with Eu(fod)$_3$ and Pr(fod)$_3$ aided the assignment of ^{13}C resonances of methyl esters of cholic acid and some of its derivatives.[154] The position of deuterium substitution in hydroxy-12β-cananine has been determined through the use of Eu(dpm)$_3$.[734] The configuration of the olefin bond in pregnadienones has been assigned on the basis of shift data with Eu(dpm)$_3$.[735]

In steroids with more than one hydroxyl group, the least sterically hindered site is usually the preferred site of complexation.[79,154,733] The hydroxyl groups of 3,7,12-trihydroxycholanate(XXXIX) bond in the order 3 > 7 ≫ 12.[79]

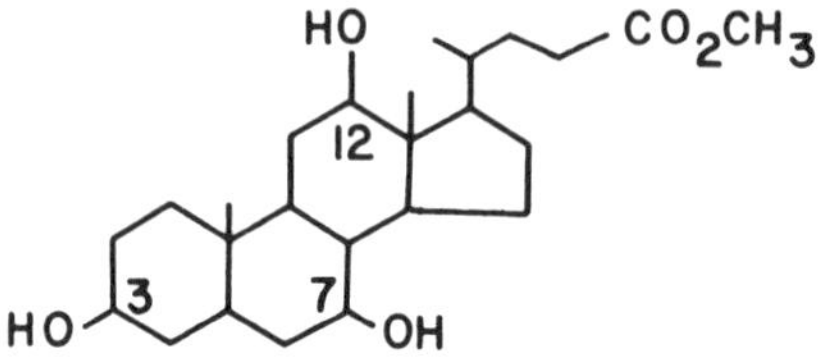

XXXIX

Another frequent occurrence with steroid derivatives is to observe complexation at more than one site. This is especially pronounced at higher concentrations of shift reagent.[295,666,667,727,729,735-737] Hinckley et al. have described a graphical procedure for assessing whether coordination of a polyfunctional substrate occurs at two sites.[737] The method assumes that the shifts exhibit only a distance dependence. This assumption is rather tenuous, and such an analysis can only be considered in a qualitative nature.

Organic-soluble carbohydrate derivatives have been studied with the aid of LSR. In most instances, all but one of the hydroxyl groups has been derivatized. Coordination with the LSR usually involves the underivatized hydroxyl group. The possibility of chelate bonding must always be considered with these substrates. The shifts in the spectra of pyranosides,[128,738] including ribo-,[739] xylo-,[739] talo-,[739] manno-,[665,679,739] psico-,[739] fructo-,[739] gluco-,[665,670,740] galacto-,[670,741] and allopyranosides[740] in the presence of LSR have been reported. The shifts in the spectra of gluco-,[127,678,742] allo-,[127,678] and ribofuranosides[741] with LSR have been reported. In certain instances, the spectra are first-order in the presence of the shift reagent.[127,665,741]

It was possible to assign aspects of the stereochemistry of unknown carbohydrate derivatives by comparing the shift gradients to glycosides of known configuration.[739,740] Data with LSR have been used to determine the position of deuterium substitution[743] and the position of substituent groups[744] in carbohydrates. Gluco- and galactopyranosides have been distinguished on the basis of coupling constants obtained in the shifted spectra, as well as differences in the magnitude of the lanthanide shift values.[670] The carbon resonances of mannopyranosyl-mannopyranoside and glucopyranosyl-glucopyranoside derivatives have been assigned by comparing the shifts with a LSR to those of simpler monosaccharide derivatives.[665]

Blocking groups — In many instances, it would be advantageous if shift reagents could be used to probe a polyfunctional substrate independently at each of its donor sites. It is often possible to probe a substrate at a hydroxyl group because of strong preferential binding. A number of derivatization schemes for hydroxyl groups that effectively reduce or prohibit complexation with a LSR have been described in the literature.[667,716,738,745-748] After carrying out such a derivatization, lanthanide shift data indicative of complexation at less preferred bonding sites can be obtained for the substrate. Derivatization of alcohols to trifluoroacetate esters,[667,736,745,748] trimethylsilyl ethers,[716,748] or *tert*-butyldimethylsilyl ethers[746-748] is usually effective at reducing complexation of the hydroxyl functional group. Procedures such as methylation and tosylation[738] are not as reliable. In a study of monofunctional trifluoroacetate esters, trimethylsilyl ethers, and *tert*-butyldimethylsilylethers, only the *tert*-butyl derivative was found to be successful at completely blocking complexation.[748]

2. Ketones

The carbonyl group is intermediate in its ability to bond to LSR. Whether or not complexation of a polyfunctional substrate occurs at a carbonyl group is highly dependent on the nature of the other donor. Groups such as phosphoryls,[603,749] hydroxyls,[154,388,666,685,696,700,711,721,735,745] and secondary amines[671,750] generally are the preferred site of complexation in polyfunctional substrates with a ketone moiety. Other carbonyl-containing groups such as esters,[79,307,751] amides,[752] and lactams[753] may compete effectively with ketones and be the preferred site of complexation with a LSR. In these instances, the two donor groups are similar, and the preferred site depends on a mixture of electronic and steric effects and varies with the nature of the substrate. An analogous result is obtained for the thionyl group.[718]

Chelate bonding involving the carbonyl group of a ketone as one of the donor groups has been observed for a number of substrates. The most common examples involve six-member rings,[91,754-763] although five-[524,752,761,764] and an eight-member chelate ring[765] have been reported. Ortho substitution of a carbonyl group and a methoxy group on an aromatic ring(I) leads to a stable bidentate complex with Eu(fod)$_3$.[91,754] With Eu(dpm)$_3$, complexation reportedly involved only the carbonyl group.[91] A similar arrangement of a carbonyl and methoxy group

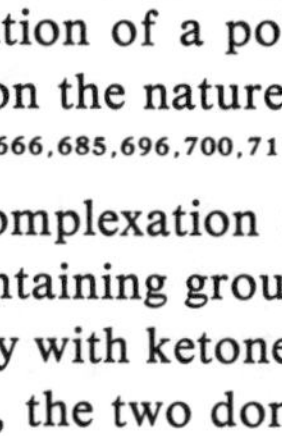

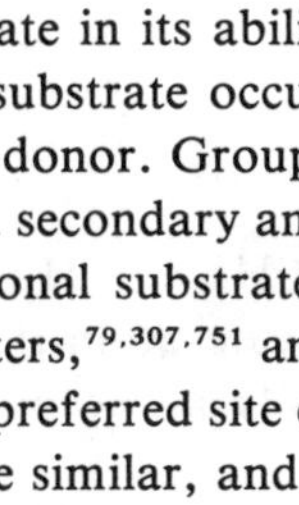

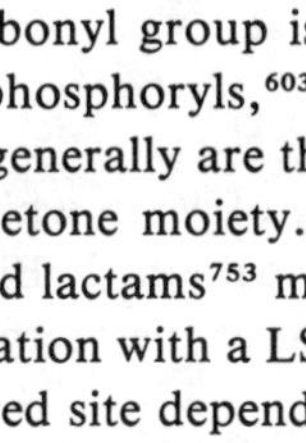

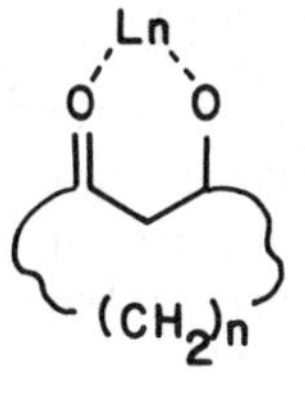

leads to chelation of quinones with a methoxy group peri to the carbonyl(II)[91,452,757-761] and alkylidene furanones(III).[762] Cycloalkanones with a methoxy group β to the carbonyl group(IV) have the potential of forming a bidentate complex with LSR.[755] In this instance, however, the size of the cycloalkane ring influences the orientation of the donor groups. Strong bidentate binding of cyclooctanone was observed. Weak bidentate binding of cycloheptanone may be present, and complexation at only the carbonyl group was noted for smaller rings.

Substitution of a methoxy group alpha to the carbonyl group of quinones reportedly results in bidentate complexation and the formation of a five-member chelate ring(V).[752,761,764]

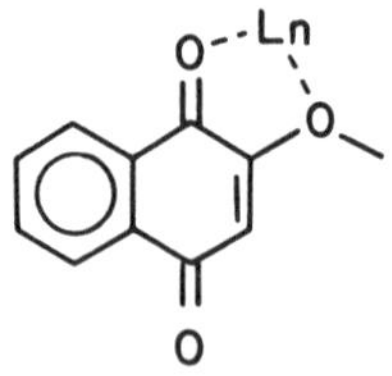

V 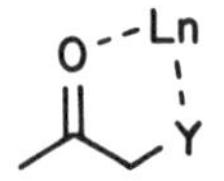   VI

The bonding of VI in which Y = $-OCH_3$, $-SCH_3$, or $-NMe_2$ was reported to occur in a bidentate fashion.[524] Bidentate complexation involving the tertiary amine and carbonyl oxygen of the acetyl group of VII has been noted.[756] A tridentate complex

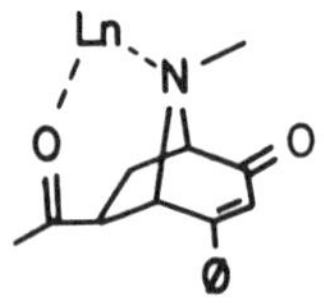

VII 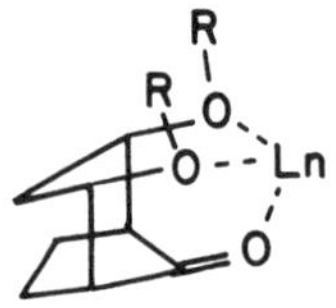   VIII

with VIII has been postulated.[763,767] The isomer of VIII that had a reversal of configuration of the two ether-substituted carbon atoms coordinated only through the carbonyl group.

The isomerism of diazoketones(IX) has been studied with Eu(fod)$_3$.[768] Complexation occurred at the carbonyl group and was stronger for the E rotamer.

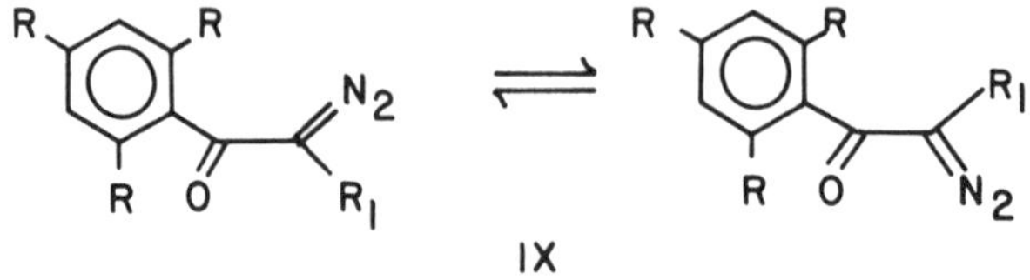

IX

The shifts in the spectra of symmetrical carotenes with Ln(dpm)$_3$ have been reported.[58,317] Independent complexation at both carbonyl groups was noted. The relative bonding of aromatic substrates with ester, ketone, and ether substituent groups with Eu(fod)$_3$ has been investigated.[754] The binding was dependent on steric and electronic effects. Bidentate bonding of ortho-substituted groups was much stronger than monodentate bonding. Aromatic ketones preferentially complexed over aromatic esters.

LSR have been employed in the study of the conformational preference of compounds of structure X(X = O, S, Se, or Te).[769-774]

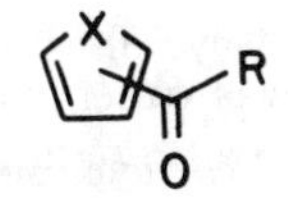

X

None of the reports noted any perturbation of the conformational preference by complexation of the shift reagent. Both the barrier to rotation and conformer populations of these substrates were determined. A conformational analysis of *trans*-4-(5-bromo-2-thienyl)-3-buten-2-one has been carried out through the analysis of lanthanide shift data.[769] The conformations of acetylbenzofurans have been studied with the aid of Eu(fod)$_3$.[332]

The configurations of substituted cyclobutanones and cyclobutenones have been assigned on the basis of lanthanide shift data.[775,776] LSR have been used to assign the stereochemistry of a number of polyfunctional substrates with five-member rings(XI) and a keto group.[393,543,762] The isomers of α-*dl-trans*-allethrin were distinguished in the presence of Eu(fod)$_3$.[393] The

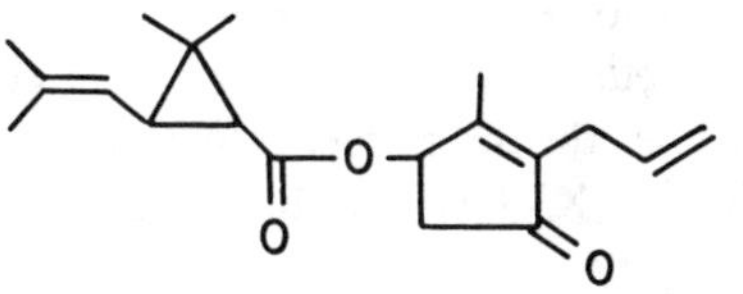

XI 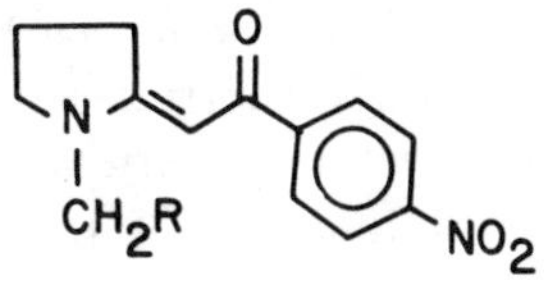   XII

stereochemistry of 1-substituted-2-α-(*p*-nitrobenzoyl)methylenepyrrolidines(XII) and -piperidines were assigned on the basis of lanthanide shift data.[543] The *cis* and *trans* isomers of alkylidene furanones were distinguished on the basis of experiments with Eu(dpm)$_3$.[762] The shifts in the spectra of XIII in the presence of Eu(fod)$_3$ and Pr(fod)$_3$ have been reported.[427]

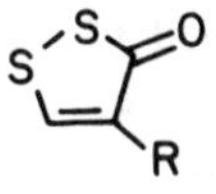

XIII 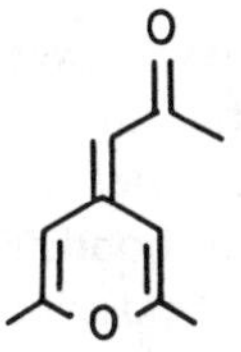   XIV

Shift data for 4-acetonylidene-2,6-dimethyl[4*H*]pyran(XIV) in the presence of Eu(dpm)$_3$ have been used to determine the conformation of the carbonyl group relative to the exocyclic double bond.[777] The spectra of cinnamic acid derivatives with cyclohexanone(XV) or cyclohexenone groups have been studied with Eu(fod)$_3$.[778]

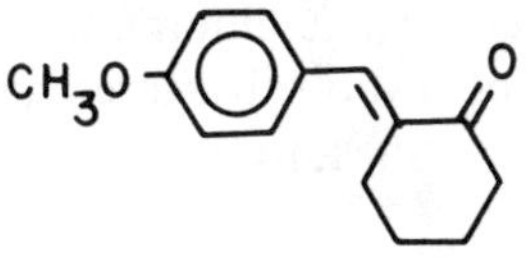

XV

The configurations of substituted 1,4-cyclohexanediones have been assigned through the use of shift data recorded with LSR.[355]

The shifts in the spectra of 3-methoxycyclohexanone, -heptanone, and -octanone with Eu(fod)$_3$ have been reported.[755] Shift data with Eu(fod)$_3$ have been used to assign the positions of substituent groups in chlorazepine-2,5-diones(XVI).[753] The structural assignment

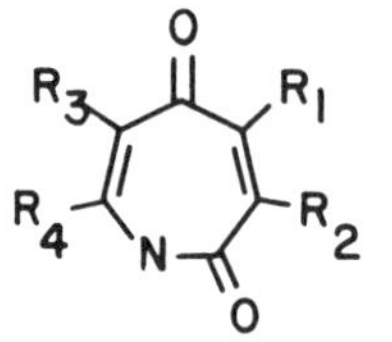

XVI 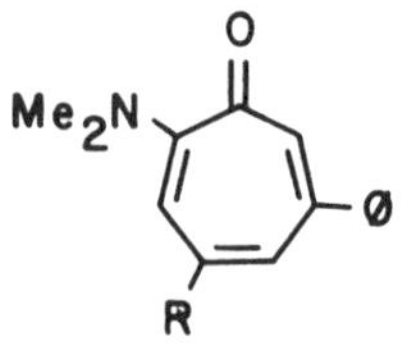   XVII

of XVII was facilitated through the use of shift data with Pr(fod)$_3$.[756]

The shifts in the ^{1}H and ^{13}C spectra of 6-methoxyphenalone, methoxyindanones, and fluorenone with Yb(fod)$_3$ have been reported.[134] The position of substituent groups in naphthoquinones has been determined through the use of Eu(fod)$_3$.[752] Complexation of these substrates was observed at the less sterically hindered carbonyl group. The spectra of a series of anthraquinones(XVIII) have been studied in the presence of Eu(fod)$_3$ and Pr(fod)$_3$.[757]

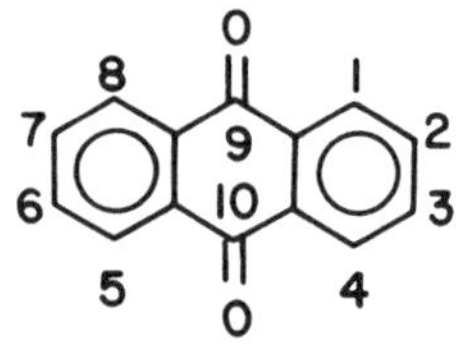

XVIII

Hydrogen bonding occurs between the carbonyl oxygen at position 9 and amine hydrogen of 1-(methylamino)anthraquinone.[757] The lanthanide shift data for this substrate were indicative of complexation at position 10.[757] The 1-dimethylamino derivative cannot form a hydrogen bond. For this substrate, the shifts implied chelate bonding of the carbonyl group at position 9 and the amine nitrogen. The 2-dimethylamino derivative cannot form a stable chelate bond and seemed to bond more favorably at the carbonyl oxygen at position 10.[757] The structure of trimethoxy averufin(XIX) was assigned on the basis of shift data with a LSR.[779]

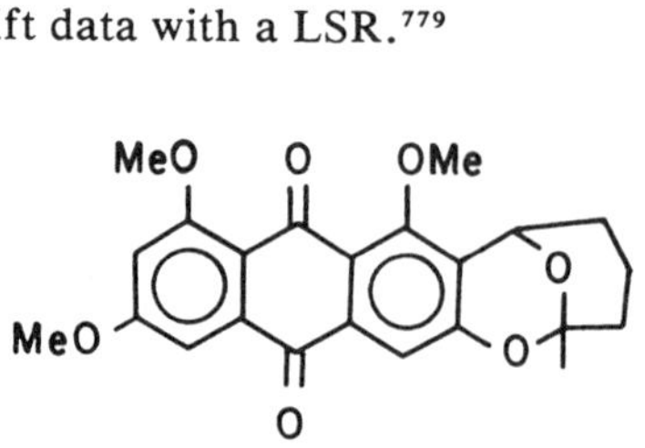

XIX

Shift data with LSR have been used to locate the position of substituent groups in xanthones.[452,759]

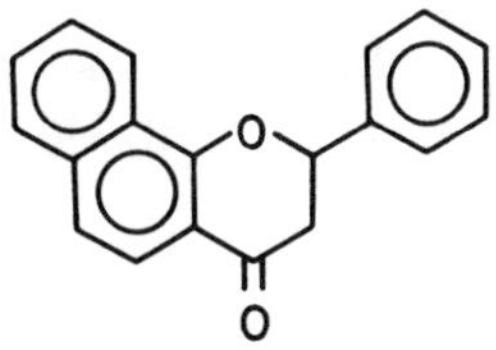

XX 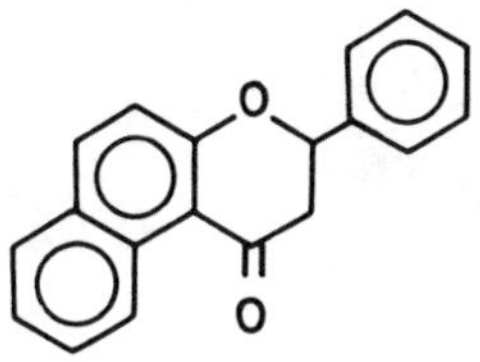   XXI 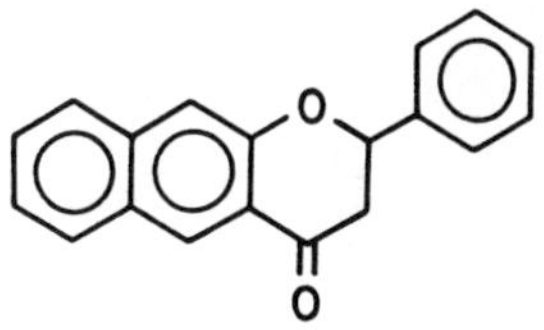   XXII

The shifts in the ^{13}C spectrum of xanthone with Yb(fod)$_3$ have been reported.[307] Flavones and isoflavones have been extensively studied with LSR.[91,134,307,759-761,764,765,780] The shifts in the ^{13}C spectrum of flavone with Yb(fod)$_3$ have been reported.[134,307] The majority of these reports, however, have used data generated with the LSR to locate the position of substituent groups.[91,759-761,764,780] The three isomers of naphthoflavone (XX, XXI, XXII) were distinguished through the use of shift data with Eu(fod)$_3$.[765]

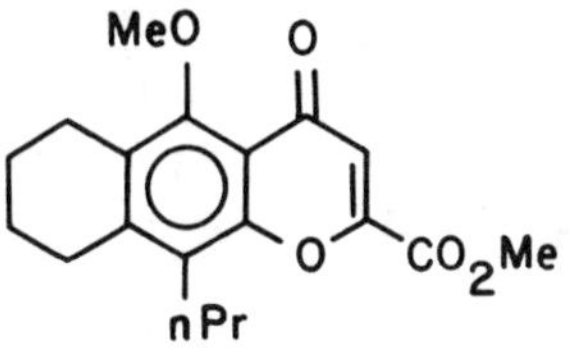

XXIII

The structure of XXIII and XXIV were assigned on the basis of shift data with a LSR.[758]

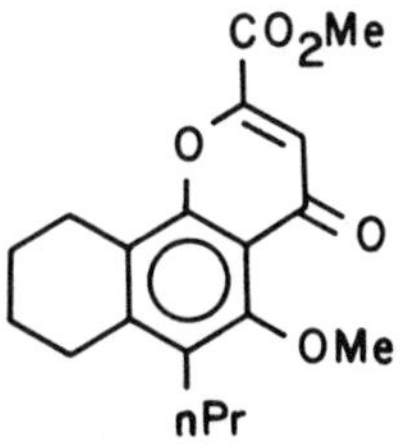

XXIV

Two diastereomers of XXV were distinguished through the use of shift data with Eu(fod)$_3$.[781] The shift data also permitted a determination of the ring stereochemistry for XXV.

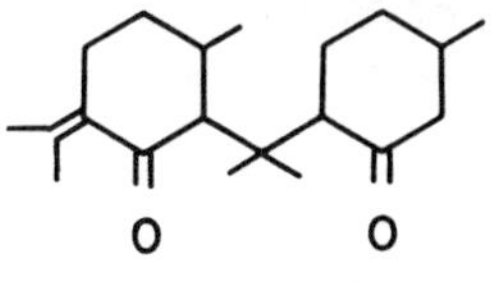

XXV 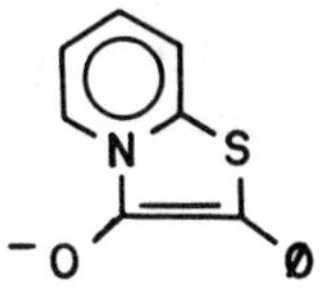   XXVI

The planarity of XXVI and its derivatives was determined through an analysis of shift data with Eu(dpm)$_3$.[782] The structure of XXVII was

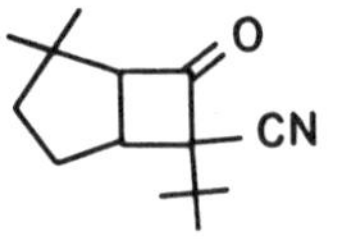

XXVII 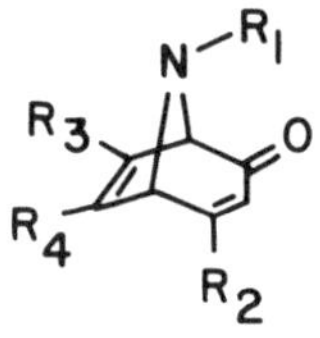   XXVIII

assigned with the aid of LSR.[366] The structures of derivatives of XXVIII were determined through the use of shift data with Eu(fod)$_3$ and Pr(fod)$_3$.[756,783] The bridgehead hydrogens of bicyclo[3.3.0]octane-3,7-dione were shown to occupy a *cis* configuration by analysis of the spectrum in the presence of a LSR.[784] The configurations of the alkoxy groups in 2,4-dialkoxy[3.3.1]nonan-9-one were determined with the aid of Eu(fod)$_3$.[763,767] Shift data with a LSR have been used to facilitate the structural assignments of 1,3-oxazolidin-2-ones(XXIX).[785]

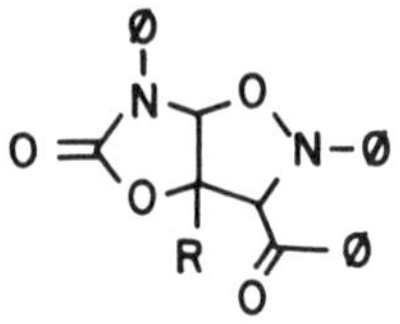

XXIX 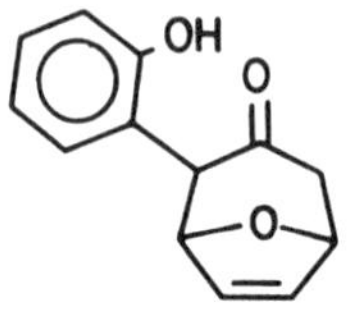   XXX

Complexation of XXIX reportedly involved the oxygen atom of the benzoyl group. The structural assignment of XXX was aided by the use of shift data with Eu(fod)$_3$.[694] This report did not discuss the site of complexation of XXX.

The conformation of XXXI was determined through the analysis of lanthanide shift data.[786] The shifts indicated preferential

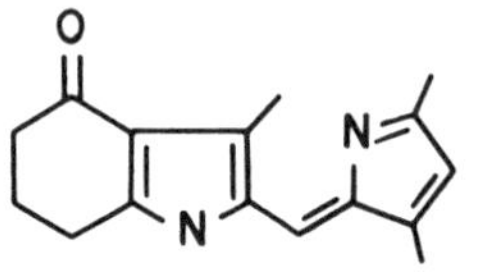

XXXI 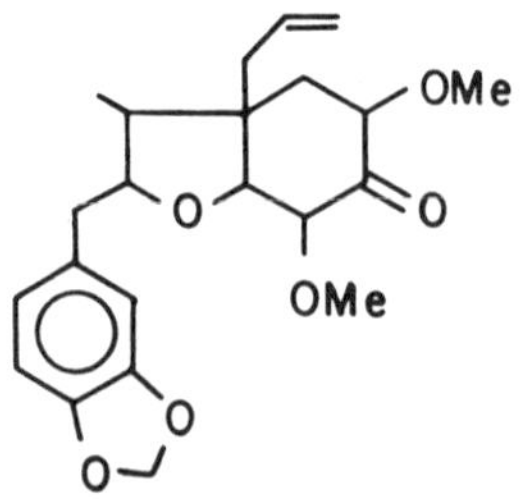   XXXII

complexation at the carbonyl group of XXXI. Lanthanide shift data was used to assign the structure of canellin B(XXXII).[787] Shift data with Pr(fod)$_3$ was employed in an attempt to assign the absolute configurations of the benzofuranoid neolignins XXXIII and XXXIV.[788] Large shifts

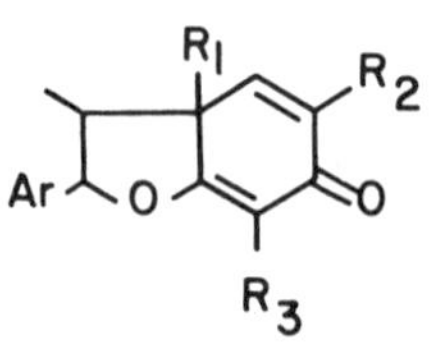

XXXIII 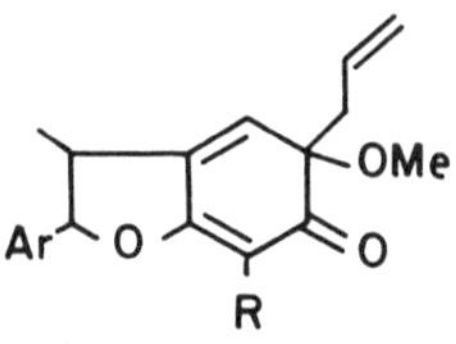   XXXIV

were noted, but the configuration at C-5 could not be assigned. LSR were used to distinguish the two isomers of XXXV and to analyze and assign the configuration of the four isomers of XXXVI.[789] Mixtures of the four isomers of XXXVI were also resolved and quantified in the

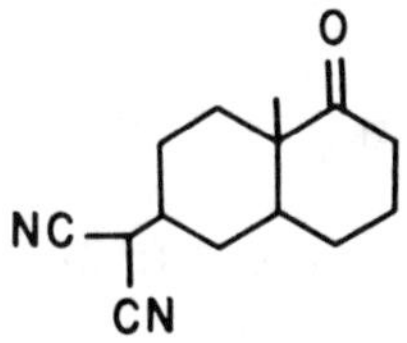

XXXV

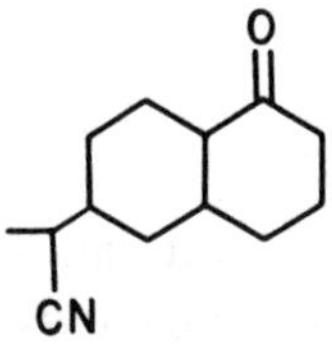

XXXVI

presence of Eu(fod)$_3$.[789] The stereochemistry of 3-thiadecalone and 4-thiadecalone was assigned on the basis of shift data with Eu(dpm)$_3$.[790] The shifts in the spectra of pyrazolotropones(XXXVII,XXXVIII) with Eu(dpm)$_3$ have been used to distinguish hydrogenated from dehydrogenated analogs and to locate the position of double bonds.[791]

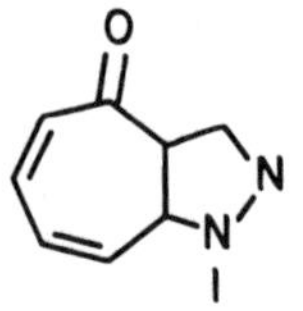

XXXVII

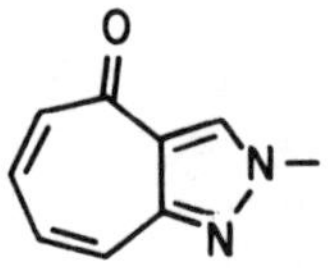

XXXVIII

The structure and conformation of 7,11-diphenyl-2,4-diazaspiro[5.5]undecan-1,3,5,9-tetraones(XXXIX) have been determined through the application of Eu(dpm)$_3$.[792]

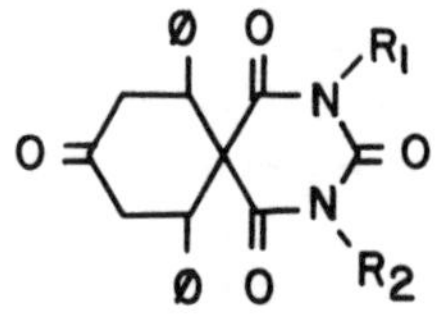

XXXIX

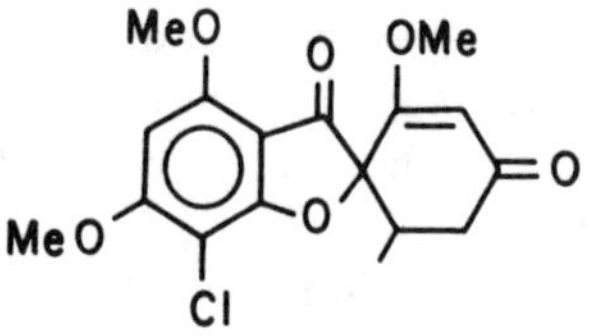

XXXX

The conformation of griseofulvin(XXXX) was assigned with the aid of shift data with Eu(dpm)$_3$.[793] Coupling constants obtained in the shifted spectra, and the study of partially deuterated analogues, facilitated the determination of the conformation.

The stereochemistry of the methoxy group in XXXXI was assigned on the basis of lanthanide shift data.[388] Shift data with Eu(fod)$_3$ were used to determine the structure of the adduct(XXXXII) of N-phenylmaleimide and 2,3,4,5-tetraphenylcyclopentadienone.[794]

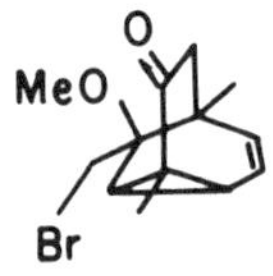

XXXXI 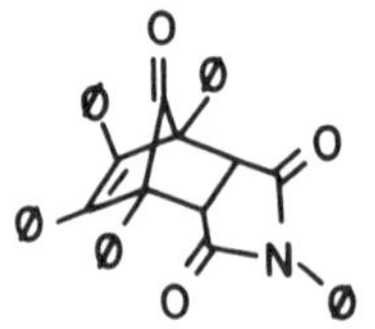 XXXXII 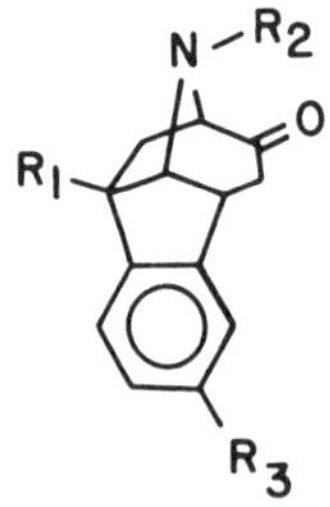 XXXXIII

The structural assignment of XXXXIII was facilitated through an analysis of the spectrum with LSR.[795] The *endo-syn* and *endo-anti* isomers of 5,6-dimethyl-3-thio-3-phenyl-3-phosphatricyclo[5.3.2.0²,⁶]-dodeca-4,8,11-trien-10-one (XXXXIV) were distinguished using shift data obtained with Eu(fod)$_3$.[749] Complexation of XXXXIV reportedly involved the carbonyl group. For the derivative of XXXXIV, in which the sulfur atom was replaced with an oxygen atom, complexation occurred through the oxygen atom of the phosphoryl group. Shift data with Eu(dpm)$_3$ permitted the distinction of XXXXV from XXXXVI and the assignment of the ring conformations.[796]

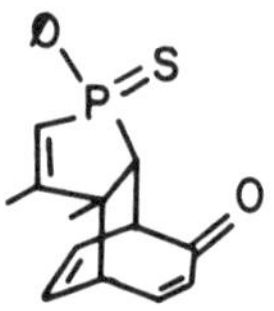

XXXXIV 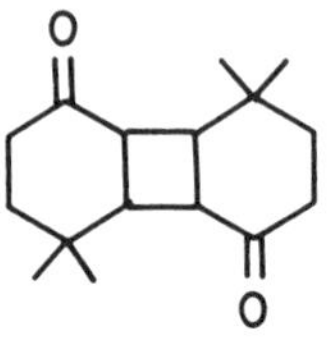 XXXXV 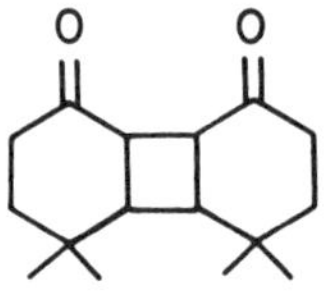 XXXXVI 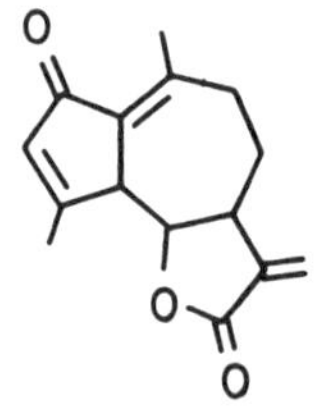 XXXXVII

The structural assignment of XXXXVII was facilitated through the use of LSR.[797]

Studies of polyfunctional adamantanones with LSR have been the focus of several reports.[718,719,721,798] The shifts in the spectra of 2,4-adamantanedione and 2,6-adamantanedione with LSR reflected an additive effect of the equal, but independent, complexation of the shift reagent at both carbonyl groups.[719] The effects of substituent groups on the shifts in the spectra of 4-substituted-2,6-adamantanediones with LSR have been reported.[798]

The spectra of polyfunctional steroids with one or more carbonyl groups have been analyzed in the presence of LSR.[79,295,307,666,734,737,745,746,748,799-802] Many of these reports have discussed factors that influence the site of complexation.[79,295,666,737,745,746,748,799] Hydroxyl groups generally bind in preference to keto groups; however, if the hydroxyl group is sufficiently hindered,[79] or has been blocked through derivatization,[745,746,748] coordination at a ketone site will occur. Steric considerations influence the bonding when two or more carbonyl groups are present in a steroid.[799] Increased conjugation of a carbonyl group reduces its ability to bond to a LSR.[666] A number of studies have noted alterations in the degree of association of donor sites as the ratio of shift reagent to substrate is varied.[295,666,737,799] This observation necessitates that the spectra of polyfunctional steroids be recorded over a range of shift reagent concentrations.

Data with LSR have been used to determine the conformation of rings close to the preferred site of complexation.[734,799,802] The assignment of methyl resonances can be facilitated through the use of LSR.[800] The assignment of ¹³C resonances of certain steroids has been confirmed through the use of LSR.[801] LSR have been used to distinguish an androsta-4-ene-3-one from two isomeric dieneones and a trieneone.[666] Identification and quantification of each of the four steroids in mixtures was also performed.

The dimerization of chlorophyll a has been studied with the aid of Eu(fod)$_3$.[803] The substrate reportedly complexed at the carbonyl group of ring 5. This was demonstrated by studying two magnesium free derivatives, methyl pheophorbide and methyl bacteriopheophorbide a. It was reported that the shift reagent altered the conformational population distribution of the dimer.

Blocking groups — On occasion it might be advantageous to eliminate or reduce complexation of a keto group so as to facilitate study about another donor site. Conversion of a carbonyl group to its ethylene thioketal(XXXXVIII) effectively prevents complexation with a LSR.[745]

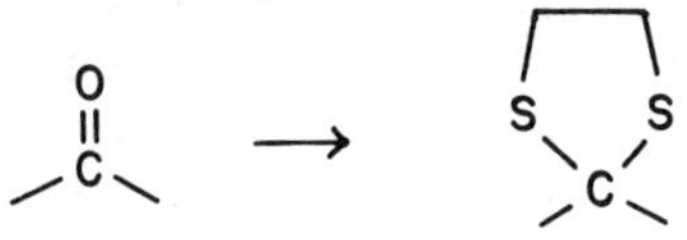

XXXXVIII

3. Esters

The complexation of LSR with esters occurs at the carbonyl oxygen. Ester groups are of intermediate basicity and in polyfunctional substrates alcohols,[79,154,295,674,685,703,704,706,711,713,715,723,725,728,729,732,736,778,804] primary amines,[410,805] secondary amines,[806] nitrogen oxides,[807] and phosphoryl groups[95,808] are often the preferential site of complexation. In polyfunctional substrates with an ester group and an amide[805,809-818] or lactam group,[809,812,819] the LSR usually bonds at the carbonyl oxygen atom of the amide. Substrates in which preferential complexation occurs at an epoxy group,[745,820] ketone,[756,778,801,803] thionyl group,[821] or heterocyclic nitrogen atom[822,823] have been observed. A cyclic ether oxygen was the site of complexation with Eu(dpm)$_3$ in anhydrogalactopyranoside derivatives with ester functional groups.[670] Of course, any functional group is subject to electronic and steric effects that may influence the binding preference when compared to an ester group.

Chelate bonding involving one or more ester functional groups has been noted with a number of substrates.[675,725,754,756,824-826] Six-member chelate rings were observed for o-methoxymethylbenzoate(I)[754] and esters of 2,3-diphenyl-3-hydroxypropionic acid(II).[675] The nitrogen atom and carbonyl oxygen of III reportedly complexed in a chelate fashion with LSR.[756]

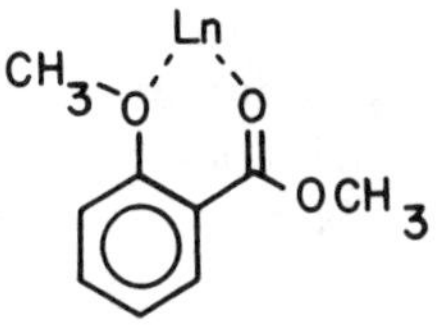

I 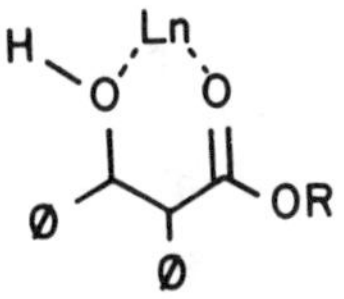 II 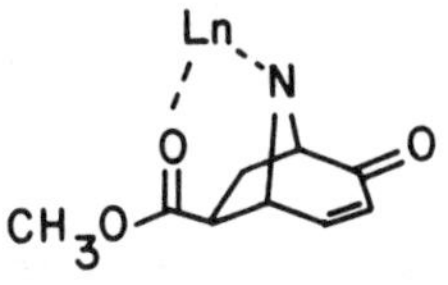 III

A seven member-chelate ring was reported for the complex between Eu(fod)$_3$ and IV.[824] Alteration of the configuration of the bridging carbon (V) resulted in tridentate bonding.

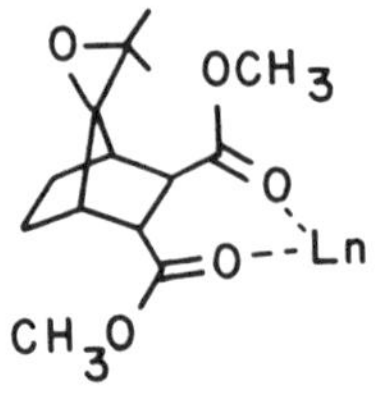

IV 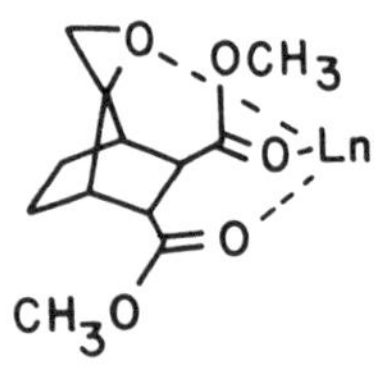   **V**

Certain derivatized carbohydrates have the proper orientation of substituent groups so that stable bidentate complexation occurs.[825,826] The bonding of the macrocycle VI was dependent on the location of substituent groups.[725] If R_2 and R_3 or R_6 and R_7

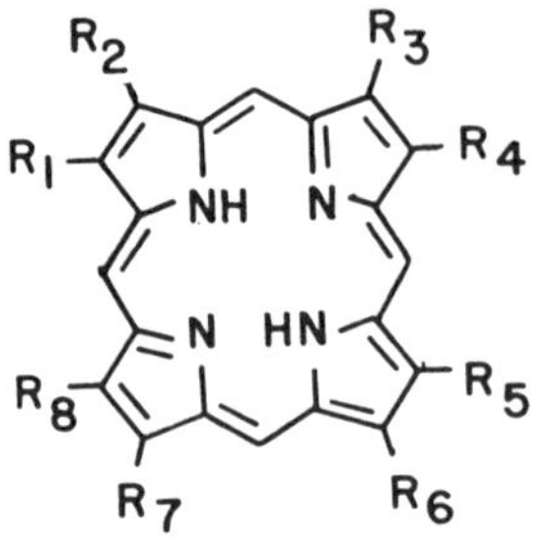

VI

were $-CH_2CH_2CO_2CH_3$ groups, bidentate complexation involving the two carbonyl oxygen atoms with the LSR was indicated. This would correspond to a 14-member chelate ring. LSR were used with these substrates to help determine the position of substituent groups.

LSR have been used in the study of esters of fatty acids[166,397,447,684] and triglycerides.[827-829] The complexation of polyfunctional fatty acid methyl esters with LSR is dependent on the concentration of shift reagent. Studies of polyfunctional derivatives with LSR often provide more information than the study of monofunctional derivatives.[447,684] Shift data with $Eu(fod)_3$ have been used to determine the position of deuterium substitution in dimethyl-1,7-heptanedioate.[397] LSR can be used to distinguish isomeric triglycerides.[828,829] Systematic differences in the shifts of the end groups and center substituent of triglycerides are observed in the presence of LSR.[828,829] The triglycerides can be analyzed individually or in mixtures.

Shift data with LSR have been used to assign the configuration of polyfunctional olefins with ester groups.[405,830,831] The configuration of the olefin bond of the dimethyl ester of L-γ-ethylideneglutamic acid was assigned on the basis of lanthanide shift data.[832]

Structural analyses of the juvenile hormones methyl-3,11-dimethyl-10,11-*cis*-epoxy-7-ethyl-2-*trans*- and methyl-10,11-epoxy-2-*trans*-6-*trans*-farnesoate were aided through the use of $Eu(fod)_3$.[833] Complexation at both the epoxy and ester sites was noted with these substrates. The configuration of the olefin bonds in VII and VIII were assigned on the basis of shift data with $Eu(fod)_3$.[714]

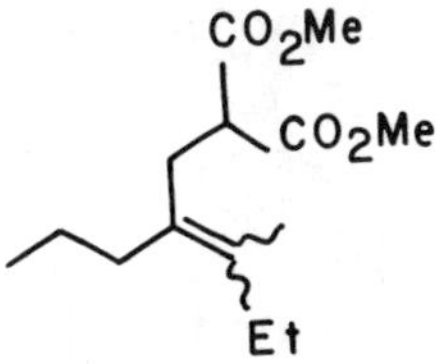

VII, VIII structures

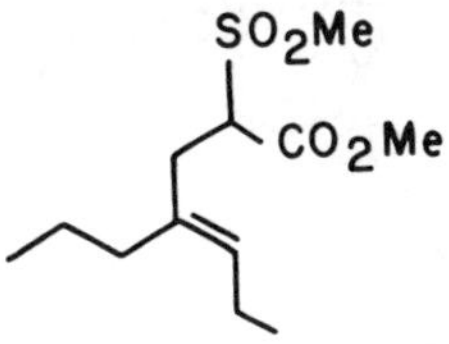

The dimethyl ester of 1-butene-2,4-dicarboxylic acid and the trimethylester of 1-hexene-2,4,6-tricarboxylic acid were used as models for studies of polymers with LSR.[834] The acetate derivatives of substituted phenols have also been used as model substrates for studies of polymers with LSR.[412] It was observed that ester groups on aliphatic side chains associated more strongly with LSR than ester groups attached to aromatic rings.[412]

The shifts in the spectra of substituted methyl benzoates with LSR have been shown to correlate with inductive effects.[410] The conformations of tetrasubstituted-biphenyls have been evaluated through the use of shift data with Eu(fod)$_3$ and Pr(fod)$_3$.[689] Shift data with Eu(dpm)$_3$ aided the structural assignment of IX.[835]

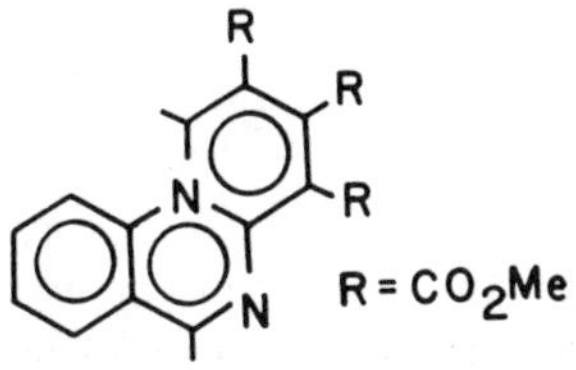

IX, 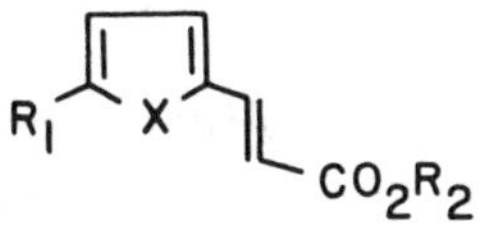 X, 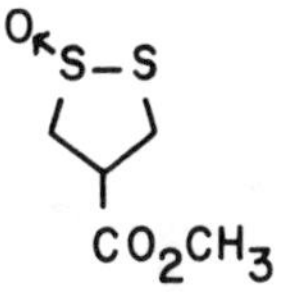 XI structures

The conformations of furyl and thienyl derivatives of X have been studied in the presence of Eu(dpm)$_3$.[769] The stereochemistry of the sulfinyl group in *syn*- and *anti*-1,2-dithiolane-4-methylcarboxylate-1-oxide(XI) was determined through the use of Eu(dpm)$_3$.[836] The conformations of ethyl *cis*- and *trans*-2-hydroxy-1-cyclopentane carboxylate, as well as the cyclohexane, -heptane, and -octane derivatives, were determined using shift data with Eu(fod)$_3$.[698] An analogous study of the conformations of the corresponding trichloroacetylcarboxamide derivatives was also performed.[698] The stereochemistry of pyrrolidines and piperidines with ester substituent groups have been assigned on the basis of shift data with LSR.[543] The epimers of 2,3-diphenyl-5-carboxymethylisoxazolidine(XII) were distinguished in the presence of Pr(fod)$_3$, and the stereochemistry of each was assigned.[837] Shift data with LSR have

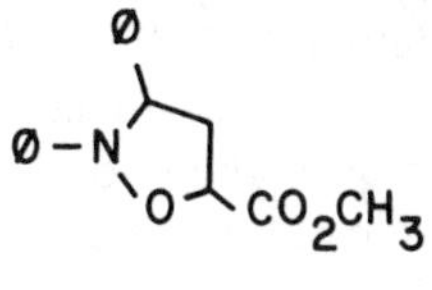

XII, 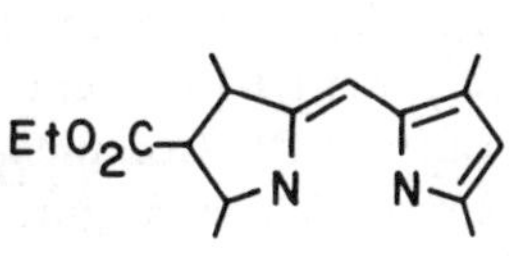 XIII structures

been used to determine the conformations of pyrromethenes(XIII).[786]

The assignment of the configuration of peracetylated pinitol (XIV) was facilitated through the use of shift data with Eu(dpm)$_3$.[838] The *cis* and *trans* isomers of XV and its hydrogenated derivative were distinguished in the presence of Eu(dpm)$_3$.[839] Binding of the

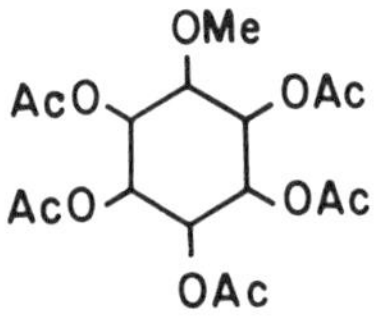

XIV 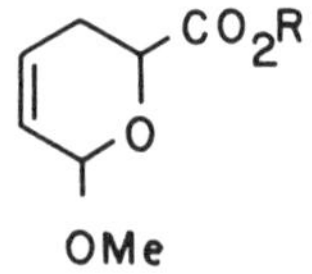   XV 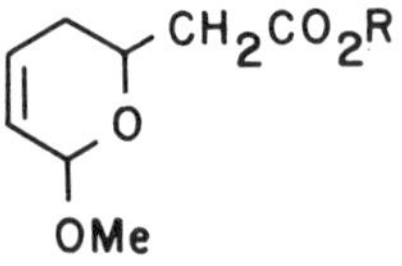   XVI

trans isomer was stronger due to more favorable steric effects. The distinction of cis and trans XVI with LSR was less pronounced than for XV because steric hindrances in the cis isomer were reduced.

Shift data with LSR have been used to assign the resonances of 1-(x-benzo[b]thienyl)ethyl acetate.[840,841] The shifts in the spectrum of ethylindole-2-carboxylate with a LSR have been fit to the dipolar shift equation.[842] Data collected with LSR permitted the assignment of the cis or trans configuration of model polyfunctional esters, but did not reliably permit the assignment of the cis and trans isomers of XVII.[830]

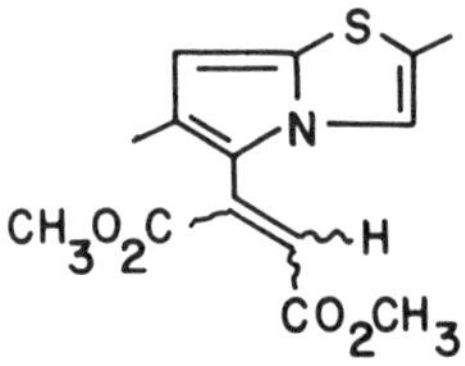

XVII

The variability of steric factors for different rotational conformers of XVII was believed to inhibit the results obtained with the LSR. The configurations of methyl groups of tocopherol acetates have been assigned on the basis of shift data with Eu(dpm)$_3$.[407,409]

Shift data obtained with Eu(fod)$_3$ have been used to aid the structural assignment of polyfunctional bicyclo[2.2.1]heptadienes.[824] The configuration and conformation of a disubstituted bicyclo[3.3.1]octane (XVIII) has been determined through studies with Eu(dpm)$_3$.[705] It was reported that complexation of the shift reagent did not alter the conformation of XVIII.

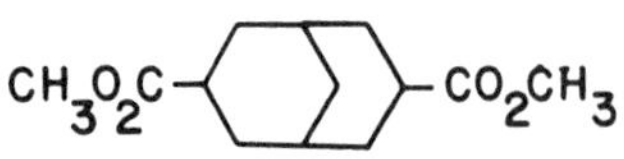

XVIII 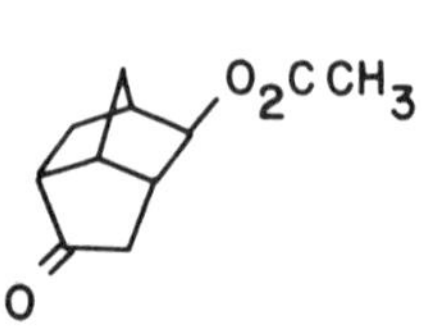   XIX 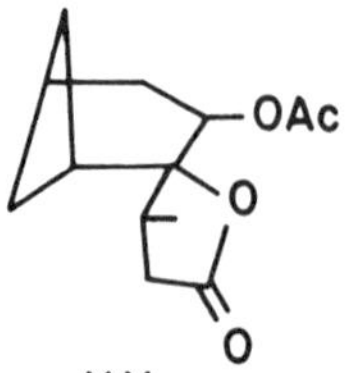   XX

The spectra of a number of polyfunctional tricyclic compounds with one or more ester groups have been recorded with LSR.[714,751,758,798,843,844] Eu(dpm)$_3$ was used to distinguish the cis-endo and cis-exo isomers of 3,5-disubstituted nortricyclenes.[843] The structural assignment of the acetate of 9-exo-hydroxy brendan-4-one (XIX) was facilitated through the use of Eu(fod)$_3$.[751] The site of complexation was reportedly the carbonyl group of the ester. The stereochemistry of the methyl group in XX was assigned with the aid of shift data with Eu(fod)$_3$.[714] Approximately equal bonding at both the ester and lactone carbonyl oxygens was noted. Shifts in the spectrum of XXI

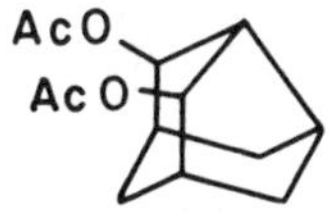

XXI

with Eu(fod)$_3$ were used to confirm the equivalence of the two acetate groups.[844] Independent and approximately equal complexation of the three carbonyl groups of an acetate derivative of an adamantanedione was observed.[798] XXII and XXIII have been distinguished on the basis of shift data with Eu(fod)$_3$.[758]

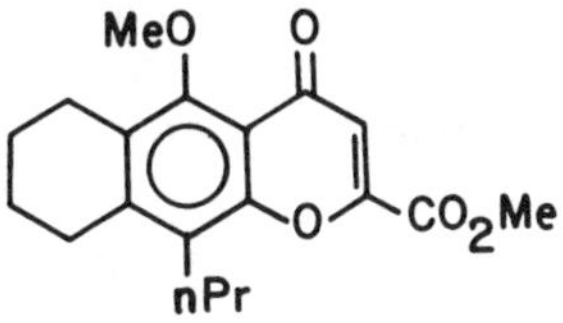

XXII 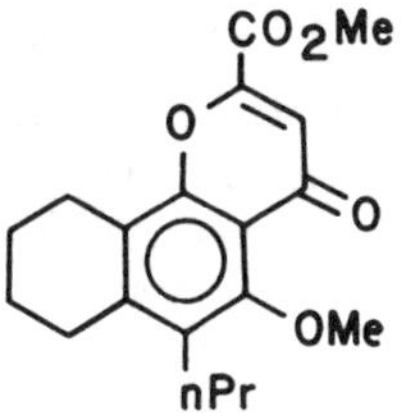   XXIII

Shift data with Eu(fod)$_3$ have been used to assign the structures of isomers of XXIV.[845] The assignment of the structure of reserpine (XXV) was facilitated through the use of shift data with LSR.[846,847] The two independent studies of reserpine with LSR indicate the difficulty that can be encountered in determining the site

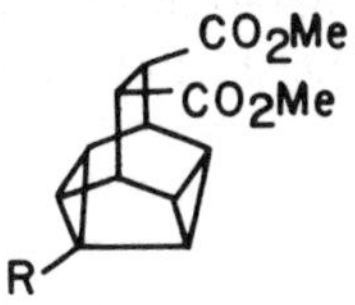

XXIV 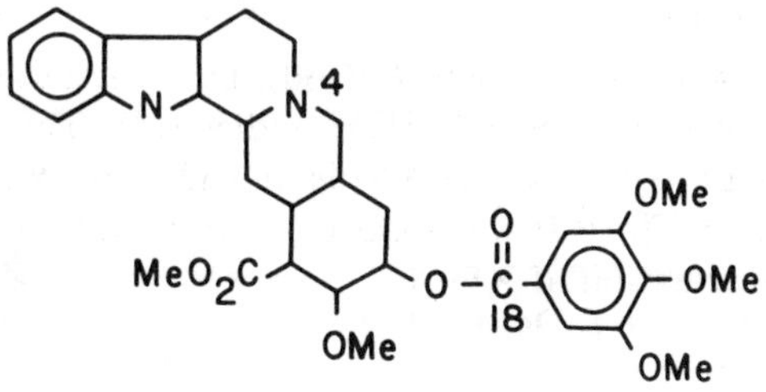   XXV

of complexation of polyfunctional substrates. One report claimed that appreciable complexation occurred at the nitrogen atom at the 4-position and the carbonyl oxygen at C-18.[846] The second report claimed that chelate bonding of the ortho methoxy groups was the principal bonding interaction.[847] Both reports employed Eu(fod)$_3$ as the shift reagent. Partially deuterated derivatives were studied to help in assigning the resonances.[846] The shift data permitted assignment of the stereochemistry and certain conformational aspects of the C and E ring.[846]

The complexation between polyoxadioxo[n]paracyclophanes(XXVI) and Eu(dpm)$_3$ and Eu(fod)$_3$ has been assessed.[394,848]

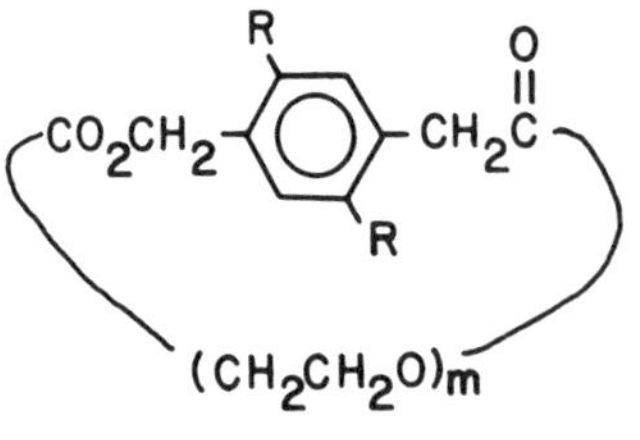

XXVI

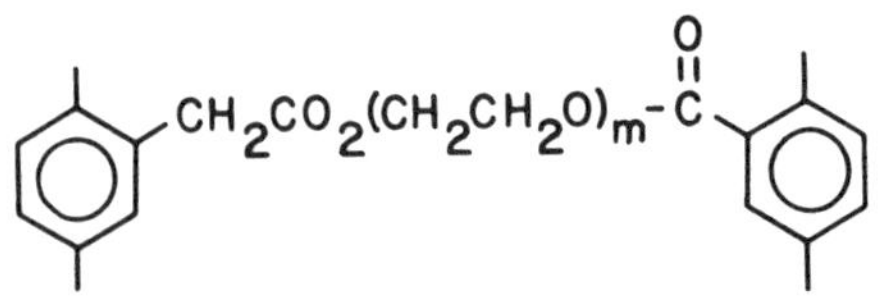

XXVII

Preferential complexation occurred at the carbonyl group of XXVI and XXVII when m = 2. For derivatives with longer chains, adjacent ether oxygens formed a stable chelate bond with the shift reagent. The ether oxygen of the ester group was not involved, however, in the chelate bonding.

Polyfunctional steroids that bond through an ester group have been the focus of studies with LSR.[79,307,729,736,800] An ester group has been observed to complex in substrates with a ketone moiety provided steric hindrance is present near the ketone.[79,307] Methyl esters bond more strongly than trifluoromethyl esters, and derivatization to trifluoromethyl esters has been used to block alcohol functionalities.[736] Shift data with Eu(dpm)$_3$ has been employed in the assignment of methyl resonances of meliacins.[729]

Effects of lanthanide shift reagents on the NMR spectra of carbohydrate derivatives have been described in several reports. These include studies of gluco-,[665,738,741,805,806,826,849-853] galacto-,[805,806,849,851,852,854] and mannopyranosides.[665,806,851] Steric factors influence the site of complexation of the LSR and differences in the bonding site allowed distinctions to be made among the gluco-, galacto-, and mannopyranoside derivatives.[806,849,851] More specifically, preferred complexation occurs at the 6-acetoxy group of glucopyranosides, but at the 4-position of galactopyranosides.[806] Studies of arabino-,[826,852] gluco-,[127,128,678] and allofuranosides[127,128] with LSR have been described. The shifts in the spectra of arabino- and xylopyranosides[826,852,855] with LSR have been reported.

Shift data with LSR can be used to distinguish alpha and beta anomers of glucopyranosides,[851] galactopyranosides,[851] mannopyranosides,[851] arabinopyranosides,[826] and arabinofuranosides.[826] Shift data with a LSR can also be used to identify a sugar as either a pyranoid of furanoid derivative.[826] Certain mixtures of carbohydrate derivatives have been resolved or quantified in the presence of a LSR.[852,855] The shifts in the spectra of octaacetylated trehalose and octaacetylated sucrose in the presence of LSR have been described.[825]

Most of the reports on carbohydrate substrates have discussed and specified the site of complexation. Despite the complexity of these substrates in regards to the multiple number of bonding sites, a surprising degree of specificity in complexation has been noted.[127,128,678,738,741,805,806,825,826,849-851,854] Hydroxyl groups bond more strongly than ester groups unless sterically hindered.[665] The amide functional group of amino sugars bonds preferentially over esters and, as a result, amino sugars exhibit considerably larger shifts in the presence of LSR.[805,806] The carbonyl oxygen of ester groups bonds preferentially over ether groups.

4. Lactones

The preferred site of complexation of polyfunctional substrates with a lactone group is highly dependent on the nature of the other donor groups. Studies of polyfunctional lactones with LSR are somewhat limited, but it has been noted that hydroxyl,[307,687,697,714-716,729] epoxy,[428,797] ester,[729] and ketone[797] groups often complex in preference to lactones. Complexation is also favored at heterocyclic nitrogen atoms[856] and at ortho-substituted methoxy groups in which a chelate bond forms with the shift reagent.[307,716,857] It must be kept in mind, however, that particular electronic and steric effects within a substrate may alter these general observations.

Data with LSR have been used to determine the position of substituent groups on coumarins.[429,716,857,858] An effective method for comparing different substituent patterns is to select a reference proton and calculate a ratio of the lanthanide shifts for each resonance of the substrate.[716] Distinct patterns for the different positions and isomers are noted.

The diastereomers of I have been distinguished through the use of LSR.[859]

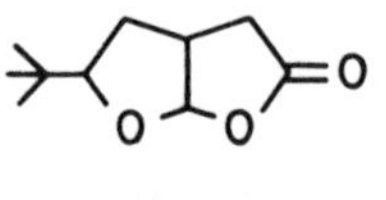

I

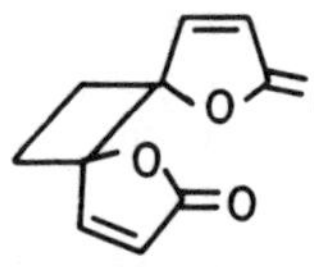

II

The shifts in the spectra of anemonin(II) and tetrahydroanemonin with Eu(fod)$_3$ were used to confirm the *trans* ring arrangement.[784] The two donor groups are identical, and it was noted that equal, but independent, complexation occurred at the carbonyl oxygen atoms. The structural assignment of r-1,C-2-diphenyl-t-3-methoxycyclobutane-1,2-diolcarbonate (III) was facilitated through the use of Eu(fod)$_3$.[860]

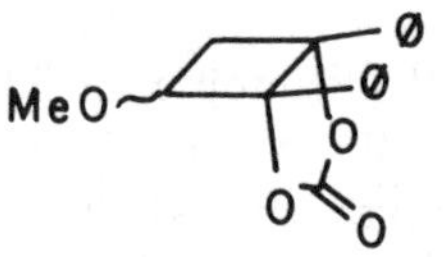

III

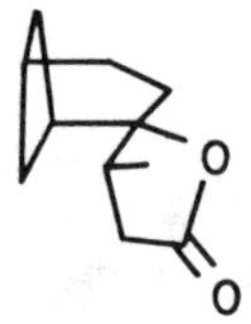

IV

The diastereomers of III that differed in the configuration at the alkoxy-substituted carbon could also be distinguished and assigned on the basis of the lanthanide shift data. The stereochemistry of the methyl group in the lactone ring of IV was assigned on the basis of shift data with Eu(fod)$_3$.[714] Approximately equal complexation of the shift reagent at both the ester and lactone carbonyl oxygen atoms of IV was noted.

The conformation of the B ring of 3β-hydroxy-β-homocholestan-7,8α-lactone (V) was determined through studies with a LSR.[667] Appreciable complexation reportedly occurred at both the hydroxyl and lactone group. Trifluoroacetylation of the hydroxyl group limited the complexation to only the lactone moiety.

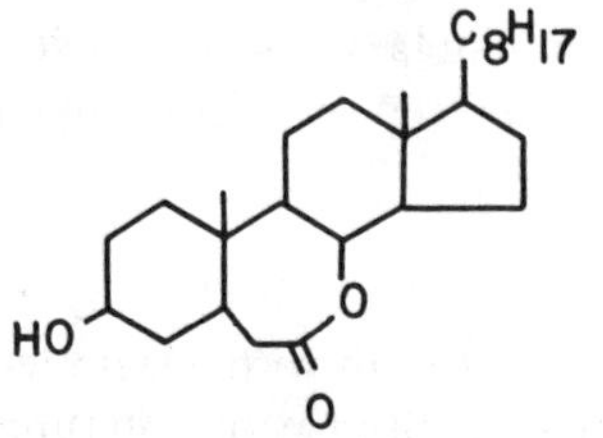

V

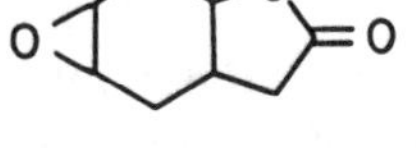

VI

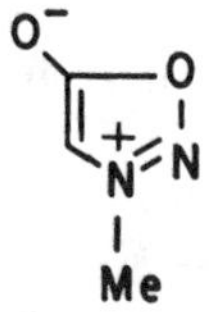

VII

The shifts in the spectra of 4,5-dihydroxyoctanedioic acid *bis* lactone and the dodecane analog with Eu(dpm)$_3$ have been reported.[428] The isomers of VI that differ in the configuration of the epoxide ring were distinguished in the presence of Eu(dpm)$_3$.[428] The epoxide oxygen was the preferred site of complexation, although significant association of the lactone site was indicated. The shifts in the ^{17}O NMR spectrum of *N*-methyl sydnome(VII) in the presence of Ln(dpm)$_3$ chelates have been reported.[189] The shifts were indicative of complexation at the exocyclic oxygen atom.

5. Aldehydes

The relative ability of aldehyde groups to compete with other functional groups for coordination sites on LSR was assessed in a study of 4,4'-methylenedicinnamic acid derivatives.[778,804] The ordering that resulted from this study is as follows.

$$-OH \gg CO_2CH_3 \simeq -\overset{O}{\overset{\|}{C}}H > -O_2CH_3 > OCH_3$$

LSR have been employed in the study of the conformational preference of substrates of structure I in which X = O,[343,344,435,771-773] S,[343,344,434,435,771-773] N,[771,786] Se, and Te.[772]

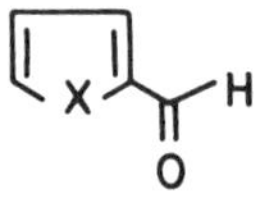

Significant complexation shifts were noted for these substrates,[343] and it has been recommended that complexation shifts be accounted for before attempting conformational analyses.[434] This is done by measuring the shifts with a La(III) complex. It has also been reported that the conformational preference of I was not altered on complexation with the shift reagent.[771,772]

Analogous studies of the conformational preferences of formylmethylenthiopyrans have been performed.[861] The conformational preferences of 1-methyl-2-imidazole carboxaldehyde (II)[862] and III[863] have been assessed with the aid of LSR.

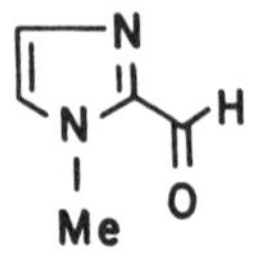

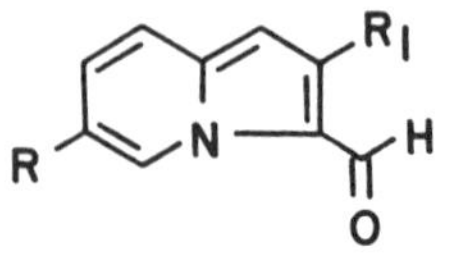

The configuration of the olefin bonds of substrates of the formula (CH$_3$)$_2$N(CH)$_n$O, in which n = 2, 3, 5, and 7, have been determined through the use of Eu(dpm)$_3$.[864-866] The shifts in the spectra of symmetrical carotenoids with Eu(dpm)$_3$ have been reported.[58]

Studies of methoxybenzaldehydes,[436,867] hydroxybenzaldehydes,[688] hydroxynaphthaldehyde,[688] and *p*-dimethylaminobenzaldehyde[868] with LSR have been described. The position of substituent groups in methoxybenzaldehydes was assigned on the basis of shift gradients and by comparing shift data of compounds with unknown to known configurations.[867] The orientation of the aldehyde group relative to the ring in methoxybenzaldehydes has been assessed on the basis of studies with LSR.[436,867]

6. Acid Chlorides

The relative binding strength of acid chlorides with Eu(fod)$_3$ has been assessed in a study of cinnamic acid derivatives.[778] Complexation occurs through the carbonyl group of the acid chloride and the following order was reported for these derivatives.

$$-OH > -\overset{O}{\underset{\|}{C}}CH_3 \simeq -\overset{O}{\underset{\|}{C}}H > -\overset{O}{\underset{\|}{C}}Cl > -O\overset{O}{\underset{\|}{C}}CH_3 > OCH_3 > CN$$

7. Ethers

Most ethers exhibit small association constants with LSR and are, therefore, among the least competitive donor groups in polyfunctional substrates. Polyfunctional substrates that bond to LSR through an ether oxygen tend to fall into specific categories. One is when the other donor is also a weak base such as a nitrile[507,869,870] or sulfide[770,871,872] group. Aliphatic methoxy groups or cyclic ethers will coordinate in preference to these two donor groups. Epoxides as a class exhibit rather strong binding to LSR and have been noted to bond in preference to esters,[447,715,820,833] lactones,[428,715,797] and tertiary-substituted amines.[873] Epoxides have even been observed to compete for binding in substrates containing a hydroxyl group.[715] The third situation in which complexation through an ether oxygen has been observed is when the other group is sterically hindered. Such reasoning has been used to explain the preferential binding of cyclic ethers over a hydroxyl group[668-670] and tertiary amine.[874,875]

Chelate bonding involving one or more ether groups has been reported for a variety of substrates. Examples with alcohols,[306] esters,[824] and ketones[91,452,524,752,755,757-765,767] have already been described. Polyglycoldimethyl ethers, which have repetitive $-CH_2CH_2O-$ units, form stable chelate bonds with LSR(I).[135,848,876-880] The chelate bond in 1,2-dimethoxyethane was found to be more stable than that in a flexible compound such as 1-methoxy-2-n-octyloxyethane.[135] A similar

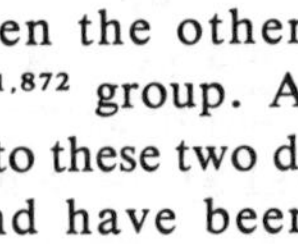

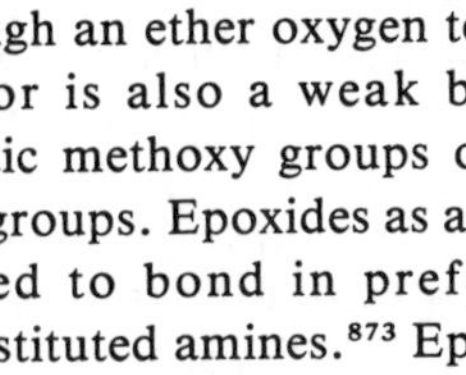

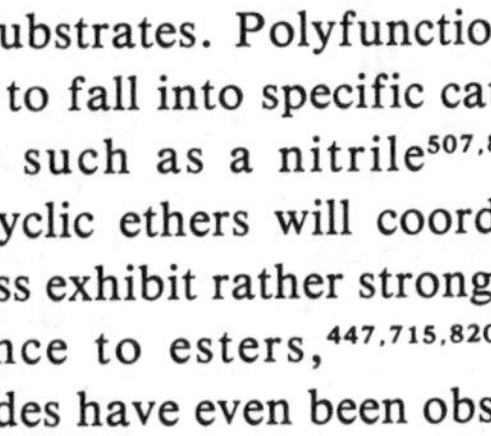

chelate interaction often explains the shifts and broadening observed in the spectra of crown ethers in the presence of LSR.[823,881-883] A stable chelate bond is formed in substrates containing ortho substitution of two methoxy groups on an aromatic ring(II).[135,307,672,757,847,857,884-886] The bidentate complexation of o-dimethoxy aromatics is reportedly more stable than that for glymes.[135] The chelation of ortho methoxy groups has been observed to compete favorably for coordination sites on the LSR with esters,[884] lactones,[307,857] ketones,[757] tertiary amines,[672,885,886] and to a lesser degree secondary amines.[885] A similar arrangement of oxygen atoms results in bidentate complexation between Pr(fod)$_3$ and eusiderin B(III).[884]

Bidentate complexation of derivatized disaccharides with LSR has been noted.[679,887] Specific configurations of the potential donor groups are necessary for the formation of chelate bonds. Benzylidine anilines (IV) form a bidentate complex with Eu(fod)$_3$.[569]

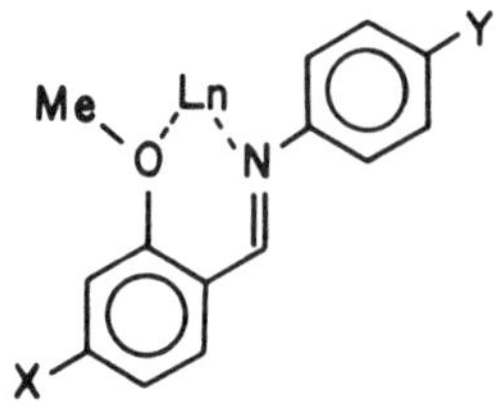

IV 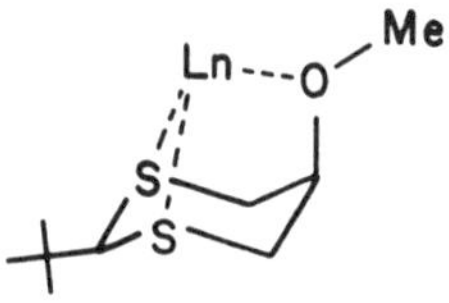   V

The large shifts for the *cis* isomer of V in the presence of a LSR were believed to result from tridentate complexation.[888] The *trans* isomer does not have the proper orientation of the donor groups to coordinate in a tridentate manner, and small shifts are observed in the spectrum with the shift reagent.

The spectra of symmetrical carotenoids with ether functional groups have been studied with Eu(dpm)$_3$.[58] The shifts were indicative of equal, but independent, complexation at the two ether sites. Shift data with a LSR have been used to assess the conformation of monoheptylglycol ether.[682] Studies of glymes with LSR have been used to demonstrate the presence of contact shifts[876,877,880] and 2:1 substrate-shift reagent complexes.[878] It was also demonstrated in a study of glyme-6 that the shifts of the resonances in the presence of a LSR were the weighted sum of components for each particular binding site.[879] The effects of LSR on the spectra of crown ethers have been described.[823,848,881-883,889-893]

The stronger association of aliphatic compared to aromatic ethers was noted in a study of bifunctional cinnamic acid derivatives.[778,804] A similar conclusion was reached in a study of $o,o,o'o'$-tetrasubstituted diphenyls in which it was found that Ph-CH$_2$-O-CH$_3$ derivatives associated more strongly than Ph-O-CH$_3$ analogs.[689] The shifts in the spectra of methoxy benzenes with LSR have been reported, and only in the case of ortho-substitution were substantial shifts observed.[847] The shifts in the ^{1}H and ^{13}C spectra of 1,5- and 2,6-dimethoxynaphthalene with Yb(fod)$_3$ have been reported.[134]

Shift data with Pr(fod)$_3$ were used to assign the structures of licarin B(VI) and eusiderin B(III).[884]

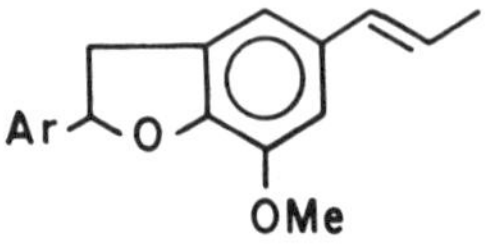

VI 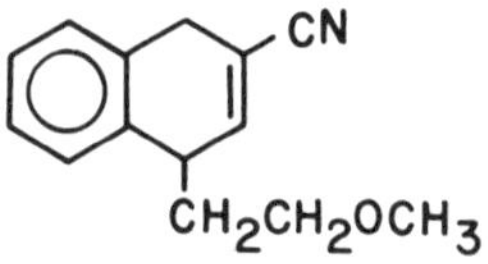   VII 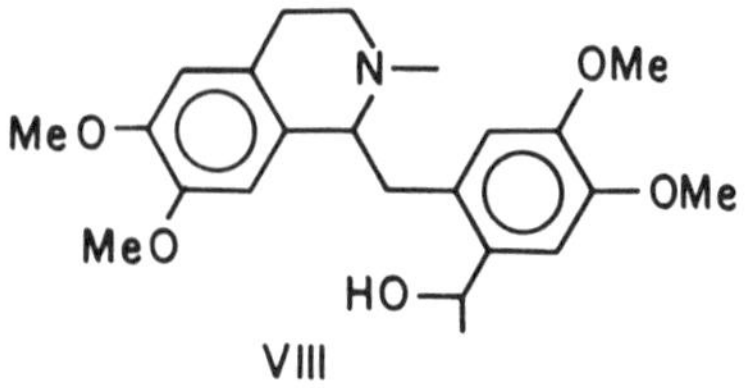   VIII

Shift data with Eu(dpm)$_3$ were used to assign the structure of VII.[870]

The *cis*- and *trans*- isomers of 2-methyl-5-*tert*-butyl-1,3-dioxane were distinguished on the basis of shift data with Eu(dpm)$_3$.[875] The shifts in the spectrum of VIII with Eu(dpm)$_3$ were used to resolve and assign the two diastereomers.[672] Competitive complexation between the hydroxyl and ortho-methoxy groups was observed.[672]

LSR have been employed in studies of 2-alkoxytetrahydropyrans.[839,894,895] Complexation occurred through the ring oxygen for all substrates. The resonances of the axial and equatorial protons were resolved in the presence of the shift reagent, and the conformations of the substrates were assigned.[894,895]

LSR have been utilized in conformational analyses of 6,8-dioxabicyclo[3.2.1]octanes.[896] The preferred site of coordination was the oxygen atom at position 6. Difficulty was encountered, however, in distinguishing certain configurations on the basis of lanthanide shift data. The ability to assign conformations of these substrates on the basis of only lanthanide shift data was also questioned.[896] The distinction of the *syn* and *anti* configurations of IX and X was achieved through the use of shift data with a LSR.[897]

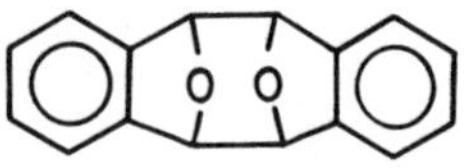

IX 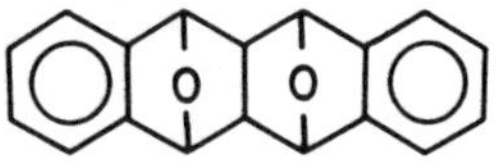   X

The *syn* isomer of X formed a chelate bond with the shift reagent. The *anti* isomer formed a weak monodentate complex and much smaller shifts were obtained in the spectrum. Shift data with Eu(dpm)$_3$ have been used to determine the amount of XI and XII present in mixtures.[872]

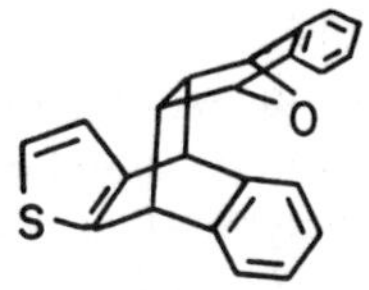

XI 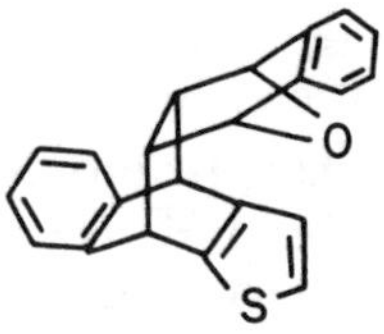   XII

Certain assignments in the spectra of pentacyclodecanes (XIIIa,XIIIb) were confirmed on the basis of shift data with a LSR.[275]

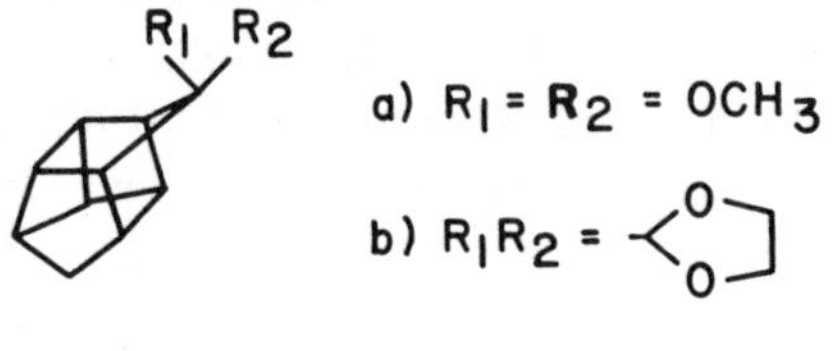

a) R$_1$ = R$_2$ = OCH$_3$

b) R$_1$R$_2$ =

XIII

The *exo* and *endo* configurations of bicyclo[3.1.0]hexylmorpholines(XIV,XV) (n = 4, 5, 8, and 9) were assigned using shift data with Eu(fod)$_3$.[874] The configurational assignment

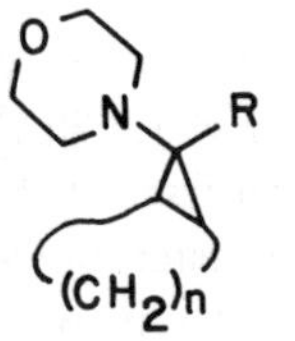

XIV 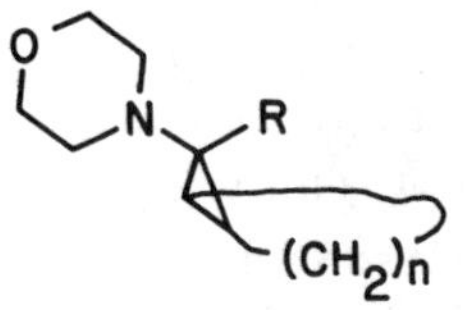   XV

when R was either a hydrogen atom or methyl group was straightforward. When R was a cyano or methoxy group, the *exo* and *endo* isomers were not readily distinguished in the presence of the LSR. It was proposed that complexation of the shift reagent occurred at the cyano or methoxy substituent groups in addition to the ring oxygen.⁸⁷⁴ The reduced specificity of the bonding eliminated the shift differences between the *exo* and *endo* isomers that were observed with derivatives that bonded only through the ring oxygen.

The stereochemistry of the carbinol carbon of XVI was determined with the aid of lanthanide shift data.⁶⁶⁸ The shift data were also useful in analyzing the conformation of these substrates. Shift data with

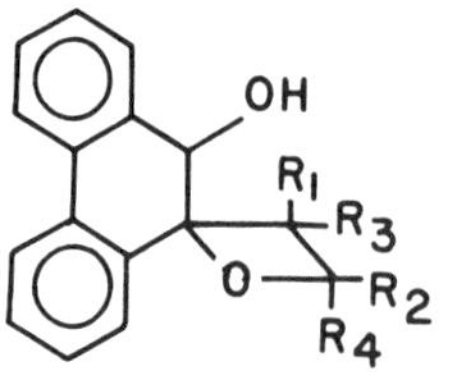

XVI

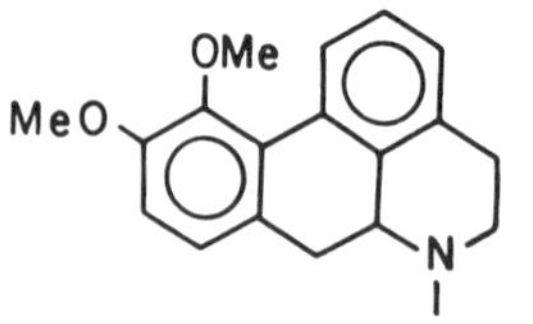

XVII

Eu(fod)₃ have been used to determine the position of deuterium substitution in XVII.⁸⁸⁶

The equatorial methoxy group of 3,3-dimethoxycholestane was the preferred site of complexation with Eu(dpm)₃.⁸⁷¹ The corresponding ethyleneketal derivative associated only at the 3-β position. The resonances for the diastereomers of 3β-(1'-methoxyethoxy)cholest-5-ene were resolved in the presence of Eu(dpm)₃.⁸⁹⁸ The spectra of the diastereomers of 3β-tetrahydropyranyloxy-5α-cholestane, 3β-tetrahydropyranyloxycholest-5-ene and 3β-(1'-methoxyethoxy)-5α-cholestane were also resolved using Eu(dpm)₃.⁸⁹⁸

LSR have been applied to the study of derivatized carbohydrates.⁶⁷⁰,⁶⁷⁹,⁷³⁸,⁸⁸⁷ A variety of these substrates bond in a bidentate fashion, and the particular arrangements of donor atoms that form the most stable chelate bonds have been discussed.⁶⁷⁹,⁸⁸⁷ The anhydro group of 3,6-anhydroglucopyranosides and 3,6-anhydrogalactopyranosides bonds preferentially over other ether and ester groups within these substrates.⁶⁷⁰ The shifts in the presence of a LSR have been used to assign many of the methoxy resonances in per-*O*-methylated aldohexosylaldohexoses as well as methylated monomer species.⁸⁸⁷

LSR have been employed in the study of juvenile hormones with epoxide groups.⁸³³ Significant complexation reportedly occurred at both the epoxide and ester groups in these substrates, and structural analysis was facilitated on the basis of lanthanide shift data. A similar finding with regards to bonding was noted in a study of methyl *cis*- and *trans*-9,10-epoxy stearate with Eu(fod)₃.⁴⁴⁷ The epoxy group exhibited preferential binding when compared to the carbomethoxy group; however, a significant amount of complexation at the carbonyl oxygen was noted. This was especially true at higher concentrations of shift reagent. The degree of bonding of the epoxy group in the *trans* isomer was less than in the *cis* isomer because of increased steric hindrance. The shifts in the spectrum of XVIII, which is a component of many epoxy resins, with Eu(dpm)₃ and Pr(dpm)₃ have been reported.⁸⁹⁹

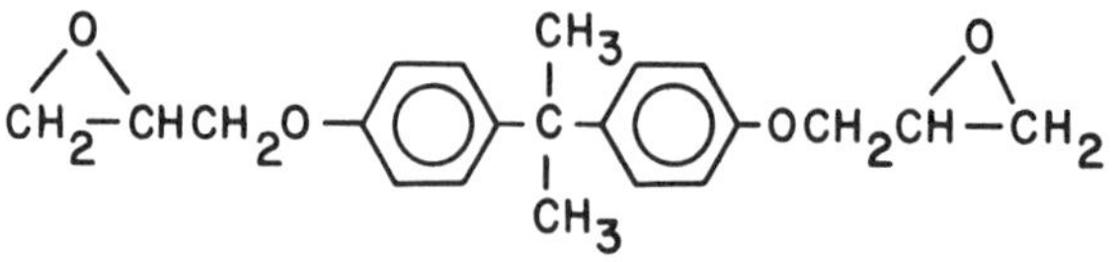

XVIII

The shifts in the ^{1}H and ^{2}H spectra of XIX with Eu(fod)$_3$ have been used to determine the configuration and position of deuterium substitution.[820] The configuration of the epoxy group in XX was assigned on the basis of shift data with Eu(dpm)$_3$.[428]

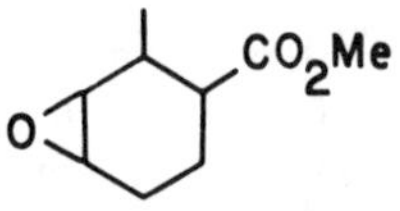

XIX 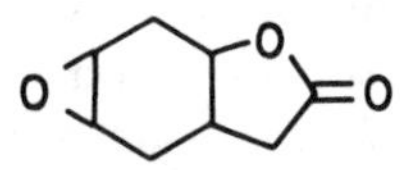   XX

8. Peroxides

The autooxidation product of methyl linoleate was studied with the aid of Eu(fod)$_3$.[166] Association of the shift reagent took place at the peroxide group rather than the ester group.

9. Amines

The association constants or primary, secondary, and tertiary amines are so markedly different that each class will be individually discussed.

a. Primary

As a result of favorable steric and electronic effects, aliphatic primary amines exhibit strong binding to LSR. Almost all polyfunctional substrates with a primary aliphatic amine would be expected to complex at the amine nitrogen. It has been reported that aminoethyldiphenyl phosphine complexes with Eu(dpm)$_3$ through the phosphine group.[415] This report is almost certainly in error since the phosphine group is a soft Lewis base and has little affinity for the lanthanide ion of the tris chelates. Aromatic primary amines associate less strongly than aliphatic primary amines because of electronic effects. Preferential complexation of a heterocyclic nitrogen atom over an aromatic amine has been observed in some polyfunctional substrates.[822,900] Appreciable complexation was noted at both the amine nitrogen and carbonyl oxygen of 3- and 4-amino methyl benzoate with Eu(fod)$_3$.[410] Aromatic amines are the preferred site of complexation in substrates with cyano and methoxy groups.[461] It has been reported that complexation of I with Yb(dpm)$_3$ occurred at the cyano nitrogen rather than the amine nitrogen.[901] This was explained

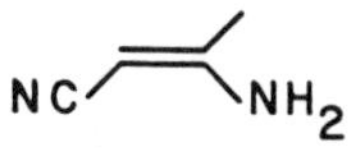

I 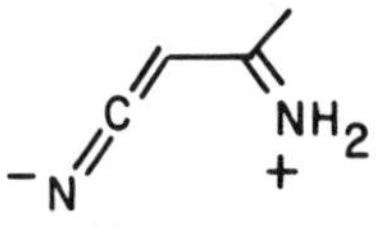   II

by assuming an appreciable contribution of the resonance structure II. The shifts in the spectrum of IIIb with a LSR were

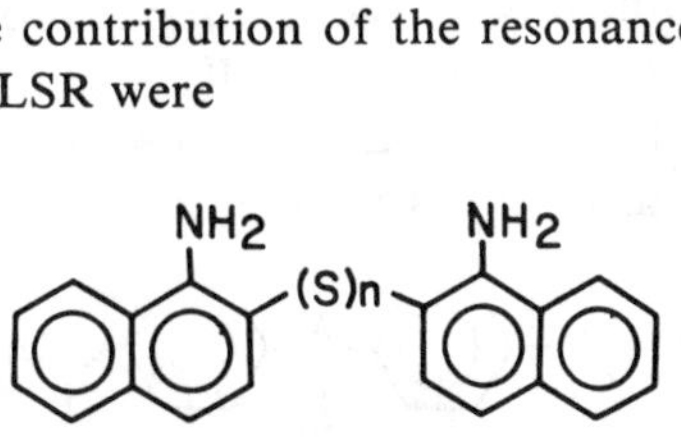

a) n = 2
b) n = 3

III

significantly larger than those for IIIa.⁹⁰² It was postulated that the configuration of IIIb allowed for bidentate binding of the amine groups.⁹⁰² The lanthanide shift data were used to establish the position of the sulfur bridge in these substrates.

LSR have been used to distinguish a variety of amphetamine isomers.⁸⁸⁵ The stereoisomers (*cis-cis, cis-trans, trans-trans*) of bis(4-aminocyclohexyl)methane were distinguished on the basis of shift data with Eu(dpm)₃.⁹⁰³ It was also possible to resolve and quantify mixtures of the three isomers through the use of the shift reagent. The equatorial amine of the *cis-trans* isomer exhibited a higher degree of association than the axial amine group.

The structural assignment of IV was facilitated through the use of shift data with Eu(fod)₃.⁹⁰⁴ Measurement of the NOE in the shifted spectrum allowed further structural information to be obtained from the study with the shift reagent. Shift data with a LSR permitted the distinction of the *exo* and *endo* isomers of V.⁴⁸⁵

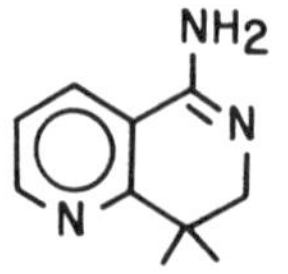

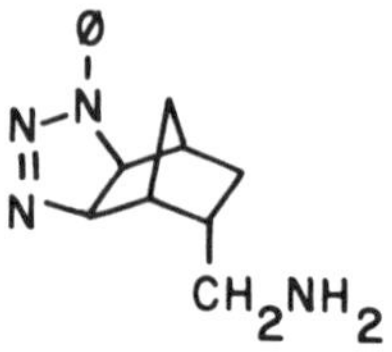

b. Secondary Amines

The increased steric hindrance of secondary amines compared to primary amines accounts for a significant reduction in the complexation of polyfunctional substrates at secondary amine sites. Other functional groups at which complexation of LSR occurred in preference to secondary amines include hydroxyls,⁶⁸¹,⁶⁹³ phosphoryls,⁹⁰⁵ ketones,⁷⁵²,⁷⁵⁷,⁷⁸⁶ aldehydes,⁷⁸⁶ lactams,⁸¹⁹,⁹⁰⁶ esters,⁷⁸⁶,⁸³¹,⁸³²,⁸⁴²,⁸⁴⁶ and heterocyclic nitrogen atoms.¹²²,⁹⁰⁷,⁹⁰⁸ The chelate bonding of ortho-substituted methoxy groups effectively competes with secondary amines.⁸⁴⁷ The influence of electronic and steric effects on binding within a particular substrate can be demonstrated for bicyclobenzo[*b*]-1,4-diazabicyclo[3.2.1]octane (I).⁹⁰⁹ The shifts in the spectrum of I in the presence of the LSR were indicative of bonding exclusively at the bridgehead nitrogen. Complexation at the secondary amine would be predicted on the grounds of steric effects. The electronic effects in I favor complexation at the tertiary amine.

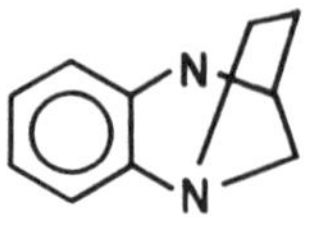

The shifts in the spectrum of 2[{[5-(dimethylamino)-methyl-2-furanyl]thio}ethyl]amino-2-methylamino-1-nitroethene (II) with a LSR were indicative of a chelate bond involving the nitro group and secondary amine *cis* to the nitro group.⁹¹⁰

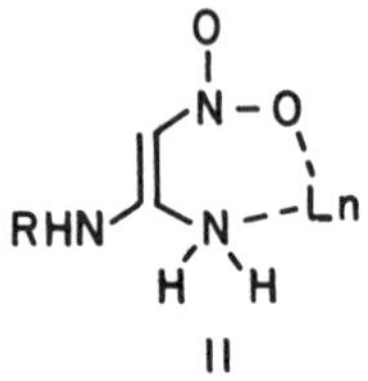

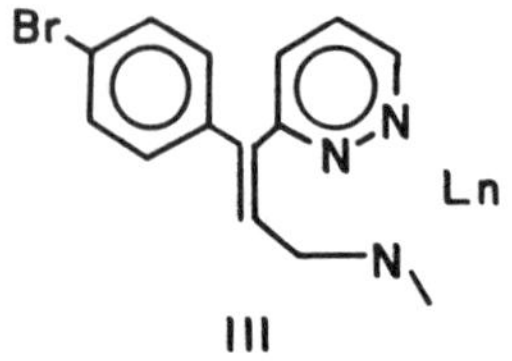

Bidentate coordination involving the secondary amine and heterocyclic nitrogen atom was proposed to explain the anomalous shifts in the Z isomer of 3-(4-bromophenyl)-3-pyridylallylmethylamine (III).[908] The configuration of the double bond of III and other derivatives was assigned through the use of LSR.

The shifts in the spectrum of IV in the presence of Eu(dpm)$_3$ facilitated the assignment of the configuration of the double bond.[750] The

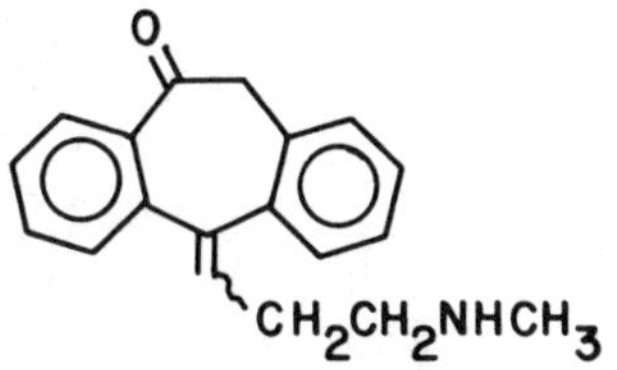

IV

assignment was not as readily performed using shift data with Eu(fod)$_3$. The shifts with Eu(fod)$_3$ were indicative of appreciable complexation at both the amine and ketone and lacked the specificity necessary to distinguish the two configurations. Eu(dpm)$_3$ exhibited a higher degree of specificity for the amine nitrogen.

Polyfunctional substrates with a secondary amine and cyclic ether have been noted to preferentially bond through the amine nitrogen.[770,911] Data with LSR have been used to assign the configurations (axial vs. equatorial) and conformations of 3-methyl- and 4-methyl-2-methylaminotetrahydropyran.[911]

The structure about the secondary amine nitrogen in aristoteline(V) was established through the use of LSR.[912]

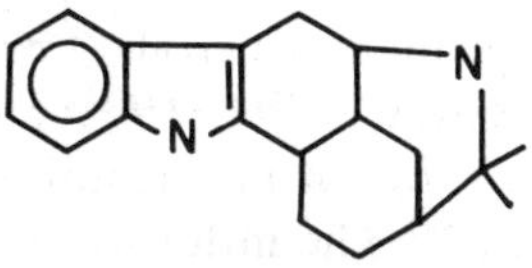

V 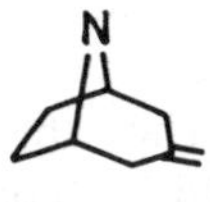   VI

The conformation of nortropinone (VI) and nortropine were determined by fitting the shift data with Eu(dpm)$_3$ to the dipolar shift equation.[671] Preferential complexation reportedly occurred at the secondary amine site in polyfunctional substrates with secondary alcohol or ketone groups.[671] Complexation of the LSR did not alter the conformation of these substrates. Shift data with Eu(dpm)$_3$ have been used to assign resonances in the spectra of diazacyclophanes.[913] The chemical shift values in the unshifted spectra were obtained by extrapolation of the lanthanide shift data and were used to assess ring current effects.

c. Tertiary Amines

Tertiary amine groups in polyfunctional substrates seldom bind to LSR because of pronounced steric hindrance. Preferential complexation of alcohols,[671,672,695,708,709,734] primary amines,[485] phosphoryls,[905] sulfinyls,[914] ketones,[543,734,752,756,757,771,783,792,795,798] aldehydes,[771,862-866,868] esters,[543,830,837] lactones,[785] lactams,[915-918] heterocyclic nitrogen atoms,[88,122,822,919] imines,[676,920,921] and ethers[873-875] over tertiary amines has been noted in studies with LSR. The chelate bonding of ortho-substituted methoxy groups is stronger than that of tertiary amines.[885,886] Only in instances in which the other donor group has unfavorable electronic or steric effects,[307,484,873,909] or is a relatively weak donor group such as an ether[873,911] or phosphine,[415] is complexation at a tertiary amine observed.

Participation of a tertiary amine group in chelate bonding of a polyfunctional substrate has been reported for several compounds. A five-member chelate ring was reportedly formed in the complex between Eu(fod)$_3$ and I.[524] The derivative of anthraquinone with a dimethylamine

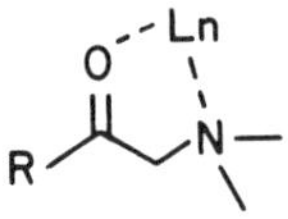

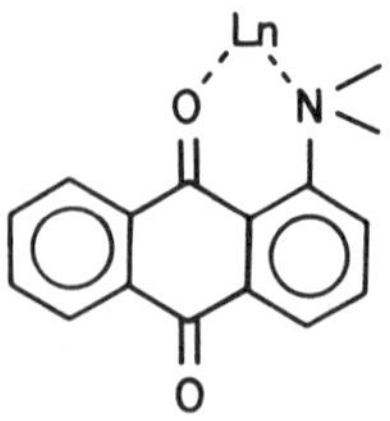

group peri to one of the carbonyl groups(II) bonded in a bidentate manner with Eu(fod)$_3$ and Pr(fod)$_3$.[757] The derivative with a methylamine group peri to the carbonyl did not associate as a bidentate donor because of hydrogen bonding between the carbonyl group and the amine hydrogen. A six-member chelate ring was observed in the complexation of III with Eu(fod)$_3$.[756]

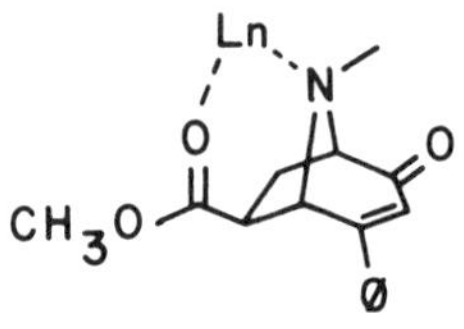

Shift data with Eu(dpm)$_3$ have been used to distinguish the *cis* and *trans* isomers of 3-methyl sulfonylcyclobenzaprine(IV).[484] The configurations and conformations of substituted 2-dimethylaminotetrahydropyrans have been determined through the analysis of shift data with Eu(dpm)$_3$.[911] An assessment of the ring current effects in diazacyclophanes was facilitated through a study with Eu(dpm)$_3$.[913]

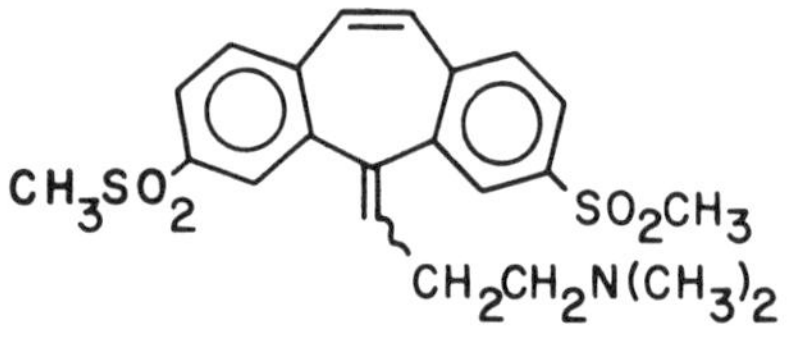

The complexation of Eu(dpm)$_3$ with tropinone(V) reportedly occurred to a significant extent at both donor sites.[671] All of the proton resonances

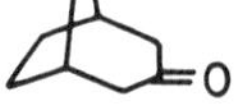

of bicyclobenzo[*b*]-1,4-diazobicyclo[3.2.1]octane(VI) were assigned on the basis of shift data with Eu(fod)$_3$.[909] The values of the *geminal* and *vicinal* coupling constants of VI were also obtained from the shifted spectra.

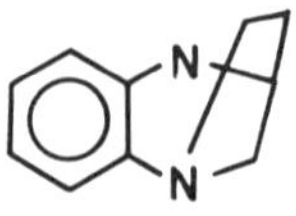

10. Nitrogen Heterocycles

The degree to which heterocyclic nitrogen atoms in polyfunctional substrates bind to LSR is highly dependent on the degree of steric hindrance about the heterocyclic nitrogen and the nature of the other functional group. Other functional groups that have been observed to associate in preference to heterocyclic nitrogen atoms include alcohols,[307,687,706] primary amines,[485,904] ketones,[786,791,795,803,862] esters,[725,786,835] aldehydes,[862] amides,[805,882,922,923] lactams,[916,924,925] and nitrogen oxides.[926,927]

Chelate bonding of substrates with one or more heterocyclic nitrogen atoms has been reported for a number of compounds. The interaction between Eu(fod)$_3$ and I was believed to be a chelate bond involving the pyridine nitrogen and oxygen atom.[927] The shifts in the spectrum of

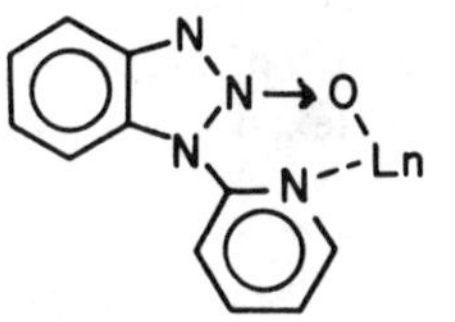

II with Eu(dpm)$_3$ were indicative of complexation at the carbonyl group of the amide moiety.[922] With Eu(fod)$_3$, however, a chelate bond involving the carbonyl oxygen and ring nitrogen was indicated by the shift data. The bond between 3-(4-bromophenyl)-3-pyridylallyl-N-methylamine (III) and Eu(fod)$_3$ was reportedly bidentate in nature.[908]

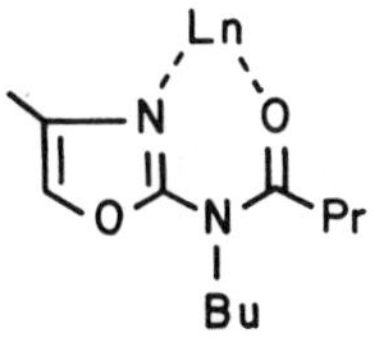

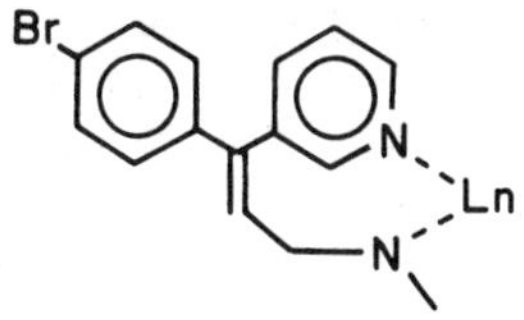

Certain polycyclic nitrogen heterocycles have an arrangement of nitrogen atoms suitable for stable chelate bonding. Bidentate complexation of 1,8-naphthyridine (IV) with Eu(dpm)$_3$ and Eu(fod)$_3$ has been noted.[495,500,928] A similar arrangement of nitrogen atoms explains the strong association of 1,4,5-triazanaphthalene and 1,3,5,8-tetraazanaphthalene with LSR.[495] The substrates 1,10-phenanthroline[495,500,929] and 2,2′-bipyridine[500,928-930] form especially stable chelate complexes with LSR. An idea of the stability of these complexes can be gained by considering the observation that the shifts with Eu(dpm)$_3$ in the spectrum of 1,10-phenanthroline were essentially identical in CDCl$_3$ and acetone-d$_6$.[929] In addition, no shifts are observed in the spectrum of ethanol that has been added to solutions of Eu(dpm)$_3$ with 2,2′-bipyridine or 1,10-phenanthroline.[929]

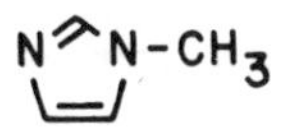

LSR have been used to study N-methylimidazoles (V)[465,466,919,931,932] and N-methylbenzimidazoles.[822] Complexation always occurred at the nitrogen atom that was not methylated. Excessive steric hindrance, such as that in N-methyl-2,4-diphenylimidazole, reduces the complexation to the extent that no shifts are noted in the spectra with

Eu(dpm)$_3$ or Eu(fod)$_3$.[931] The specificity of the bonding site permits the distinction of the 2,5- and 2,4-diphenyl-substituted isomers. The shifts in the ^{14}N spectrum of N-methylimidazole with Eu(dpm)$_3$, Yb(dpm)$_3$, Dy(dpm)$_3$, and Dy(fod)$_3$ have been reported.[465,466] The shifts in the ^{13}C spectrum of N-methylimidazole with LSR have been reported.[932] The contact contribution with Eu(III) was large enough to cause an upfield shift for one of the carbon resonances.

N-methylbenzimidazoles with carboxyethyl, cyano, nitro, and amine substituents on the aromatic ring coordinated to LSR through the N-3 position.[822] The shift data were used to determine the position of substituent groups in these substrates.

The effects of LSR on the spectra of triazole derivatives have been described.[927,932] Complexation of 1-methyl-1,2,3-triazole (VI) occurred predominantly at the N-3 position, whereas 1-methyl-1,2,4-triazole (VII) complexed at the N-4 position.[932] Both of these

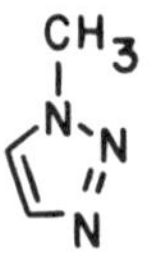

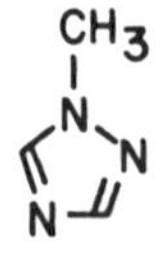

observations are predicated on the grounds of binding at the least sterically hindered site. The shifts for 1-(2-pyridyl)benzotriazole (VIII) with Eu(fod)$_3$ indicated some degree of complexation at all four nitrogen atoms.[927] The 3-oxide

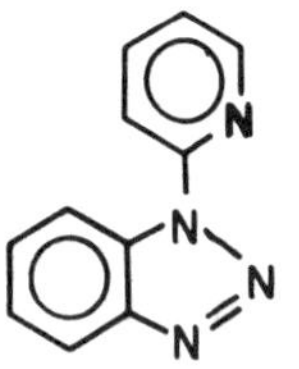

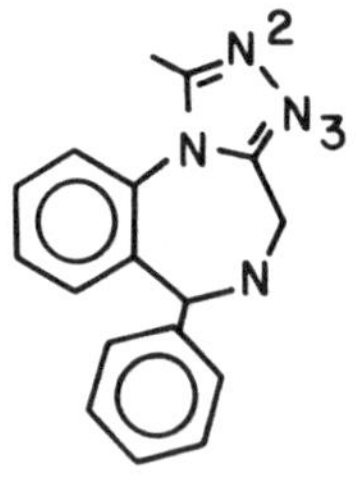

derivative of VIII bonds preferentially at the oxygen atom. The 2-oxide derivative, as previously discussed, appears to bond in a bidentate manner.

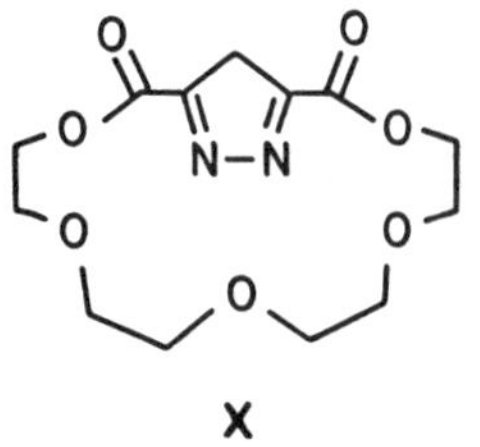

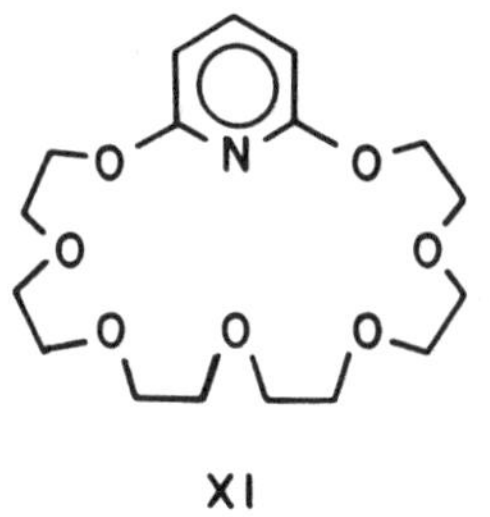

Five-member heterocyclic rings with oxygen or sulfur atoms in addition to a nitrogen atom bond to LSR through the nitrogen atom.[502] The shifts in the ^{15}N spectrum of triazolobenzodiazepines (IX) with Yb(dpm)$_3$ have been reported.[933] The magnitude of the shifts was indicative of complexation at the N-2 and N-3 positions. The shifts in the spectra of X with Eu(fod)$_3$ were indicative of binding at a variety of sites with a preference, however, for the heterocyclic nitrogens.[823] For a somewhat analogous pyridine derivative (XI), it was difficult to determine an exact site of complexation, but, the largest shifts were observed for the methylene groups furthest from the pyridine ring.[882]

Factors that influence the binding of pyridine substrates with LSR have been described.[900,908,923] The bonding of 2-, 3-, and 4-aminopyridine with chelates of dpm and fod occurred almost exclusively at the ring nitrogen.[900] The substrates 2-, 3-, and 4-cyanopyridine also complexed at the ring nitrogen.[489] The magnitude of the shifts for these substrates was dependent on both electronic and steric effects and were largest for the four-substituted, and smallest for the two-substituted derivatives. The effects of steric hindrance on binding preference can be demonstrated from a study of substituted nicotinamides (XII).[923] If there were no ring substituents in XII,

XII

complexation of the shift reagent was observed at the ring nitrogen. Substitution at the two- and six-position on the ring resulted in complexation at the carbonyl group. The conformations of these nicotinamides were determined through an analysis of the lanthanide shift data. The geometry of the L-S complex was fixed by fitting the shift data for the pyridine ring. The conformation of the side chain was then determined using this geometry. A determination of the conformation of nicotine was facilitated through the use of LSR.[88]

The effects of steric hindrance on binding of polyfunctional substrates with a heterocyclic nitrogen atom was also demonstrated in a study of 3-(4-bromophenyl)-3-pyridylallylamines (III).[908] When R_1 and R_2 were methyl groups, complexation at the pyridine ring was observed. When R_1 was a hydrogen atom and R_2 was a methyl group, a bidentate complex with Eu(fod)$_3$ was proposed. Increasing the steric bulk of R_2 by replacing the methyl group with an ethyl or n-propyl group reduced complexation at the secondary amine and enhanced complexation at the heterocyclic nitrogen.

The shifts in the spectra of pyridazine (XIII), pyrimidine (XIV), and pyrazine (XV) with Eu(dpm)$_3$ have been reported.[495]

XIII   XIV   XV

The spectra of substituted pyrazines in the presence of LSR have been used to assign resonances, determine coupling constants, and distinguish isomers.[934,935] Partially deuterated derivatives were analyzed in some instances to facilitate assignment of the resonances. In methylpyrazine, preferential complexation was observed at the nitrogen atom meta to the methyl group.

The complexation of naphthyridines (IV, XVI, XVII) with LSR has been studied.[495,500] For XVI and

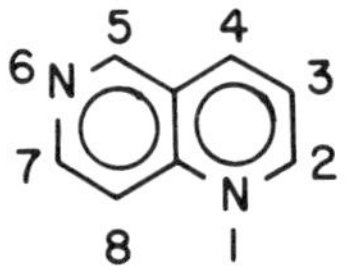

XVII, complexation occurred almost exclusively at the isoquinoline type nitrogen atom. For IV a strong chelate bond was observed. By studying mixtures of monofunctional substrates, Nagawa was able to determine the relative percent of complexation with Eu(fod)₃ at each nitrogen atom in the polyfunctional substrates.[500] The binding site in methyl-1,5-naphthyridines was influenced by steric effects of the methyl group.[500] Substitution of the methyl group at the 2-position reduced the complexation at N-1. Substitution at the 4-position reduced the complexation at N-5.

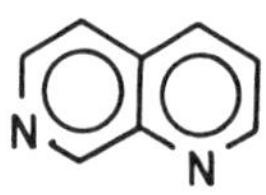

The complexation of benzodiazines (XVIII-XXI) with LSR preferentially occurred at the isoquinoline type nitrogen.[495]

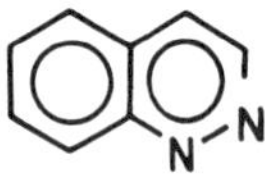

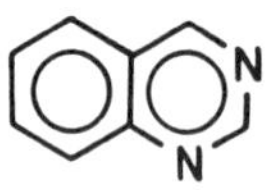

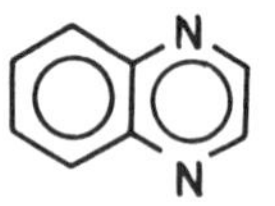

The shifts in the spectra of XVIII and XXI, which have isoquinoline type nitrogen atoms, were smaller than expected.[495] Binding of polyazanaphthalenes to LSR occurred through an isoquinoline type nitrogen unless the compound had two nitrogen atoms with the same configuration as those in 1,8-naphthyridine(IV). In the latter case, chelate bonding of these two nitrogen atoms was observed. The binding between pyridazines (XXII) and Eu(fod)₃ or Pr(fod)₃ occurred principally at N-1, although a slight complexation at N-5 was noted.[936]

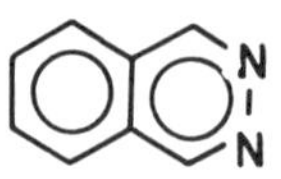

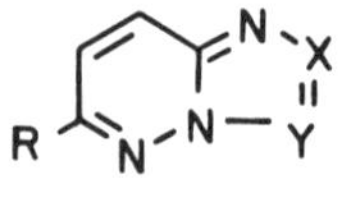

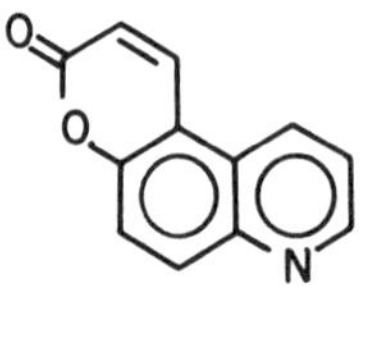

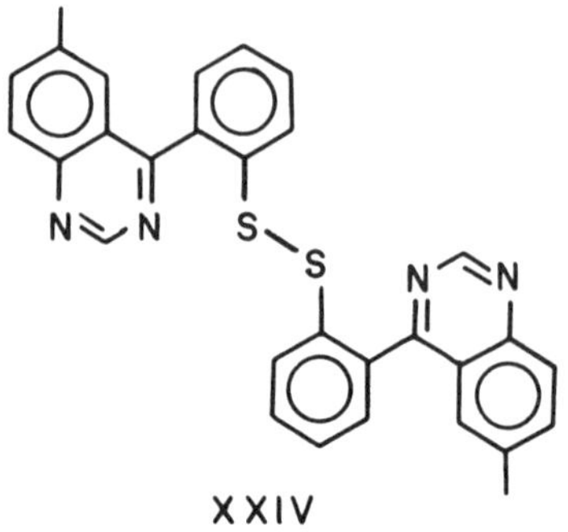

Shift data with Eu(fod)₃ have been used to assign the position of the substituent group in cyano- and nitroisoquinolines.[506] The structure of 3-H-pyrano[3,2-f]-quinolin-3-one (XXIII) has been confirmed on the basis of shift data with Eu(dpm)₃.[856] Association of the shift reagent at the ring nitrogen was confirmed by evaluating the magnitude of the shifts in the spectrum of coumarin. The structure of XXIV was assigned by comparing the shifts in the spectrum in the presence of Eu(fod)₃ to those of model compounds with known configurations.[937]

The conformation of 1-[2-(1,3-dimethyl-2-butenylidene)hydrazino]phthalazine (XXV) was determined by fitting observed lanthanide shift data and relaxation data to calculated values.[907] The conformation of

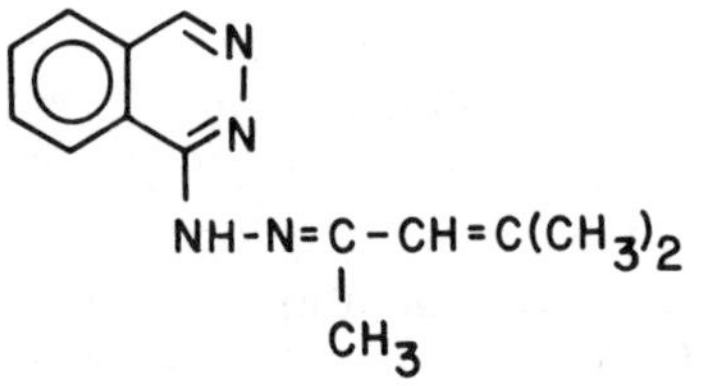

XXV

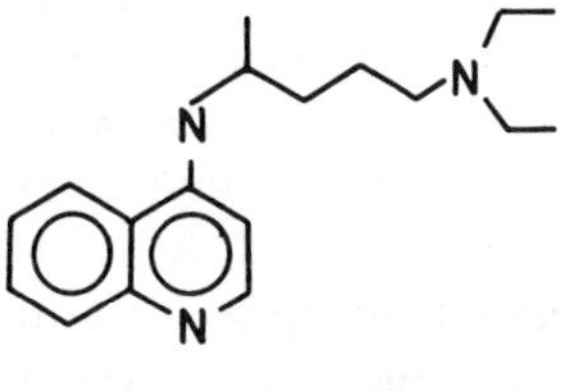

XXVI

chloroquine (XXVI) has been determined by fitting observed shift data with Pr(dpm)₃ to calculated values.[122]

The site of association of methyldiazaphenanthrenes with LSR has been assessed.[500] Complexation occurred at both nitrogen atoms, but was dependent on steric effects of the methyl groups. Shift data with Eu(fod)₃ have been used to determine the position of the substituent group in XXVII.[938]

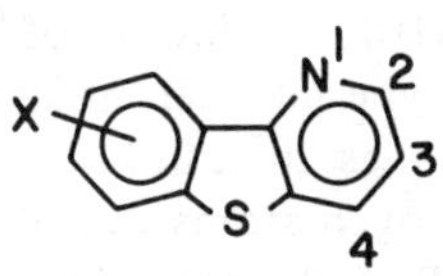

XXVII

Derivatives of XXVII with the nitrogen atom at the 2-, 3-, and 4-position were also studied. Steric factors influenced the association constants of the substrates with the LSR, and the following order was reported: N2=N3 >> N4 >> N1.

The percent of the *cis* and *trans* configurations of *N*-(4-aza-9-fluorenylidene)amines(XXVIII) has been determined through the use of Eu(dpm)₃.[939,940]

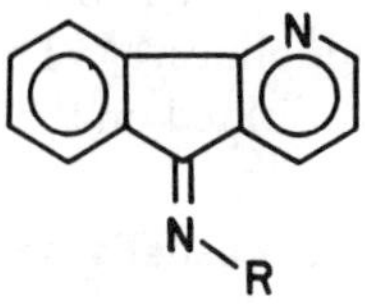

XXVIII

The assignment of resonances in the spectra of 3,6-diphenylimidazo[1,2-b]-as-triazine(XXIX) and a methyl-substituted derivative were confirmed through the application of Eu(fod)₃.[941] The ³¹P resonances of cyclotriphosphazene (XXX) were resolved in the presence of a LSR.[942]

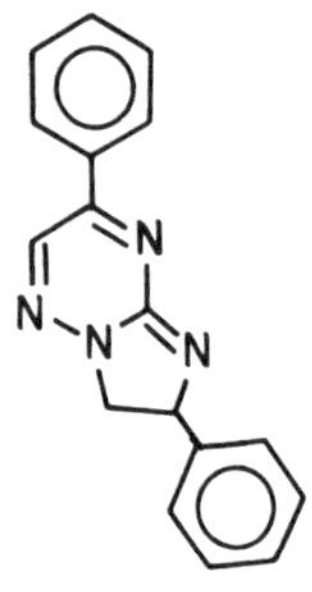

XXIX 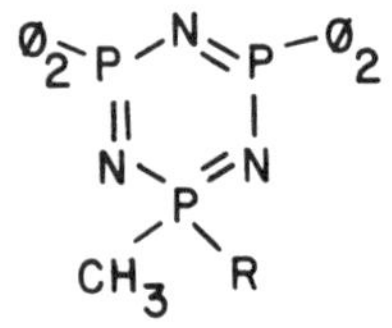   XXX

11. Nitriles

Polyfunctional substrates seldom bond to LSR through a nitrile group. Polyfunctional substrates in which hydroxyl,[255,366,507] ketone,[756,775,776,789] ester,[410,543,824,831] primary amine,[461] imine,[920] and heterocyclic nitrogen[489,506,822] groups bond in preference to a nitrile group have been noted. Most polyfunctional substrates that reportedly bond to a LSR at a nitrile group had an ether functionality as the second binding site.[507,778,869,870,874] The bonding between I and Yb(dpm)$_3$ reportedly involved the nitrile nitrogen atom.[901] This rather anomalous result was explained on the basis of a significant contribution of the resonance structure II.

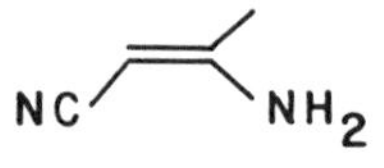

I 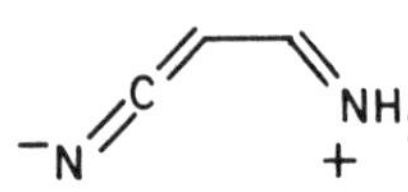   II

The nitrile group in phenoxybutyronitrile[507,869] and a derivative of cinnamic acid(III)[778] bonds in preference to the aromatic ether. In polyfunctional substrates with a nitrile and aliphatic or cyclic ether group.

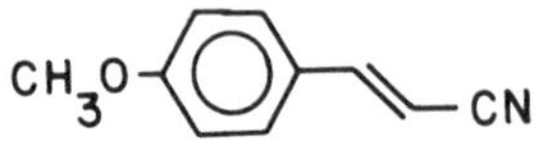

III 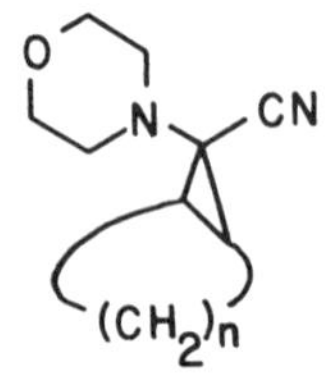   IV

appreciable complexation of the LSR is usually observed at both sites. Such a finding has been reported for 3-methoxypropionitrile, 3-ethoxypropionitrile, 3,3-tetramethylenedioxydipropionitrile,[507,869] and bicyclo[3.1.0]hexylmorpholines (IV).[874] The shifts in the spectrum of β,β'-oxydipropionitrile in the presence of LSR indicated that the principal mode of interaction was through the nitrile groups.[507] These studies noted a distinction, however, in the bonding of Eu(dpm)$_3$ and Eu(fod)$_3$ with oxynitrile substrates.[507,869] Eu(fod)$_3$ exhibited a greater preference for the nitrile group than Eu(dpm)$_3$.

The aromatic and vinyl resonances of V were resolved in the presence of Eu(dpm)$_3$ and permitted assignment of the structure.[870] This report did not discuss the site of bonding.

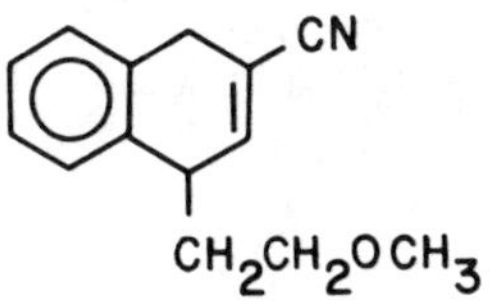

12. Amides

Amide functional groups are of intermediate basicity and are the preferred site of complexation in a number of polyfunctional substrates. Phosphoryl[520,943] and alcohol[740] groups associate in preference to amide groups. Preferential coordination at a ketone group over an aromatic amide has been reported.[752] The complexation of LSR with substituted nicotinamides was dependent on the position of the substituent groups.[882,923] Complexation took place at the pyridine nitrogen except in derivatives with a 2,6-disubstituted ring.

Severe broadening in the spectrum of N-butyl-N(4-methyloxazol-2-yl)-2-methylpropionamide (I) in the presence of Eu(fod)$_3$ was considered evidence for chelate bonding of the carbonyl oxygen and ring nitrogen.[922] The spectrum

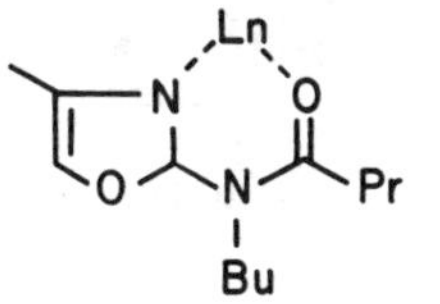

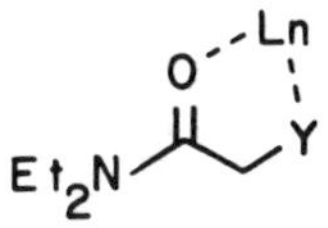

of I with Eu(dpm)$_3$ was not significantly broadened, which implied that bonding took place only at the carbonyl oxygen. The complexation between LSR and II in which Y = −OMe or −NMe$_2$ reportedly involved bidentate coordination of the substrates.[524] LSR were employed in an attempt to determine the conformations of II; however, complexation with the LSR caused considerable alteration of the conformations.

Barriers to rotation and conformational preferences of polyfunctional substrates with amide groups have been determined through analyses with LSR. These include compounds of structure III in which X=O,[528,772,773,944,945] S,[528,772,773] Se,[772] and Te.[772] It

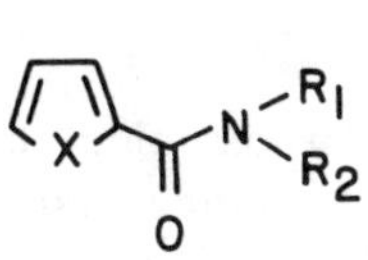

has been reported that complexation of the LSR with III did not change the conformational preference.[772] The conformations of substituted nicotinamides have been determined through the use of shift data with LSR.[882,923] The conformations of 1-acetylimidazole, -pyrazole, -1,2,4-triazole, and -tetraazole have been determined on the basis of shift data with Eu(fod)$_3$.[862]

Eu(fod)$_3$ has been employed in assigning the conformations of symmetrical diamides.[522,528] Resonances were observed for all three of the possible rotamers of IV in the NMR spectrum with Eu(fod)$_3$. Analogous substrates with a −CH=CH− and

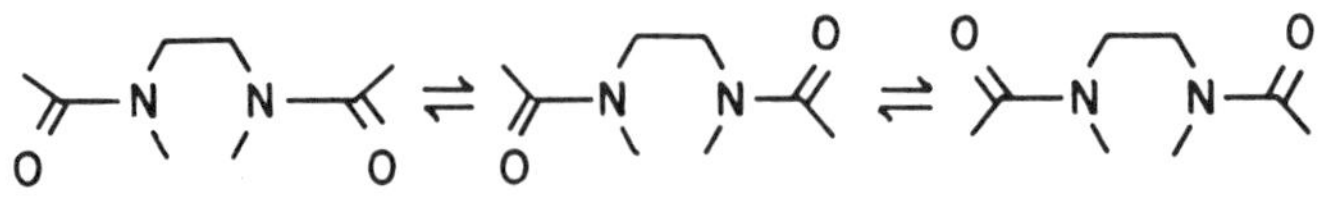

−CH=CH−CH=CH− group between the two amide groups were also studied.[528] The cis and trans isomers of V and VI were distinguished using LSR.[528] The rotamers of

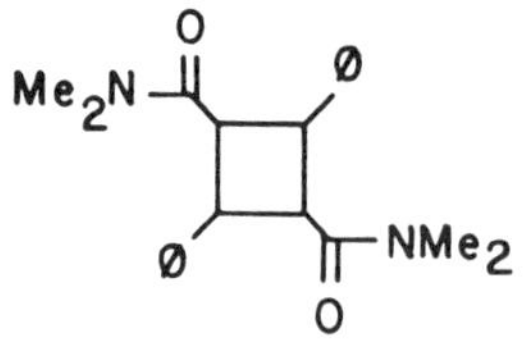

VII and VIII were distinguished on the basis of shift data with a LSR.[522]

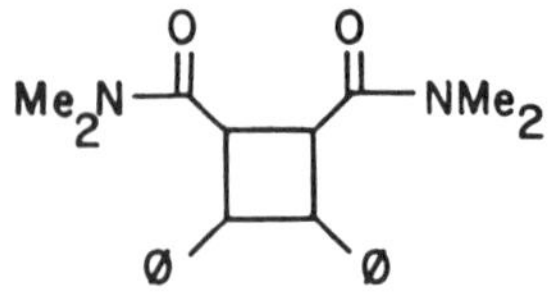

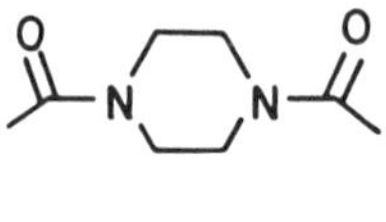

The geometry of cyclopentanes, cyclohexanes, and bicyclo[2.2.2]octanes(IX) with two amide substituent groups were determined through the use of shift data with a LSR.[522,946] These substrates were employed as models in a study of polyamides.[946] The conformations of trichloroacetyl carboxamide derivatives of ethyl cis- and trans-2-hydroxy-1-cyclopentane, -hexane, -heptane, and -octane carboxylates (X) were assigned on the basis of shift data with Eu(fod)$_3$.[698]

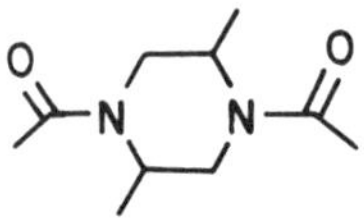

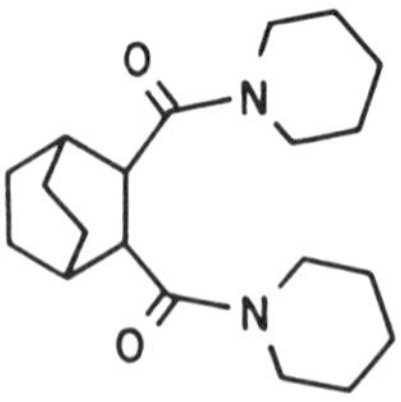

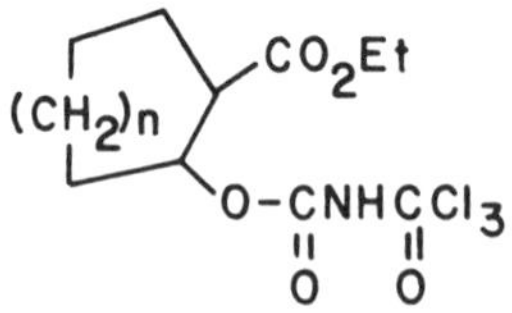

The shifts in the ^{1}H[129] and ^{13}C[947] spectrum of piperine (XI) with Eu(dpm)$_3$ have been reported. Analysis of the ^{13}C spectrum with the shift reagent facilitated the assignment of all of the resonances. Slow rotation about the N−C=O bond was observed in the presence of the shift reagent.[129]

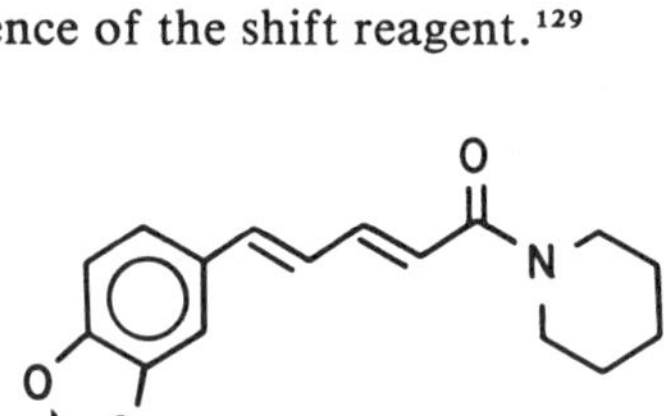

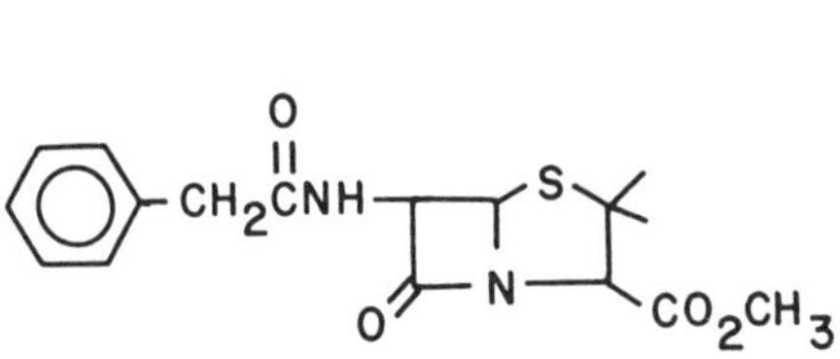

The spectra of penicillin-6-methyl ester (XII),[809,812] penicillin benzylester, and 1-oxadethiapenicillins[810] have been analyzed in the presence of LSR. The data with the shift reagent was used to assign the configuration about the carbon with the ester substituent.[810] The stereochemistry of methyl 7-(2-thienylacetamide)-3-methylenecepham-4-carboxylate (XIII) was assigned on the basis of shift data with Eu(fod)$_3$.[817]

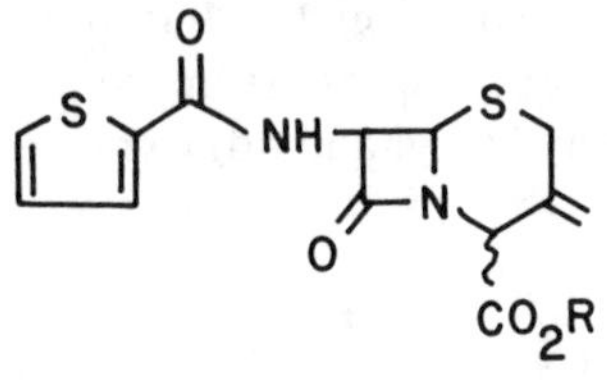

XIII

Derivatized carbohydrates with an amide functional group reportedly bond at the oxygen atom of the amide.[805,806] In α and β-D-glucoseamine pentaacetates the amide site was reportedly the preferred site of binding by four to five times over that of the ester groups.[805] The spectra of the α and β-anomers of peracetylated 2-acetamido-2-deoxy derivatives exhibited significant differences in the presence of Eu(fod)$_3$.[806] The shifts in the spectrum of N-acetylcytidine triacetate (XIV) in the presence of a LSR were indicative of complexation at the amide group.[805] The spectrum with the shift reagent permitted the interpretation and assignment of the resonances.[805]

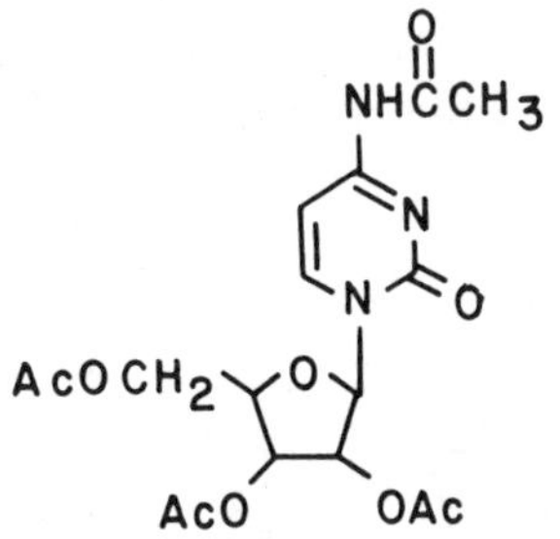

XIV

LSR have been applied in the study of derivatized tripeptides, dipeptides, and amino acids.[813-818] The stereochemistries of N-benzyloxycarbonyl-L-phenylalanyl-L-alanine methyl ester and the D-alanine derivative were confirmed through the use of shift data with Eu(dpm)$_3$.[813] The *cis-trans* isomerism resulting from slow rotation of the N—C=O bond in *t*-butoxycarbonyl α-amino acid esters has been studied with the aid of Eu(fod)$_3$.[814,816] Resonances for the two rotamers were observed in the presence of the shift reagent; however, complexation of the shift reagent altered the population of the rotamers. Extrapolation to zero concentration of LSR enabled a determination of the population of each rotamer in the uncomplexed form. It was also reported that complexation of the shift reagent increased the barrier to rotation for conversion of the *E* to *Z* isomer, but did not alter the rotational barrier of the reverse process.[814] The shifts in the ^{15}N NMR spectra of N-benzyloxycarbonyl-glycine-leucine-leucine methyl ester and N-benzyloxycarbonyl-alanine-leucine-leucine methyl ester with Eu(dpm)$_3$ and Dy(fod)$_3$ have been reported.[818]

13. Lactams and Imides

Most polyfunctional substrates with one or more lactam groups that have been studied with LSR coordinated at the oxygen atom of the lactam. Exceptions include the preferential association of the amide group of penicillins[809,810,812] and similar substrates.[817] The oxygen atom of the benzoyl group of I was reportedly the preferred site of complexation with chelates of fod.[785] Appreciable association of both the secondary amine and lactam oxygen of furo[2.3-e]-1,4-diazepin-5-ones (II) was observed with Eu(dpm)$_3$.[906] The shifts in the spectrum of III in the presence of Eu(fod)$_3$ were indicative of chelate bonding of the hydroxyl and adjacent carbonyl group.[673]

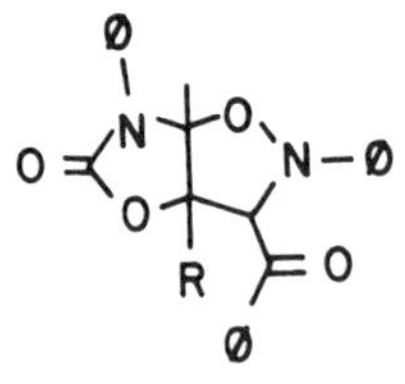

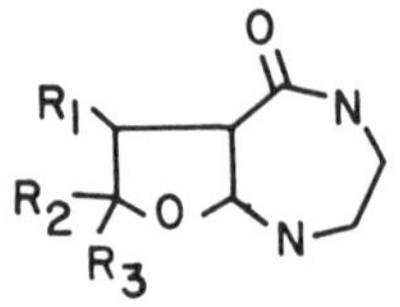

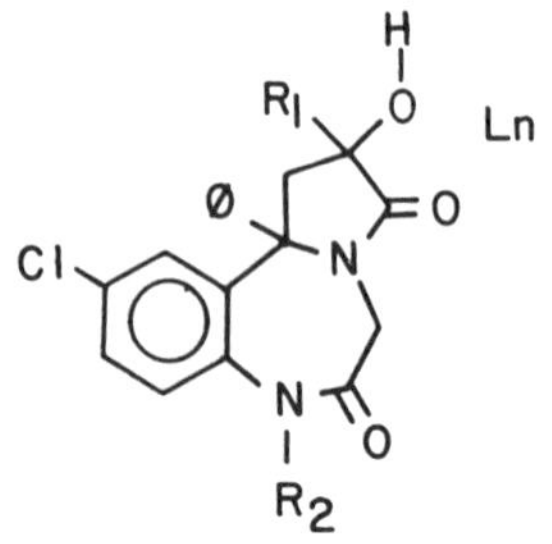

Monocyclic lactams with four[541,809,812] five-,[123,819,917,948] six-,[949-951] seven-,[753,916,924,925] and eight-[69] member rings have been studied with the aid of LSR. The stereochemistries of derivatives of IV have been assigned on the basis of lanthanide shift data.[541] Several derivatives of IV had R_1 or R_2 groups with diastereotopic methylene hydrogens, and these were frequently resolved in the presence of the shift reagent.[809,812]

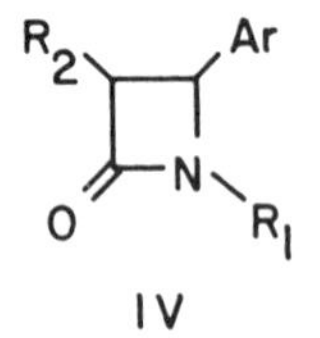

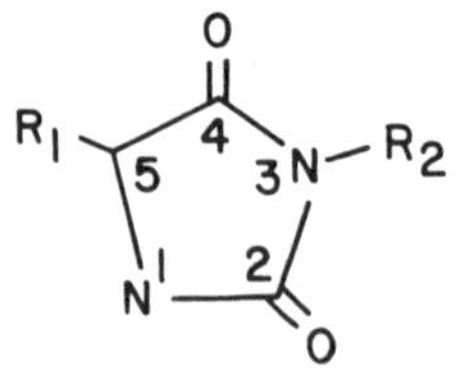

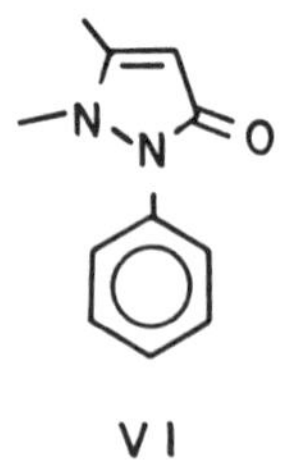

Data collected with LSR have been used to aid in the determination of the conformations of hydantoins (V).[123] Complexation of these substrates with the LSR reportedly involved the less hindered C-2 carbonyl oxygen. The shifts in the spectrum of antipyrine (VI) with Pr(fod)$_3$ have been reported.[917] The isomers of VII were distinguished on the basis of shift data with a LSR.[819] The shifts in the spectra of VII were indicative of coordination at the oxygen atom of the lactam.

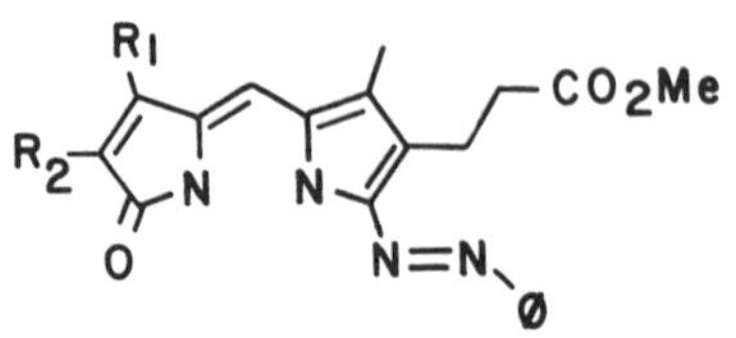

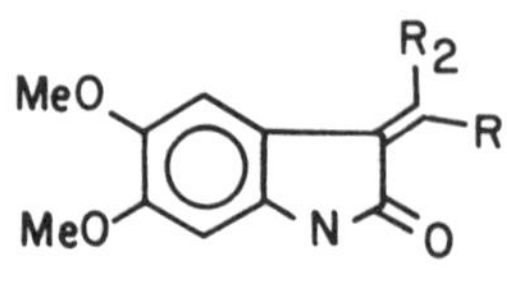

Stereochemistries about the double bond of 5,6-dimethoxyisatylidenes (VIII) were assigned on the basis of shift data with Eu(fod)$_3$.[948] Complexation at the carbonyl group, rather than the ortho-substituted methoxy groups, was indicated by the shift data.

The configurations and conformations of 2'-substituted-1-(1,3-dioxan-5-yl)uracils (IX) were determined through the analysis of shift data with Eu(fod)$_3$.[951]

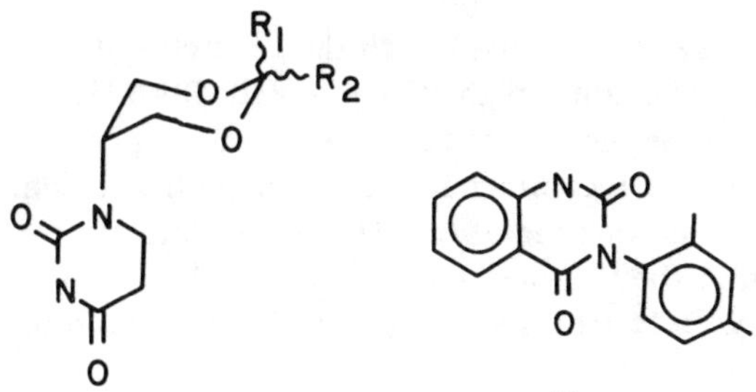

IX   X

The positions of substituent groups in 1-ethyl-1,6-dihydro-2,4-dimethoxy-5-methyl-6-oxopyrimidine and 1,5-diethyl-1,6-dihydro-2,4-dimethoxy-6-oxopyrimidine were determined through the application of LSR.[950] Each position was identified by a distinct shift gradient. Most of the resonances in the ^{1}H spectrum of 3-(2',4'-dimethylphenyl)-1H,3H-quinazoline-2,4-dione (X) were assigned with the aid of shift data with Eu(fod)$_3$.[949] Based on steric and electronic effects, binding of X was expected at the oxygen atom of C-2. The shift data was indicative of that conclusion.

The position of substituent groups in chloroazepine-2,5-diones (XI) has been assigned on the basis of shift data with Eu(fod)$_3$.[753] The stereochemistries

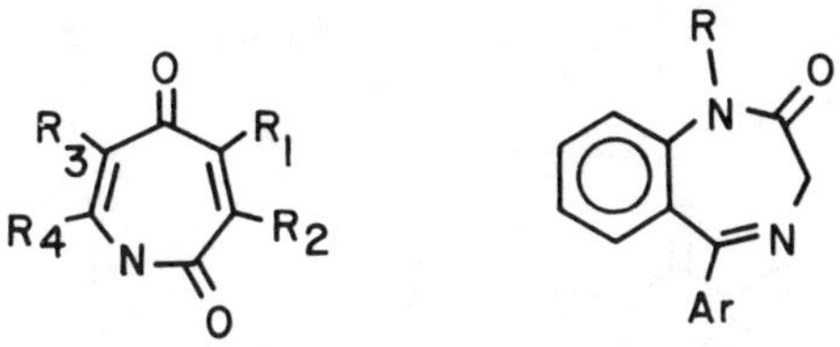

XI   XII

and conformations of 1,4-benzodiazepine-2-ones (XII) have been determined through the use of LSR.[916,924,925] The assignment of resonances in the spectrum of XIII was facilitated with Eu(dpm)$_3$ and Pr(dpm)$_3$.[69] The spectra of furo[2.3-e]-1,4-diazepin-5-ones (II) were first-order in the presence of Eu(dpm)$_3$.[906]

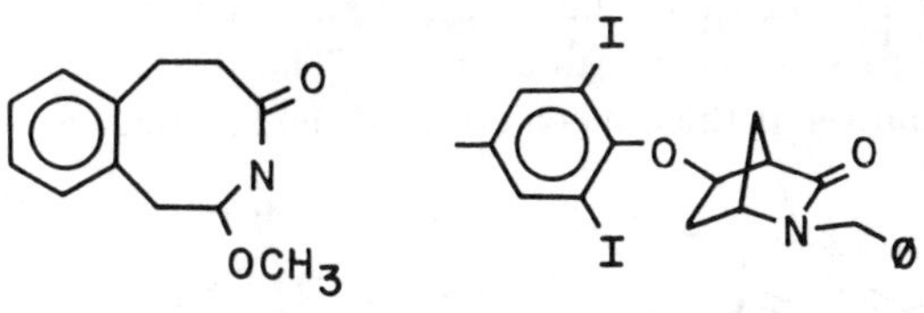

XIII   XIV

Conformational analyses of bicyclic lactams XIV,[952] XV,[953] and XVI[953,954] have been performed by fitting observed to calculated lanthanide shift data.

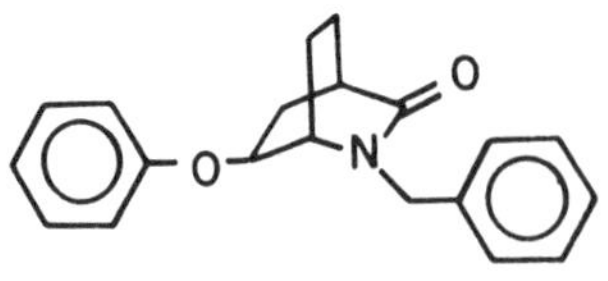

XV 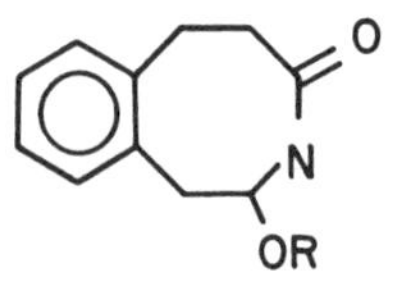 XVI

The rigid bicyclic ring system was used to fit the geometry of the L-S complex, and the conformation of the side chain was then determined. Both ^{1}H and ^{13}C lanthanide shift data were used to increase the size of the data set and improve the accuracy of the fit. The use of shift ratios was recommended to eliminate the constant of the dipolar shift equation and permit easier mixing of the ^{1}H and ^{13}C data.[953]

The configuration at the C-3 position of 1,3-oxazolidin-2-ones (XVII) was assigned through an analysis with Eu(fod)$_3$ and Pr(fod)$_3$.[785] The two isomeric

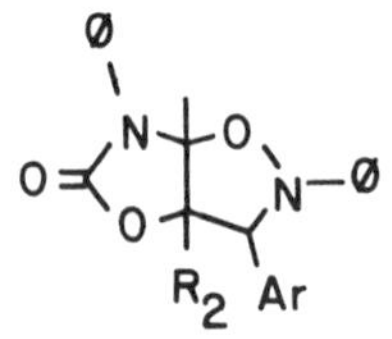

XVII 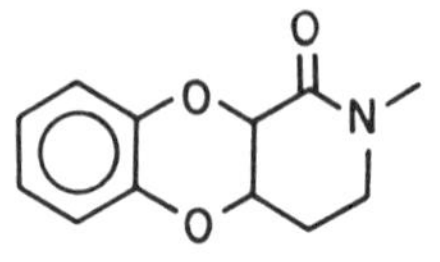 XVIII

tricyclic 1,4-benzodioxans (XVIII) were distinguished using shift data obtained with Eu(fod)$_3$.[955] The coupling constants of the *trans* isomer, which is the more rigid of the two, did not change in the presence of the shift reagent. Changes were noted, however, for the coupling constants of the *cis* isomer. On the basis of shift data with a LSR, the diastereomers of XIX were distinguished individually and in mixtures.[956] Configurational and conformational

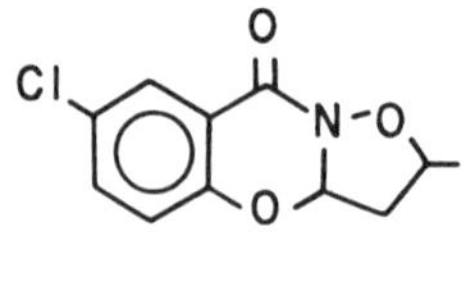

XIX 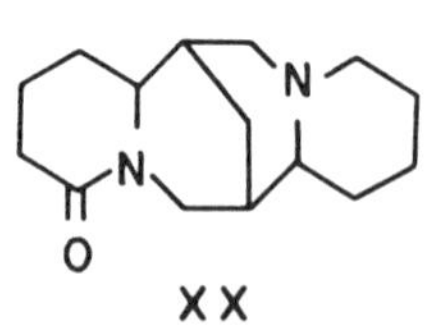 XX

studies of lupanine (XX) have been performed with the aid of LSR.[915,957] Both NMR and IR data indicated that the conformation of XX was unperturbed by complexation of the shift reagent.

LSR were employed in a study of the conformation of phthalazino[2.3-b]-phthalazine-5,12-(14H,7H)-dione (XXI).[958]

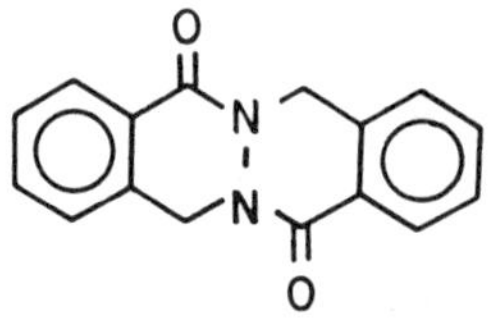

XXI 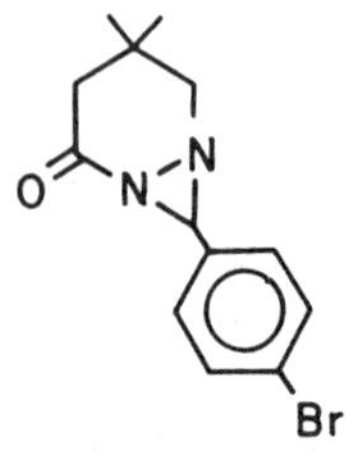 XXII

The configuration of a 1,5-diazabicyclo[3.1.0]hexan-2-one(XXII) was assigned on the basis of shift data with Eu(fod)$_3$.[918]

A determination of the conformations of cyclodipeptides was facilitated through the use of shift data with Eu(fod)$_3$.[959] LSR have been used to study aspects of the solution conformation of naphthomycin[726] and valinomycin.[811]

Complexation of XXIII with LSR occurred at the bridging ketone group rather than the carbonyl groups of the imide.[794] This was unequivocally demonstrated by

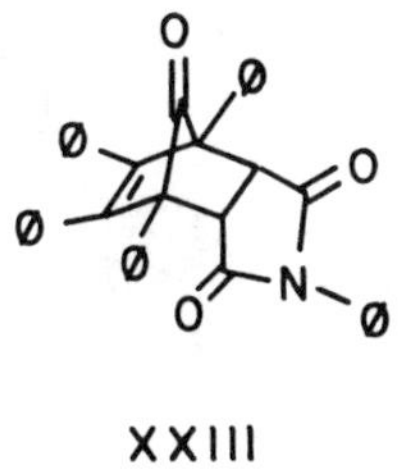

XXIII

recording the shifts in the spectrum of the compound with a methylene group at the bridging position.

14. Nitrogen Oxides, Nitrones, and Azoxys

The N—O group exhibits a strong association with LSR and, even though limited in scope, it is not surprising that studies of polyfunctional substrates with nitrogen oxide, nitrone, and azoxy groups bond at these sites. Pyrimidine N-oxides (I,II) bond to LSR through the oxygen atom, and I and II have been distinguished on the basis of lanthanide shift data.[550,926]

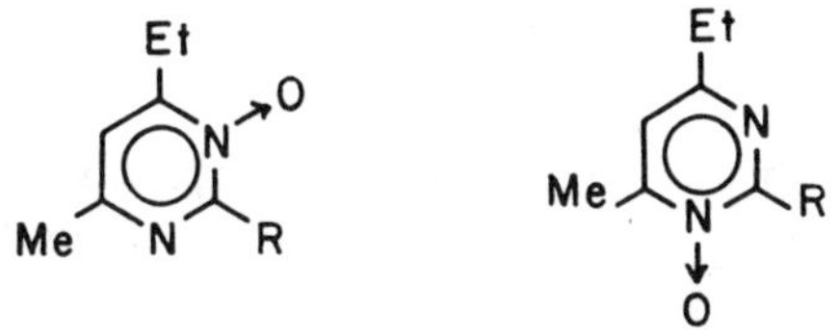

I II

The isomers III and IV were distinguished in the presence of Eu(fod)$_3$.[807] The geometries of the

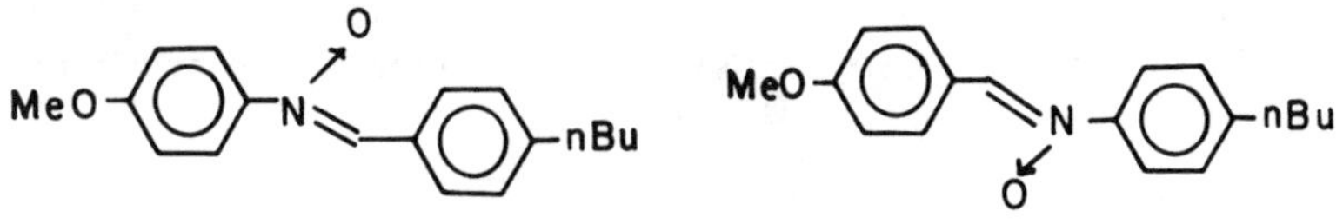

III IV

analogous azoxy isomers with a N=N—O group were assigned in an experiment with Eu(fod)$_3$.[112,807] Mixtures of the two azoxy isomers were also resolved and quantified.[807] Certain of the resonances of the azoxy derivatives of III exhibited "wrong way" shifts with Eu(fod)$_3$. This observation was believed to result from a change in the sign of the angle term of the dipolar shift equation. No shifts were observed in the spectra of the azoxy compounds with Eu(dpm)$_3$.[112]

The derivatives of III and IV in which the n-butyl groups were replaced with acetyl groups were also distinguished in the presence of Eu(fod)$_3$.[807] The shifts in the spectra of these substrates were indicative of some complexation at both the carbonyl group and the N—O group. A distinct difference has been reported for the complexation of V and VI with Eu(fod)$_3$.[927] The shifts in the spectrum of V indicated bidentate complexation of the oxygen atom and pyridine nitrogen. Coordination of VI seemed to occur exclusively at the oxygen atom.

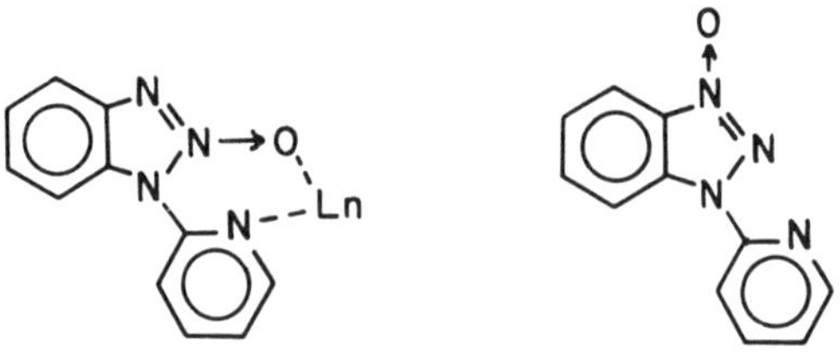

V   VI

15. Nitros

Nitro groups are among the least effective functional groups in bonding to LSR, and polyfunctional substrates seldom associate through a nitro group. One exception appears to occur with substrates of structure I.[910] The shifts in the spectrum of I with Eu(fod)$_3$

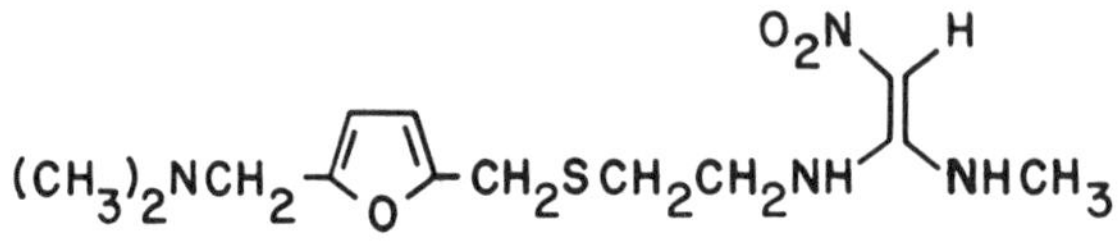

I

and Pr(fod)$_3$ were indicative of some degree of complexation at the nitro group. The shifts with Eu(dpm)$_3$ were more indicative of coordination at the tertiary amine site. Model substrates with an olefin configuration identical to that of I also appeared to exhibit appreciable coordination at the nitro group with chelates of fod. It was postulated that a bidentate complex, as shown in II, was the most probable bonding situation.

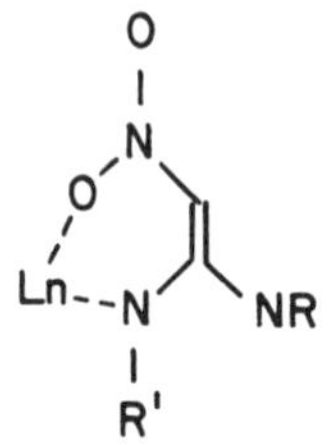

II

16. Oximes

LSR have been employed in only a few studies of polyfunctional substrates with an oxime group. Conformational analyses of I have been carried out with the aid of Eu(dpm)$_3$[960] and Eu(fod)$_3$;[774] however, the ability to distinguish different conformations based on the lanthanide shift data was limited.[774] Severe broadening was observed in the spectrum of I with Eu(dpm)$_3$. The *cis* and *trans* rotamers of the *anti* configuration were distinguished using LSR. The shifts for the *syn* isomer were much smaller, however, and the *cis* and *trans* isomers could not be distinguished.[960]

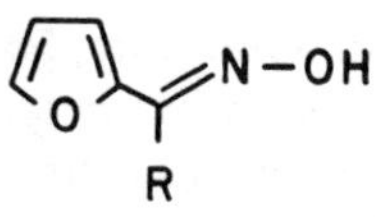

Carbohydrate derivatives with an oxime functional group have been studied with the aid of Eu(dpm)$_3$.[961] It was possible to determine certain aspects of the conformation of these substrates and to distinguish the *syn* and *anti* isomers through an analysis of the lanthanide shift data.

17. Azos, Azines, and Azides

Polyfunctional substrates containing azo (–N=N–), azine (=N–N=), and azide (N$_3$) groups have been the focus of only a limited number of studies with LSR. The azo unit is not particularly effective in competing for coordination to a LSR. As a result, the exocyclic oxygen in N-methyl syndrome (I) bonds in preference to the azo unit.[189] The carbonyl group of a lactam[819] and in diazoketones,[768] has been noted to bond in preference to an azo unit.

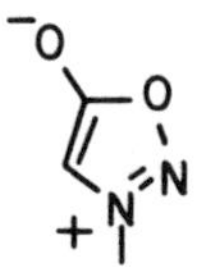

The configurations of the R$_1$ and R$_2$ groups in 2-alkylidenehydrazono-3-methyl-2,3-dihydrobenzothiazoles (II) were determined through the application of Eu(fod)$_3$.[676] If the R groups had no other donor atoms suitable for complexation with the shift reagent, complexation was noted at one of the azine nitrogens. In the case of the Z-configuration (III), the shifts were indicative of binding at the β-nitrogen. The E-configuration (IV) reportedly coordinated through the alpha nitrogen. The derivative in which R$_2$ was a phenyl substituent with a hydroxyl group at the ortho position (V) seemed to form a chelate bond involving the β-nitrogen and hydroxyl oxygen.

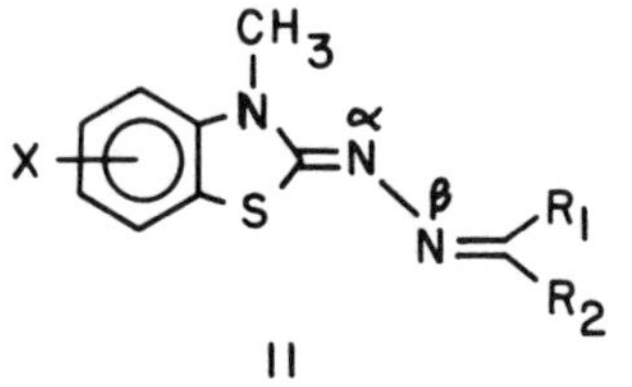

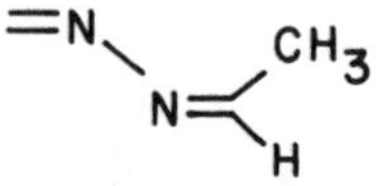

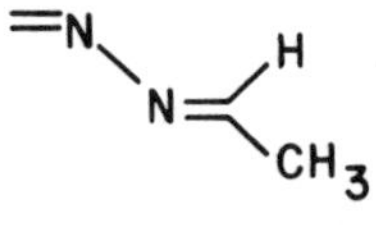

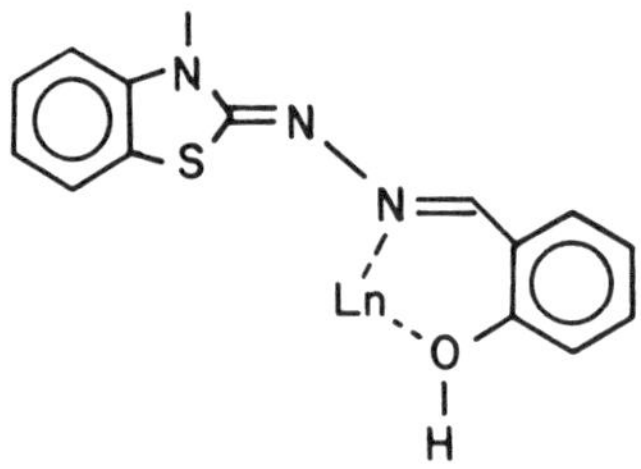

Azide units form only weak complexes with LSR. The principal site of coordination for VI was the oxygen atom of the lactam, although some degree of complexation at the azide group was indicated.[541] The effects of Eu(fod)$_3$ on the spectra of the *meso-* and *d,l*-diastereomers of 1,2-diazido-1,2-di-*tert*-butylethane have been described.[962] No shifts were observed for the *meso* isomer, whereas the spectrum of the *d,l*-isomer did exhibit shifts with Eu(fod)$_3$. The difference was ascribed to steric effects of the *tert*-butyl groups. In the *d,l*-isomer, the *t*-butyl groups have a configuration that permits a stable bidentate complex between the substrate and shift reagent. Such a favorable arrangement of the donor groups does not occur in the *meso* derivative.

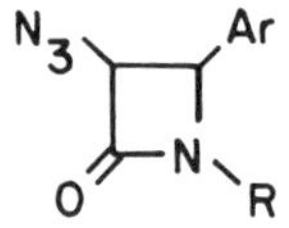

18. Imines

The imine group (=N—) does not form particularly strong complexes with LSR. The coordination of 7-chloro-5-phenyl-2,3-dihydro-1*H*-1,4-benzodiazepines (I) with a LSR occurred through the imine nitrogen.[921] A lactam analog (II), however, complexed principally through the carbonyl group.[921] Preferential complexation of an endocyclic pyridine nitrogen over an exocyclic imine group has been reported with *N*-(4-aza-9-fluorenylidene)arylamines.[939,940]

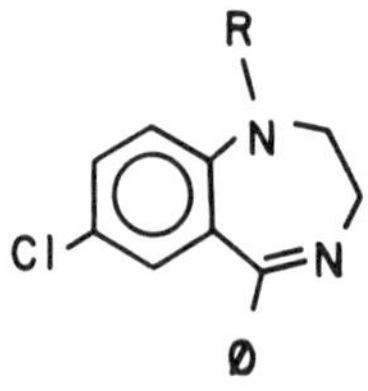

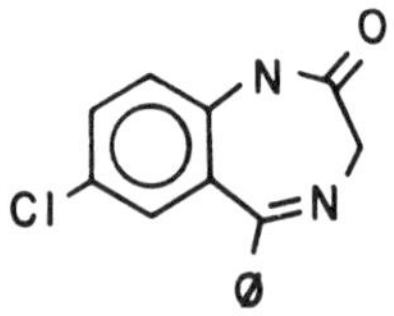

The *cis* and *trans* isomers of benzylidene anilines (III) have been distinguished through the use of Eu(fod)$_3$.[569] For the derivative in which R was a methoxy group, the shifts were indicative of bidentate complexation involving the imine nitrogen and ether oxygen.

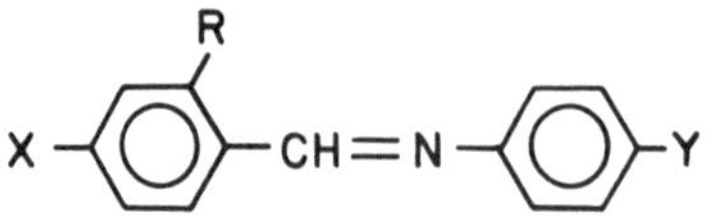

19. Substrates with a Sulfur Atom

Sulfides, sulfoxides, and sulfones encompass the functional groups with a sulfur atom present in the majority of polyfunctional substrates studied with LSR. Polyfunctional substrates with thiol, P=S, and C=S groups have been the focus of a few studies. The sulfide group associates weakly with LSR. Only in cases of chelate bonding,[888] or in polyfunctional substrates with just sulfide groups,[913] is complexation at sulfide groups observed. The cis isomer of I coordinated in a tridentate manner with Eu(fod)$_3$.[888] The trans isomer of I, which did not have the proper alignment of the donor groups for tridentate binding, exhibited weak complexation with the LSR.

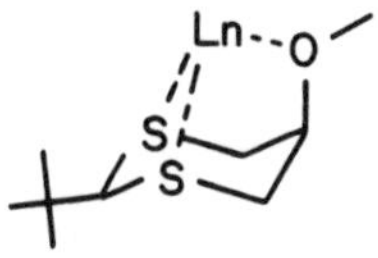

I

Suitable shifts were observed in the spectra of dithiacyclophanes in the presence of Eu(fod)$_3$-d$_{27}$,[913] Eu(dpm)$_3$ was judged unacceptable for studies of these substrates because of poor solubility and Lewis acidity.

The complexation between asparagusic acid methyl ester (II) and Eu(dpm)$_3$ and Eu(fod)$_3$ involved only the ester group.[119] The shifts in the spectrum of dihydroasparagusic acid (III), however, were indicative of some complexation at both the ester and thiol groups.[119]

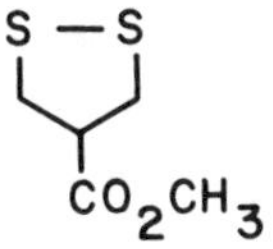

II 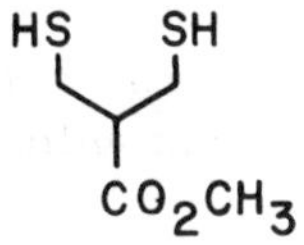 III 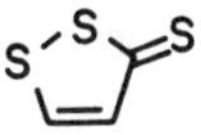 IV

The P=S[749] and C=S[427,944] groups complex only weakly, if at all, to LSR. No shifts were observed in the spectra of 1,2-dithiole-3-thione (IV) in the presence of Eu(fod)$_3$ and Pr(fod)$_3$.[427] The analogous carbonyl derivative exhibited small, but measurable shifts.

The thionyl group is of intermediate basicity and in polyfunctional substrates has been reported to bond with LSR in preference to ether,[584,914] sulfide,[585,821,836] ester,[821,836] and tertiary amine[914] groups. Severe broadening was observed in the spectrum of V in the presence of Eu(dpm)$_3$ and Pr(dpm)$_3$ and was ascribed to bidentate complexation.[674]

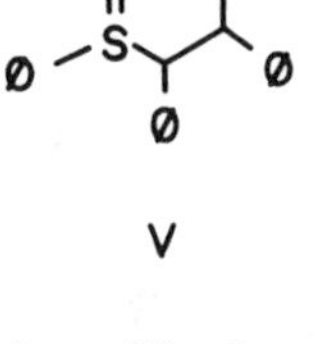

V 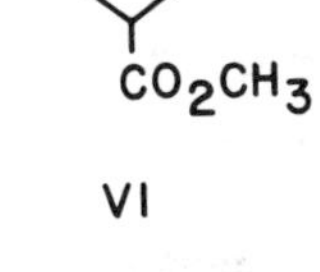 VI

The orientation of the sulfinyl group relative to the ester group in syn- and anti-1,2-dithiolane-4-methylcarboxylate-1-oxide (VI) has been assigned on the basis of shift data with a LSR.[821,836] The configuration of neothiobinupharidine (VII) was determined through an analysis of the spectrum with Eu(fod)$_3$.[914] The conformations of

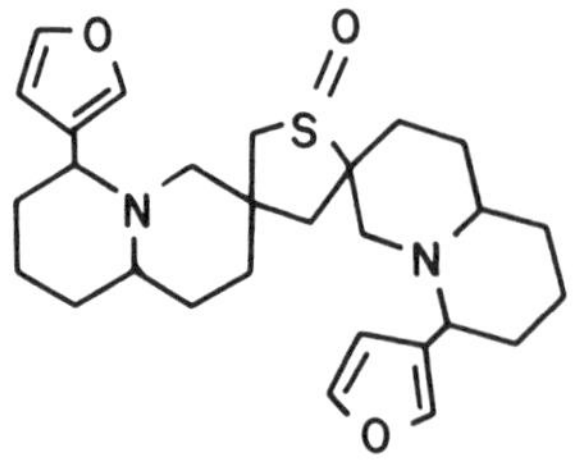

VII 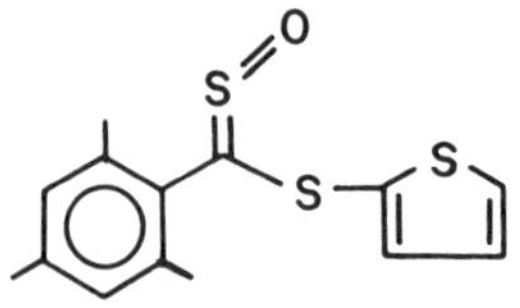   VIII

3,3'-disubstituted diphenylsulfines[584] and 4,4'-disubstituted diphenylsulfines[585] were determined through the use of shift data with Eu(dpm)$_3$. A similar analysis was performed for VIII.[585] The *meso* and *racemic* forms of *bis*(phenylsulfinyl)methane were distinguished in the presence of a LSR.[963]

The sulfone group is intermediate in its ability to compete for coordination sites on a LSR. Hydroxyl groups bond in preference to sulfone groups.[674,704,718,740] Preferential complexation of a tertiary amine over an aromatic sulfone moiety has been observed.[484] Certain derivatized carbohydrates with tosyl and ether groups reportedly bond to shift reagents at the tosyl group.[738] Sulfone groups have been observed to preferentially complex in polyfunctional substrates with sulfide[964] and ether[574] groups. The bonding between Pr(fod)$_3$ and aromatic substrates with methoxy and sulfone groups occurred at the S=O group furthest removed from steric hindrances.[574]

The four geometrical isomers of IX were distinguished using shift data with Eu(fod)$_3$.[964]

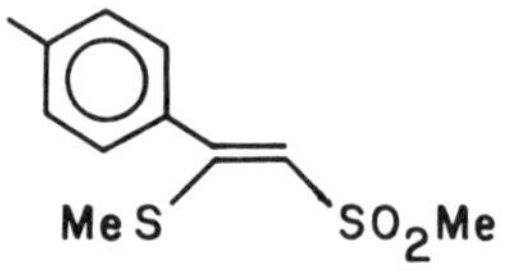

IX

20. Substrates with a Phosphorus Atom

The phosphoryl group exhibits a high association constant with LSR. It is the preferred site of complexation in substrates with hydroxyl,[603] amide,[520,943] ketone,[603,749] ester,[95,808] amine,[905] and ether[603] functional groups. Increased complexation at the site other than the phosphoryl group is often observed at higher shift reagent concentrations.[95,603] The shifts in the spectrum of diethyl-2-aziridinyl phosphonate (I) with Eu(dpm)$_3$ were indicative of either complexation at the nitrogen atom or a bidentate complex involving the nitrogen atom and phosphoryl oxygen.[905]

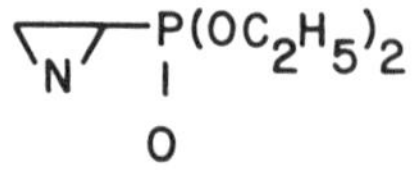

I

The two nitrogen invertomers of I were distinguished on the basis of the shift data.

The two isomers of 8-phenyl-8-oxo-8-phosphabicyclo[3.2.1]-octan-3-ones that differed in the configuration about the phosphorus atom were distinguished using Eu(dpm)$_3$.[603] The shifts in the spectra of II and III with LSR have been reported.[965]

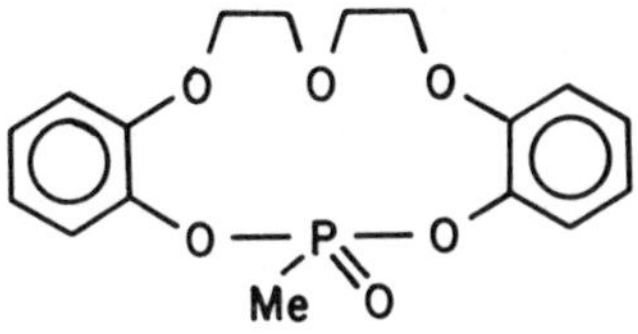

II

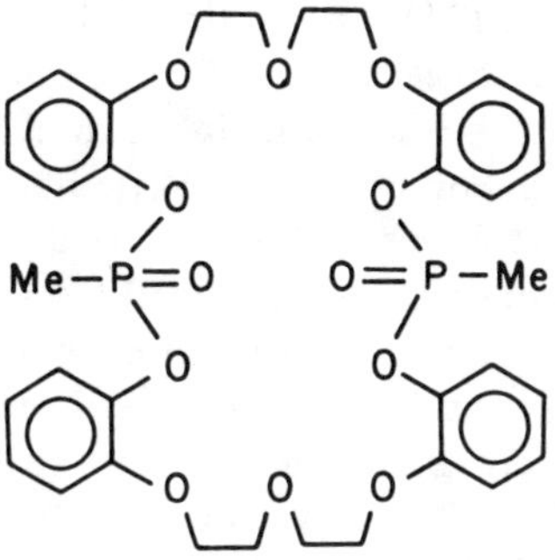

III

The resonances of the nonequivalent ethoxy groups of α-aminophosphonic acid esters (IV) were resolved in the presence of LSR.[943]

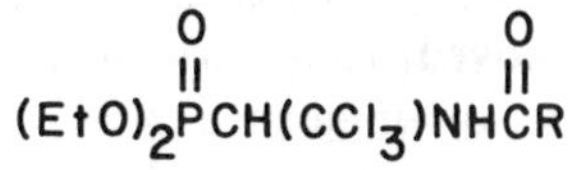

$(EtO)_2\overset{O}{\overset{\|}{P}}CH(CCl_3)NH\overset{O}{\overset{\|}{C}}R$

IV

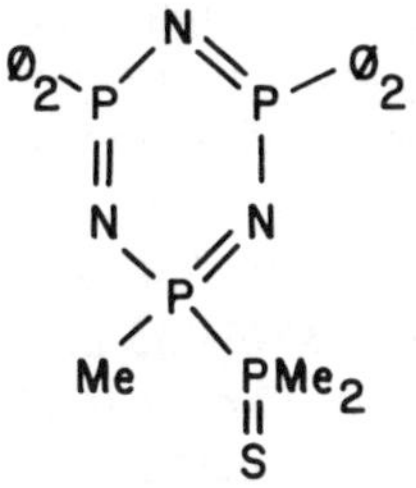

V

The P=S group exhibits a weak complexation with LSR. Coordination of a polyfunctional substrate with an amide and P=S group occurred through the carbonyl group of the amide.[520] The ^{31}P resonances of cyclotriphosphazene (V) were resolved in the presence of LSR.[942]

E. Polymers

A number of organic-soluble polymers have been studied with the aid of lanthanide tris chelates. LSR have been used to resolve the methylene resonances[966] and assess the content of *trans-gauche-trans* conformation present in polyethylene glycols (PEG).[967] Complexation occurred preferentially at the terminal hydroxy groups, but also took place to some extent at the ether oxygens.[966,967] It was reported that the shift reagent did not alter the conformation of PEG.[967] LSR have been used to determine the molecular weight of polypropylene glycols.[968] The shift reagent preferentially bonds at the terminal hydroxy groups, and resolution of the end group methyl resonances was obtained. Three overlapping methyl doublets were resolved in the spectrum of polypropylene oxide in the presence of Eu(fod)$_3$ and Pr(fod)$_3$.[969] Resolution of end and center methylene resonances in formaldehyde oligomers was noted in the presence of Eu(dpm)$_3$.[966] The distinction results from competition of the end and center oxygen atoms for coordination to the shift reagent.

Copolymers of formaldehyde and ethylene oxide have been studied with LSR.[966,970,971] The shift reagent enhanced the resolution of methylene resonances and permitted aspects of the sequence to be determined.[970,971] The copolymerization of ethylene and propylene oxide with n-butanol was studied with the aid of Eu(dpm)$_3$.[972] It was possible to determine that the ethylene residues were closer to the terminal n-butoxy groups. The orientation of the isopropylidene group in the propylene derivative was also assessed on the basis of the shift data.[972]

The isotactic, heterotactic, and syndiotactic triads of poly(vinylmethylether) were determined on the basis of lanthanide shift data.[969,973] Resolution of the methoxy resonances of the triads was obtained in the spectrum with the shift reagent. The shifts in the spectrum of poly(vinylethylether) with LSR were smaller than the methyl compound.[969] This was believed to result from steric effects.

Poly(methylmethacrylate) (PMMA) has been the focus of a number of studies with LSR.[974-979] The methoxy resonances of the isotactic, heterotactic, and syndiotactic triads of atactic PMMA were resolved in the presence of a LSR.[976,979] It was also possible to determine the pentad tacticity of PMMA.[979] The conformations of isotactic[974,975,978] and syndiotactic[975,978] PMMA have been assigned on the basis of lanthanide shift data. It was shown that the nature of the solvent influenced the conformation of the polymer.[975] The resolution of the triad sequences of poly(tert-butylmethacrylate) was not significantly improved, however, in the presence of Eu(fod)$_3$.[977]

The spectra of copolymers of styrene[980] and chloroprene[981] with methacrylate have been examined in the presence of LSR. Aspects of the sequence distribution could be determined through analysis of the shifted spectra. With the chloroprene derivative, six methoxy signals, which were assigned to the six pentad sequences, were observed in the presence of the LSR.[981]

Resolution of the isotactic, heterotactic, and syndiotactic triads of poly(vinylacetate) was observed with Eu(fod)$_3$ and Pr(fod)$_3$.[969] The spectra of vinyl acetate copolymers with vinyl chloride[982,983] and ethylene[984] have been studied with LSR. It was possible to resolve and assign the diad and triad sequence of these polymers through analysis of the shifted spectra. Similar findings were obtained with vinylchloride-vinyl propionate copolymers.[982,983] In certain instances, superior results were obtained by monitoring the shifts in the ^{13}C NMR spectra.[982,983] Shift data with Eu(fod)$_3$ has been used to deduce the monomer distribution and determine the relative concentration of the triad sequences in the terpolymer prepared from vinyl chloride, vinyl acetate, and ethylene.[983]

The shifts in the spectra of acetylation products of phenol-formaldehyde resins with a LSR facilitated assignment of the aromatic hydrogen atoms.[412] Through this study, it was possible to determine the degree of polymerization. The conformation of the repeating unit in the homopolymer D-poly-β(−)-hydroxybutyrate was determined by fitting shift data with Eu(fod)$_3$.[985,986] Signals from the different sequences of poly(ethyleneterphthalate) and poly(ethyleneisophthalate) were distinguished on the basis of shift data with a LSR.[987] It has been reported that the diad, triad, and tetrad signals of copolyterphthalate were resolved with Eu(fod)$_3$ and Pr(fod)$_3$.[988] A study of the equilibrium and rate constant of the polymerization of dialkyladipates and glycols was facilitated through the application of LSR.[989]

The shifts in the ^{13}C spectrum of the natural, isotactic, and atactic forms of poly(β-methyl-β-propiolactone) with Eu(dpm)$_3$ have been used to assign the diad tacticity.[990] Larger shifts were observed for the signals of the *meso* diad than for the *racemic* diad. The percentage of *meso* and *racemic* diad configuration in poly[β-(2-cyanoethyl)β-propiolactone] was determined using Eu(dpm)$_3$.[990]

Shift data with Eu(fod)$_3$ has been used to determine the concentration of triads in isoprene-nitrile[991] and butadiene-nitrile[992] rubbers. The microtacticity of atactic polyvinyl pyridine was assessed with the aid of Eu(dpm)$_3$ and Pr(dpm)$_3$.[993] The conformations of stereoregular polyamides(I) were assigned on the basis of a study with Eu(fod)$_3$.[946]

I

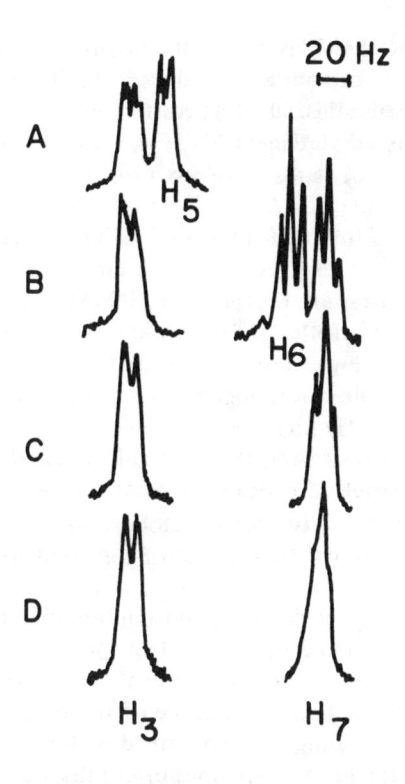

FIGURE 4. The metamorphosis of the H_3 and H_7 resonances of quinoline with increasing Eu(dpm)$_3$/quinoline ratio: (A) 0.193, (B) 0.276, (C) 0.378, (D) 0.588. (Reprinted with permission from Reuben, J. and Leigh, J. S., Jr., *J. Am. Chem. Soc.*, 94, 2789, 1972. Copyright 1972, American Chemical Society.)

No shifts were observed in the spectra of methylphenyl- and methylhydrogen polysiloxane with Eu(fod)$_3$ and Pr(fod)$_3$.[969]

IV. INFLUENCE OF LSR ON COUPLING CONSTANTS

One common application of LSR is to obtain values of coupling constants from the more disperse spectrum. Caution must be advised, however, since the coupling constants of the free substrate are not always the same as those of the complexed substrate. Changes in the coupling constants of a substrate upon complexation with a LSR arise from two mechanisms.

The first, and most common, occurs when the conformation of the substrate is perturbed upon complexation. In this situation, the dihedral angle, or time averaged dihedral angle, between coupled protons is altered. The second mechanism occurs because the lanthanide ion is a Lewis acid and withdraws electron density from the donor.[353] This mechanism explains the changes in coupling constants observed on complexation of camphor, a rigid substrate, with a LSR.[353] The relative change of coupling constants on complexation has been used to demonstrate that the chelates with fod are stronger Lewis acids than those with dpm.[80,353]

A second phenomena that is sometimes observed with paramagnetic lanthanide chelates is a loss of coupling in the shifted spectrum.[71] This is most likely the result of chemical exchange spin decoupling.[71] An example is shown in Figure 4 for the H_5 and H_7 resonances of quinoline.[71] The multiplet nature of the resonances collapses as the concentration of shift reagent is increased.

A method to determine the signs of coupling constants that employs LSR has been described.[994] The sign of a coupling constant can be determined in second or higher order spectra. The procedure described by Roth seems to contradict every other application of LSR in that the shift reagent is used to coalesce the resonances of the substrate and convert a first-order spectrum into a second-order spectrum.

V. SEPARATION OF DIASTEREOTOPIC PROTONS

The resonances for diastereotopic protons coincide in the spectra of many compounds. In the presence of a LSR, different geometric terms (r and Θ) may occur for diastereotopic protons causing a pronounced difference in chemical shifts. The coupling constants obtained in the resolved spectra often provide useful information in the determination of molecular conformation.

Examples include the resolution of the methylene resonances of amphetamine[478] and of the ethyl group in 2-ethylcyclohexanone.[358] In the latter study, it was also found that the methyl resonances of 2-isopropylcyclohexanone were resolved in the presence of the shift reagent. The methyl resonances of the isopropyl group of the hemithioethylene ketal of methyl isopropyl ketone were resolved in the presence of a LSR.[898] Diastereotopic protons of the substituent groups of β-lactams have been resolved in the presence of LSR.[809,812] Diastereotopic resolution in substrates with ester[404] and epoxide[447] functional groups has been noted.

A number of reports have described enhanced resolution of diastereotopic protons of acyclic alcohols in the presence of LSR.[170,173,684,995,996] No substantial change of the coupling constants was observed in the presence of the shift reagent, implying that no conformational perturbation occurred upon complexation.[173] For many of the acyclic alcohols, resolution of more than one set of diastereotopic protons was obtained.[170,173] Resolution of the diastereotopic protons α and β to the sulfinyl group of di-n-butylsulfoxide was achieved using a LSR.[573]

VI. SECONDARY DEUTERIUM ISOTOPE EFFECT

LSR have been applied on numerous occasions in determining the site and percentage of deuterium substitution. The first observation of a secondary deuterium isotope effect with LSR was reported by Smith et al.[85] In a study of verbanol(I),

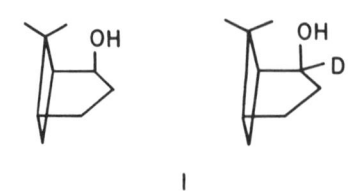

I

a doubling of all of the resonances was observed in the proton NMR spectrum with Eu(dpm)$_3$, py$_2$.[85] The doubling of resonances was caused by differences in the shifts of the spectra of 4-deuterioverbanol and verbanol in the presence of the shift reagent.

Secondary deuterium isotope effects with LSR have since been noted for other alcohols,[86,184,203,997] glymes,[876,877,998] aldehydes,[85,999] ethers,[999] amines,[999] and pyridine derivatives.[1000,1001] In each instance, the shifts for the deuterated derivative were larger than those for the nondeuterated compound. Increasing the deuterium substitution increased the shifts as evidenced by the ordering of the shifts for a series of ethanol derivatives: $CH_3CD_2OH > CH_3CDHOH > CH_3CH_2OH$.[86] At high shift reagent concentrations, under the absence of competition, a coalescence of the resonances of the deuterated and nondeuterated substrates is commonly observed.[997,998] Deuterium substitution apparently increases the association constant between the substrate and shift reagent. The mechanism that causes the larger association constant in deuterated derivatives is not known.

One possibility is that the increased inductive effect of deuterium relative to hydrogen increases the base strength of the donor. A second is that the hydrogen atoms geminal to the donor group form a hydrogen bond with oxygen atoms of the chelate ligand.[85,86,184] Such a hydrogen bond would be stronger for the deuterium isotope. A third possibility is that the difference results from steric effects. This reasoning has been used to explain the differences in the shifts between methylpyridines and deuterated derivatives in the presence of LSR.[1000,1001]

The secondary deuterium isotope effect, while perhaps of not much practical utility, represents an interesting example of the subtle differences in substrates that can be distinguished in the presence of LSR.

VII. CHEMICALLY INDUCED DYNAMIC NUCLEAR POLARIZATION (CIDNP)

Applications of LSR in CIDNP studies have been described in the literature.[1002,1003] The reason for employing LSR in CIDNP studies is the same as in conventional NMR, to remove accidental degeneracy of signals. There are a number of complications, however, that limit the use of LSR in CIDNP studies. LSR quench the CIDNP signals, so must be employed at low concentrations.[1002,1003] LSR cannot be employed in photochemically initiated reactions because they are efficient quenchers.[1003] One final caution involves the fact that the concentration of products is continually increasing in a CIDNP experiment, while the concentration of shift reagent remains constant. As a result, the locations of the signals vary over the course of the experiment.[1002,1003] The application of LSR in CIDNP experiments was demonstrated by studying the decomposition of a mixture of benzoyl propionyl peroxide and the m-chloro derivative.[1002] The overlap of the methylene signals of ethyl benzoate and m-chloroethyl benzoate was removed in the presence of the shift reagent.

Chapter 3

STUDIES OF CHIRAL SUBSTRATES WITH LANTHANIDE TRIS CHELATES

Lanthanide tris chelates with optically pure ligands are powerful tools for the study of optically active hard Lewis bases. If an optically pure compound interacts with a pair of optical isomers, two diastereomers result, and these may exhibit different chemical shifts in an NMR spectrum. The differences in chemical shifts are often accentuated when the optically pure compound is a lanthanide chelate. Separation of the resonances of optical isomers under these circumstances results from differences in the geometric term of the pseudocontact shift equation. That is, a particular proton of a pair of optical isomers complexed to a lanthanide shift reagent has a different r and/or Θ value in the *dextro* and *levo* enantiomer.

A second mechanism that would also result in enantiomeric resolution of the resonances of optical isomers in the presence of a chiral LSR is if the association constants for the *dextro* and *levo* enantiomers are different. In this case, the resonances for the enantiomer with the larger association constant would exhibit greater shifts. Since this mechanism requires competition between the substrates for coordination sites on the LSR, it is more apt to be operative at low shift reagent concentrations. As will be pointed out, most studies that have assessed the mechanism that caused enantiomeric resolution with chiral LSR have attributed it to geometric differences.

The first report that described the utility of chiral LSR for enantiomeric resolution was by Whitesides and Lewis in 1970.[4] The Eu(III) complex with the ligand 3-(*tert*-butylhydroxymethylene)-*d*-camphor was employed in this report. Improved chiral LSR have since been developed and will be described in the section on shift reagent selection.

The major use of chiral LSR is in the determination of optical purity or enantiomeric excess. The interaction between chiral LSR and enantiomers cannot be defined with enough certainty to permit assignment of absolute configuration. Substrate to substrate variability reduces the ability to assign absolution configurations through the study of known compounds to all but the most similar of derivatives. Procedures for assigning the absolute configurations of certain classes of compounds have been developed. Interestingly, these procedures rely on the use of achiral LSR. Since they involve the distinction of chiral substrates, however, they are described in this chapter.

I. SHIFT REAGENT SELECTION

A. Metal

Criteria governing the selection of a particular lanthanide ion for use in chiral LSR are much the same as previously described for achiral applications. Since the enantiomeric resolution of a substrate in the presence of a chiral LSR is frequently quite small, broadening must be kept to a minimum. For this reason, chelates with europium are the obvious first choice. Acceptable levels of broadening are often obtained with chelates of praseodymium and ytterbium, and these may find use in certain applications.

B. Ligands

Since Whiteside's first report involving chelates with the ligand 3-(*tert*-butylhydroxymethylene)-*d*-camphor, H(*t*-cam),[4] a variety of chiral ligands have been employed in shift reagents. Many of these ligands are derivatives of camphor(I). R groups employed in these ligands,

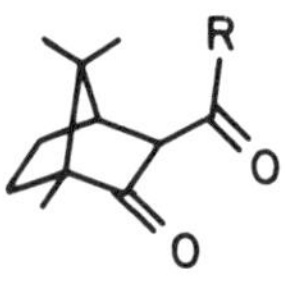

I

in addition to the *tert*-butyl group,[4-7,1004] include methyl,[6] phenyl,[6] pentafluorophenyl,[1005] trifluoromethyl,[5,7,1006] pentafluoroethyl,[1007] heptafluoropropyl,[5,6,1006,1008] formyl,[1006] carbethoxy,[1006] *d*- and *l*-fenchyl(II),[7] and III.[1004] It is the

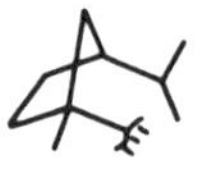

II 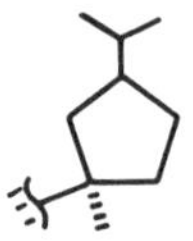   III

concensus of these reports that the $-CF_3$ derivative, [3-(trifluoroacetyl)-*d*-camphor], [H(facam)], and $-C_3F_7$ derivative (3-(heptafluorobutyryl)-*d*-camphor, [H(hfbc)] are the most effective of the ligands derived from camphor.[5-7,1006] It is worth noting that a number of shorthand notations in addition to "facam" and "hfbc" are used in the literature to denote these ligands. In comparing chelates of hfbc and facam, the shifts are usually larger with chelates of hfbc than facam, but the degree of enantiomeric resolution exhibits no consistent pattern.[1006] The chelates of hfbc are more effective at resolving diastereotopic protons than those of facam.[1006]

Other cyclic ketones such as *l*-menthone,[7] carvone,[1006] pulegone,[1006] and *d*-nopinone(IV)[7,1005] have been derivatized to β-diketone ligands. The only ones of these worthy of special mention are the trifluoroacetyl-[7] and heptafluorobutyryl-[1005] derivatives of nopinone. Goering et al. noted that the chelates of 3-heptafluorobutyryl-*d*-nopinone were superior to those of hfbc or facam.[1005] McCreary et al. rated chelates of 3-trifluoroacetyl-*d*-nopinone as superior to those of facam.[7]

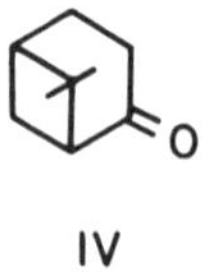

IV

Chelates of the ligands 3,7-dimethyl-3,7-diphenyl-4,6-nonanedione[1006] and 3,7-dimethyl-4,6-nonanedione[1009] have been evaluated as chiral LSR. Chelates of β-diketone ligands containing substituted cyclopentyl groups have also been evaluated.[1004] None of these examples offers any unique advantages over other available chiral LSR.

Kawa compared the enantiomeric resolution in the spectrum of 1-phenylethylamine with Eu(hfbc)$_3$ to that with the Eu(III) complex with di(perfluoro-2-propoxypropionyl)methane (V).[1010]

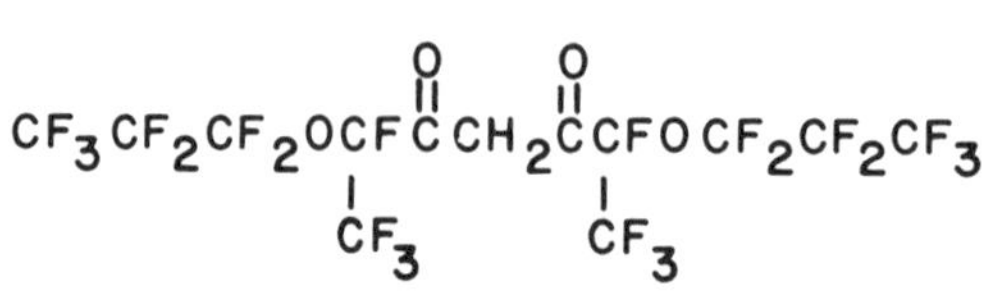

$$CF_3CF_2CF_2OCFCCH_2CCFOCF_2CF_2CF_3$$
$$\quad\quad\quad\; |\;\;\;\;\;\;\;\;\;\;\;\;\;\;\; |$$
$$\quad\quad\quad CF_3\quad\quad\;\; CF_3$$

V

The chelate with V produced superior enantiomeric resolution. This ligand is also of special interest because it has only the hydrogen resonance of the methine position to overlap with the spectrum of the substrate.

A final set of chiral ligands deserving mention were described in a paper by McCreary et al.[7] These ligands consisted of various campholyl and fencholyl derivatives of methane. In particular, complexes with the ligand d,d-dicampholylmethane(VI), (H[dcm]) were especially effective at bringing about enantiomeric resolution in the spectra of substrates. Only

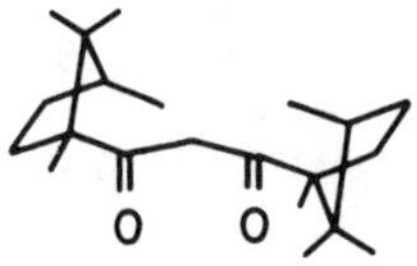

VI

a few workers have utilized Eu(dcm)$_3$, but all of the information available indicates its superiority over other chiral LSR.

To summarize, lanthanide complexes with the ligands facam, hfbc, trifluoroacetyl- and heptafluorobutyryl-d-nopinone, di(perfluoro-2-propoxypropionyl)methane, and especially dcm merit special note as effective chiral shift reagents. The chelates with facam and hfbc are by far the most widely used. In many respects, this is more a measure of their ease of preparation and widespread commercial availability rather than superior effectiveness. Based on optical resolution alone, Eu(dcm)$_3$ is clearly the reagent of choice. Students working in my laboratory have reproduced the literature preparation of H(dcm) with no difficulty.[7] In addition, H(dcm) and the Eu(III) complex with dcm recently have become available commercially.

II. EXPERIMENTAL TECHNIQUES

A. Solvent

The shift difference between enantiomers ($\Delta\Delta\delta$) in the presence of chiral LSR is typically small, necessitating the use of noncoordinating solvents. Goering et al. have noted larger $\Delta\Delta\delta$ values in CCl$_4$ compared to CDCl$_3$.[1006] McCreary et al. reported that larger $\Delta\Delta\delta$ values were obtained in pentane, 1,1,2-trichloro-1,2,2-trifluoroethane, and CCl$_4$ compared to CFCl$_3$, CS$_2$, benzene, CHCl$_3$, and CH$_2$Cl$_2$.[7] One limitation to the use of pentane as the solvent is that the hydrogen resonances of pentane overlap with regions of the spectrum from about 1 to 3 ppm.

B. Preparation and Purification

Procedures for the preparation of chiral LSR have been described in the first reports describing their use. The procedures will not be elaborated on here. An improved access to chelates with facam has been reported.[1011] In this process, a homogeneous exchange is carried out between Ba(facam)$_2$ and a lanthanide salt. The Ba(facam)$_2$ is prepared by adding Ba^{2+} to the reaction mixture, and isolation of the ligand is not necessary. This procedure is probably applicable with most, if not all, chiral ligands.

The purity of chiral LSR has not been an issue discussed in the literature as much as the purity of achiral analogues. Suffice it to say that impurities in the shift reagent or solvent, such as water and other scavengers, will reduce the shift values and enantiomeric resolution. Commercially obtained chiral LSR should at least be dried *in vacuo* over P$_4$O$_{10}$ for 24 hr and then stored over P$_4$O$_{10}$. A better procedure, however, is to dry the chelates at elevated temperatures (100°C) *in vacuo* over P$_4$O$_{10}$. This is most conveniently carried out in an abderhalden.

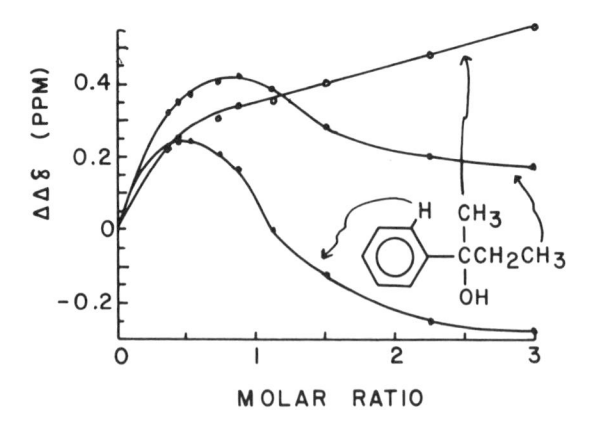

FIGURE 1. Plots of ΔΔδ vs. molar ratio for the designated protons of 2-phenyl-2-butanol in the presence of 0.3 M Eu(hfbc)$_3$. (Reprinted with permission from Goering, H. L., Eikenberry, J. N., Koermer, G. S., and Lattimer, C. J., J. Am. Chem. Soc., 96, 1493, 1974. Copyright 1974, American Chemical Society.)

C. Procedures for Using Chiral LSR

Chiral LSR are best employed by obtaining a series of spectra as increasing amounts of the shift reagent are added to a solution of the substrate. There are several reasons for such an approach. The first is that a series of spectra permit a better distinction between enantiomeric resolution and resolution of resonances of chemically different protons. In many cases, it is advantageous to first record the shifts in the spectrum of a substrate with an achiral shift reagent. The spectrum with an achiral shift reagent is usually easier to assign. The spectrum with an achiral LSR also permits an assessment of the relative shifts of the resonances. For polyfunctional substrates, the preferred site of complexation can often be determined more readily with an achiral LSR.

The second reason for recording a series of spectra with chiral LSR is that the degree of enantiomeric resolution sometimes exhibits a marked variability with L:S ratios. A small degree of enantiomeric resolution at a low L:S ratio will not necessarily improve at higher L:S ratios. An excellent example of this phenomenon can be seen by the shift data presented in Figure 1 for 2-phenyl-2-butanol with Eu(hfbc)$_3$.[1006] The values of ΔΔδ for the individual resonances are plotted vs. the L:S ratio. The results for the proton *ortho* to the substituent group are especially interesting. At low ratios, enantiomeric resolution is observed for this resonance. As the ratio is raised, the resonances gradually coalesce. Increasing the ratio further causes a reversal of the magnitude of the shifts and enantiomeric resolution is once again present. The authors believe this observation results from changes in the stoichiometry of the shift reagent-substrate complex.[1006] The relative shifts of the resonance in the *dextro-* and *levo-* enantiomer are different in the LS and LS$_2$ complexes. A similar behavior has been noted for other substrates.[7]

As the concentration of shift reagent is increased, it is not uncommon to observe an insoluble material.[7] This could either come from insoluble oxides or hydroxides in the shift reagent or by a reaction of the chelate with impurities in the solvents. The broadening that results from this solid material may be enough to eliminate or impair the enantiomeric resolution in the spectrum. Any insoluble material should therefore be removed by filtration prior to recording the spectrum.

A second caution in using chiral LSR is not to confuse the resolution of hydrogen resonances on a prochiral carbon with enantiomeric resolution.[7,1006] Chelates with hfbc are especially effective at resolving diastereotopic protons.[1006]

The temperature of the sample has also been shown to influence the magnitude of $\Delta\Delta\delta$.[6,7] Lower temperatures generally enhance the value of $\Delta\Delta\delta$. For weakly bonding substrates such as 2-nitrobutane, 2-butanethiol, and 2-cyanobutane, negligible values of $\Delta\Delta\delta$ were noted at ambient temperatures. The magnitude of $\Delta\Delta\delta$ became appreciable, however, at -50 to $-75°C$.[7] For substrates that bond strongly such as amines, very little improvement in $\Delta\Delta\delta$ was obtained on lowering the temperature.[7] Raising the temperature has proved advantageous in some studies with chiral LSR.[1012,1013] These have typically involved substrates for which severely broadened spectra were obtained with the shift reagent at ambient probe temperatures. The broadened spectra may be the result of a strong chelate interaction of the substrate and LSR. Raising the temperature speeds up the exchange of the substrate and results in sharper peaks. Enantiomeric resolution was then observed in the spectrum.[1012,1013]

A procedure to quantify a mixture of enantiomers when only partial enantiomeric resolution is obtained in the shifted spectrum has been described.[1014] The method requires fairly rigorous control of certain experimental conditions to be accurate. Its use was demonstrated on a variety of substrates with different functional groups.

III. ASSIGNMENT OF ABSOLUTE CONFIGURATION

Two general methods for assigning the absolute configurations of optical isomers using LSR have been reported in the literature. The first involves a study of the circular dichroism spectrum of a chiral substrate in the presence of an achiral LSR.[1015-1019] Characteristic Cotton effects that can be correlated to absolute configuration are noted in the spectrum. The second procedure is to derivatize the *dextro* and *levo* enantiomers with an optically pure derivatizing agent. Two diastereomers result, and resonances of the diastereomers may be resolved in the NMR spectrum. In many cases, addition of a lanthanide ion enhances the resolution of the resonances of the diastereomers.[1020] The relative magnitude of the shifts of the two diastereomers in the presence of the shift reagent can often be correlated with the absolute configuration. The most common derivatizing agent is Mosher's reagent,[1021] α-methoxy-α-trifluoromethylphenylacetic acid (MTPA), but others have been used as well.

A. Circular Dichroism Spectra

The circular dichroism spectra of certain substrates with $Pr(dpm)_3$ exhibit Cotton effects that can be correlated with the chirality of the substrate. While a wide variety of compounds complex with $Pr(dpm)_3$, this method will not work for all substrates and is most suited to bifunctional compounds such as cyclic and acyclic α-diols,[1015-1018] α-hydroxyamines,[1018,1019] α-aminoketones,[1019] and α-hydroxymonoacetates.[1015] Configurational assignment has also been achieved for certain monoamines.[1019] The success with the bifunctional substrates is believed to result because of strong chelate bonding with the shift reagent.[1018]

Reports by Nakanishi and Dillon[1015,1018] provide the best description of the method and include a discussion of the variables and limitations involved. The solvents n-hexane, CCl_4, and $CDCl_3$ have been found to be suitable for this method, but must be rigorously dried before use.[1015,1018]

B. MTPA Esters

Diastereomers resulting from the derivatization of primary carbinols,[696,1022-1024] secondary carbinols,[696,1024-1035] amino acids,[1036] biaryls,[1037-1040] and hydroxycarboxylic acid esters[1025,1041] with MTPA have been studied with the aid of LSR. The different shifts for the diastereomers in the presence of an achiral lanthanide shift reagent frequently permits assignment of the absolute configuration of the original substrate.

Yashuhara and Yamaguchi studied the MTPA esters of 12 primary carbinols with a chiral center at the C-2 position.[1022] The methoxy resonances of the diastereomers were resolved in the presence of a LSR, and larger shifts were observed for the (R) derivative. The differences in lanthanide shift values were believed to result from differences in the association constants caused by steric effects of the diastereomers. Yamaguchi and Mosher[1024] and Reich et al.[1023] recorded the spectra of MTPA derivatives of primary carbinols with Eu(fod)$_3$ and were able to determine the enantiomeric purity of the carbinols. The resolution of the signals of the diastereomeric esters with Eu(fod)$_3$ was larger than the resolution of the carbinol enantiomers with chiral LSR. The absolute configurations of the C-3' and C-5 position of (4R, 5S, 3'R)-3,3-dimethyl-4-(3'-hydroxybutyl)-5-hydroxymethylcyclohexan-1-one(I) were assigned by studying MTPA derivatives with the aid of LSR.[696]

I

It was possible to individually block each hydroxyl group by selective acetylation and to assign the configuration about the other by preparing the MTPA ester.

Yamaguchi et al. studied the spectra of the MTPA esters of 32 secondary carbinols with Eu(fod)$_3$.[1026] The methoxy signal was the most convenient to monitor for diastereotopic resolution, and in general, larger shifts were observed for the ester resulting from the R enantiomer. The different shifts of the diastereomers in the presence of the LSR were believed to be due to different association constants resulting from steric effects. Since the steric bulk of the substituent groups on the chiral carbon is also used to determine the R or S configuration, it is essential that one be able to rank these substituent groups in order of steric bulk. Ambiguities in this ordering will cause ambiguities in determining which diastereomer should exhibit the larger shifts in the presence of the LSR. Yamaguchi et al. reported that 4 examples of the 32 studied did not follow the general correlation scheme.[1026] Other violations to the general rule that the shift of the (R)-methoxy resonance in the presence of a LSR is greater than that of the (S)-methoxy resonance have been noted.[1034,1035] These include the compounds cis-3-tert-butylcyclohexanol, trans-carveol, cis-3-methylcyclohexanol, and alpha- and beta-tetralol.[1034] The lanthanide shift data for the MTPA derivatives of trans-3-tert-butylcyclohexanol exhibited the expected behavior.[1034]

Studies of MTPA derivatives of 2-substituted,[1031,1035] 3-substituted,[1027,1031] and 4-substituted cyclohexanols,[1028,1029] bicyclic and tricyclic secondary carbinols,[1032] and steroids[1033] have been carried out with the aid of LSR. The conformations of MTPA derivatives of cis- and trans-4-tert-butylcyclohexanols were assigned on the basis of lanthanide shift data. The data indicated the presence of two rotamers in solution.[1028,1029] These reports cautioned against using models involving only one rotamer to predict the relative shifts of the diastereomers in the presence of a LSR. They also found no evidence for a chelate bond involving the ester carbonyl and methoxy oxygen atoms, as stated in another report.[1022]

A number of reports have advocated monitoring the shifts in the ^{19}F spectra of the MTPA derivatives with LSR,[1030-1032,1035] since better resolution of the signals of the diastereomers was observed in the ^{19}F spectrum. The ^{19}F data exhibit the same general trend as the ^{1}H data in that larger shifts are noted for the diastereomer prepared from the R-alcohol.[1031,1032,1035] The configuration of an enantiomerically pure substrate can be assigned by preparing derivatives from (R)–$(+)$ and (S)–$(-)$-MTPA and recording the shifts with Eu(fod)$_3$.[1026]

The absolute configuration of the carbinol carbon in 2- and 3-hydroxy carboxylic acid methyl esters was assigned on the basis of lanthanide shift data of MTPA derivatives.[1025] The substrate reportedly coordinated through both ester carbonyl groups; however, this did not interfere with assignment of the configuration. A wide range of substrates was studied and larger shifts were consistently obtained for the diastereomer of the S-carbinol.[1025] The configuration of methyl-13-hydroxy-9-*cis*-11-transoctadecadienoate was assigned on the basis of shift data for the MTPA derivative with Eu(fod)$_3$.[1041]

The absolute configuration of amino acids have been assigned by analysis of shifts with a LSR in the spectrum of the MTPA derivative of amino acid methyl esters.[1036] Derivatization of amino acid methyl esters with MTPA converts the amine group to an amide. Preferred complexation of the shift reagent reportedly occurs at the carbonyl group of the amide functionality. Using (S)–$(-)$-MTPA as the derivatizing agent, larger shifts were always noted for the derivative of the (S)-amino acid unless the substituent group on the alpha carbon was either a benzyl or substituted benzyl group.[1036] Other chiral centers in the amino acid did not interfere with the assignment of the configuration of the alpha carbon. It was not necessary to have both enantiomers to assign the absolute configuration.

The spectra of MTPA derivatives of axial chiral biaryls have been recorded in the presence of LSR.[1037-1040] These include biphenyl, binaphthyl, bianthryl, and bridged biphenyl derivatives. For each substrate, either a hydroxy or amine group was derivatized with MTPA. The data with the LSR were suitable for determining optical purity or assigning absolute configuration.[1037,1038,1040] Assigning the absolute configuration involved a consideration of the relative shifts of the methoxy resonances of the diastereomers and of steric effects in the derivatives. For amide derivatives, unacceptable broadening was observed with Eu(fod)$_3$, and the use of Eu(dpm)$_3$ was recommended.[1039]

C. Camphanate Esters

Derivatization of chiral primary[1042-1045] and secondary carbinols[1046] with $(-)$-ω-camphanic acid chloride (I) produces diastereomers whose spectra can be resolved in the presence of LSR.

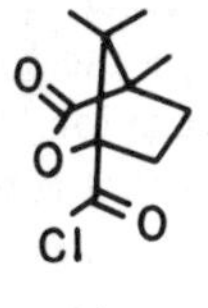

I

Resolution of the diastereomers in the unshifted spectrum is sometimes observed, but is enhanced through the use of a LSR. The methyl resonances of the camphanoyl group are particularly useful signals to monitor.[1042] Characteristic differences in the magnitude of the shifts for the derivatives of the R and S enantiomers are noted.

The absolute configurations of *trans*-6-hydroxymethyl-2-methoxy-5,6-dihydro-2H-pyran(II),

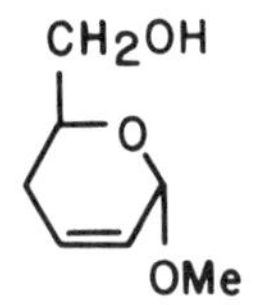

II

2-phenyl-1,2-dideuterioethanol,[1043] 1-deuterio-1-octanol,[1045] and other alpha-deuterated primary alcohols have been assigned by analyzing their camphanate esters with LSR.[1044] Jurczak and Konowal have reported the results of studies of a wide range of camphanate esters of primary and secondary carbinols with Eu(fod)$_3$.[1042,1046]

D. Preparation of Amides

Optically pure amines such as (R)-phenylethylamine can be used to prepare diastereomers from chiral acid chlorides[1047] and carboxylic acids.[1048] Chiral carboxylic acids have also been derivatized to diastereomers with (+)-sec-butylamine.[1048] An optically pure carboxylic acid such as O-methylmandelic acid can be used to prepare diastereomers from chiral amines. Addition of a LSR to the resulting mixture of diastereomers enhances the resolution, and the different shifts can be used to determine optical purity[1048] and assign absolute configurations.[1047] The absolute configuration of 2-[4-(1-oxo-2-isoindolinyl)-phenyl]propionic acid(I) was assigned by analyzing the spectrum of the derivative with (R)-phenylethylamine with Eu(fod)$_3$.[1047] In this work, the magnitude of the lanthanide shifts was compared to those in the spectra of similar derivatives of known configurations.

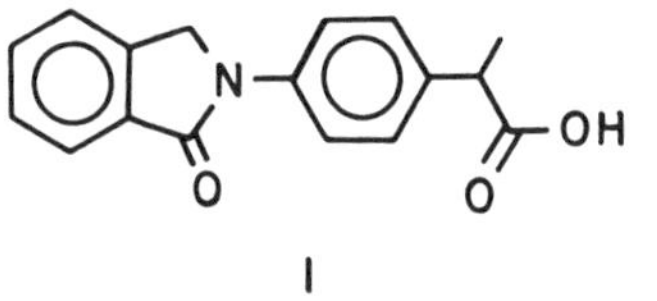

I

E. Other Derivatives

Furukawa et al. have described a procedure for assigning the absolute configuration of 3-alkylpropanols deuterated at the 2- or 3-position.[1049,1050] The procedure involves the preparation of a 1-phenyl-3-alkylpropanol derivative. In the aryl derivative, the methylene hydrogen atoms at C-2 and C-3 are diastereotopic, and resolution of the four protons is observed in the presence of Eu(fod)$_3$.

RCH$_2$CH$_2$CH$_2$OH $\longrightarrow$ RCH$_2$CH$_2$CHOH

The validity of the method was demonstrated by studying a series of compounds with known configurations.[1049] The method has been applied in a study of deuterium incorporation into the polyketide 2-n-hexyl-5-n-propylresorcinol by bacteria.[1050] This study employed ^{2}H NMR in the analysis.

IV. DETERMINATION OF OPTICAL PURITY

A. Enantiomeric Resolution with Achiral LSR

Several reports have noted enantiomeric resolution in the spectra of chiral compounds in solutions containing an achiral LSR and an optically pure additive.[7,477,1051-1053] This observation was first reported by Jennison and Mackay[1053] using Eu(fod)$_3$ and Eu(dpm)$_3$ in the chiral solvents (−)-2,2,2-trifluorophenylethanol and (+)-1-phenylethylamine. Enantiomeric resolution in the spectra of 1-phenylethylamine, amphetamine, methyl alanate, methylphenyl sulfoxide, and certain oxadiazines(I) was noted.

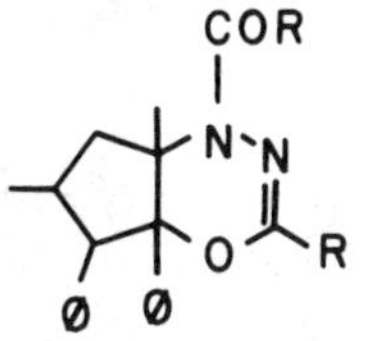

I

The enantiomeric resolution in the spectra of these substrates was at best comparable to that with Eu(hfbc)$_3$.[1053]

Two mechanisms may explain this process. The first is that the optically pure additive (A) complexes with the achiral LSR(L) to create a chiral LSR. This chiral species now complexes with the substrate(S) to bring about enantiomeric resolution.

$$L + A = LA$$
$$LA + S = LAS$$

This mechanism requires the formation of a 1:2 shift reagent-donor complex. The second explanation is that the chiral substrate and optically pure additive associate in some fashion, and the association constants of the enantiomers with the additive are different. The enantiomer that interacts less favorably with the chiral additive is more available for complexation with the LSR and exhibits larger shifts.

McCreary et al.,[7] and Ajisaka and Kainosho[477] favored the first mechanism in studies of amines. Ajisaka observed enantiomeric resolution in the spectra of partially resolved alkylamines such as α-phenylethylamine and α-(2-thienyl)ethylamine in the presence of Eu(fod)$_3$.[477] Whitesides observed enantiomeric resolution in the spectrum of α-phenylethylamine mixed with (R)-N-methyl-1-phenylethylamine and Eu(fod)$_3$.[7] Conclusive evidence for the first mechanism was obtained in a study of isopropylamine.[7] Enantiomeric resolution of the methyl resonances of isopropyl amine was observed with Eu(fod)$_3$ and (R)-N-methyl-1-phenylethylamine. Since these methyl groups are enantiotopic by internal comparison, resolution requires the interaction with an optically pure environment, and the second mechanism cannot be operable.

Pirkle and Sikkenga favored the second mechanism in studies of sulfoxide(II)[1051] and lactone(III)[1052] enantiomers with perfluoroalkyl carbinol additives and Eu(fod)$_3$.

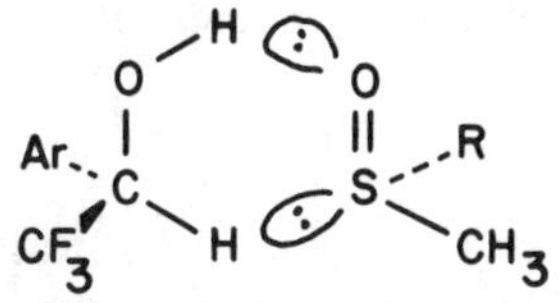

II 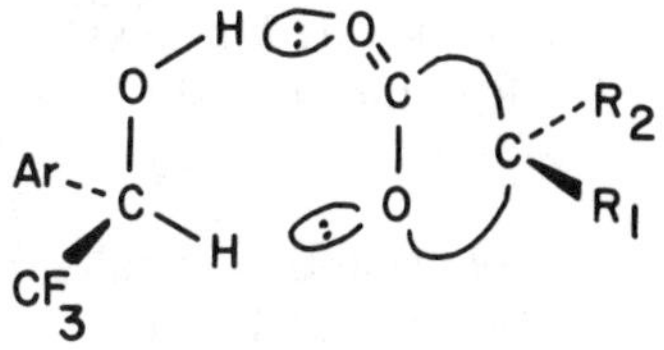   III

The complex between Eu(fod)₃ and the substrates is stronger than the complex between the additive and the substrates. At low concentrations of shift reagent, the Eu(fod)₃ "strips" away the more weakly hydrogen-bonded sulfoxide, and larger shifts are observed in its spectrum.[1051] At high concentrations of Eu(fod)₃, all of the substrate is essentially "stripped" from the additive, and enantiomeric resolution is no longer observed in the spectrum. Enantiomeric resolution was noted in the spectra of dimethylsulfoxide-d₃, methyl-*tert*-butylsulfoxide, methyl-*p*-tolylsulfoxide, and methyl-*p*-chlorophenylsulfoxide.[1051] Enantiomeric resolution was not observed in the spectra of simple chiral lactones exhibiting the interaction in III.[1052] For certain nitrated lactones (IV), however, enantiomeric resolution was noted in solutions with the chiral additive and Eu(fod)₃. This observation may reflect a third

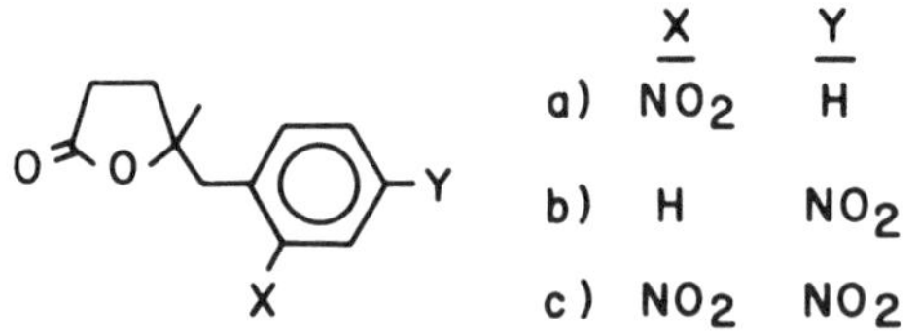

	X	Y
a)	NO₂	H
b)	H	NO₂
c)	NO₂	NO₂

IV

interaction between the substrate and additive involving the nitro group of the substrate and aromatic ring of the additive. This would increase the likelihood of observing differences in the association constants between the substrate and chiral additive.

These observations, while interesting, appear to be of little practical utility since the enantiomeric resolution is usually less than that with chiral LSR.[7,1053] Pirkle and Sikkenga have pointed out, however, that in instances in which the second mechanism is operative and the interaction between the chiral additive and substrate is well understood, the relative shifts with a LSR may permit the assignment of absolute configurations.[1051]

B. Applications of Chiral LSR
1. Alcohols

Enantiomeric resolution has been observed in the spectra of a wide variety of chiral substrates with a hydroxyl group in the presence of chiral LSR. On occasion, however, better enantiomeric resolution has been observed in the spectrum of the corresponding methyl ester.[1006,1054-1056] If direct study of a hydroxyl compound with a chiral LSR is not successful at achieving enantiomeric resolution, conversion to the corresponding ester is recommended.

The spectra of a number of derivatives of I have been recorded with chiral LSR.[1057-1060] The spectra of derivatives in which R is an acyclic aliphatic group,[1057-1059] cyclic aliphatic group,[1057] or aromatic

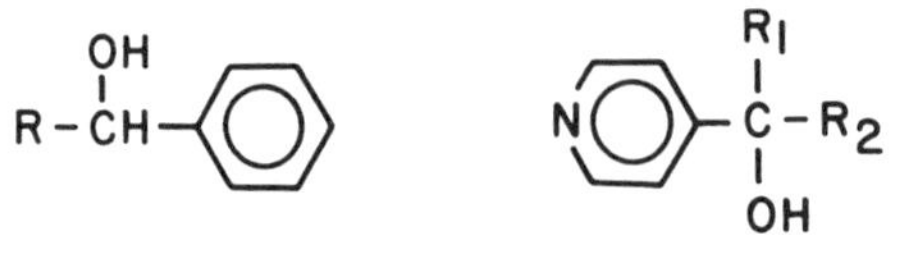

I
II

group,[1060] have been enantiomerically resolved. Hofer determined the T_1 values for the enantiomers of ephedrine (R = −CH[CH₃]NH[CH₃]) and methyl ephedrine (R =

—CH[CH$_3$]N[CH$_3$]$_2$) in solution with Gd(dcm)$_3$.[1058] Significant differences were observed in the T$_1$ values for the enantiomers. This observation implies that the geometry of the shift reagent-substrate complex is different for the *dextro* and *levo* enantiomers. The values observed for a series of benzhydrols (R = *p*-substituted phenyl group) were related to geometric differences between the shift reagent-substrate complexes, and these geometric differences correlated with electronic effects of the substituent group.[1060] Rules for assigning the absolute configuration of benzhydrols based on lanthanide shift data were presented in these reports. Enantiomeric resolution was not observed in the spectra of 1-pyridylalkanols(II) with Eu(facam)$_3$.[1061]

The enantiomeric purity of 2-phenylethanol,[1043,1062,1063] 2-phenylethanol-1,1,2-d_3,[1043,1062,1063] and 2-phenylethanol-1,2-d_2[1043] have been determined through the use of Eu(dcm)$_3$. The enantiomeric resolution with Eu(dcm)$_3$ was better than that with Eu(hfbc)$_3$[1063] or Eu(facam)$_3$.[1062] Enantiomeric resolution was not obtained in the spectrum of 2-phenylethanol-2-*d* with either Eu(facam)$_3$ or Eu(dcm)$_3$.[1062] The enantiomeric purities of 1-deuterio alcohols (III) in which R was a *t*-butyl, *i*-propyl, *n*-propyl, phenyl, adamantyl, or trichloromethyl group have been determined using Eu(hfbc)$_3$.[1023]

R—CHD—OH

III

In all cases, larger shifts were noted for the (R) configuration of III in the presence of the shift reagent.

Enantiomeric resolution in the spectra of β-hydroxy esters[1012,1064] and β-hydroxyketones[1009] has been achieved with chiral LSR. Severe broadening was observed in the spectrum of IV with Eu(facam)$_3$ at ambient probe temperatures.[1012] Heating the sample to 95°C resulted in a sharpening of the spectrum, and the CH$_3$ resonances were enantiomerically resolved.

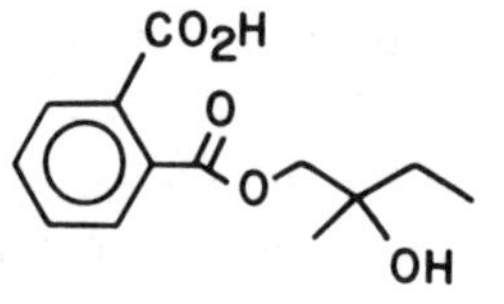

IV

Severe broadening has also been observed in the spectrum of the methyl ester of 2-hydroxy-2-methylbutanoic acid with Eu(facam)$_3$ and Eu(hfbc)$_3$ at ambient probe temperatures.[1065] The broadening eliminated the possibility of observing enantiomeric resolution in the spectrum. It is likely that the severe broadening in the spectra of these substrates is the result of bidentate complexation.

Chiral LSR have been applied in a study of the hydroxylated products(V) of 9-oxo-13(*cis*- and *trans*-)prostenoic acids.[1534] The enantiomeric purity

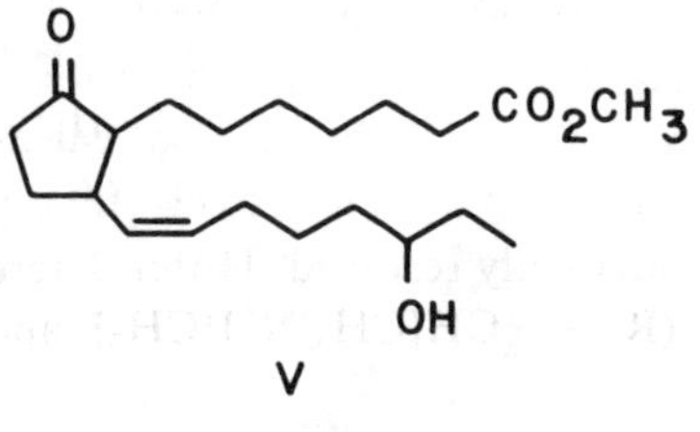

V

of lactones has been determined by first converting the lactone to its corresponding diol and then recording the spectrum of the diol with Eu(facam)$_3$.[1067] The procedure reportedly worked for all ring sizes and for substituted lactones. Enantiomeric resolution has been obtained in the spectra of 1,2-diglycerides and monoglycerides with Eu(hfbc)$_3$.[1068]

Enantiomeric resolution was observed in the ^{1}H and ^{2}H spectra of 2-[^{2}H$_3$]methyl-2-methylbutanol with Eu(hfbc)$_3$.[1069] The optical purity of VI was determined with the aid of Eu(hfbc)$_3$.[1069] The optical purity of linalool (VII) was assessed through analysis of the ^{1}H and ^{13}C spectra with Eu(hfbc)$_3$.[1070]

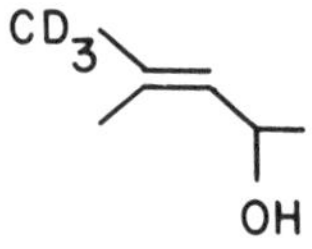

VI 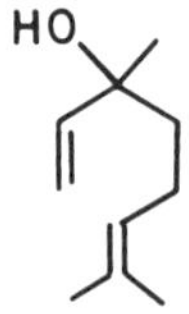 VII 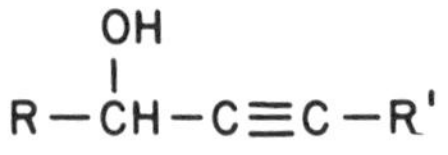 VIII

The optical purities of alkynes with structure VIII have been determined with the aid of Eu(dcm)$_3$.[1071]

The spectra of chiral allenic alcohols (IX,X) exhibited enantiomeric resolution in the presence of Eu(hfbc)$_3$.[1072]

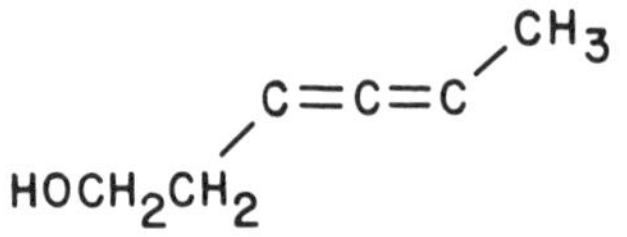

IX 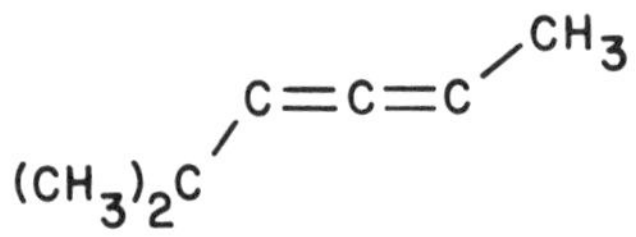 X

The methylene hydrogens of IX are diastereotopic and were resolved in the shifted spectrum. The methyl groups of X attached to the carbinol carbon are diastereotopic, and four methyl singlets were obtained in the presence of the chiral shift reagent.

Optical purities of terpin-1-en-4-ol[1073] and XI[1074] have been determined using chiral LSR. The optical isomers of

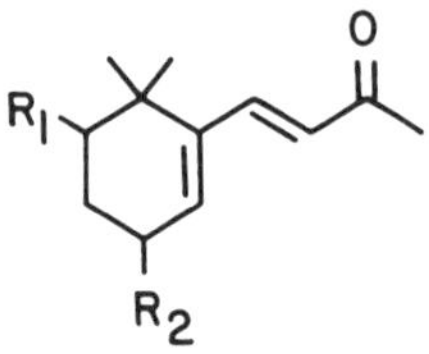

XI XII

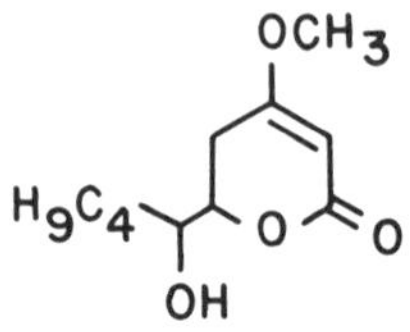

3,3,5-trimethyl-2-cyclohexen-1-ol were not resolved in the presence of Eu(facam)$_3$.[1075] Enantiomeric resolution was obtained in the spectra of XII and XIII with Eu(facam)$_3$,[1076] and Eu(hfbc)$_3$,[1070] respectively. The optical purity of XIV was determined through the use of a chiral LSR.[1077] The optical purities of cis- and trans-3-tert-butyl- and 3-isopropyl-6-methylenecyclohexanol were assessed using Eu(facam)$_3$.[1078]

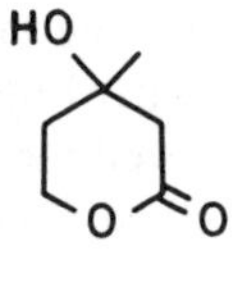

XIII 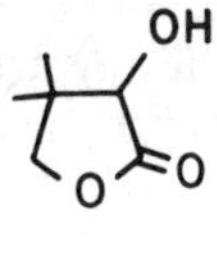   XIV

The enantiomeric purities of chiral multicyclic carbinols have been determined through the use of chiral LSR.[1079-1084] These include a bicyclo[3.2.0]heptane(XV)[1079] and XVI.[1080]

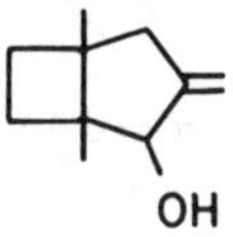

XV 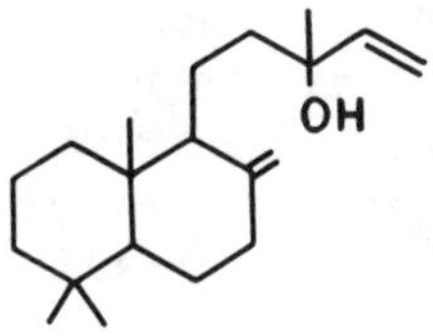   XVI

Optical resolution in the ^{13}C and ^{31}P spectra of XVII with Eu(facam)$_3$ was used to confirm the chirality of this substrate.[1081]

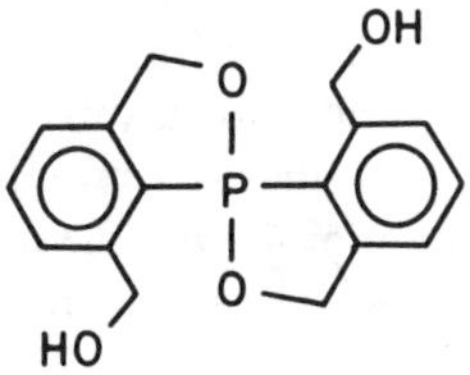

XVII 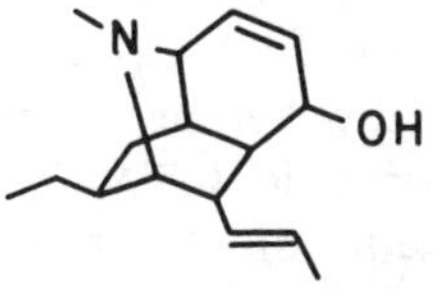   XVIII

The absolute configuration of cannivonine b (XVIII) was assigned by comparing the spectrum with Eu(facam)$_3$ to that of a similar compound of known configuration.[1082] The optical purity of *trans*-4,5-dihydroxy-4,5-dihydroaldrin was determined through the use of Eu(hfbc)$_3$.[1083] The optical purities of inokosterone-3,22,26-triacetate (XIX) and the 2,3,22-triacetate derivative have been determined with the aid of Eu(facam)$_3$.[1084]

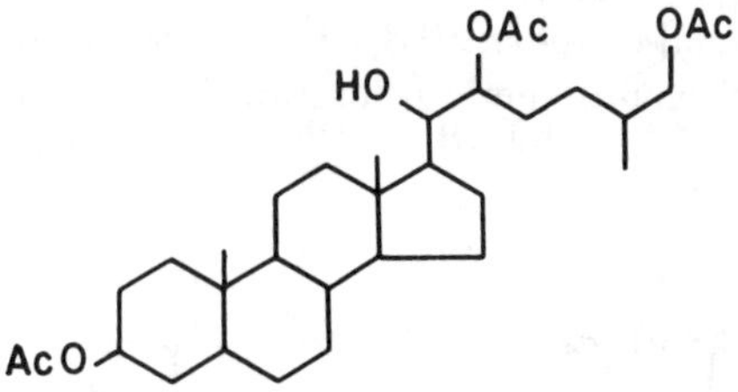

XIX

Certain compounds exhibit chirality because rotation about one or more bonds is slow on the time scale of NMR spectroscopy. Perhaps the most well-known example of this occurrence is with *o,o,o'o'*-tetra-substituted biphenyls(XX).

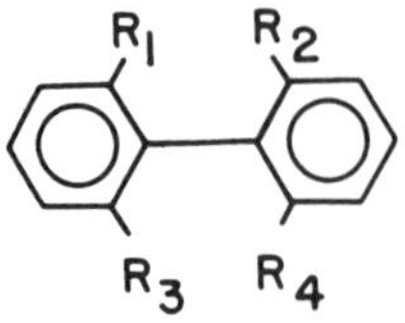

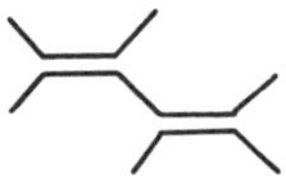

140   NMR Shift Reagents

XX

XXI

Derivatives in which one or more of the R groups is a hydroxymethyl group have been enantiomerically resolved through the application of chiral LSR.[1055,1085] Butadienes with slow rotation about the C_2–C_3 bond are chiral.[1086,1087] Chiral LSR have been used to resolve the enantiomers of butadiene derivatives and to determine the barriers to rotation. It was reported that complexation of the shift reagent did not alter the rotational barrier of these substrates.[1086,1087]

Hydrogen atoms that are enantiotopic by internal comparison can be resolved in the presence of chiral LSR. Such protons become nonequivalent in an optically pure environment. Resolution of protons enantiotopic by internal comparison has been observed in the spectra of benzylalcohol,[1006,1088] 2-propanol,[1088,1089] 2-methyl-2-butanol,[1089] 1,3-diglycerides,[1068] and cis-4,5-dihydroxy-4,5-dihydroaldrin with chiral LSR.[1083]

2. Ketones

Enantiomeric resolution has been obtained in the spectra of a variety of substrates that bond to a chiral LSR through a ketone group. The optical purities of 2-phenylmercaptoethyl methyl ketone, 2-phenylmercaptopropyl phenyl ketone, and α-methyl-β-phenylmercaptopropiophenone were determined by monitoring the methyl signals in the presence of Eu(hfbc)$_3$.[1090] Enantiomeric resolution was observed in the spectrum of I with Pr(hfbc)$_3$.[1091] The

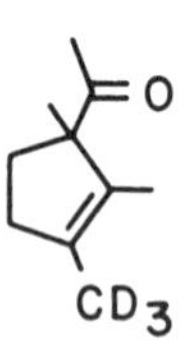

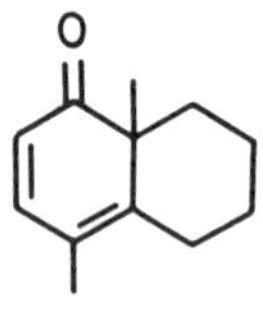

I

II

optical purities of the bicyclo compounds II and III were determined through the use of Eu(facam)$_3$.[1092] The signal of the methyl group closest to the carbonyl group exhibited the most pronounced enantiomeric resolution in these substrates.[1092] The optical purity of IV

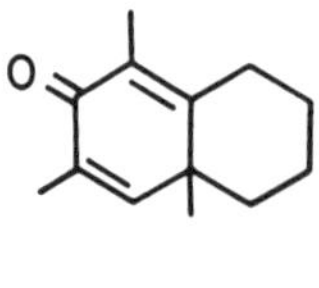

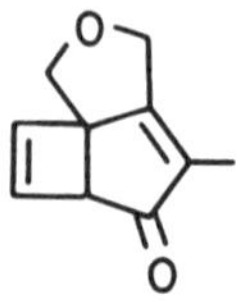

III

IV

was determined with the aid of Eu(hfbc)$_3$.[1093]

Shift data with Eu(facam)$_3$ has been used to distinguish *threo*- and *erythro*-3,4-dimethylhexa-2,5-dione.[1094] Separate CH$_3$CH– signals were observed in the spectrum of the *threo* isomer in the presence of the chiral LSR. Decoupling the methyl resonances permitted the determination of the CH—CH coupling constant and assignment of the configuration. Results with the *erythro* isomer were inconclusive as the enantiotopic 3- and 4- protons were not resolved.[1094]

The compounds benzo[9]annulenone[1095] and naphtho[9]annulenone(V)[1096] exhibit helical chirality because of steric conjestion. This has

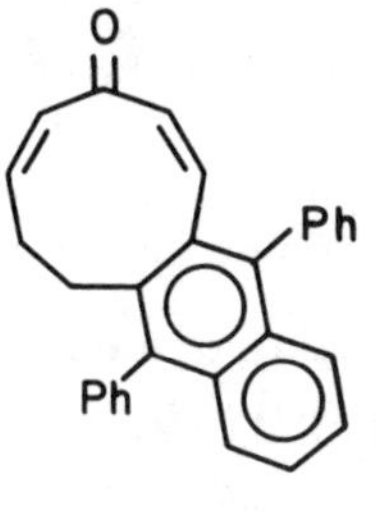

V

been unequivocally demonstrated by recording the spectra of the two compounds in the presence of Eu(dcm)$_3$.[1095,1096] Enantiomeric resolution of the *dextro* and *levo* isomers was observed with the shift reagent.

Other compounds that exhibit chirality by virtue of steric conjestion include 2,2'- and 3,3'-dimethylbianthrones (VI).[1097]

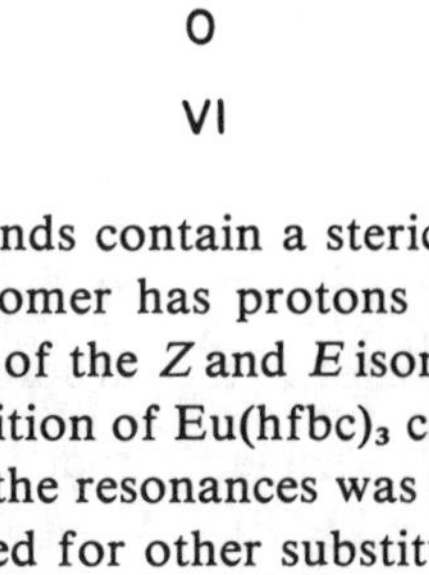

VI

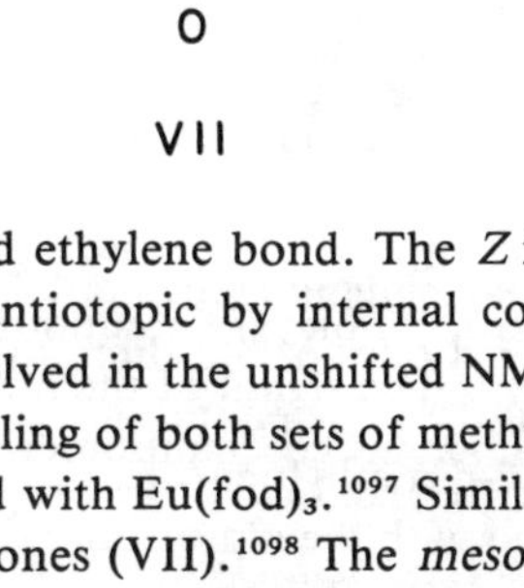

VII

These compounds contain a sterically crowded ethylene bond. The *Z* isomer is chiral while the *E* isomer has protons that are enantiotopic by internal comparison. The methyl signals of the *Z* and *E* isomers are resolved in the unshifted NMR spectrum of mixtures. Addition of Eu(hfbc)$_3$ causes a doubling of both sets of methyl signals. Such a doubling of the resonances was not observed with Eu(fod)$_3$.[1097] Similar observations have been noted for other substituted bianthrones (VII).[1098] The *meso* and (+/−) isomers of VII were distinguished in studies with chiral LSR. The resonances of the methine protons of the (+) and (−) isomers were resolved with a chiral LSR, but the methine protons do not couple and appear as singlets. The methine protons of the *meso* isomer are anisochronous and exhibit coupling when resolved in a chiral environment. These differences have been used to assign the stereochemistry of a number of derivatives of VII.[1098]

The spectra of substrates with ketone functionalities that are chiral by virtue of slow rotation about a bond have been enantiomerically resolved with the aid of chiral LSR.[1099,1100] These include substituted flavones(VIII) in which the bond between the two phenyl rings exhibits slow rotation.[1099]

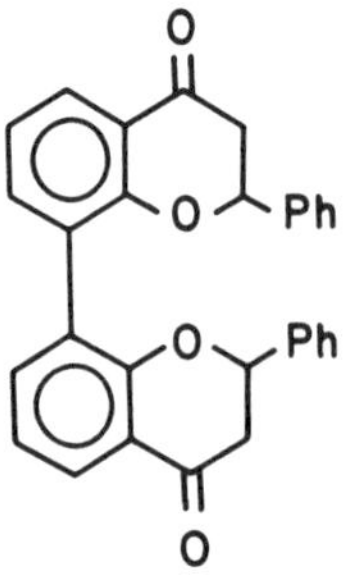

The rotational barriers in pivalophenones have been determined through the application of chiral LSR.[1100] It was reported that complexation of the shift reagent did not alter the rotational properties of these substrates.

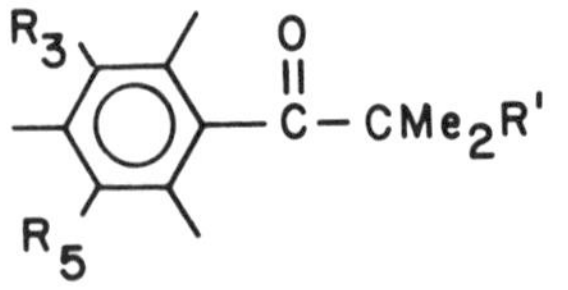

3. Esters

The spectra of chiral compounds with ester functionalities frequently exhibit enantiomeric resolution in the presence of chiral LSR. Some workers have reported that the enantiomeric resolution observed in the spectrum of a carbinol substrate in the presence of a chiral LSR is enhanced on acetylation.[1006,1054-1056] The methyl singlet of the ester group of the resulting derivative is particularly useful to monitor for enantiomeric resolution.

A number of acyclic esters of structure I in which the carbon alpha to the carbonyl group is chiral have been enantiomerically resolved in the presence of chiral LSR.[1062,1065,1101,1102]

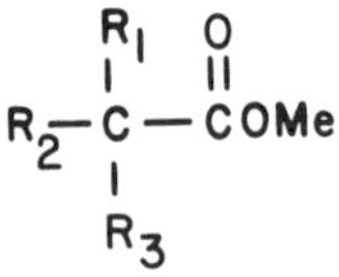

The R groups have included aromatic[1062,1101,1102] and aliphatic[1065] moieties. The optical purity of the ethyl ester of 3,7-dimethyloctanoic acid has been determined through the use of Eu(hfbc)$_3$.[1103] The enantiomeric purities of the methyl esters of 3,7-dimethyloctanoic acid, 3,7-dimethyloct-6-enoic acid, and 3,7,11-trimethyldodecanoic acid have been determined with the aid of Eu(dcm)$_3$.[1104,1105] The 15-(R) and 15-(S) configurations of II were distinguished in the presence of Eu(hfbc)$_3$, and the optical purity was determined.[1106]

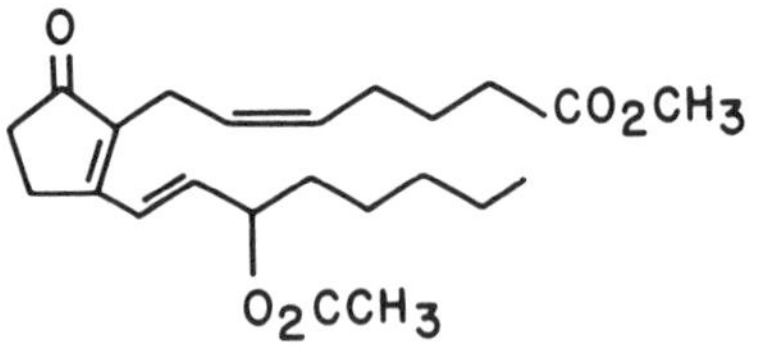

The optical purities of certain triglycerides have been determined with the aid of Eu(hfbc)$_3$.[1068] The spectra of derivatives with a long and short chain at the 1- and 3-positions, as in 1,2-diacetyl-3-stearoyl-sn-glycerol and 2,3-diacetyl-1-stearoyl-sn-glycerol, exhibited the most pronounced optical resolution. The analysis of the optical purity of monoglycerides with chiral LSR was improved if the substrate was first converted to a di-trimethylsilylether or, better yet, a diacetate derivative.[1054] For diglycerides, the most pronounced enantiomeric resolution was obtained in the spectrum if the hydroxy group was derivatized to a trimethylsilylether.[1054] A scheme for determining the optical purity of triglycerides whose spectra were not enantiomerically resolved with chiral LSR was described in this report. The method involved an epoxidation, hydrolysis, and silylation of the triglyceride. The product was then subjected to analysis with a chiral LSR.[1054]

Methods employing chiral LSR for determining the optical purity of methyl esters of amino acids have been described.[1107-1109] In two of these reports, the amine group was either acetylated[1108] or trifluoroacetylated prior to analysis with the shift reagent.[1107] Successful resolution of the derivatives of alanine,[1108] norvaline, norleucine, and methionine[1107] was demonstrated. The results with Eu(hfbc)$_3$ were judged superior to those with Eu(t-cam)$_3$, Eu(facam)$_3$, and Yb(facam)$_3$.[1107]

Ajisaka et al. observed enantiomeric resolution in the spectra of nine methyl esters of amino acids with chiral LSR.[1109] The site of bonding was not discussed in this report. A correlation of the relative shifts with absolute configuration was noted; and it was postulated that this trend could be applied, with caution, to amino acid methyl esters of unknown configuration.

Enantiomeric resolution of the methoxy resonances was observed in the spectra of eight allenic esters(III) and diesters in the presence of Eu(hfbc)$_3$.[1110]

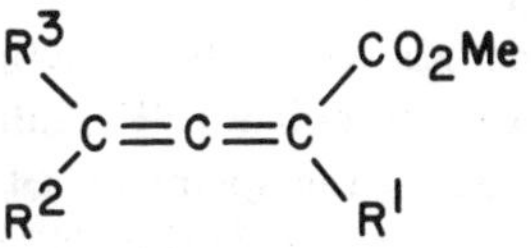

III

Substrates with ester functionalities containing rings of three,[1013,1101] four,[820,1111] five,[1017,1112] and six[820,1113] members have been studied with chiral LSR. The optical purity of methyl (+)-(1S:2R)-2-phenylcyclopropane carboxylate was determined with the aid of Eu(facam)$_3$.[1013] At ambient probe temperatures, the spectrum with the shift reagent was severely broadened. Raising the temperature of the sample to 100°C resulted in a reduction of the broadening, and enantiomeric resolution was observed in the spectrum. With the aldehyde and carboxylic acid derivatives, severe broadening in the presence of the shift reagent precluded the determination of optical purities.[1013] The optical purity of the epoxide IV was determined using chiral LSR.[1101]

Portions of the spectra of the *d,l*-isomers of (1S,2S)-(+)-dimethyl *trans*-1,2-cyclobutanedicarboxylate were resolved in the presence of Eu(hfbc)$_3$.[820] Chiral LSR have also been used to distinguish the *meso(cis)* from *dl(trans)* isomer of dimethyl 3,3,4,4-tetramethyl-1,2-cyclobutane dicarboxylate.[1111] The *meso* isomer has protons that are enantiotopic by internal comparison. These protons were resolved in the presence of the chiral LSR, and the fact that they were coupled enabled the assignment of the *meso* configuration.[1111] Two signals were also observed in the spectrum of the *dl* isomers with the shift reagent, but there was no coupling.

The optical purity of cyclopentenes V[1017] and VI[1112] were determined through the use of chiral LSR. The enantiomeric purities

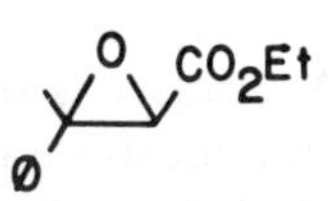

IV 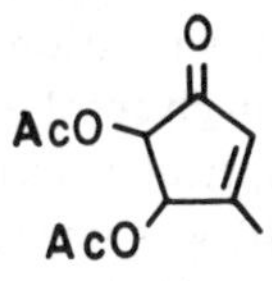   V 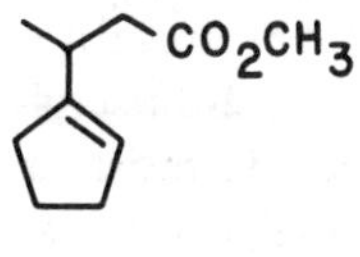   VI

of (3R,4S)-(−) methyl cis-3-methylcyclohexene-4-carboxylate,[820] VII,[1113] and VIII[1056] have been determined with chiral LSR. The

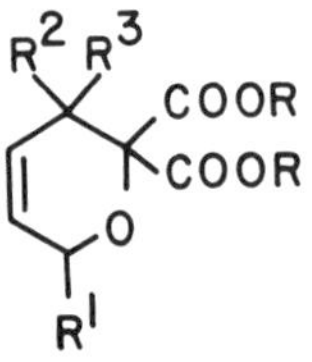

VII

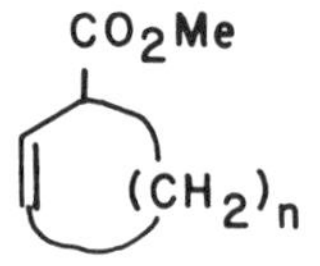

VIII

enantiomeric resolution in the spectrum of VIII with Eu(facam)$_3$ was found to be more favorable than that observed in the spectrum of the corresponding alcohol.[1056]

Chiral LSR have been used to determine the optical purities of bicyclo[3.1.0]hexane,[1114] bicyclo[4.1.0]heptadiene,[517] and bicyclo[3.2.1]octene[1115,1116] derivatives. Mixtures of d,l-cocaine have been quantified with the aid of Eu(facam)$_3$.[1117] The methyl signal of the ester group was convenient to monitor in such an analysis. Splitting of the ester methyl resonances of ^{1}H-benzo[c]quinolizine-1,2,3,4-tetracarboxylate (IX) was observed as a result of enantiomeric resolution with Eu(facam)$_3$.[835] In this study, it was noted that one set of the

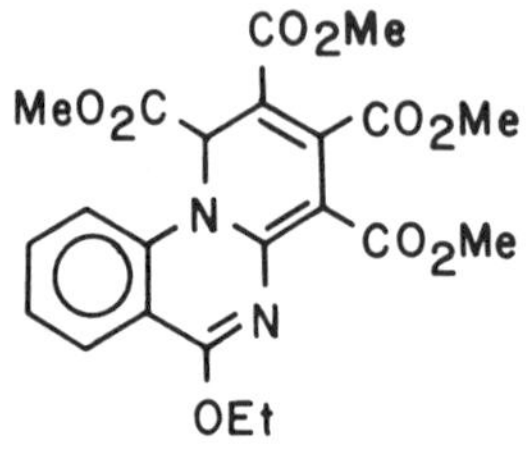

IX

X

resonances gradually disappeared on standing. It was postulated that the shift reagent facilitated a ring opening and rearrangement to the most stable enantiomer. The optical purities of methyl tricyclo[3.3.0.0^{3,7}]octane-2-carboxylate (X)[1118] and D$_3$-trishomocubanol acetate (XI)[1119] have been determined through the use of Eu(facam)$_3$. Eu(facam)$_3$ has also been employed in a determination of the optical purity of XII.[1084]

XI

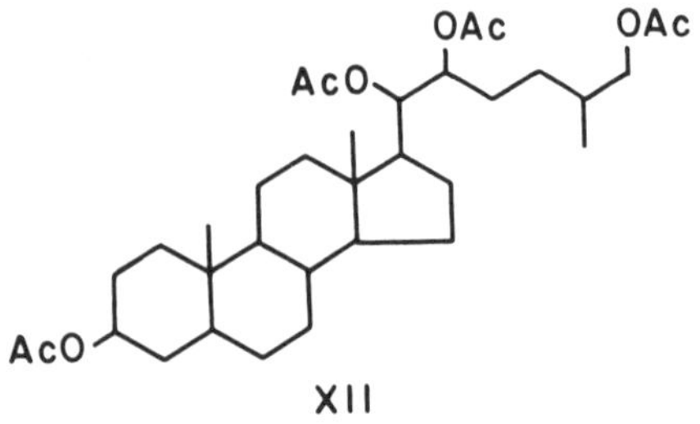

XII

The spectra of several substituted biphenyls with ester functional groups have been recorded in the presence of chiral LSR.[1055,1085,1120,1121] The enantiomeric resolution observed in the spectra of these substrates confirmed the chirality that results from slow rotation about the biphenyl bond. The enantiomeric resolution in the spectra of certain biphenyl derivatives with hydroxyl groups was less than that in the spectra of the corresponding methyl esters.[1055]

4. Lactones

The spectra of chiral lactones are not often enantiomerically resolved in the presence of Eu(facam)$_3$.[1067] A procedure for determining the optical purity of lactones that involves a derivatization to the corresponding diol, and analysis of the diol with Eu(facam)$_3$, has been described.[1067] Before carrying out such a derivatization scheme, however, direct analysis should be attempted since enantiomeric resolution has been obtained in the spectra of some lactones with chiral LSR.

The optical purities of 2-phenyl-3,1-benzoxathian-4-one (I) and its corresponding sulfone have been determined with Eu(hfbc)$_3$.[1122]

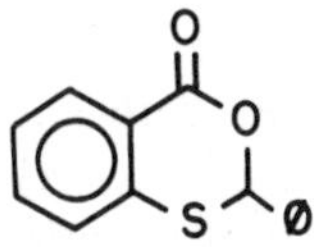

I 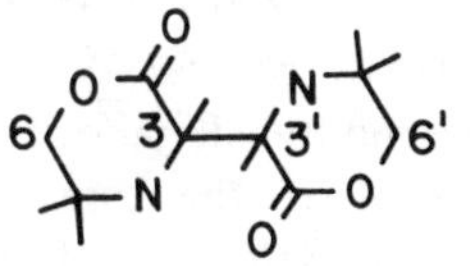 II

Both substrates reportedly complexed through the carbonyl oxygen. The enantiomeric resolution in the spectrum of I was better than that of the sulfone derivative. The *meso* and *dl* isomers of II were assigned on the basis of a study with Eu(facam)$_3$.[1123] A doubling of the 3,3' and 6,6' resonances was observed for the *dl* isomer in the presence of the shift reagent. No such doubling was obtained in the spectrum of the *meso* isomer.

5. Aldehydes

There are only a few reports describing the application of chiral LSR to aldehydes,[6,1013,1124,1125] and most of these have achieved limited success. The enantiomers of α-phenylpropionaldehyde were not resolved in the presence of Eu(facam)$_3$.[1124] Enantiomeric resolution has been obtained in the spectrum of 2-phenylpropionaldehyde, however, with Eu(hfbc)$_3$.[6] The spectrum of 2-phenylcyclopropane carboxaldehyde was too severely broadened in the presence of Eu(facam)$_3$ to detect the presence of enantiomeric resolution.[1013] The optical purity of the corresponding ester was successfully determined using chiral LSR. Chiral LSR were used to resolve enantiomers of I that result from slow rotation about the ring-nitrogen bond and to determine the barrier to rotation.[1125]

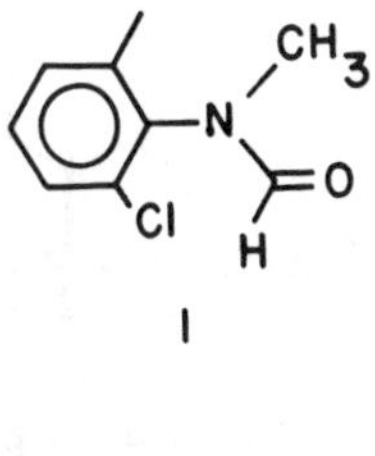

I

6. Carboxylic Acids

Solutions of carboxylic acids and Eu(facam)$_3$ or Eu(hfbc)$_3$ do not form a precipitate, and carboxylic acids may be studied directly with chiral LSR. A procedure for determining the enantiomeric purities of derivatives of 2-methyl-5,6-dihydro-α-pyran-6,6-

dicarboxylic acid has been described.[1113] Through ozonolysis, the starting material was converted into a lactic acid. An analysis of the resulting lactic acid with Eu(hfbc)₃ was then compared to analyses of lactic acids with known configurations.

The spectrum of 2-phenylcyclopropane carboxylic acid was broadened in the presence of Eu(facam)₃ to the point that any enantiomeric resolution could not be detected.[1013] The spectrum of the corresponding ester was not broadened, however, and enantiomeric resolution was observed.

7. Ethers

Enantiomeric resolution has been obtained in the spectra of a number of ethers with chiral LSR. The optical purity of propylene oxide has been determined with Eu(facam)₃.[1077] Chiral LSR have been used to assign the configuration of *cis*(*meso*) and *trans*-(*dl*)-2,3-butylene oxide.[1089] The protons of the *meso* isomer that are enantiotopic by internal comparison were resolved in the presence of the chiral LSR and were coupled to each other. For the *d,l* isomer, the enantiomers were resolved, but no coupling of these protons was observed. The *meso* stereochemistry of dieldrin (I) has been demonstrated through an analysis of the spectrum in the presence of Eu(facam)₃.[1089] The H₃,H₆ and

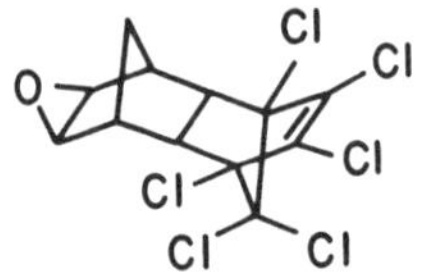

I

H₄,H₅ pairs, which are enantiotopic by internal comparison, are resolved in the presence of the shift reagent. Vicinal coupling between H₄ and H₅ confirmed the *meso* configuration.

The optical purities of 2-methyl-1-*p*-tolylsulfonyl-1,2-epoxybutane and -propane were determined by recording spectra with Eu(facam)₃ at 270 MHz.[1126] The 2-isopropyl-1-*p*-tolylsulfonyl-1,2-epoxypropane and 2-ethyl-1-*p*-tolylsulfonyl-1,2-epoxybutane derivatives were not enantiomerically resolved in the presence of Eu(facam)₃.[1126] Poor results were also obtained in attempts to determine the enantiomeric purities of II and III with Eu(facam)₃.[1127] The use of

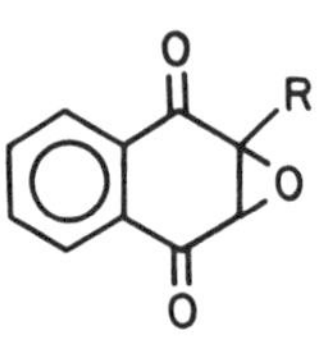

II 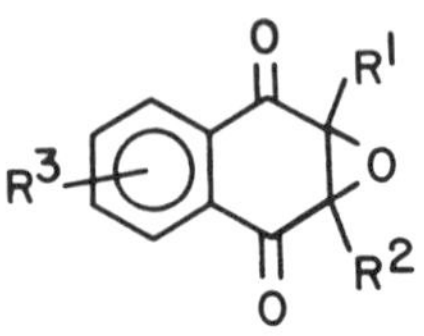   III

Eu(dcm)₃, however, significantly improved the enantiomeric resolution.

The enantiomeric purities of 6,7-dimethoxy-2-benzonorbornenone (IV) and similar derivatives have been determined with Eu(facam)₃, Eu(hfbc)₃, *tris*(3-pentafluorobenzoyl-*d*-camphorato)europium(III), and *tris*(3-heptafluorobutyryl-*d*-nopinato)europium(III).[1005]

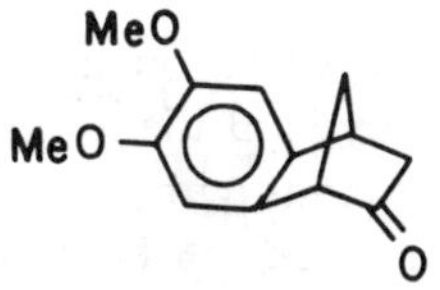

IV

These substrates coordinated to the shift reagent in a chelate manner through the *ortho*-substituted methoxy groups. The optical purities of (+)-2-deuterio-3,7-dimethyl-7-methoxymethylcycloheptatriene and thermolysis products of this compound were determined with the aid of Eu(hfbc)$_3$.[1128]

Different signals for the optical isomers of 2,2'-dimethoxy-6,6'-dimethoxymethyl[1055] and 2,2'-difluoro-6,6'-dimethoxybiphenyl[1120] that result from slow rotation about the biphenyl bond have been observed in the presence of chiral LSR.

8. Amines

The enantiomeric resolution in the spectrum of 1-phenylethylamine has been measured in almost all reports of new chiral LSR. The best results have been obtained with Eu(dcm)$_3$.[7] The T$_1$ values of 1-phenylethylamine and *N,N*-dimethyl-1-phenylethylamine with Gd(dcm)$_3$ have been measured, and significant differences between protons of the two enantiomers were observed.[1058] This observation implies that the enantiomeric resolution in the presence of the chiral shift reagent is the result of geometric differences of the two shift reagent-substrate complexes.

Amphetamine (I) has been the focus of several studies with chiral LSR.[885,1066,1129-1131] A comparison of the enantiomeric resolution in the spectrum of I with Eu(dcm)$_3$, Eu(hfbc)$_3$, and Eu(facam)$_3$ has

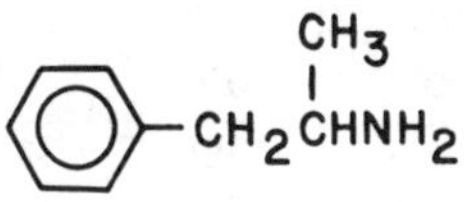

I 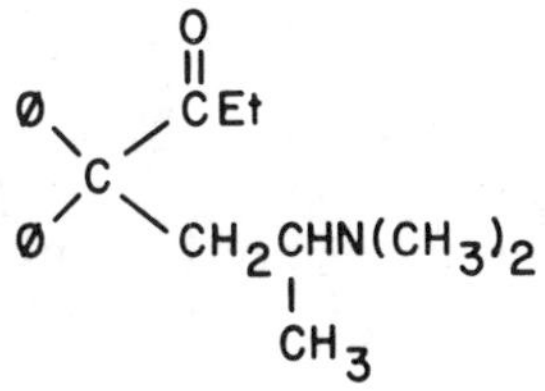   II

been reported.[1066] Baseline resolution of the methyl resonances was obtained only in the presence of Eu(dcm)$_3$. Partial enantiomeric resolution in the spectrum of 2,5-dimethoxyamphetamine in the presence of Eu(facam)$_3$ has been reported.[885] The optical purity of ephedrine,[1130,1132] norephedrine,[1133] and pseudoephedrine[1130] has been determined with chiral LSR. The optical purity of methadone(II) has been determined with the aid of Eu(hfbc)$_3$.[1134] The enantiomeric resolution of the *N*-methyl signals was monitored.

The methyl resonances of 2-aminopropane, which are enantiotopic by internal comparison, were resolved in the presence of Eu(facam)$_3$.[1089] A procedure for determining the optical purity of penicillamines (III) has been reported.[1135] The penicillamine is derivatized to the cyclic derivative IV, which is then analyzed with the chiral shift reagent.[1135] Complexation of IV with Eu(hfbc)$_3$ reportedly occurs at the secondary amine. The ^{13}C spectra of β-amino alcohols were not enantiomerically resolved in the presence of Eu(hfbc)$_3$; however, the spectra of the corresponding oxazolidine derivatives(V) were enantiomerically resolved.[1136] The substrates reportedly bonded to the shift reagent through the nitrogen atom.

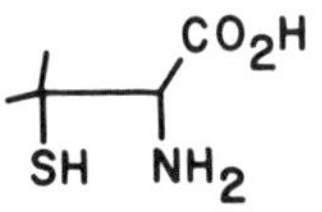

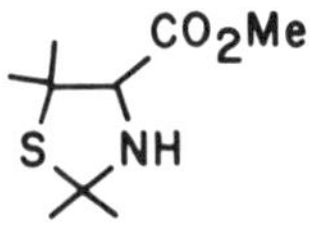

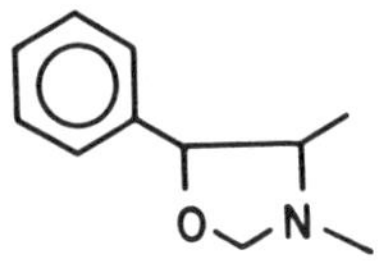

The optical purities of isoquinoline alkaloids including landanosine,[1082,1137] papaverine,[1082] N-methylpavine, tetrahydropavine, tetrahydropalmatine, salsolidine, and glaucine(VI)[1137] have been determined with chiral LSR.

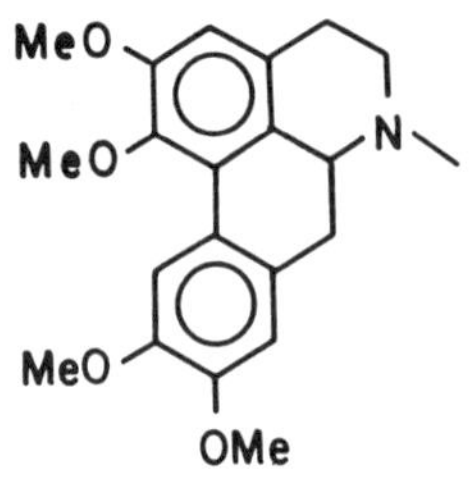

These reports did not speculate on the bonding site however, the *ortho*-substituted methoxy groups are likely bonding locations. The enantiomeric purity of methorphan (VII) has been determined through the use of chiral LSR.[1138] Complexation of the shift reagent reportedly occurred at the nitrogen atom, but the best resonance to monitor for enantiomeric resolution was that of the methoxy group.

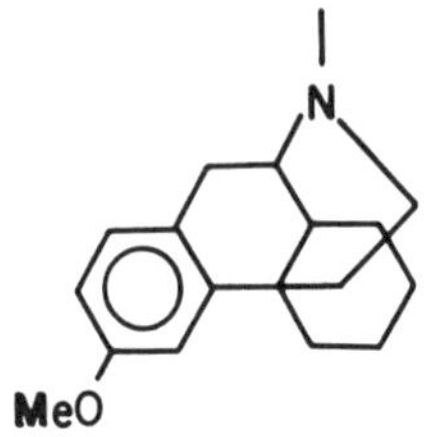

The study of several biphenyl substrates with amine functional groups has been facilitated with Eu(hfbc)$_3$.[1121] These substrates are chiral by virtue of slow rotation about the biphenyl bond.

9. Nitrogen Heterocycles

The optical purity of 2-*sec*-butylpyridine was successfully determined with Eu(facam)$_3$,[1139] however, the spectrum of the 4-*sec*-butyl derivative exhibited no enantiomeric resolution.[1061] The failure to enantiomerically resolve the spectra of 1-(4-pyridyl)-1-phenylethane and 1-(2-pyridyl)-1-phenylpropane with Eu(facam)$_3$ has also been noted.[1061] The optical purity of tetramisole(I) was determined through the use of Eu(hfbc)$_3$.[1140] Complexation at the

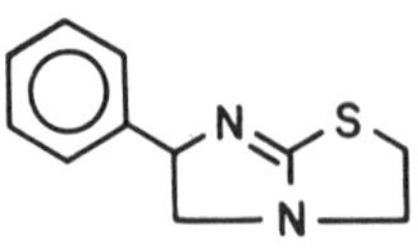

heterocyclic nitrogen was indicated, and complete enantiomeric resolution of the adjacent CH resonance was observed.

10. Amides

The barrier to rotation in 2,6-disubstituted benzamides (I) has been studied with the aid of Eu(hfbc)$_3$.[1141] These substrates are

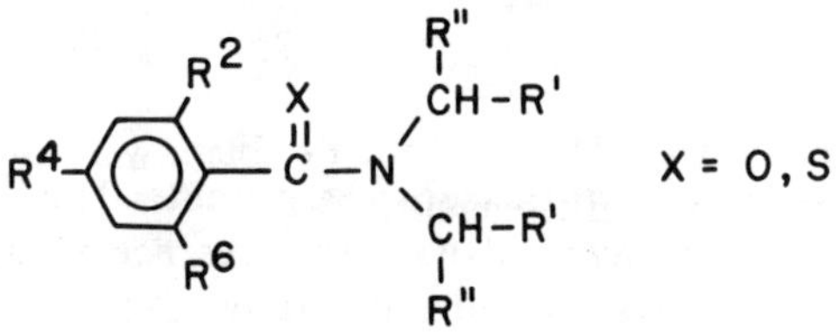

chiral by virtue of slow rotation. The substrates complex at the X atom, and better results were obtained for the amides compared to the corresponding thiamides.

11. Lactams

Derivatives of barbituric acid (I) are optically active, and several reports have described the use of chiral LSR with these substrates.[1142-1145] These studies have described

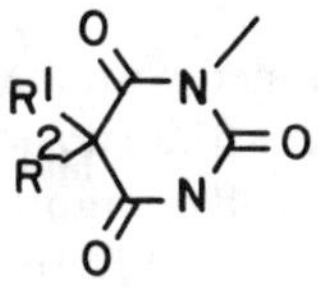

the determination of the optical purity of mephobarbital (R_1 = Ph, R_2 = Et),[1142] hexobarbital (R_1 = cyclohexenyl), R_2 = Me)[1142,1143] methohexital (R_1 = allyl, R_2 = 1-methyl-2-pentynyl),[1142,1144] 1-methyl-5-butyl-5-phenylbarbital,[1143] 1-ethyl-5-(1'-cyclohexenyl)-5-methylbarbital,[1143] and thiamylal[5-allyl]-5-(1-methylbutyl)-2-thiobarbituric acid.[1145]

The spectra of somewhat analogous compounds including

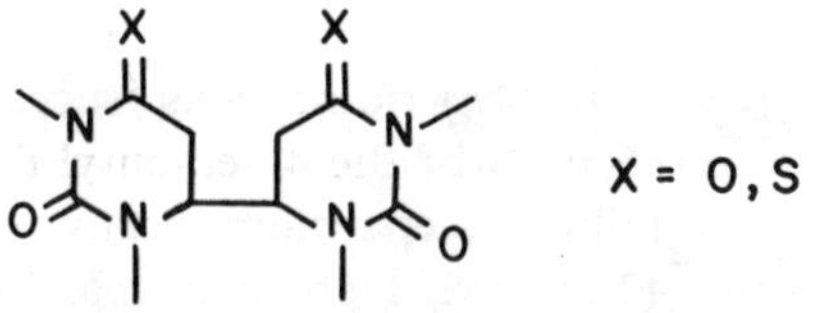

3-ethyl-3-phenyl-2,6-piperidinedione,[1146] 3-ethyl-5-phenyl-2,4-imidazolidinedione,[1147] and tetrahydropyrimidines(II)[1148] have been enantiomerically resolved with chiral LSR. The *dl* and *meso* isomers of II were distinguished on the basis of shift data with Eu(facam)$_3$.[1148] The optical purities of tetrahydrofuranyl-5-fluorouracil derivatives have been determined with Eu(facam)$_3$.[1149] The optical purities of III[1150] and IV[1151] have been determined with the aid of Eu(hfbc)$_3$ and Eu(t-cam)$_3$, respectively.

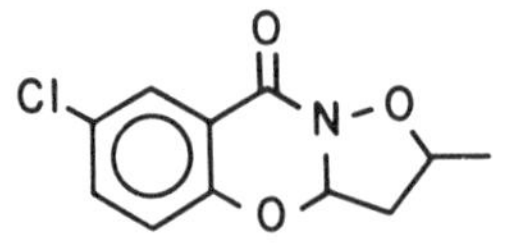

There are two chiral centers in III, and four doublets were observed for the methyl group in the spectrum with the shift reagent.[1150]

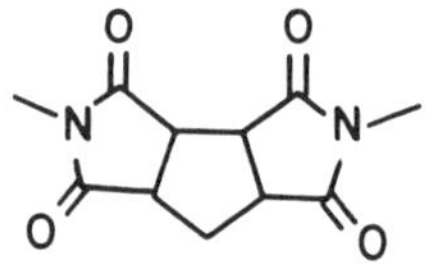

12. Nitrosoamines

Nitrosoamines of structure I are chiral because of slow rotation of the ring-nitrogen bond. The rotational barriers

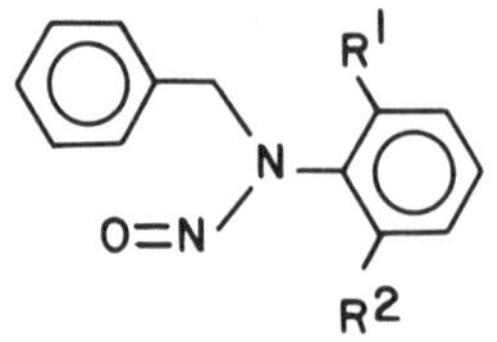

of these substrates have been determined with the help of chiral LSR.[1125,1152,1153] It was concluded that complexation between the substrate and LSR did not alter the rotational properties. This was unequivocally demonstrated through an analysis of the derivative in which $R_1 = CH_3$ and $R_2 = Cl$.[1152] In this case, the methylene protons are diastereotopic and are resolved in the unshifted spectrum. For this substrate, the barrier to rotation could be determined without a shift reagent. The results obtained with no LSR, racemic Eu(hfbc)₃, and d-Eu(hfbc)₃ were essentially identical, indicating that no alteration of the rotational properties occurred on complexation.

13. Organoborons

Compounds of type I and II exhibit axial chirality if the R groups are different.

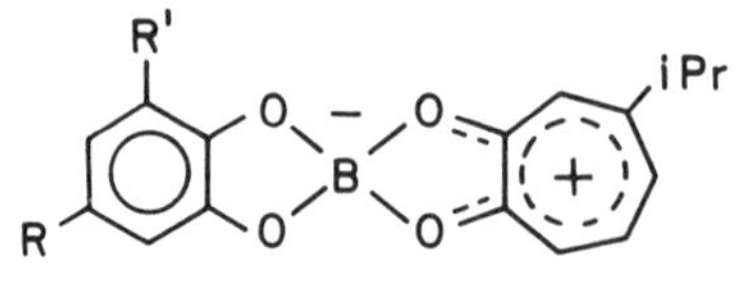

Optical isomers of I and II have been resolved and quantified using chiral LSR.[1154] Complexation reportedly involved the oxygen atoms of the uncharged ring.

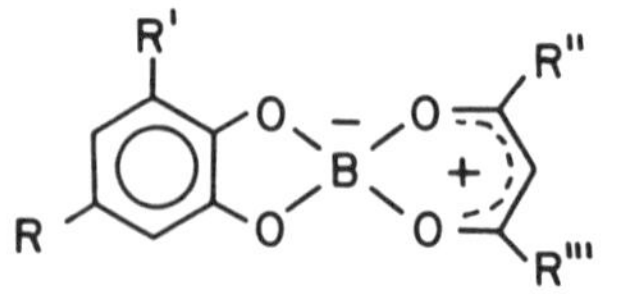

14. Sulfur

Chiral substrates with sulfur-containing functional groups such as sulfoxides,[6,7,963,1004,1006,1008,1089,1155-1160] sulfines,[1159] sulfones,[1006] sulfoximes,[1155] sulfilimines,[1155] sulphinamides,[1155,1161] and thiocarbonyls[629] have been studied with chiral LSR. Except for the thiocarbonyl compounds, all of these classes of compounds bond to the LSR through an oxygen atom.

The methyl groups of dimethylsulfoxide are enantiotopic by internal comparison and are resolved in the ¹H and ¹³C NMR spectra with Eu(facam)₃ and Yb(facam)₃.[1089]

Enantiomeric resolution has been observed in the spectra of methyl p-tolyl sulfoxide,[1158] methyl phenyl sulfoxide,[1158] and benzyl methyl sulfoxide[6] with chiral LSR. Eu(facam)$_3$ and Eu(hfbc)$_3$ have been used to resolve enantiomers of β-ketosulfoxides, α-cyanosulfoxides, β-enaminosulfoxides, β-oximinosulfoxides, and β-sulfonylsulfoxides.[1155] Enantiomeric resolution was not observed in the spectrum of bis-(phenylsulfinyl)methane with Eu(t-cam)$_3$.[963] The optical purities of N-[2-(allylsulfinyl)benzyl] N-methylmethanamine (I)[1157] and II[1160] were determined with Eu(facam)$_3$. There was no discussion of the bonding site in the latter report. Deshmukh et al. have reported that

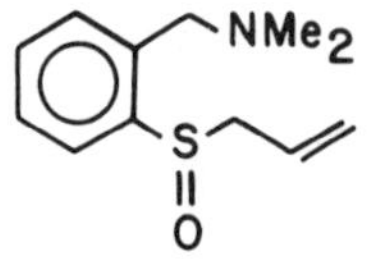

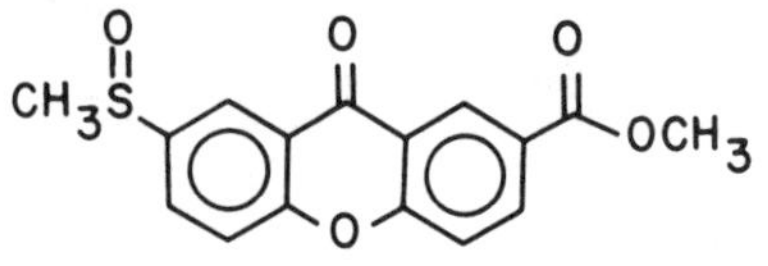

the enantiomeric resolution in the spectra of certain sulfoxides is greater with (R)-(−)-N-(3,5-dinitrobenzoyl)-α-phenylethylamine than with Eu(hfbc)$_3$.[1156] The chiral amine interacts with sulfoxides via hydrogen bonding.

Unsymmetrical sulfines of structure III are chiral by virtue of hindered rotation about the ring-carbon bond. Symmetrical derivatives of III with restricted rotation have

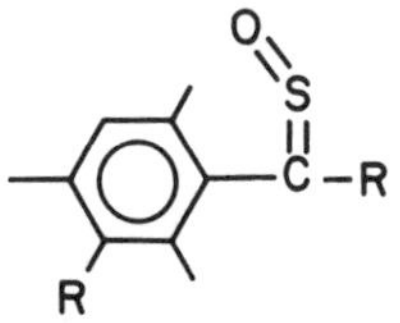

a prochiral center and contain protons that are enantiotopic by internal comparison. Shift reagents with the facam and hfbc ligands have been employed to resolve enantiotopic protons in both unsymmetrical and symmetrical derivatives.[1159] The barrier to rotation in certain of these substrates was determined through studies with LSR. It was reported that complexation of the shift reagent did not alter the barrier to rotation.[1159]

The spectra of chiral N-alkylidenesulphinamides have been recorded with chiral LSR.[1155,1161] The optical purity of IV was determined with Eu(facam)$_3$.[1161] No enantiomeric

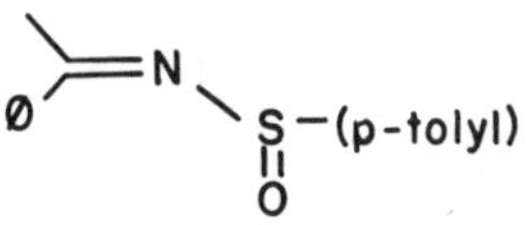

resolution was noted in the spectra of substrates of structure IV in which the methyl group was replaced with an ethyl, isopropyl, or phenyl group.[1161]

Enantiomeric resolution has been observed in the spectra of dibenzylsulfone, 2-octylphenylsulfone, 1-adamantyl-4-(1-pentenyl)sulfone, and phenyl-1-(3-methyl-2-butenyl)sulfone in the presence of Eu(hfbc)$_3$.[1006] Resolution of the methyl groups of V, which are enantiotopic by internal comparison, was observed using Eu(hfbc)$_3$.[629]

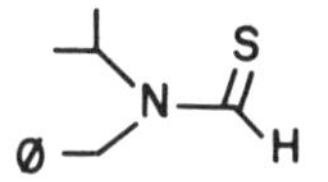

V

15. Phosphorus

The spectra of chiral substrates with the P=O group have been enantiomerically resolved with chiral LSR. The optical purity of allenes of structure I was determined using Eu(t-cam)$_3$.[1162]

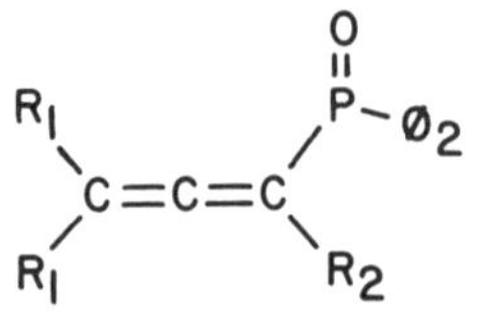

I 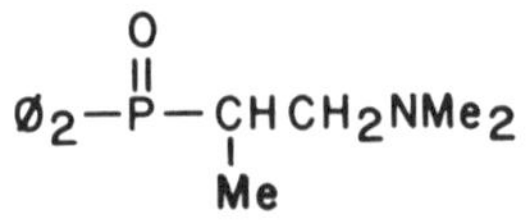 II

The enantiomeric purity of II was determined with Eu(dcm)$_3$.[1102] The spectra of the compounds 1,2,2-trimethylpropylmethylphosphonofluoridate (III), isopropylmethylphosphonofluoridate, ethyldimethylphosphoramidocyanidate (IV), and ethyl-2-diisopropylaminoethylmethylphosphonothioate (V) were recorded with chiral LSR.[1163] Spectra obtained with

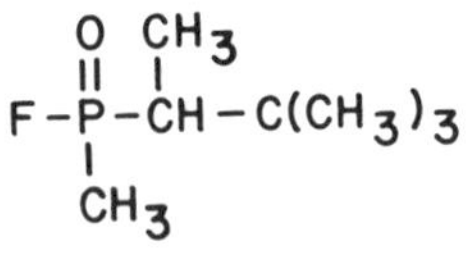

III 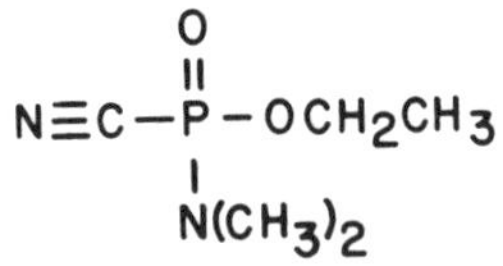 IV

Eu(fod)$_3$ were used to facilitate assignment of the resonances. The enantiomers resulting from chirality about the phosphorus atom were resolved for each substrate. III has two chiral centers, and resonances for all four enantiomers were observed in the spectrum with the chiral shift reagent.[1163]

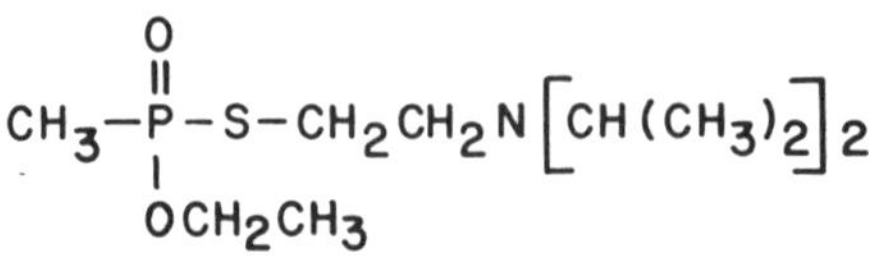

V

The optical purity of VI, a cyclophosphoramide, was determined through an analysis of the ^{1}H and ^{31}P spectra with Eu(facam)$_3$.[1164] The chirality of VII was demonstrated by the observation of enantiomeric resolution in the spectrum with Eu(facam)$_3$.[1165]

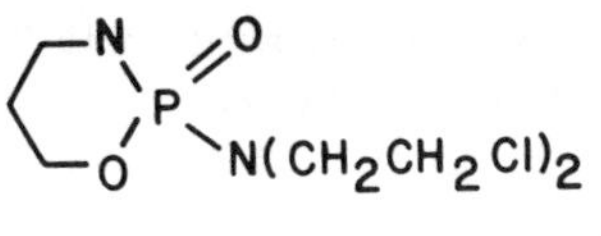

VI 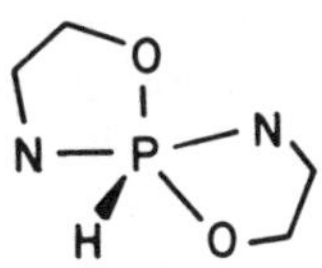   VII

16. Metal Complexes

Chiral LSR have been employed in studies of the enantiomeric purity of metal complexes. The metal complexes that have been studied contained functional groups in the ligands that coordinated to the lanthanide ion of the shift reagent. Both organometallic and metallo-organic species have been the focus of reports.

Racemic resolution was observed in the spectrum of tris(3-nitroso-2,4-pentanedionato)cobalt(III) (I) with Eu(hfbc)$_3$, and served to demonstrate that the complex had facial, rather than meridional, geometry.[1166] The spectra of the enantiomers of

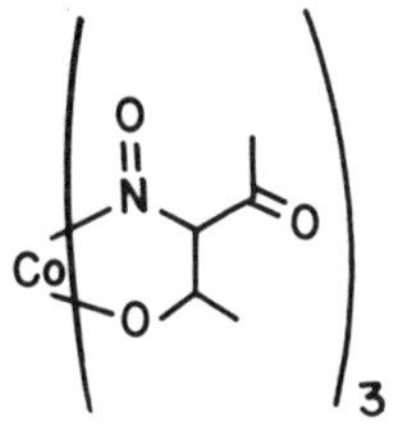

I

a Ni(II) complex with the ligand prepared from the Schiff's base condensation of 2,4-pentanedione and 1,2-diaminopropane were resolved with Eu(hfbc)$_3$.[634] The complex with an analogous ligand prepared from 2,3-diaminobutane can have either a *meso* or *dl* configuration depending on the configuration of the starting amine. In each case, a doubling of resonances was observed in the presence of chiral LSR.[634] With the *meso* configuration, the two sets of resonances result from resolution of protons that are enantiotopic by internal comparison.

The spectra of complexes of the formula (CO)$_3$ML in which M = Fe or Mn and L had the structure II,[1167] III,[1168] IV,[1169,1170] or V[1171] have been recorded with chiral and achiral LSR

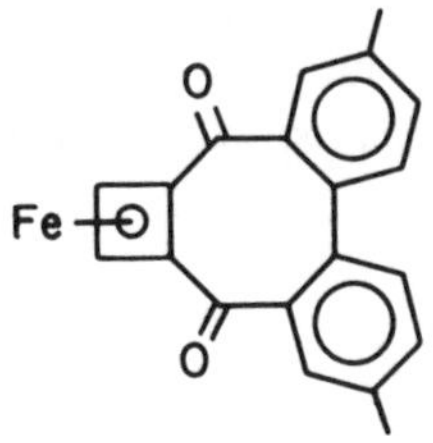

II 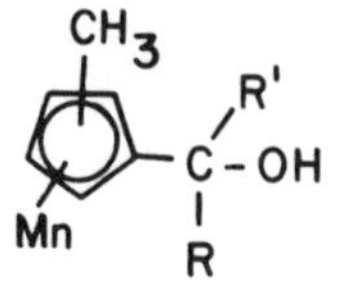   III 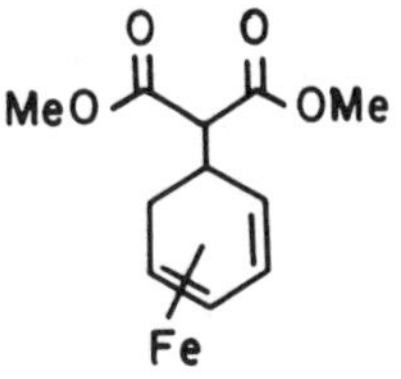   IV 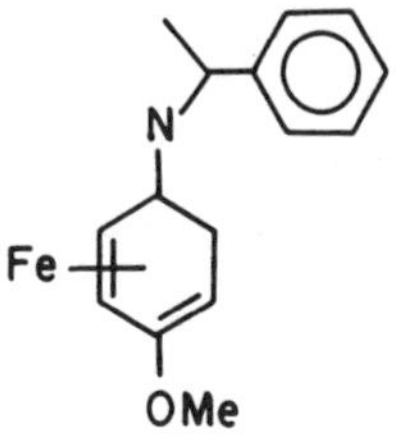   V

The enantiomeric purities of the Fe complexes with II and IV have been determined with Eu(facam)$_3$. The carbonyl oxygens of the dimethyl malonate group of IV reportedly coordinated to the shift reagent in a chelate manner.[1169,1170] The spectra of the diastereomers (V) resulting from the reaction of (−)-1-phenylethylamine with (−)-tricarbonyl[(1,2,3,4-*n*-2-methoxy-1,3-cyclohexadiene]iron were resolved in the presence of Eu(fod)$_3$.[1171]

Enantiomeric resolution was observed in the spectra of dimethylfumarate and dimethylmaleate iron tetracarbonyl complexes in the presence of Eu(facam)$_3$.[1172] The cyclopentadiene proton resonances of $(n^5\text{-}C_5H_5)Fe(CO)(CN)(PPh_3)$ and $(n^5\text{-}C_5H_5)Fe(CO)(PPh_3)(O\text{-}CH_2CN)$ were enantiomerically resolved in the presence of $Yb(facam)_3$.[1236] No enantiomeric resolution was obtained in the spectrum of $(n^5\text{-}C_5H_5)Fe(CO)(PPh_3)(O\text{-}CH_2CH_2CN)$ with $Yb(facam)_3$.[1236] This was believed to be the result of the weak interaction of the CN group with the lanthanide ion. Enantiomeric resolution has been observed in the spectra of complexes of the formula $(n^5\text{-}C_5H_5)PtR_1R_2R_3$ with $Eu(facam)_3$.[1173] The two methyl groups in the complex with R_1 = acetyl and $R_2 = R_3$ = methyl are enantiotopic by internal comparison. The resonances for these two methyl groups were resolved in the presence of the shift reagent.

Mixtures of enantiomers of $[(C_5H_5)MX\{P(O)(OCH_3)_2\}\{P(OCH_3)_3\}]$ in which M = Co or Rh and X = I or CH_3 have been resolved through the use of $Eu(facam)_3$.[1174] The degree of enantiomeric resolution was dependent on X and the metal. The Ni analog of this complex was not enantiomerically resolved with $Eu(facam)_3$. Complexes of the formula $[PtCl_2(L\text{-}L)olefin]$ in which L-L was biacetyl-bis-N,N-dimethyl hydrazone and the olefins were prochiral ligands such as fumarodinitrile, acrylonitrile, acrolein, and methyl acrylate, were enantiomerically resolved in the presence of $Eu(hfbc)_3$.[1175] The enantiomeric purity of VI has been determined with the aid of $Eu(facam)_3$.[1176]

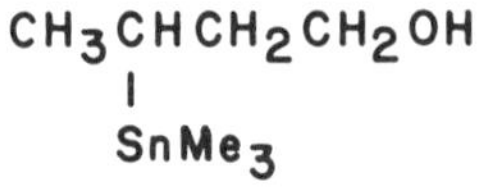

$$CH_3CHCH_2CH_2OH$$
$$|$$
$$SnMe_3$$

VI

17. Organic Salts

Resolution of enantiotopic protons in iminium (I) and imidazolium (II) salts has been observed in the presence of $Eu(hfbc)_3$.[629] The methylene groups of II are enantiotopic by internal comparison and were resolved in the spectrum with the shift reagent. The methyl groups

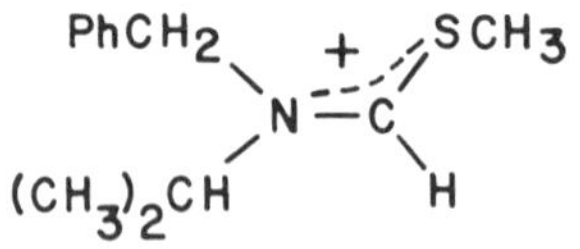

I

II

of the *E* isomer of I were resolved with $Eu(hfbc)_3$, but resolution of the methyl groups was not obtained for the *Z* isomer.

Chapter 4

BINUCLEAR LANTHANIDE(III)-SILVER(I) SHIFT REAGENTS

Lanthanide tris chelates are ineffective as NMR shift reagents for soft Lewis bases such as olefins, aromatics, halogenated compounds, and phosphines. One approach to developing shift reagents for these classes of compounds was demonstrated by Evans et al.[8] In this study, a binuclear complex formed in solution from a lanthanide tris chelate and silver heptafluorobutyrate was used to shift the spectra of olefins. The silver ion bonded to the substrate, and the paramagnetic lanthanide ion was responsible for the shifts in the spectrum. The shifts were rather small, however, and the method was applied in only a few instances. Meyers used a mixture of Eu(facam)$_3$ and silver trifluoroacetate to determine the optical purity of methylene(2-methyl)cyclohexane.[1177] Dambska and Janowski used Pr(fod)$_3$ with silver trifluoroacetate to resolve xylene isomers.[1178,1179]

In 1980, it was reported that certain silver β-diketonates, when used in conjunction with lanthanide tris β-diketonates, resulted in more effective binuclear shift reagents than either silver trifluoroacetate or heptafluorobutyrate.[9] The use of these shift reagents with olefins, aromatics, halogenated compounds, and phosphines has been described. The binuclear shift reagents are also suitable for certain oxygen- and nitrogen-containing compounds that bond weakly to the lanthanide tris chelates, but contain functional groups to which silver effectively bonds. The optical purity of chiral substrates can be determined by employing chiral lanthanide tris chelates with either achiral or chiral silver β-diketonates.

A simplified set of equilibria representative of the processes that occur with the binuclear reagents is shown in Equations 1 and 2. Both equilibria are rapid on the NMR time scale at ambient probe temperatures.[1180]

$$Ag(\beta - dik) + Ln(\beta - dik)_3 = Ag[Ln(\beta - dik)_4] \qquad (1)$$

$$Ag[Ln(\beta - dik)_4] + \text{substrate} = (\text{substrate})Ag[Ln(\beta - dik)_4] \qquad (2)$$

The exact nature of the binuclear complex remains unknown; however, evidence indicates that an ion pair is formed between Ag$^+$ and a lanthanide tetrakis chelate anion.[1180] Further evidence for the formation of an ion pair is provided by the recent application of binuclear complexes as shift reagents for organic salts.[630] In this application, the species Ln(β-dik)$_4^-$ is used to shift the spectra of organic cations.

I. SHIFT REAGENT SELECTION

A. Metal

The criteria for the selection of the appropriate metal for binuclear applications parallel those already described for the tris chelates. The direction of shifts, as well as the relative magnitude and broadening, is the same as with the tris chelates. The only significant difference is that, whereas europium is most commonly employed with the tris chelates, ytterbium is recommended in binuclear applications.[1180] The silver bridge in the binuclear shift reagents lengthens the distance between the substrate and lanthanide ion relative to that observed with the tris chelates. With binuclear reagents containing Eu(III), the resulting shifts are often too small to be of much practical utility.[1180] Yb(III) causes downfield shifts that are larger than those with Eu(III). In

Table 1
LIGANDS

Name	Abbreviation	Structure	
I 2,4-pentanedione	H(acac)	$R_1 = -CH_3$	$R_2 = -CH_3$
II 1,1,1-trifluoro-2,4-pentanedione	H(tfa)	$R_1 = -CF_3$	$R_2 = -CH_3$
III 1,1,1,5,5,5-hexafluoro-2,4-pentanedione	H(hfa)	$R_1 = -CF_3$	$R_2 = -CF_3$
IV 2,2,6,6-tetramethyl-3,5-heptanedione	H(dpm)	$R_1 = -C(CH_3)_3$	$R_2 = -C(CH_3)_3$
V 6,6,7,7,8,8,8-heptafluoro-2,2-dimethyl-3,5-octanedione	H(fod)	$R_1 = -C(CH_3)_3$	$R_2 = -C_3F_7$
VI 4,4,4-trifluoro-1-(2-thienyl)-1,3-butanedione	H(tta)	$R_1 = $ 2-thienyl	$R_2 = -CF_3$
VII 4,4,5,5,6,6,6-heptafluoro-1-(2-thienyl)-1,3-hexanedione	H(hfth)	$R_1 = $ 2-thienyl	$R_2 = -C_3F_7$
VIII 4,4,4-trifluoro-1-(2-furyl)-1,3-butanedione	H(tfb)	$R_1 = $ 2-furyl	$R_2 = -CF_3$
IX 4,4,5,5,6,6,6-heptafluoro-1-(2-furyl)-1,3-butanedione	H(hfh)	$R_1 = $ 2-furyl	$R_2 = -C_3F_7$
X 4,4,4-trifluoro-1-phenyl-1,3-butanedione	H(ptb)	$R_1 = $ phenyl	$R_2 = -CF_3$
XI 4,4,5,5,6,6,6-heptafluoro-1-phenyl-1,3-hexanedione	H(phd)	$R_1 = $ phenyl	$R_2 = -C_3F_7$
XII 4,4,4-trifluoro-1-(2-naphthyl)-1,3-butanedione	H(tnb)	$R_1 = $ 2-naphthyl	$R_2 = -CF_3$
XIII 4,4,5,5,6,6,6-heptafluoro-1-(2-naphthyl)-1,3-hexanedione	H(hnh)	$R_1 = $ 2-naphthyl	$R_2 = -C_3F_7$

binuclear applications, the broadening with Yb(III) is within acceptable limits, and coupling is retained in the shifted spectra. The best upfield binuclear reagents are those involving Pr(III).

For halogenated substrates, the shifts with the binuclear reagents are considerably smaller, and acceptable shifts are only obtained by using chelates of Dy(III).[1180] Dy(III) is noted for causing the largest shifts of all of the lanthanides. For applications with organic salts, the distance between the substrate and lanthanide ion is reduced, and europium reagents provide acceptable shifts.[630]

B. Ligand

Selection of the best ligand for binuclear reagents is complicated by the fact that the ligand of both the lanthanide tris chelate and silver β-diketonate influence the magnitude of the shifts. The selection of chiral binuclear LSR will be discussed in Section IV. The structures, names, and abbreviations for the achiral ligands evaluated in binuclear shift reagents are shown in Table 1. Lanthanide tris chelates with ligands I through XIII have been studied. The concensus of these reports is that only the chelates of fod are effective in binuclear applications.[83,1181-1183] The silver complexes of ligands I through XIII have been evaluated with Yb(fod)$_3$ and other chelates of Yb(III).[83,1180-1183] The best shifts were observed with the combination of Yb(fod)$_3$ and Ag(fod). Fewer combinations of shift reagents with Pr(III) have been evaluated, although the results vary considerably from those with Yb(III).[1180,1183] Pr(fod)$_3$ is the tris chelate of choice; however, the silver complexes with tta and tfa, when paired with Pr(fod)$_3$, result in larger shifts than the reagent with Ag(fod).[1180,1183] The combination of Pr(fod)$_3$ and Ag(tfa) is recommended as the first choice for an upfield shift reagent. Ag(fod) and Ag(tfa) are available from commercial sources.

The reasons why changing only one ligand of the binuclear complex causes such a pronounced variation in the effectiveness of the shift reagent are not known. A number of possible explanations may be offered.

Lanthanide chelates with bulky ligands such as dpm may not be able to form the tetrakis chelate anion believed to be the active shift reagent species.[1181,1182] This reasoning cannot explain the ineffectiveness, however, of many other binuclear reagents. Another possibility is that the association constant between binuclear complexes and substrates varies with the nature of the ligand. This mechanism seems unlikely, based on the complexation shifts recorded in the spectrum of 1-hexene with La(fod)$_3$-Ag(fod), La(fod)$_3$-Ag(tfa), Lu(fod)$_3$-Ag(fod) and Lu(fod)$_3$-Ag(tfa).[1180] The complexation shifts were essentially identical for all four diamagnetic reagents.[1180] Since La(III) is close in size to Pr(III), and Lu(III) is close in size to Yb(III), pronounced differences in the association constant across the series of lanthanides should be reflected in the complexation shifts.

It is also possible that the geometry of the shift reagent-substrate complex changes with both the ligands and the size of the lanthanide ion.[83] Any changes in the geometry would be reflected in the distance and angle terms of the dipolar shift equation. The magnetic susceptibility of the lanthanide ion, which is incorporated into the constant term of the dipolar equation, might also vary with the nature of the ligands. Finally, Horrocks' and Sipe's proposal for explaining the ineffectiveness of certain tris chelates may be extended to certain binuclear reagents.[110] Binuclear complexes with small ligands such as acac, tfa, and hfa may exhibit a large number of configurational isomers in solution. The shifts in the spectrum of the substrate would then average out to zero.[1181]

II. EXPERIMENTAL TECHNIQUES

A. Solvent

The solvents that can be employed with the binuclear reagents are more limited than with the tris chelates. The principal criteria are that a solvent must not bond to either the lanthanide ion or the silver ion. Either situation results in a significant reduction in the shifts in the spectrum of a substrate.[1180] Only small shifts were observed in the spectra of substrates in benzene, acetone, acetonitrile, and dimethylsulfoxide.[1180] The silver ion reacts with carbon disulfide to produce an insoluble material, rendering this solvent unsuitable for use.[1180] Chloroform, carbon tetrachloride, cyclohexane, and pentane[1184] are suitable for use with the binuclear reagents. The magnitudes of the shifts were found to exhibit the following trend with solvent: pentane > cyclohexane > CDCl$_3$ ~ CCl$_4$.[1180,1184] The improvements in pentane are especially pronounced for those substrates that bond only weakly to the binuclear reagents. Typical values for the improvements in the shifts in pentane are for organochlorides (100%), organobromides (10 to 20%), aromatics (50%), and olefins (15%).[1184] The pentane resonances overlap with the region of the spectrum from 1 to 4 ppm and cannot be used with substrates whose shifted spectrum has peaks in this region.

B. Shift Reagent Preparation and Purification

Preparation and purification of lanthanide tris chelates to be used in binuclear applications takes on the same dimensions as previously discussed in Chapter 2. Once again the degree of purity is dependent on the particular interest of the user. Quantitative fitting of the dipolar shift equation requires rigorous purification, whereas qualitative studies often require less stringent purification.

Preparation and purification of the silver β-diketonate presents certain additional problems that may ultimately limit the use of binuclear reagents in quantitative studies.[9,1180] The silver β-diketonates are light sensitive and therefore should be stored at all times in light-proof containers. The shelf life varies among the different silver β-diketonates and even exhibits some batch to batch variability. We have found that Ag(fod) and Ag(tfa) can be stored for at least a year without unacceptable degradation. The light sensitivity of the silver β-diketonates is accentuated in solution. This presents no special problems in running spectra; in fact, the spectra of solutions containing a binuclear reagent and substrate that were covered to exclude light did not change over a period of 4 days.[1180] What is found, however, is that recrystallization of the silver β-diketonates usually results in a material no more pure than the original product. Attempts at sublimation resulted in the decomposition of the silver β-diketonate.[1183] For these reasons, the silver β-diketonates have been used without further purification. After preparation, the silver β-diketonates should be stored *in vacuo* over P_4O_{10} for 24 hr and then stored over P_4O_{10} until the time of use.

C. Use of Binuclear Reagents

Addition of a binuclear reagent to solution usually results in the formation of some insoluble material. The exact nature of this material is unknown, but it is believed to result from either impurities in the silver β-diketonate or a reaction of the silver with impurities in the solvent. The procedure we follow is to weigh out the appropriate amount of lanthanide chelate, silver β-diketonate, and substrate into a test tube or NMR tube. Solvent is added, and the tube is stoppered and shaken vigorously for 2 min. The mixture is centrifuged and the supernatant is decanted into an NMR tube for analysis. During this process, the solutions are covered with aluminum foil to exclude light.

III. STRUCTURE AND THEORY

Little detail is known about the structure and theoretical basis of the shift mechanism of the binuclear reagents. Evidence indicates that a 1:1 complex between the lanthanide tris chelate and silver β-diketonate is formed in solution.[141] The structure of the resulting complex is unknown, although indications are that an ion pair is formed between the silver cation and a lanthanide tetrakis chelate anion.[630,1180] Such structures are known for binuclear lanthanide-cesium complexes with β-diketonate ligands.[1185,1186] Solid state crystal structures of binuclear lanthanide(III)-silver(I) β-diketonate complexes have yet to be reported.

The principal evidence for the formation of an ion pair came from a study that attempted to reproduce the shifts obtained in the spectrum of toluene with Eu(fod)$_3$-Ag(fod) using other combinations of europium and silver complexes.[1180] It was found that mixing K[Eu(fod)$_4$] and AgBF$_4$ in solution resulted in a precipitate that was presumably KBF$_4$. The shifts in the spectrum of toluene with the remaining lanthanide species, presumably Ag[Eu(fod)$_4$], were essentially the same as those with Eu(fod)$_3$-Ag(fod). Recent work has shown that organic cations can be exchanged for the silver ion in the binuclear reagents.[630] If the organic salt contains a halide counterion, silver halide precipitates from solution leaving the ion pair R$^+$[Eu(fod)$_4$]$^-$. Mass spectra of Ag[Ln(fod)$_4$] show the presence of the ion Ag[Ln(fod)$_4$]$^{+\,[141]}$ and Ag[Ln(fod)$_3$]$^+$.[141,1181]

If an ion pair is formed in solution, the exact mechanism by which the silver cation and lanthanide tetrakis chelate anion associate is still uncertain. The silver could bond to oxygen atoms of the chelate ligands,[1185] pi electrons of the chelate rings,[1187] or the methine carbon.[1188] Each of these situations has been observed in other bimetallic β-diketonate complexes. One final possibility is a simple electrostatic attraction of the two ions.

The stoichiometry of the binuclear reagent-substrate complexes has been the focus of only limited study. Offermann has reported that 1:1 stoichiometry is observed between a wide range of alkene substrates and Ag(fod) in CDCl$_3$.[1189] The substrates in this study included a variety of olefins with markedly different steric and electronic effects.

The shift mechanism of the binuclear reagents is expected to be described by either the extended or simplified form of the dipolar equation. Since the binuclear shift reagent-substrate complex undergoes rapid intra- and intermolecular exchange at ambient probe temperatures,[141,1180] it is reasonable to assume that the criteria necessary for "effective" axial symmetry are met.[11,148,1190,1191] This would permit the simplified equation to be employed with these reagents. Contact shifts are expected to be minimized with the binuclear reagents because of the diamagnetic silver ion bridging the substrate and lanthanide.[83] Two workers have reported the presence of contact shifts with the binuclear reagents.[1192,1193] Audit used the method of internal ratios to postulate a contact mechanism for the shifts in the proton spectrum of cis-2-hexene.[1192] This method requires uniform geometry of the shift reagent-substrate complex for all of the lanthanides. It seems unlikely that the binuclear shift reagent-substrate complexes are isostructural for all members of the series when one considers that $Pr(fod)_3$-$Ag(tfa)$ was more effective than $Pr(fod)_3$-$Ag(fod)$, whereas the opposite was observed with binuclear reagents with $Yb(fod)_3$.

Dambska and Janowski,[1193] employing the method described by Reilley et al.,[1194] reported the presence of contact shifts in the ^1H spectra of 1-hexene, styrene, and ethylbenzene. It was reported that the contact shifts were more pronounced for ethyl benzene. This method requires that the binuclear complexes be isomorphous so that each lanthanide ion has identical crystal field parameters. Once again the influence of the ligands on the effectiveness of the binuclear reagents implies that isomorphous structures are not observed. Until further work is done, the exact shift mechanism of the binuclear reagents remains unknown.

IV. APPLICATIONS

A. Olefins

Several reports have described the application of binuclear LSR to olefins.[8,1180,1181,1183,1192,1195-1197] Shifts in the ^1H spectrum of 1-hexene[1180,1192] and ^{13}C spectrum of 2-methyl-2-butene with binuclear shift reagents have been reported.[1195] The binuclear shift reagents can be used to distinguish cis and trans olefins.[1180,1192,1196,1197] The spectrum of a mixture of a cis and trans olefin exhibits a characteristic dependence on the concentration of shift reagent. At low L:S ratios, all of the resonances for the cis isomer are shifted further. This results on account of steric effects that cause the association constant between the cis isomer and silver to be larger than that for the trans.[1198] At low L:S ratios, the cis isomer competes more effectively for coordination sites on the shift reagent. As the L:S ratio is raised above one, a reversal of the order of the shifts is observed, and all of the resonances for the trans isomer exhibit larger shifts. This concentration dependence of the relative magnitude of the shifts has been noted for cis-trans-2-hexene,[1180,1192,1196] 2-octene,[1180] and 4-nonene.[1199]

The reason why the resonances of the trans isomer are shifted further at high L:S ratios is not clear. At such a ratio, any competitive bonding effects must be removed. The most likely explanation would seem to involve differences in the geometric term of the dipolar equation (distance and angle values) between the cis and trans olefin. Relaxation data for cis- and trans-2-hexene with $Gd(fod)_3$-$Ag(fod)$ indicated that distances alone could not explain the different shifts for the two isomers.[1192]

Cyclic[1183,1195] and multicyclic olefins[8,1181,1195] have been the focus of studies with binuclear LSR. The shifts in the ^{13}C spectrum of methylcyclohexene with $Yb(fod)_3$-$Ag(fod)$ have been reported.[1195] The shifts in the ^1H spectrum of α-pinene, β-pinene, delta-3-carene, and camphene with Yb(fod)$_3$-Ag(fod) have been reported.[1181] In each example, the shifted spectrum was more informative than the unshifted spectrum. The shifts in the ^{13}C spectra of α-pinene, norbornene, and *syn*-(I) and *anti*-(II) sesquinorbornene with Yb(fod)$_3$-Ag(fod) have been reported.[1195]

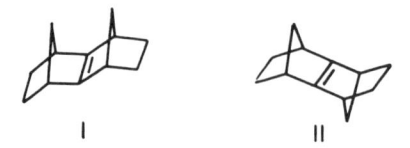

The shifts in the ^{1}H and ^{13}C spectra of the multicyclic compounds indicated that the silver bonded preferentially to one side of the olefin bond.[1181,1195] Such a preference was the result of steric effects. The preferential bonding enhanced the selectivity of the shifts, providing more interpretative value to the shifted spectra.

B. Alkynes

Alkynes bond to silver and can therefore be studied with the aid of binuclear LSR. The shifts in the spectrum of 2-octyne with Yb(fod)$_3$-Ag(fod) have been reported.[1180] The *t*-butyl resonance of the fod ligands overlapped with certain resonances in the region from 1 to 3 ppm. This problem was eliminated by recording the spectrum with Dy(dfhd)$_3$-Ag(fod). The *t*-butyl resonance of the one fod ligand in this combination was upfield of TMS.

C. Aromatics

Binuclear LSR have been used to shift the spectra of monocyclic[9,1178,1179,1182,1200,1201] and polycyclic[9,1182] aromatic hydrocarbons. The bonding of aromatic substrates to silver(I) is considerably weaker than that of olefins,[1202] but adequate shifts are observed in the spectra of aromatic substrates with the binuclear reagents. The location of silver binding to aromatic substrates is predicated by steric effects. For a compound such as toluene, the silver preferentially bonds to the ring at positions removed from the methyl group. As a result, the proton *para* to the methyl group exhibits the largest shift in the presence of a binuclear LSR.[9,1178] The spectrum of toluene in the presence of Yb(fod)$_3$-Ag(fod) or Pr(fod)$_3$-Ag(tfa) is first order at 90 MHz.[9]

The ^{1}H spectra of other methyl benzenes can be converted to first-order spectra using the binuclear reagents.[1178,1182,1201] Binuclear reagents have also been used to quantitate mixtures of methyl benzenes.[1178,1179,1182] The methyl resonances of mixtures of ethylbenzene, *o*-, *m*-, and *p*-xylene, are fully resolved in the presence of binuclear reagents permitting quantitation of the isomers.[1182] The spectrum of unleaded gasoline in the presence of Yb(fod)$_3$-Ag(fod) exhibits unique methyl resonances for toluene; *o*-, *m*-, and *p*-xylene; 1,2,3-trimethylbenzene, 1,2,4-trimethylbenzene, 1,3,5-trimethylbenzene, and 1,2,4,5-tetramethylbenzene.[1182]

The shifts in the ^{13}C spectra of toluene and *o*-xylene with Eu(fod)$_3$-Ag(fod) have been measured.[1182] The resonances of carbon atoms near the preferred site of complexation exhibited upfield shifts in the presence of the binuclear reagent. Complexation of silver causes upfield shifts in ^{13}C spectra,[1203,1204] and for toluene and *o*-xylene, the upfield complexation shifts are probably larger than the downfield shifts caused by the europium.[1182] The complexation shifts can be recorded with La(fod)$_3$-Ag(fod) or Lu(fod)$_3$-Ag(fod).

The shifts in the ^{1}H spectra of benzene, *o*-xylene, and 1,3,5-trimethylbenzene with Pr(fod)$_3$-Ag(fod) were compared to the shifts when these substrates were complexed with Cr(CO)$_3$.[1201] The shifts in the spectra of the complexes with Cr(CO)$_3$ were larger with the symmetrical compounds benzene and 1,3,5-trimethylbenzene. For o-xylene, larger shifts were observed with Pr(fod)$_3$-Ag(fod). These differences may reflect the fact that Cr(CO)$_3$ bonds over the center of the aromatic ring,[1201] whereas Ag(I) is known to bond over the carbon-carbon bonds of the ring.[1202]

A secondary deuterium isotope effect with binuclear shift reagents has been noted in the ^{1}H NMR spectra of p-xylene/p-xylene-d_6 and benzene/benzene-d.[1200] For the p-xylenes, resolution of the aromatic signals of the deuterated and nondeuterated derivatives was observed. The shift was larger for the resonance of the deuterated derivative. Complete resolution of the signals for benzene and benzene-d was not obtained; however, a shoulder attributable to the derivative of lower concentration was noted. Once again a larger shift was observed for the deuterated derivative.

The shifts in the spectra of p-terphenyl[9] and fused ring polycyclic aromatic hydrocarbons have been reported.[1182] Selective shifts were only observed in the spectra of polycyclic aromatics with pronounced steric encumbrances. The spectra of naphthalene, phenanthene, anthracene, pyrene, 1,2:5,6-dibenzanthracene, and benzo[a]pyrene were shifted in the presence of binuclear LSR; however, all of the resonances shifted essentially the same distance.[1182] Selective shifts were noted in the spectra of 1-methylnaphthalene, acenaphthene, 1-methylphenanthrene, 9-methylphenanthrene, and 2-methylanthracene.[1182] For each of these substrates, the ring substituents result in more specificity in the location of silver binding. Unique resonances were observed for almost all of the aromatic hydrogens of these substrates in the presence of Yb(fod)$_3$-Ag(fod).

D. Halogenated Compounds

The binuclear shift reagents can be used with primary chlorides, -bromides, and -iodides and secondary chlorides.[83,1180] Secondary bromides and -iodides and tertiary chlorides, -bromides, and -iodides react with the binuclear reagents, presumably producing a silver halide.[1180] For a series of halides such as 1-chloro-, 1-bromo-, and 1-iodobutane, the largest shifts were observed in the spectrum of the iodo derivative.[1180] Since iodine is the softest Lewis base of the halogens, the relative magnitude of the shifts probably reflects differences in the association constants between silver and the substrates.

The shifts in the spectra of halogenated compounds in the presence of the binuclear shift reagents are small in comparison to those obtained in the spectra of most other substrates with LSR. As a result, the shifts observed with chelates of Pr(III) and Yb(III) are not of much practical utility. In order to obtain useful shifts in the spectra of organohalides with the binuclear reagents, chelates of Dy(III) must be employed.[1180,1184] Complexes with Dy(III) cause upfield shifts and are noted as being the most powerful of the lanthanide shift reagents. Complexes with Dy(III) also cause the most broadening in the spectra of substrates. The broadening in the spectra of organohalides in the presence of binuclear reagents with Dy(III) is within acceptable limits, although, most of the coupling is lost. Sharpening of the signals was noted, however, in decoupling experiments.[1180]

Shifts in the ^{1}H spectra of 1-bromopentane, 1-chloropentane, 1-iodohexane, and chlorocyclohexane with Dy(fod)$_3$-Ag(fod) have been reported.[1180] The shifts in the spectra of aromatic halides such as chloro-, bromo-, and iodobenzene with the binuclear reagents have also been reported.[1180] The shifts were small for each of the aryl halides. The lanthanide shift data for chloro- and bromobenzene were indicative of bonding to the aromatic ring. For iodobenzene, the silver preferentially bonded to the iodine atom.

E. Phosphines

As described in Chapter 2, phosphines bond weakly to lanthanide tris chelates. Silver, on the other hand, binds effectively to phosphines.[1205] The ^{1}H spectrum of triphenylphosphine in the presence of Pr(fod)$_3$-Ag(tfa) is first-order at 90 MHz.[1182] The magnitude of the shifts was indicative of complexation of the silver at the phophorus atom; a conclusion that was confirmed by the ^{31}P spectrum. The ^{31}P spectrum of triphenyl phosphine (0.15 M), Pr(fod)$_3$ (0.2 M) and Ag(fod)(0.2 M) at 22°C consists of two doublets.[1182] The doublets are the result of ^{107}Ag and ^{109}Ag coupling to the phosphorus. Increasing the concentration of triphenylphosphine higher than that of the shift reagent causes exchange of the substrate to occur and the ^{31}P resonance collapses to a singlet. Coupling between silver and phosphorus was also observed in the ^{31}P spectrum of triphenylphosphite in the presence of Pr(fod)$_3$-Ag(fod).

F. Polyfunctional Substrates

The term polyfunctional substrate is used in this chapter to denote substrates with either two or more olefin bonds or a phenyl ring and an olefin bond. It is generally recognized that olefin bonds associate more strongly with silver than do aromatic rings.[1202] For this reason, the compounds styrene[1192] (III), 4-methylstyrene,[8,1180] and indene[1192](IV)

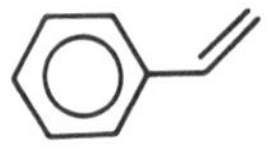

III 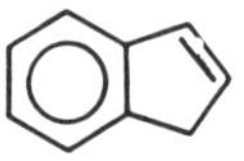 IV

complex to silver preferentially at the olefin bond. The magnitude of the shifts in the spectra of polyfunctional substrates in the presence of LSR frequently exhibit nonlinear behavior with shift reagent concentration. This behavior is usually indicative of significant complexation at a second site at higher shift reagent concentration. The shifts in the spectrum of styrene[1192] and 4-methylstyrene[1180] indicated bonding of the shift reagent at the aromatic ring at high shift reagent concentrations.

For polyfunctional substrates with two or more olefin bonds, factors such as steric effects, strain effects, and localization of the pi electrons determine the preferred site of complexation.[8,1180,1189,1192,1206,1207] Shift[8,1180,1192] and relaxation data[8,1197]

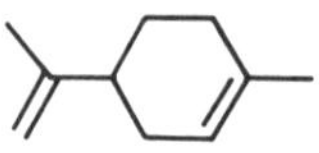

V 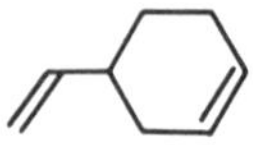 VI

have been used to show that the binuclear reagents complex with limonene(V) at the exocyclic olefin bond. This bond is less sterically hindered than the ring olefin. For 4-vinyl-1-cyclohexene(VI), ^{1}H and ^{13}C shift data indicated a significant degree of complexation at both olefin bonds.[1180] Complexation of a binuclear reagent with (cis,trans)-hexa-2,4-diene reportedly occurred at the less hindered cis olefin.[1192] Only weak complexation, and no useful spectral simplification, was noted in the spectrum of 1,6-diphenylhexatriene with a binuclear shift reagent.[1208]

Independent, but equal, complexation at the exo faces of the carbon-carbon double bonds of norborna-2,5-diene was observed with a binuclear reagent.[1192,1208,1209] The shifts in the spectrum of 5-methylenenorborn-2-ene(VII) with binuclear lanthanide shift reagents indicated that complexation occurred at the exo face of the intracyclic olefin bond.[1192] The

VII

exocyclic olefin is less sterically hindered, but silver bonding at the intracyclic bond causes a greater relief of strain.[1192]

The spectrum of 3,7-dimethylenebicyclo[3.3.1]nonane (VIII) was recorded with binuclear reagents involving Eu(fod)₃ and AgNO₃, AgClO₄, Ag(OAc), and Ag(fod).[1210] The shifts with Ag(fod) were far superior to the other combinations. It was postulated that, for all combinations, the substrate complexed in a bidentate manner with the silver.

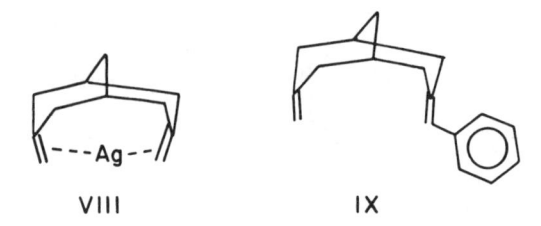

VIII   IX

A marked difference in the shifts was noted in the spectrum of 3-methylene-7-benzylidenebicyclo[3.3.1]nonane(IX) with the binuclear reagents Eu(fod)₃-AgNO₃ and Eu(fod)₃-Ag(fod).[1210] The shift data with AgNO₃ was indicative of bidentate bonding similar to that with VIII. With Ag(fod), it appeared that the shift reagent bonded to the *exo* face of the 3-methylene group.[1210]

The spectra of polypentenylene, polyoctenylene, and polybutadiene exhibited improved resolution in the presence of binuclear shift reagents.[1211] From the shifted spectra, it was possible to distinguish the configurations of the olefin bonds and assign resonances to specific diads and triads. The content and distribution of *cis* and *trans* units in polybutadiene and *trans*-polypentenamer was determined using Eu(fod)₃-AgNO₃.[1212] It was also possible to resolve *cis-cis, cis-trans, trans-cis,* and *trans-trans* diad configurations using the binuclear shift reagent.

G. Multifunctional Substrates

The term multifunctional substrate is used in this chapter to denote a compound that has both a hard Lewis base (oxygen or nitrogen atom) and a soft Lewis base. An obvious question is whether the binuclear reagents are specific enough to probe soft Lewis bases in multifunctional substrates. If so, one could employ lanthanide tris chelates to obtain a series of shifted spectra reflecting bonding at one particular site; and then, a binuclear complex to obtain a second series of spectra for a second site within the compound. Although it is somewhat dependent on the properties of particular substrates, hard Lewis bases tend to disrupt the formation of the binuclear complex by bonding directly to the lanthanide tris chelate in solution. The shifts in the spectrum of a multifunctional substrate in the presence of a binuclear reagent most often reflect a certain degree of bonding at both donor sites.[1192,1199,1210,1213,1214]

This has been observed with 1-hydroxymethyl-3,7-dimethylenebicyclo[3.3.1] nonane,[1210] 1,2-dimethoxybenzene,[1192] and linalool(X).[1199]

X

Shift data with Eu(fod)$_3$-Ag(fod) facilitated the structural determination of permethyl ether derivatives of compounds extracted from mushrooms.[1213] The shifts were indicative of complexation of the shift reagent at certain methoxy and olefin groups.

Binuclear reagents have been recommended for use with certain substrates that associate strongly with lanthanide tris chelates and, as a result, have severely broadened spectra.[1214] Examples include the compounds isooxazole(XI), thiazole(XII), benzothiazole, and 1-methylimidazole(XIII). Each of these

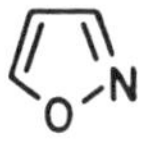

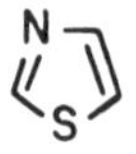

has a C=C or C=N bond at which the binuclear shift reagent can associate. The shifts with the binuclear reagent are considerably smaller than with the tris chelates, but the absence of broadening results in spectra of more interpretative value.[1214]

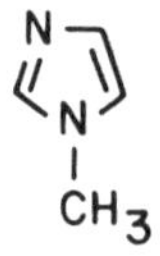

Binuclear reagents have also been shown to be effective for certain compounds with a nitrogen or oxygen atom that, by virtue of either electronic or steric effects, associate weakly with the tris chelates.[1192,1197,1201,1208,1209,1214,1215] These compounds must have a functional group to which silver can bond. Examples include nitrogen-containing compounds such as pyrrole(XIV), indole(XV), N-methylindole,[1201,1208]

azobenzene,[1214] benzonitrile,[1214] and cyanine dyes.[1215] For azobenzene, the largest shifts in the ^{1}H spectrum in the presence of Pr(fod)$_3$-Ag(tfa) were noted for the protons *ortho* to the azo group.[1214] A similar finding was noted for benzonitrile.[1214] These observations imply that the silver bonds to the −N=N− bond in azobenzene and the −C≡N bond in benzonitrile rather than the phenyl rings. For aliphatic nitriles, the shifts with the tris chelates were larger than with the binuclear complexes.[1214]

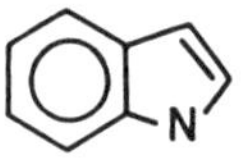

The binuclear reagents reportedly bond to the C=N bond of benzylidene(anilines) (XVI).[1214]

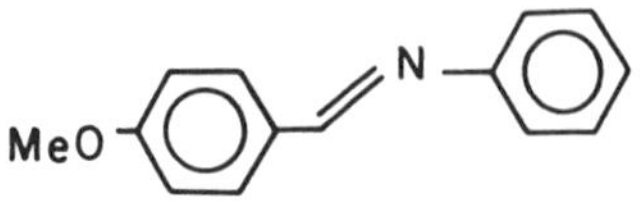

Larger shifts were noted in the spectrum of the *trans* isomer. The application of binuclear LSR to these compounds is limited, however, because of a slow *trans-cis* isomerization that occurs in the presence of the shift reagent.

The shifts in the spectra of 1-methoxybenzene,[1192] benzofuran,[1192,1197,1208] methyl benzofurans,[1192] benzoxazole(XVII),[1208]

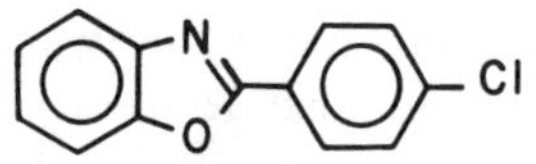

XVII

and 7-t-butoxynorbornadiene[1209] were larger with binuclear reagents than with lanthanide tris chelates. The shift data for 7-t-butoxynorbornadiene indicated that complexation of the silver occurred at the *exo* face of the olefin bond near the butoxy group. A chelate bond involving the carbon-carbon double bond and ether oxygen was proposed for this substrate.[1209]

The spectra of sulfur-containing compounds such as thiophene,[1197] benzothiophene,[1192,1208] and 2-ethylbenzothiophene[1192] are shifted in the presence of the binuclear reagents. The shift data indicate that these substrates complex to silver through the sulfur atom.[1192,1197]

H. Organic Salts

The species $Ln(fod)_4^-$, formed in solution from $Ln(fod)_3$ and $M(fod)$ ($M = K^+$ or Ag^+) has been found to be an effective shift reagent for organic salts.[630,1216] Reactions 3 and 4 are representative of the proposed equilibria.

$$Ln(fod)_3 + M(fod) = M[Ln(fod)_4] \qquad (3)$$

$$M[Ln(fod)_4] + R^+X^- = R[Ln(fod)_4] + MX(s) \qquad (4)$$

If the resulting MX salt is insoluble in the solvent, the system is forced in favor of the ion pair involving an organic cation (R^+) and a lanthanide tetrakis chelate anion. The shifts in the spectrum of R^+ are considerably larger with $Ln(fod)_4^-$ than with $Ln(fod)_3$. The applicability of $Ln(fod)_4^-$ as a shift reagent for ammonium,[630] sulfonium,[1216] and isothiouronium salts[1216] has been demonstrated. The ability of a species of the formula $Ln(\beta\text{-dik})_4^-$ to cause shifts in the spectrum of an organic cation was actually noted as far back as 1968 by Burkert et al.[1217] The chemical shifts for the piperidinium cation(pip) in the complexes $[Pr(dbm)_4]$pip (dbm = 1,3-diphenyl-1,3-propanedione) and $[Pr(benzac)_4]$pip (benzac = 1-phenyl-1,3-butanedione) were reported.

The procedure for using the binuclear reagents for organic salts is much the same as described in the experimental section of this chapter. In a test tube or NMR tube is placed the appropriate amount of $Ln(fod)_3$, $M(fod)$, substrate, and solvent. The tube is stoppered and shaken vigorously for a period of time. The time allowed for dissolution of the substrate and shift reagent is dependent on the solubility of the substrate. Sparingly soluble organic salts are gradually solubilized because of the precipitation of the metal salt with the counterion. Periodic shaking over the course of 1 hr is recommended for salts that are only sparingly soluble in CCl_4 or $CDCl_3$.[1216]

The shifts in the spectra of organic salts in the presence of $Ln(fod)_4^-$ are larger than those of olefins and aromatics in the presence of the binuclear reagents. This finding reflects the absence of the silver bridge and closer association of the organic cation and lanthanide ion. For most substrates, the shifts in the spectrum with $Eu(fod)_4^-$ are of sufficient magnitude.[630,1216] The choice of M in the M(fod) species depends on the anion of the organic salt. For halide counterions, the shifts when Ag(fod) were used were larger than those with K(fod).[630] For tetrafluoroborate salts, the results with K(fod) were superior.[1216] These differences are presumably the result of solubility differences of the salts formed from K^+ or Ag^+ and halide or BF_4^- ions.

Examples of ammonium salts the spectra of which were effectively shifted with Eu(fod)$_4^-$ include N-methylnicotinium iodide, N-ethylquinolinium iodide, diethylamine hydrobromide, dimethylamine hydrochloride, and [3-(dimethylamino)propyl]trimethylammonium iodide.[630] Only small shifts were observed for tetra-n-butyl ammonium iodide and diphenylamine hydrochloride in the presence of Eu(fod)$_4^-$. This is probably the result of the steric hindrance of these substrates. Adequate shifts were obtained in the spectrum of tetra-n-butylammonium iodide in the presence of Yb(fod)$_4^-$. For diphenylamine hydrochloride, however, only small shifts were observed with Yb(fod)$_4^-$.

Sulfonium salts, the spectra of which were effectively shifted with Eu(fod)$_4^-$, include trimethylsulfonium iodide, S-methyltetrahydrothiophenium iodide and -tetrafluoroborate, n-butyldimethylsulfonium iodide and -tetrafluoroborate, di-n-butylmethylsulfonium iodide and -tetrafluoroborate, benzyldimethylsulfonium tetrafluoroborate, and S-methylthianaphthenium tetrafluoroborate(XVIII).[1216] The species Yb(fod)$_4^-$ were necessary to obtain adequate shifts in the spectrum of ethyldiphenylsulfonium tetrafluoroborate. Resolution of certain diastereotopic hydrogens for the n-butyldimethyl- and di-n-butylmethylsulfonium cations was observed in the presence of the shift reagent.[1216] The four different ring protons of the methyl tetrahydrothiophenium cation(XIX) were resolved in the presence of Eu(fod)$_4^-$.[1216]

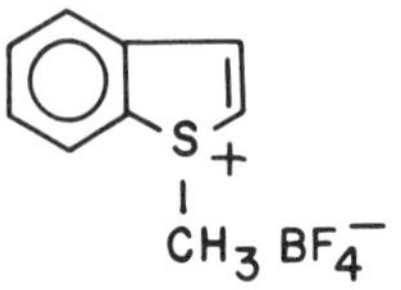

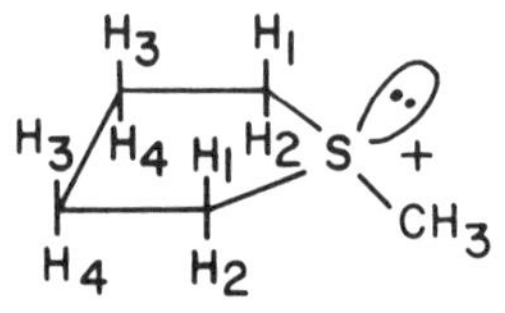

It is not known whether the shift reagent associates preferentially on the side of the ring with the methyl group or the lone pairs of electrons.

A finding of these studies was that the shifts in the spectrum of a sulfonium cation in the presence of Eu(fod)$_4^-$ were considerably larger than those for the corresponding sulfide in the presence of Eu(fod)$_3$. Since sulfides are readily converted into sulfonium salts,[1218-1227] a means of studying these compounds with NMR shift reagents is realized. Certain reactive alkyl halides can be converted into sulfonium salts by reacting them with methyl sulfide.[1218-1220] The shifts in the spectrum of the salt in the presence of Eu(fod)$_4^-$ are larger than those in the spectrum of the alkylhalide in the presence of binuclear shift reagents. Since many alkylhalides are not readily converted into sulfonium salts, however, a general scheme for analyzing alkyl halides is to convert them into their corresponding isothiouronium salts.[1228-1230]

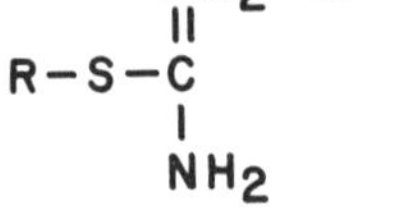

Isothiouronium salts (XX) are prepared by the reaction of the alkylhalide with thiourea.[1228-1230] Large shifts were observed in the spectra of n-butylisothiouronium chloride, n-butylisothiouronium iodide, n-pentylisothiouronium bromide, n-nonylisothiouronium bromide, benzylisothiouronium chloride, and sec-butylisothiouronium chloride and -bromide[1216] in the presence of Ln(fod)$_4^-$. It is recommended that Yb(fod)$_4^-$, rather than Eu(fod)$_4^-$, be used as the shift reagent since the thiouronium center at which the shift reagent associates serves to remove the alkyl portion of the compound from the lanthanide ion. The resonances of the diastereotopic methylene hydrogen atoms in sec-butylisothiouronium chloride and -bromide were resolved in the presence of Yb(fod)$_4^-$.[1216] Enantiomeric resolution was observed in the spectrum of sec-butylisothiouronium bromide with the species Eu(facam)$_3$(fod)$^-$ formed from Eu(facam)$_3$ and Ag(fod).

I. Chiral Binuclear Reagents

Chiral binuclear reagents can be used to determine the enantiomeric purity of chiral olefins[83,1177,1181,1183,1210,1231-1235] and aromatics.[1235] The reagents are formed from a chiral lanthanide tris chelate and either a chiral or an achiral silver β-diketonate. The degree of enantiomeric resolution is highly dependent on the silver β-diketonate employed in the binuclear reagent.[83,1181,1183,1210] The particular resonances that are enantiomerically resolved also vary with the nature of the silver reagent.[1181] As with the chiral tris chelates, experimentation with a variety of chiral binuclear reagents may be necessary to achieve and optimize enantiomeric resolution.

Lanthanide tris chelates with the ligands 3-(trifluoroacetyl)-d-camphor, H(facam), and 3-(heptafluorobutyryl)-d-camphor, H(hfbc), have been used successfully in binuclear applications. Chelates with the ligands d,d-dicampholylmethane, H(dcm), and 1,1-difencholylmethane, H(dfm), have been evaluated in binuclear applications.[83] The Pr(III) complexes with dcm and dfm were tested in combination with ten silver β-diketonates.[83] For each combination, no shift or enantiomeric resolution was noted in the spectrum of the substrate. It was postulated in this study that the dcm and dfm ligands may be too sterically hindered to permit formation of the tetrakis chelate anion believed to be the active shift reagent species.

More than ten silver β-diketonates have been employed with Pr(III) and Yb(III) chelates of facam and hfbc in an effort to determine the most effective chiral binuclear reagent.[83,1181,1183] Initial screening of the different combinations was carried out using camphene and 3-methyl-1-pentene. The best enantiomeric resolution with binuclear reagents of Yb(III) was achieved with Yb(hfbc)$_3$-Ag(fod), Yb(facam)$_3$-Ag(tta), and Yb(facam)$_3$-Ag(facam).[83] The combination of Yb(facam)$_3$ and Ag(facam) is particularly effective at bringing about enantiomeric resolution in the spectra of chiral substrates, but suffers from the problem of having limited solubility in CDCl$_3$.[1181] Precipitation of this binuclear reagent was observed at concentrations above 0.05 M. Another problem with this reagent is that the shelf life of Ag(facam) varies, and it does not seem to be as stable as many of the other silver β-diketonates. For upfield shift reagents, Pr(hfbc)$_3$-Ag(tnb), Pr(hfbc)$_3$-Ag(ptb), Pr(facam)$_3$-Ag(tfb), and Pr(facam)$_3$-Ag(hfh) were judged most effective.[83]

Partial enantiomeric resolution of the terminal vinyl protons of methylene(2-methyl)cyclohexane was observed with Eu(facam)$_3$ and silver trifluoroacetate.[1177] Slight enantiomeric resolution was obtained in the spectrum of 3-methylene-7-benzylidenebicyclo[3.3.1]nonane with Eu(hfbc)$_3$ and AgNO$_3$.[1210,1231] No noticeable enantiomeric resolution was observed in the spectrum of this substrate with Eu(hfbc)$_3$-Ag(fod) or Eu(hfbc)$_3$-Ag(hfbc). The difference in effectiveness was believed to result from different binding interactions of the shift reagent with the substrate. Chelate bonding was proposed for AgNO$_3$ and monodentate bonding for Ag(fod) and Ag(hfbc).[1210] The binuclear reagent formed from Eu(fod)$_3$ and silver d-camphor-10-sulfonate has been used to resolve the spectra of enantiomers of bicyclo[3.3.1]nonane ring systems.[1232]

Enantiomeric resolution was obtained in the NMR spectrum of XXI through the use of Yb(hfbc)$_3$-Ag(fod).[1233]

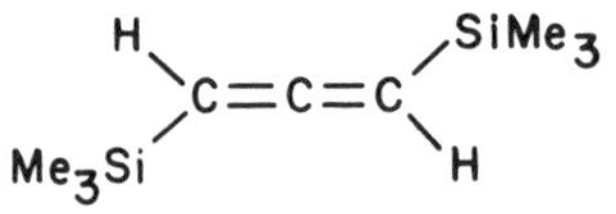

$$\underset{Me_3Si}{\overset{H}{>}}C=C=C\underset{H}{\overset{SiMe_3}{<}}$$

XXI

The olefin protons were completely resolved in the presence of the shift reagent.

The extent of enantiomeric resolution in the ¹H spectra of α-pinene, camphene, limonene, 3-methylcyclohexene, 4-vinylcyclohexene, epi-β-santalene(XXII), *trans*-menthene, and bornene with Eu(hfbc)₃-Ag(fod), Pr(hfbc)₃-Ag(fod) and Yb(hfbc)₃-Ag(fod) has been measured.[1234,1235] Enantiomeric resolution has also been noted in the ¹³C spectra of α-pinene, camphene, and bornene with the same binuclear reagents.[1234,1235] The resolution was sufficient to permit quantitative determination of enantiomeric excesses for each compound. It was also

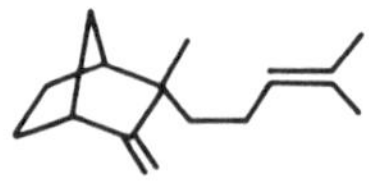

XXII 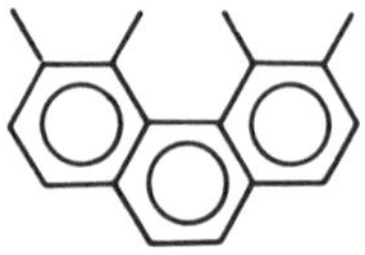   XXIII

reported that the binuclear reagent based on Yb(III) provided the best enantiomeric resolution.[1235] The spectra of norbornadiene, a compound which has protons that are enantiotopic by internal comparison, did not exhibit enantiotopic resolution in the presence of binuclear reagents.[1234]

A small degree of enantiomeric resolution was observed in the spectrum of 3,4,5,6-tetramethylphenanthrene(XXIII) with Yb(hfbc)₃-Ag(fod).[1235]

Chapter 5

THEORY OF LANTHANIDE SHIFT REAGENTS

The theoretical basis of the shifts resulting in the spectra of substrates with LSR has been widely discussed and debated in the literature. It is easy to understand the interest in this area when one considers the potential power of applying lanthanide shift data to structural determination. In its most elegant form, the shifts recorded with LSR could provide a tool for solution structural analysis as powerful as X-ray crystallography is for solid state structural analysis. Unfortunately, this lofty goal has never been realized and probably never will. Many of the problems arise from the fluxional nature of molecules in solution. The problems inherent in using shift data with LSR to arrive at conformations of molecules in solution will be discussed in this chapter.

I. SHIFT MECHANISM

A. Pseudocontact and Contact Shifts

Paramagnetic metal ions such as the lanthanides cause shifts in the NMR spectrum of a substrate by two possible mechanisms. A third mechanism, the complexation shift, is also present, but is not the result of paramagnetic properties and will not be discussed here. The two mechanisms are the pseudocontact (also through-space or dipolar) shift and contact (also scalar) shift. A contact shift results if there is a finite probability of finding the unpaired electrons of the metal at the nucleus whose NMR spectrum is being recorded. The pseudocontact shift is a result of a dipole-dipole interaction between the magnetic moment of the electrons of the metal and the nuclear spin of the nucleus of interest.

The contact shift is difficult to quantify or predict by an equation and requires a detailed evaluation of the molecular orbitals involved. The pseudocontact shift is more readily described for a particular nucleus and depends on the location of the nucleus relative to the metal ion. Since the unpaired electrons of the lanthanide ions reside in 4f orbitals, which are shielded by filled 5s and 5p orbitals, the shift mechanism with LSR is principally pseudocontact in origin.

B. Pseudocontact Shift Equation

An equation that describes the pseudocontact shift in the NMR spectrum of a nucleus in the presence of an anisotropic metal ion was first derived by McConnell and Robertson in 1958.[10] The McConnell-Robertson equation, which was derived for transition metals with partially filled 3d orbitals, related the dipolar shift to anisotropy in the g-tensor. This treatment predicted a temperature dependence of T^{-1}. Bleany used anisotropy in the susceptibility, rather than the g-tensors, to account for dipolar shifts with LSR.[1237,1238] and predicted a T^{-2} dependence for all lanthanide ions except Eu^{3+} and Sm^{3+}. Bleany's derivation assumed incorrectly that the crystal field splittings for the lowest J state were small compared to kT. This and other criticisms of Bleany's theorem were raised by Horrocks[111,1239] who employed ligand field splittings to derive an equation with a more complicated temperature dependence. Horrocks' equation predicted a large temperature dependence on T^{-1}. Horrocks' approach, however, requires a thorough knowledge of the crystal field and is not practical for many applications.[1240]

Golding and Pyykko[1241] included the higher order crystal field parameters omitted by Bleany and observed deviations of approximately 20% from Bleany's theorem.

The temperature dependence was still accounted for by a T^{-2} term. McGarvey extended Bleaney's method of calculation to include a T^{-3} term and reported a 10 to 20% improvement in accuracy.[1240] Bovee et al. measured the temperature dependence of the shifts for the 1:1 adduct of quinuclidine with chelates of fod.[1242] The data fit Bleaney's theorem provided that terms of T^{-3} and higher were included.[1242] Stout and Gutowsky reported that lanthanide shift data could be fit to a T^{-2} dependence with only a negligible T^{-3} component.[1243] Hill et al. reported a T^{-2} dependence for the dipolar shifts of tetraethylammonium N,N-diethyldithiocarbamatolanthanate(III) in acetonitrile.[1244] Shift data that were fit to $(T^{-1/2})$[1245] and (T^{-3})[1246] dependence have also been reported. Cheng and Gutowsky studied the shifts in the spectrum of dimethylacetamide in tetrachloroethane with Pr(fod)$_3$ in an attempt to fit the temperature dependence to one of the previous derivations.[1247] It was found that the complicated nature of the equilibria permitted adequate fits to a T^{-1}, T^{-2}, or mixed T^{-1} and T^{-2} dependence. The problems in attempting to use the temperature dependence of lanthanide shift data to test the theories proposed to explain the data were described.[1247] Other derivations of the dipolar shift equation have also been reported in the literature.[1248-1250] While most workers seem to favor Bleaney's theorem or a modification of Bleaney's theorem, there is still no clear agreement on the exact form of the temperature dependence and theoretical basis of the shifts with LSR.

The general form of the equation that describes the dipolar shifts with LSR is shown in Equation 1.[57,1248]

$$\Delta H/H = K[\chi_z - \tfrac{1}{2}(\chi_x + \chi_y)]((3\cos^2\theta - 1)/r^3) - K[\chi_x - \chi_y](\sin^2\theta \cos 2\Omega/r^3)$$

Magnetic susceptibility data for crystals of Ln(dpm)$_3$(pic)$_2$ have been measured, and the dipolar shifts evaluated from this solid state data were in satisfactory agreement with solution results.[57,111] Equation 1 is known as the extended form of the dipolar shift equation. The magnetic terms in the second expression of the equation exactly cancel each other when axial (C$_3$ or greater) symmetry is present. In this situation, the second term equals zero, and the simplified form of the pseudocontact equation is obtained.

$$\Delta H/H = K[\chi_z - \tfrac{1}{2}(\chi_x + \chi_y)]((3\cos^2\theta - 1)/r^3)$$

At least six assumptions are necessary to satisfy the requirements for applications of the simplified equation.[57] These are listed below.[57] The need for C$_3$ or greater symmetry in the donor can be avoided if a rapid interconversion between three equivalent rotamers takes place.[11,148]

1. The observed shifts used in the analysis are purely dipolar in origin.
2. Only a single stoichiometric complex species exists in solution in equilibrium with the uncomplexed substrate.
3. Only a single geometric isomer of this complex species is present.
4. This isomer is magnetically axially symmetric so that the shifts are proportional to the geometric factor: $[(3\cos^2\Theta-1)/r^3]$.
5. The principal magnetic axis has a particular, known orientation with respect to the substrate ligand or ligands.
6. The substrate ligand exists in a single conformation, or an appropriate averaging over internal motions is carried out.[57]

The data set obtained in a typical experiment with LSR is relatively small. Fitting the data to a unique solution using the extended dipolar equation is therefore difficult.

For this reason, most workers have attempted to fit lanthanide shift data using the simplified equation. Some of the problems with applying the simplified equation to lanthanide shift data will be discussed in the remainder of this chapter.

Attempting to explain lanthanide shift data by only invoking distance considerations is an over-simplification that may lead to serious errors in spectral and structural assignment. Ample evidence for the inclusion of the angle term exists in the form of "wrong way" shifts and distance dependencies of other than r^{-3} when the angle term is not included.[1251,1252]

II. SEPARATION OF DIPOLAR AND CONTACT SHIFTS

The simplified or extended dipolar equation can only be applied to nuclei that exhibit purely dipolar shifts. While the shift mechanism with LSR is largely dipolar in nature, a method to determine the presence of contact shifts is necessary. A number of methods for evaluating the presence of contact shifts have been reported. The advantages and disadvantages of these methods have been discussed in a paper by Reilley et al.[1194]

The most common methods, because of its relative simplicity,[68,165,1253-1257] is the ratio method first described by Horrocks et al. for use with transition metal shift reagents.[1253,1258] In this procedure, a reference nucleus is selected, and the ratio of the induced shift of each nucleus to that of the reference nucleus is calculated. In the absence of a contact shift, the ratio for a particular nucleus should remain constant for different lanthanide ions. Deviations in the ratios for different lanthanide ions imply the presence of a contact shift. This method is easy to perform: however, it requires that the shift reagent-substrate complexes be isostructural for the entire lanthanide series. This is assumed to be the case, but may not be true.

A second common procedure is to assume that a particular nucleus or set of nuclei exhibit shifts that are purely pseudocontact in origin. These nuclei are then used to fit the pseudocontact equation. The resulting geometry is used to calculate dipolar shifts for those nuclei suspected of exhibiting a contact shift. Any discrepancy between the calculated and observed shifts for these suspect nuclei are considered to result from a contact mechanism.[1259-1262] The usual procedure has been to assume that the shifts in the ^{1}H spectrum are purely dipolar and to evaluate the presence of contact shifts in the ^{13}C spectrum.

A third method that enables the direct determination of contact shifts is to measure the shifts in the spectrum of the substrate in the presence of the Gd(III) complex.[1191,1263-1266] Gadolinium(III), which has an f^7 configuration, has an isotropic magnetic field. The shifts in the spectrum of a substrate in the presence of Gd(III) result from complexation and contact effects. The complexation shifts can be measured with the diamagnetic La(III) or Lu(III) complexes. Subtraction of these values from the shifts with Gd(III) yields the magnitude of the contact shifts. The magnitude of the contact shifts with other lanthanide ions can then be calculated. The only problem with this otherwise simple method is that Gd(III) complexes also produce severe broadening in the spectrum of the substrate. For proton spectra, the broadening generally offsets any ability to reliably measure changes in the chemical shift.[1191] This method has been used to determine the magnitude of contact shifts in ^{13}C spectra.[1263-1266]

Transition metal chelates such as Ni(acac)$_2$ have been used to assess the contact shifts for a particular substrate, and the results have then been compared to data with LSR.[549,1259,1267] De Boer et al.[1191] and others[1260,1268] have attempted to use the different temperature dependences of contact (T^{-1}) and dipolar (T^{-2}) shifts to separate the contribution from each. Considering the difficulty in fitting lanthanide shift data to a particular temperature dependency, the lack of success of this method is not surprising.

Hill has reported that the contact and dipolar shifts in the spectrum of tetraethylammonium N,N-diethyldithiocarbamatolanthanate(III) could be separated on the basis of their different temperature dependencies.[1244] Theoretical treatments have also been used to assess the extent of contact contributions for the individual lanthanide ions.[1269-1271]

Reilley et al.[1194,1272] have described a method to separate contact and pseudocontact shifts that involves the use of calculated values of the relative theoretical dipolar and contact shifts for the various lanthanide ions. By recording the data for at least three lanthanide ions, it is possible to solve for the contact and pseudocontact contributions through the use of a nonlinear regression analysis. The method requires effective axial symmetry of the shift reagent-substrate complex and assumes that the crystal field splitting is constant for all of the lanthanide ions.

While these studies have found that contact contributions are minimal with LSR, they are present for certain nuclei and should be assessed if a structural analysis is to be performed by fitting the shifts to the dipolar equation. Contact shifts fall off rapidly from the site of complexation and are more significant in ^{13}C than ^1H spectra. The following order of the extent of contact shifts for the lanthanide ions indicates the preference of Yb(III) when contact shifts are to be minimized: Yb<Tm<Dy<Tb<Er<Ho<Nd<Eu.

Lee and Reilley have claimed that it is possible to prepare mixtures of Pr(fod)$_3$ and Eu(fod)$_3$ in which either the contact or dipolar contributions exactly cancel each other out.[1273] The appropriate mixtures would then constitute a contact only or dipolar only lanthanide shift reagent.

III. STOICHIOMETRY AND SYMMETRY

The optimum situation for application of the simplified pseudocontact equation would result if only one equilibrium involving the formation of a 1:1 shift reagent(L)-substrate(S) complex(Equation 1) was observed. Unfortunately,

$$L + S = LS \tag{1}$$

competing equilibria complicate the matter.[1274] The most important of these involve the formation of a 1:2 shift reagent-substrate complex(Equation 2) and the formation of a shift reagent dimer(Equation 3). Association of

$$LS + S = LS_2 \tag{2}$$

$$2L = L_2 \tag{3}$$

substrate molecules with the dimer may then take place. It has been shown that the lanthanide tris chelates of fod and dpm undergo rapid intermolecular ligand exchange at ambient probe temperatures.[73,1275] This exchange occurs in mixtures containing excess ligand in the presence of the lanthanide complex and in mixtures containing two different lanthanide chelate complexes. The mechanism for ligand exchange between two metals most likely involves a mixed dimer with ligand bridging groups.[73,1275] Evidence for equilibria 1, 2, and 3 has been obtained from a variety of sources including X-ray crystallographic data, low temperature NMR studies, vapor phase osmometry, IR spectroscopy, optical spectroscopy, emission titrations, mass spectrometry, and graphical analyses.

A. X-Ray Crystallography

The solid state structures of the complexes of Pr(III) with dpm[1276] and fod[1277] have been determined and are both dimeric in nature. In the anhydrous Pr(dpm)$_6$, each metal ion is surrounded by seven oxygen atoms with two oxygen atoms from the ligands forming the bridges(I).[1276]

I

Each metal ion of Pr$_2$(fod)$_6$.2H$_2$O is surrounded by eight oxygen atoms. Two oxygen atoms from the ligands and one from a water molecule form bridges between the metal ions. The complexes of Er(III) and Lu(III) with dpm were found to be monomeric species with trigonal prismatic structures.[72,73,1278] The smaller ionic radii of Er(III) and Lu(III) compared to Pr(III) probably accounts for the difference in the structures.

The solid state structures of the adducts of Eu(dpm)$_3$ with pyridine,[72,1279,1280] quinuclidine(II),[1281] dimethylformamide,[72,1282] 3,3-dimethylthietane-1-oxide(DMTO) (III),[244,1283] 1,10-phenanthroline,[72] DMSO,[72] Yb(dpm)$_3$ with DMSO,[72]

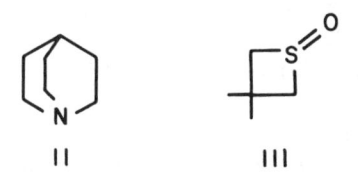

II   III

Ho(dpm)$_3$ with 4-picoline,[1284] and Lu(dpm)$_3$ with 3-methylpyridine[1285] have been elucidated. The pyridine and 4-picoline adducts exhibit 2:1 stoichiometry and have a twofold rotational axis as the highest symmetry element. The complex Eu(dpm)$_3$(DMF)$_2$ has no symmetry element higher than C$_1$.[1282] The adducts with 3-methylpyridine, 3,3-dimethylthietane-1-oxide, and quinuclidine exhibited 1:1 stoichiometry. The complex Lu(dpm)$_3$ · 3-Mepy crystallized as a capped trigonal prism with C$_{2v}$ symmetry.[1285] Eu(dpm)$_3$ · 3,3-DMTO also has C$_{2v}$ as its highest symmetry element.[244,1283] Quinuclidine, which itself has threefold symmetry, forms an adduct with Eu(dpm)$_3$ that has the threefold symmetry necessary for the application of the simplified dipolar equation.[1281]

The solid state structure of the adduct of Pr(facam)$_3$ with DMF has been determined and provides a particularly interesting example.[1286] In this case, it might be expected that the steric encumbrances of the facam ligand would be greater than those for dpm, thereby reducing the likelihood of a 2:1 complex. What was found was a dimer of the formula (facam)$_3$Pr(facam)$_3$. The oxygen atoms of the DMF molecules formed bridges between the two metal ions. Each metal atom was nine coordinate, and no evidence was found to indicate steric crowding.[1286] The ability of LSR to expand their coordination number beyond seven is also demonstrated by the structure of the adduct of Co(acac)$_3$ with Eu(fod)$_3$.[632] A 1:1 complex was found, and three oxygen atoms of the acac ligands formed bridges to the europium ion.

If the structures in solution rigorously conformed to those observed in the solid state, application of the simplified dipolar equation in studies with LSR would be invalid. What is more likely, however, is that the solid state structures represent one of a variety of configurations observed in solution. Kepert has calculated ligand-ligand repulsion energies to determine the favored structures of LS complexes in solution.[1287] Three minima were observed, and these corresponded to a capped octahedron(C_{3v}), irregular polyhedron(C_1), and a structure intermediate between a pentagonal bipyramid and capped trigonal prism(C_3).

B. Low-Temperature NMR Studies

Direct evidence for the presence of 2:1 substrate-shift reagent complexes in solution has been noted in low-temperature NMR studies performed by Evans and Wyatt,[132,1288,1289] Cramer et al.,[1290-1292] and others.[616,617] Evans and Wyatt studied the solvation numbers of a variety of lanthanide tris β-diketonates with dimethylsulfoxide(DMSO), hexamethylphosphoramide(HMPA), tetramethylurea, and triethylamine.[132] The solvation numbers were found to be dependent on the nature of the ligand, the donor properties of the substrate, the solvation of the substrate, steric repulsions between the substrate and ligand, and the radius of the lanthanide ion. Of particular interest were differences in the solvation numbers noted for complexes with fod and dpm. The stoichiometry of the adducts of HMPA with Pr(fod)$_3$ and Eu(fod)$_3$ were both 2:1, whereas the adducts of the complexes with dpm were 1:1. The only substrate of the four that formed a 1:1 complex with Pr(fod)$_3$ was triethylamine. This was also the most sterically hindered of the substrates.

Cramer et al. have investigated the adducts of 3-picoline,[1290,1291] pyridine,[1291,1292] and substituted pyridines[1292] with Eu(dpm)$_3$. The adducts exhibited 1:2 stoichiometry, and the extended dipolar equation was necessary to explain the measured lanthanide shift data. The substrate 3-picoline exhibited faster exchange with Eu(dpm)$_3$ than Eu(fod)$_3$ because of a smaller complex formation constant.[1290]

C. Vapor Phase Osmometry

Vapor phase osmometry has been used to study the extent of aggregation of chelates with dpm[139,146,156,545,1293] and fod.[139,144,156,545,1293] These studies indicated that the degree of aggregation of the fod complexes depended on the nature of the solvent,[139,156,545] the radius of the lanthanide ion[139,144,156,545] and the concentration of the complex.[1293] Nonpolar solvents lead to increased self-association. The extent of aggregation exhibited the following dependency on solvent: n-hexane>CCl$_4$>C$_6$H$_6$>CHCl$_3$.[545] Complexes of ions with a large radius such as Pr(III) exhibited more self-association than complexes with smaller ions. In no case, however, were purely monomeric complexes observed for chelates with the fod ligand. This is in distinct contrast to the chelates with dpm in which only monomeric species were indicated.[139,146,156,1293] Even in the solvent n-hexane, the complexes of Pr(III), Sm(III), Eu(III), Dy(III), Ho(III), and Tm(III) with dpm were reported to exist as monomers.[156,1293]

Feibush et al. studied chelates of facam in CHCl$_3$ by vapor phase osmometry and observed monomeric species for the smaller lanthanide ions and aggregates for larger ions.[1293] A similar finding was reported by Denning et al.[1294] The complexes of Pr(III), Nd(III), and Sm(III) with facam exhibited oligomerization in CCl$_4$, whereas the complexes with smaller lanthanide ions were monomeric. Porter et al. have studied the association of lanthanide chelates of fod with (η^5-C$_5$H$_5$)Fe(CO)$_2$(CN) in benzene.[144] The data indicated 1:2 complexes with the chelates of Eu(III), Ho(III), and Yb(III), and a mixture of 1:2 and 1:3 complexes with Pr(fod)$_3$.

D. Infrared Spectroscopy

Infrared (IR) spectroscopy has been used to determine the stoichiometry of LSR-substrate complexes.[252,1295,1296] Separate bands were observed in the IR spectrum for the free and complexed form of the substrate. Hirota and Otsuka studied the association of Eu(dpm)₃ with 2-methyl-2-propanol in CCl₄, and through the use of a Scatchard plot, determined that the complex had 1:1 stoichiometry.[1295] Kojima et al. analyzed solutions of Ln(fod)₃ and Ln(dpm)₃ with aniline.[1296] The chelates with dpm formed 1:1 complexes, whereas those with fod exhibited considerable formation of 1:2 complexes. Volka et al. reported that both 1:1 and 2:1 complexes of 1-adamantanol with Eu(dpm)₃ were observed: however, the 1:1 complex was predominant.[252]

E. Optical Spectroscopy

Changes in the intensity of the f-f absorption bands of the lanthanide ions in the presence of a substrate have been used to determine the stoichiometry of LSR-substrate complexes.[146,1297] Using Job's method, Ghotra et al. reported 1:1 stoichiometry for the complexes of Eu(dpm)₃ and Pr(dpm)₃ with pyridine, borneol, and neopentanol.[146] Optical spectroscopy can also be used to determine the symmetry of a shift reagent-substrate complex in solution.[1298] The low symmetry of complexes of the formula Ln₂(salen)₃ (salen is a ligand produced by the Schiff's base condensation of salicylaldehyde and ethylenediamine) was demonstrated by this method.[1298] Fitting of the lanthanide shift data for the ligand required the use of the extended dipolar equation.

F. Luminescence Spectroscopy

Complexes of Eu(III) with β-diketonate ligands emit an intense red luminescence characteristic of the europium ion. The intensity of the europium luminescence of tris β-diketonates increases on adduct formation because of two reasons. The first is that structural changes occur that enhance the probability of transitions. The second is that the higher coordination number of the Eu(III) reduces collisional deactivation.[1299] The enhanced luminescence in the presence of a substrate can be monitored in the form of an emission titration. Different emission titration profiles are noted for 1:1 and 1:2 adducts.[1297,1299] The alterations in the emission spectrum on adduct formation can also be used to assess the symmetry of the complex. Analysis of the luminescence spectra of adducts of substituted pyridines with Ln(fod)₃ and Ln(dpm)₃ led to the conclusion that axial symmetry was not present in these complexes.[1300]

Brittain and co-workers have used circularly polarized luminescence (CPL) to study the nature of chiral and achiral LSR-adduct complexes.[1301,1302] The CPL technique is reportedly more sensitive to structural changes in the complexes than total luminescence spectra[1301] and is particularly sensitive to steric considerations.[1301,1302] The absence of CPL in the spectra of lanthanide tris chelates with chiral camphorato ligands indicated that the ligands in these complexes were not sterically crowded.[1302] CPL spectra of N,N-dimethylformamide and DMSO complexes of lanthanide tris β-diketonates with mixed-ligands(chiral and achiral) were measured and demonstrated that the chiral ligands(facam and hfbc) exhibited greater steric effects than the achiral ligands.[1302]

G. Circular Dichroism

Nonchromophoric, optically active substrates induce circular dichroism in the electron transitions of the lanthanide ions in the tris chelates. The circular dichroism spectral changes brought about by complex formation provide a direct measure of the concentration of the resulting complexes.[1303] Andersen has used this method to study the complexation between Eu(fod)₃ and (−)-menthol in CCl₄. Under the conditions of [S] = 2[L], the principal species was the 2:1 adduct.

H. Mass Spectrometry

The high mass region of the mass spectra of Eu(fod)₃, Yb(fod)₃, and the adducts of these complexes with 1-propylamine has been analyzed.[1304] Evidence of dimeric or

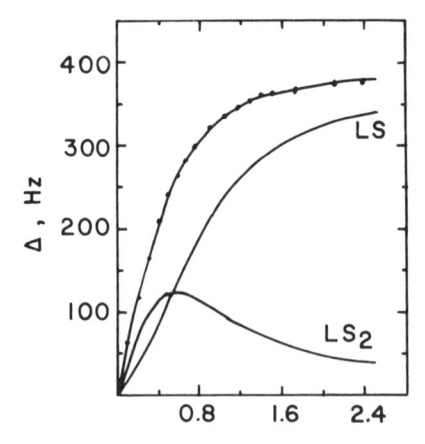

FIGURE 1. The experimental shifts of dimethylsulfoxide plotted vs. [L]/[S]. Contribution of LS and LS$_2$ are calculated with four parameters. (Reprinted with permission from Reuben, J., *J. Am. Chem. Soc.*, 95, 3534, 1973. Copyright 1973. American Chemical Society.)

higher species in the gas phase was noted in the spectra of the tris chelates. Both 1:1 and 2:1 adducts were noted in the mass spectra of the complexes with propylamine.

I. Graphical or Iterative Procedures

A variety of graphical or iterative procedures for analyzing lanthanide shift data have been described in the literature. These can be used to determine the stoichiometry, bound shifts(Δ), and equilibrium constants of the system. Many of the methods applied in studies with LSR are general methods applicable to weak molecular complexes in solution.[1305] These procedures can be divided into simple and rigorous methods. Simple methods are those that compare the magnitude of the lanthanide-induced shifts(LIS) to [L]/[S], and, by either a graphical procedure or a calculation, arrive at values for the desired constants.[68,1306-1317] These procedures generally require certain assumptions to be made or are only valid under a set of limiting conditions. Rigorous methods attempt to fit a calculated set of equilibrium constants and bound shifts to the plot of LIS vs. [L]/[S]. This method involves the use of four(K_1, K_2, Δ_1, and Δ_2) or five(K_3) parameters and was first described by Shapiro et al.[136,147,150,1318] It subsequently has been performed by other workers.[135,1275,1319,1320]

A typical diagram of the output from these rigorous fitting procedures is illustrated in Figure 1.[1319] The diagram shows the contribution to the total shift of each of the individual species in solution. Using these procedures, the presence of 1:2 adduct complexes has been noted for acetone, DMSO, 2-propanol, β-picoline,[1319] methoxy-n-butane,[135] 3-(p-chlorophenyl)-3,5,5-trimethylcyclohexanone,[136] and other cyclohexanones[147,150] with Eu(fod)$_3$ and Pr(fod)$_3$. Johnston et al.[147] reported that the fit of the data was worse if self-association of Eu(fod)$_3$(K_3) was included. De Boer et al.,[135] however, used a five parameter fit and obtained evidence for self-association of Pr(fod)$_3$ at high [L]/[S] values.

J. Conclusions and "Effective Axial Symmetry"

Almost all direct and indirect evidence indicates the formation of LS and LS$_2$ complexes in solutions of substrates with lanthanide chelates of fod. With chelates of dpm, the system tends to favor only the LS complex. The reduced solubility of the complexes of dpm compared to those of fod limits their application in quantitative solution structural analysis using the dipolar shift equation. Despite evidence that the assumptions necessary for the simplified dipolar equation are not rigorously met with LSR, success has been achieved applying the simplified equation to lanthanide shift data. Some workers have justified this by invoking the notion of "effective" or "apparent" axial symmetry.

This term was first used by Briggs et al., who demonstrated that a shift reagent-substrate system that exhibited rapid interconversion among three equivalent rotamers met the assumptions necessary for the simplified dipolar equation.[11,148] An extension of this explanation has been proposed by Horrocks.[1190] In this case, the shift reagent-substrate complex can be viewed as an ensemble of many rapidly interconverting geometrical isomers. None of these isomers possesses the necessary symmetry requirements to meet the simplified equation; however, the shifts obtained for the average observed on the NMR time scale can be explained by the simplified equation. A third explanation has been offered by de Boer et al.[1191] For many substrates, the second term of the extended dipolar equation is small relative to the first term. Omitting it from the fitting process produces errors that are no greater than other errors associated with LSR experiments. Wing et al.[244] and Horrocks et al.[57] have also reported that results obtained with the extended dipolar equation were not significantly better than those with the simplified equation. If the substrate lies in a cone with an opening angle of 40° centered at the lanthanide ion(I), omitting the second term of the dipolar equation will yield satisfactory results.[1191]

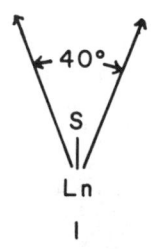

$$\underset{I}{\overset{\displaystyle \diagup\!\!\!\diagdown}{\underset{Ln}{\overset{S}{\diagdown\!\!\!\diagup}}}} \leftarrow 40° \rightarrow$$

Whichever mechanism, or mixture of mechanisms, represents the true situation, the simplified dipolar equation can often be applied to LSR studies. One must always resist the temptation to become overzealous in the degree of structural information obtained from such an experiment. The theory has yet to be totally defined, and the opportunity for the introduction of experimental error into the measured shift data is great.

IV. APPLICATION OF THE PSEUDOCONTACT SHIFT EQUATION

The typical procedure for performing structural analyses involving LSR and the dipolar shift equation is to measure a set of experimental shifts and compare them to a set of calculated shifts for proposed structures. The fitting process is most often carried out in an iterative fashion using a suitable computer program. A number of computer programs for performing such an analysis have been described in the literature.[773,1321-1327] Many of the workers whose studies are described in the following sections employ and describe other programs for such analyses. Oftentimes these programs are available upon request from the authors.

Wing et al.[1328,1329] have used a multipolar expansion to map the magnetic shielding induced by a LSR. A map of the dipolar field for an axially symmetric lanthanide-substrate complex drawn to the scale of Drieding models was provided[1328] and is shown

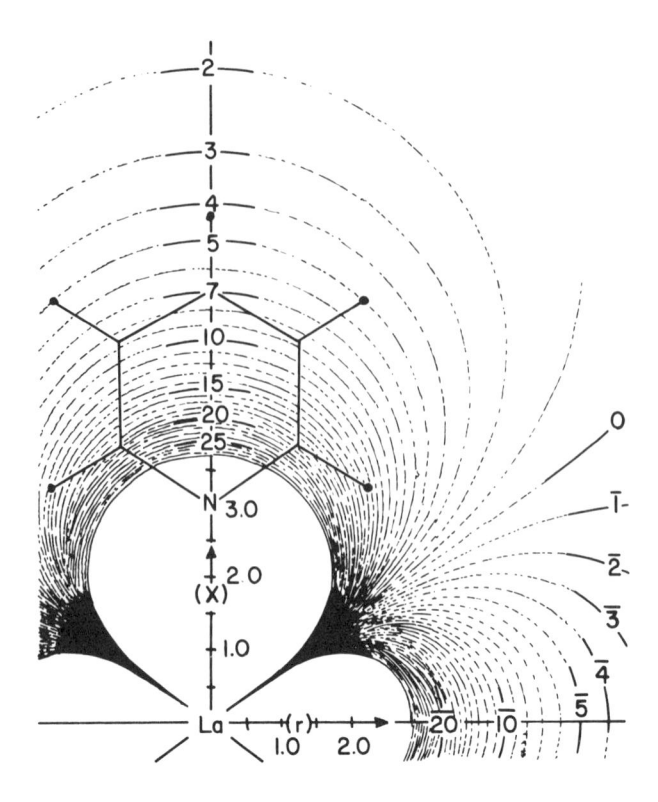

FIGURE 2. Map of the dipolar field: $((3\cos^2\theta - 1)/r^3)$ where $X = R\cos\theta$ and $R^2 = X^2 + r^2$. Pyridine has been placed on the map (the map is not drawn to scale). (Reprinted with permission from *Tetrahedron Lett.*, 4153, 1972, Wing, R. M., Early, T. A., Uebel, J. J., Copyright 1972, Pergamon Press, Ltd.)

in Figure 2. By placing the substrate appropriately on the map, the value of the $3\cos^2\theta - 1/r^3$ term can be determined. The report demonstrated how to use the map for pyridine, *cis*-4-*t*-butylcyclohexanol, and 1-adamantanol.[1328] In certain situations, this simplified approach may substitute for computer optimized fits.

The following sections describe some of the considerations inherent in performing quantitative fitting of the dipolar shift equation in studies with LSR.

A. Complexation Shifts

Shifts in the NMR spectrum of a substrate that result only through the action of complexation with the LSR are referred to as "complexation" or "complex formation" shifts. These are usually small in magnitude when compared to the shifts that result from the anisotropic magnetic properties of the lanthanide ion. Certain workers have recommended that they be accounted for when performing structural analyses using the dipolar equation.[165,343,434,435,1256,1263,1330] Complexation shifts are measured by recording the spectrum of the substrate in the presence of a diamagnetic LSR, either La(III) or Lu(III). The complex formation shift is assumed to be the same for all of the lanthanide complexes with a particular ligand. The shifts recorded with a paramagnetic metal are corrected by substracting the complexation shifts.

B. Bound Shifts

When performing a structural analysis with LSR, one is faced with the problem of how to obtain the correct value of LIS to use in fitting the dipolar equation. The shifts for a substrate in the presence of a LSR depend on variables such as temperature,[147] the magnetic field strength of the spectrometer,[1331,1332] and the absolute concentration of shift reagent and substrate.[129,200,899] The term "bound shift" is used to denote the shift of the resonance of a substrate in its complexed form. The dipolar equation describes the shifts in the NMR spectrum of a substrate in its complexed form. Bound shifts are therefore most appropriate for detailed structural fitting. The symbols Δ_1 and Δ_2 are often used to represent the bound shifts for the substrate in the LS and LS_2 complex, respectively. Since the exchange of substrate between its free and coordinated form is usually rapid on the NMR time scale, the shifts measured in the spectrum of a substrate are a time average of S, LS, LS_2 and any other species involving the substrate. In these instances, the bound shifts cannot be measured directly and must be determined by some other means.

An alternative to the use of bound shifts in structural determinations is the use of relative bound shifts.[1333-1335] Relative bound shifts are determined by calculating ratios of the LIS for the nuclei of a substrate. One nucleus of the substrate is selected as the reference for calculating the ratios. Calculations of such a set of ratios offers

$$\frac{\Delta\delta_i}{\Delta\delta_j} = \frac{k(3\cos^2\theta_i - 1)/r_i^3}{k(3\cos^2\theta_j - 1)/r_j^3}$$

two advantages. The first is that the constant (K) is canceled out and the magnetic terms for the lanthanide ion do not need to be known. Another advantage is that the concentration of shift reagent does not need to be known to a high degree of accuracy. If the slopes of LIS vs. [L]/[S] are linear for all nuclei, the ratios are independent of the concentration of LSR. Since the concentrations and ratios of the concentrations of LS and LS_2 continually vary over a plot of LIS vs. [L]/[S](see Figure 1), the use of relative bound shifts requires that the ratio of Δ_1 to Δ_2 be the same for all nuclei of the substrate. This is not always the case, and examples in which the relative bound shifts exhibited a concentration dependence have been noted.[1274,1336]

A variety of procedures for determining bound shifts have been described in the literature.[135,136,147,150,1307-1312,1318-1320,1334,1337] Some reports have compared the strengths and weaknesses of certain of these methods.[172,1336,1337] Relatively simple procedures for the evaluation of bound shifts have been described by Armitage et al.,[1307,1310,1315,1338] Bouquant and Chuche,[1311,1316] Kelsey,[1309] Williams,[1339] and Goldberg and Ritchey.[1313] These procedures employ different approaches to calculate the bound shifts, but each assumes either the formation of only a 1:1 complex or a situation that can be described by a one-step equilibrium.

The procedure described by Armitage is carried out under the conditions of [S] $\gg$ [L] and [L] = constant.[1307,1310,1315,1338] In this situation, a plot of [S] vs. 1/(LIS) results in a straight line with a slope of Δ_b[L] and a y-intercept of $-(1/K_b + [L])$. The procedure described by Bouquant involves a plot of LIS vs. $(LIS/[S])^{1/2}$ under the conditions of [L] = [S].[1311,1316] The plot is linear with a slope of $(\Delta_1/K_1)^{1/2}$ and a y-intercept of Δ_1.

A more rigorous method of obtaining bound shifts has been described by Shapiro et al. and involves a computer fit of a two-step equilibrium to a plot of LIS vs. [L]/[S].[136,147,150,1318] This procedure provides the values for Δ_1, Δ_2, K_1, and K_2. In a comparison of several of these methods, Raber and Hardee reported that the two-step fitting procedure of Shapiro was best suited for shift reagent-substrate complexes that exhibited strong binding.[1337] The conditions for strong binding are most likely to occur with chelates of fod. Chelates of fod are often preferred for structural determinations over chelates of dpm because their enhanced solubility permits study over a wider concentration range. The structural determination is then performed using Δ_1 and the LS complex rather than Δ_2 and LS_2. The data for the LS complex are fit because they are more apt to meet the assumptions of the simplified dipolar equation. In a study of complexes of the general formula $Ln(dpm)_3S_2$, Cramer et al. observed better fits of the data using the extended form of the dipolar equation.[1340,1341] Raber et al. reported that the two-step procedure was not as reliable for weakly bonding substrates, and the method of Bouquant was recommended.[1336] Weak binding is more likely to occur with chelates of dpm.

C. Location of the Principal Magnetic Axis

The location of the principal magnetic axis of the shift reagent-substrate complex is necessary to determine the values of θ to be used in fitting the dipolar equation. Solid state X-ray crystallographic and magnetic susceptibility data can be used to accurately determine the location of the principal axis. This procedure is impractical, however, and also requires the assumption that the solid state and solution structures are identical. Most workers have avoided the problem altogether by simply assuming the principal magnetic axis is colinear to the lanthanide-substrate bond.

The validity of this assumption has been discussed in the literature.[1340,1342-1345] These reports have found that the principal magnetic axis is not necessarily colinear with the lanthanide-substrate bond. The deviations have been found to be small.[1342-1344] One report specified an offset of only 0.04 to 2.40°.[1342] Even with such small deviations, however, Cramer and Maynard[1340] and Hawkes et al.[1342] reported that the location of the magnetic axis was the most sensitive of all the parameters involved in fitting the dipolar equation. Allowing for variability in the location of the principal magnetic axis complicates the fitting procedure, but should probably be done to insure more accuracy in the final results.

D. Atomic Coordinates and L-S Bond Length

Few attempts at fitting lanthanide shift data to the dipolar equation have been performed in a completely iterative fashion. Instead the measured bound shifts are compared to those calculated for a series of proposed structures for the substrate. The elimination of "unreasonable" structures based on our knowledge of the geometry of compounds shortens the time necessary for carrying out such an analysis. In order to calculate the bound shifts for a proposed structure, atomic coordinates of each nucleus are necessary. Some workers have obtained atomic coordinates by measurements of a Dreiding model of the substrate. A crude measurement of this type may be appropriate for qualitative determinations necessary in assigning spectra, but is not accurate enough for reliable fitting of the dipolar equation. Hinckley and Brumley[1346] have evaluated the effects of coordinate error on the accuracy of structures generated by fitting LIS data. In most, but not all, cases the resulting errors were small.

Raber et al.[1347] have compared five methods commonly employed for generating atomic coordinates. Three of these methods relied on data from X-ray crystallographic experiments. The fourth used standard bond lengths and angles. The fifth method employed empirical force field calculations to determine atomic coordinates. This last method was the most time-consuming of the five, but resulted in significantly better reliability in studies with LSR.[1347]

The calculated bound shifts for a substrate are dependent on the lanthanide-substrate bond length. The distance between the lanthanide and substrate can either be incorporated into the fitting process as a variable or be given a set value based on information available from X-ray crystallographic data. Cramer[1341] has reported that the lanthanide-substrate distance is not a sensitive parameter in the overall agreement of the observed and calculated shifts. In addition, it was noted that application of the simplified dipolar equation tended to give unreasonably long lanthanide-substrate bond lengths. A similar result was observed by Raber et al. in a study of adamantane-carbonitriles.[516] These findings suggest that the lanthanide-substrate bond length should not be treated as an unrestricted variable, but should only be allowed to vary within limits that are judged reasonable from available X-ray crystallographic data.

E. Significance Testing

An important aspect of any structural determination that involves fitting of the dipolar equation is to test the significance of the correlation between the observed and calculated values. The most common method of assessing confidence in studies with LSR has been the use of the Hamilton R-factor.[1348,1349] The Hamilton R-factor does not provide a direct measure of the best fit, but provides a statistical measure of the confidence with which certain structures can be rejected.

The validity of various procedures for significance testing of LIS data has been evaluated in a series of reports by Richardson et al.[1350,1351] and Li and Lee.[1352] They concluded that the Hamilton R-factor was inappropriate for use with LIS data because it was impossible to determine the number of degrees of freedom. This value is needed to meet the first Hamilton linearity assumption.[1350] It was found that repeated applications of the R-test to the same shift data produced different structural results.[1350] Three other tests of significance, the jack-knife test, the Kendall τ statistic, and Spearman's r-test, were evaluated for use with LIS data. The Kendall τ and Spearman r tests were reported to be invalid for the case of nonnormal data.[1351] The jack-knife test was recommended for use with LIS data.[1351]

F. Limitations of Configurational and Conformational Assignment

Examples of the use of LSR for configurational and conformational assignment have been presented throughout the preceding chapters. This section will only consider studies that have as their principal focus an evaluation of the limitations involved in applying LSR to configurational and conformational determinations. Suffice it to say that considerably more success is to be expected in assessing configurational detail compared to conformational detail. Most of these reports reach the conclusion that conformational information based only on the fitting of lanthanide shift data cannot stand alone.[243,261,452,1353,1354]

Sullivan used lanthanide shift data in an attempt to distinguish isomers of compounds I and II.[1353]

The shift reagent binds to the amine group of I and should not alter the conformation of the side chain. Isomers that differed about the rigid part of the molecule, such as the *exo* and *endo* diastereomers of I, were readily distinguished. The isomers that result from the diasteromeric nature of the side chain of I could not be distinguished on the basis of lanthanide shift data. The diastereomers of II that differed in the configuration of the side chain were also indistinguishable using lanthanide shift data. Sullivan's work employed Hamilton's R-factor as a test of significance. In their evaluation of significance testing, Li and Lee[1350,1352] and Richardson et al. reportedly were able to distinguish the diastereomers of I and II using more reliable tests of significance.

Hydroxy substrates present problems in studies with LSR because of the variability of the location of the lanthanide-oxygen bond.[243,261,1344,1354,1355] Schneider and Weigand have recommended using a time average of the lanthanide-hydroxy location.[243] Hinckley and Brumley were able to distinguish the *cis* and *trans* isomers of pinocarveol through fitting of LIS data.[261] Hawkes et al. have described a procedure to locate the metal in the complexation of LSR with alcohols.[1344] Davis and Willcott reported that the lanthanide shift data for *endo*-norborn-5-en-2-ol could not be fit to the *exo* isomer.[1356] Roberts et al.[1354] studied borneols as well as less rigid substrates to determine the limitations of applying lanthanide shift data in conformation analysis. The computer program employed in this study did not follow the same sequence each time it performed the fit. Repeated analyses of the same data resulted in different structures. As a result, Roberts concluded that while considerable information could be obtained from fitting of LIS data, the determination of fine details of the conformation equilibria of flexible molecules was neither simple nor particularly accurate.

Other reports[244,1315,1357,1358] have considered how to treat substrates that might exhibit free rotation. Armitage et al.[1315,1357] studied three substrates that were rigid except at the point of attachment to the shift reagent. Four situations — free rotation, free rotation over a limited range, no rotation, and jumps between minima of n-fold potential — were evaluated. Static models were judged inappropriate. In the case of free rotation, it was reportedly more difficult to get adequate fits of the data. It was also reported that no one model was reliably better for all of the substrates studied.[1315] Wing et al.[244] have also reported that averaging over molecular orientations provides more reliable results than static models. Methyl groups were treated as a static averaged point, however, in their analysis. Karnilov and Turov have discussed the limits and errors inherent in employing a one-point approximation for methyl and other freely rotating groups.[1359] Raber et al., in a study of nitrile substrates, treated methyl groups as having six equivalent rotamer positions.[515] Lenkinski and Reuben,[1358] in a study of pinacolone evaluated the fit of the data for free rotation, restricted rotation, and an averaging along the axis of rotation for the methyl and *t*-butyl groups. It was difficult to determine which method was the most appropriate; however, averaging along the axis of rotation was reportedly the least successful.

Johnston et al.[1355] have assessed the limitations of conformational analysis through the fitting of lanthanide shift data by studying a series of 3-alkyl-1-adamantane carbonitriles. These substrates bond in a linear manner with the lanthanide ion and eliminate the problem of rotation observed with substrates such as alcohols. In addition, the rigid adamantane skeleton allows optimization of the fit of the observed and calculated bound shifts. Once the rigid portion of the molecule has been fit, the conformation of the 3-alkyl group was evaluated. Excellent agreement between observed and calculated values was reported in this study.[1355]

Willcott and Davis assessed the limitations of applying LSR in conformational analysis by studying a set of substrates of varying rigidity and flexibility.[452] This report concluded that the data set in experiments with LSR were too limited to permit a study of all of the dynamic processes occurring in a flexible molecule. Difficulty was also encountered in determining the conformation of the methoxy group in III.[452] Data for the rigid portion of the molecule were used to fix the position of the lanthanide ion. A subsequent assessment of whether the methoxy group was in or out of the plane of the rings was unsuccessful. The epoxide ring of IV was used to fix the geometry of the complex with Yb(dpm)$_3$.[452] The conformation of the flexible side chain could not be determined on the basis of shift data alone. The conformation could be determined when the coupling constants obtained in the shifted spectra were used in conjunction with the lanthanide shift data.

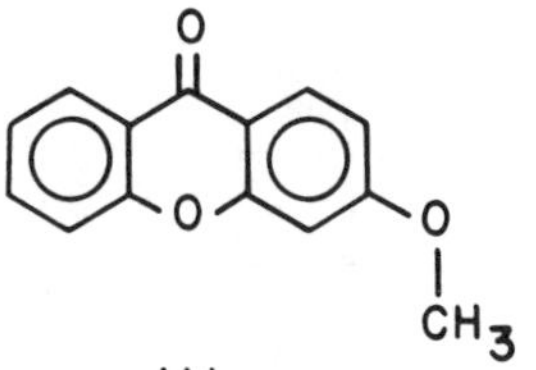

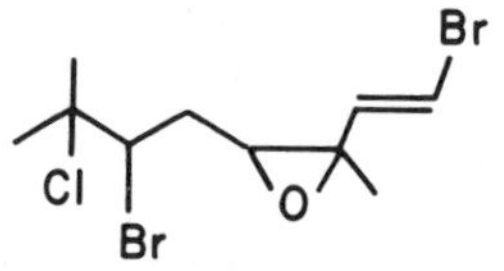

V. RELAXATION PHENOMENA WITH LSR

Paramagnetic lanthanide complexes shorten the spin lattice relaxation times of the nuclei of a substrate. The influence of the lanthanide ion on the relaxation exhibits an r^{-6} dependency. A number of workers have advocated using lanthanide relaxation data as a supplement to lanthanide shift data in structural studies employing LSR.[304,1267,1268,1358,1360-1365] Incorporation of relaxation data enlarges the limited data set obtained from ^{1}H and ^{13}C shift data and therefore should provide a more reliable fit. A second advantage is that relaxation phenomena only exhibit an r^{-6} dependency. This distance dependency rests on more solid theoretical ground than the pseudocontact shift equation[1362] and does not require any assumption as to the location of the principal magnetic axis. Uniform L-S geometry for all of the lanthanides is necessary, however, if one is to apply relaxation data in conjunction with shift data.

Relaxation data should be obtained with chelates of Gd(III). Relaxation data obtained with metals with anisotropic magnetic susceptibilities may not be described exclusively by an r^{-6} dependence.[1361,1362] Since Gd(III) usually causes excessive broadening in the spectrum of a substrate, the recommended procedure is to first use Eu(III) to shift the spectra and then measure changes in the line widths as small amounts of Gd(III) are added.[304,1362,1365,1366] Welti et al.[1366] compared the structure obtained for borneol using shift data recorded with Eu(fod)$_3$ and Pr(fod)$_3$, and relaxation data recorded with Gd(fod)$_3$. In contrast to other studies on this subject, it was reported that the results from the shift and relaxation data were not compatible. This would indicate that the lanthanide-borneol complexes were not isostructural.

Faller et al.[1363] have recommended the use of relaxation reagents such as Gd(fod)$_3$ or Gd(dpm)$_3$ as an aid in assigning the ^{13}C spectra of substrates. The relaxation times of ^{13}C resonances can be measured and related to the distance from the gadolinium ion. Faller et al.[1367] have also shown how the rapid relaxation brought about by Gd(fod)$_3$ can cause decoupling to occur in the NMR spectrum of a substrate. As an example, the ^{1}H spectrum of pyridine in the presence of Gd(fod)$_3$ is shown in Figure 3. The protons at position-2 are closest to the Gd(III) and are effectively decoupled because of the rapid relaxation. As a result, the resonance for H$_3$ collapses from a doublet of doublets to a doublet.

Johnston et al. have described a method for determining proton relaxation times in second-order spectra.[1368] The T_1 values are measured in a series of spectra with varying concentrations of shift reagent. Extrapolation to a shift reagent concentration of zero yields the T_1 values for the substrate.

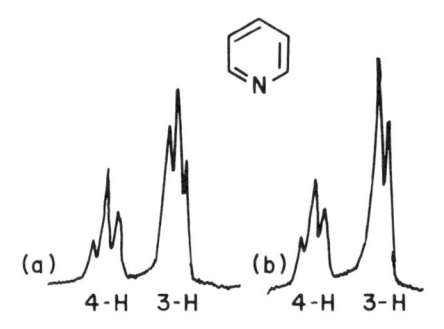

FIGURE 3. The 3- and 4-proton resonances of pyridine at 100 MHz in CDCl$_3$ in the presence of (a) 0.03 M Eu(fod)$_3$ and (b) 0.03 M Eu(fod)$_3$ and 0.0001 M Gd(fod)$_3$. (Reprinted with permission from *Tetrahedron Lett.*, 1381, 1973, Faller, J. W., LaMar, G. N., Copyright 1973, Pergamon Press, Ltd.)

Two reports have described practical applications resulting from the influence of lanthanide chelates on the nuclear Overhauser effect (NOE). Paramagnetic substances quench the NOE and Blackmer and Roberts have described how the magnitude of this quenching in ^{13}C spectra can be used to assess the site of complexation of a polyfunctional substrate.[1369] Mersh and Sanders[1370] have described the use of La(fod)$_3$ to enhance NOE growth rates for small molecules. NOE measurements are useful for determining internuclear distances; however, small molecules tumble too quickly to exhibit much enhancement. Complexation with La(fod)$_3$ results in the formation of a larger molecule and slows down the rate of tumbling.

Chapter 6

WATER-SOLUBLE LANTHANIDE SHIFT REAGENTS

Before Hinckley's report of an organic soluble LSR, several workers had already recognized that lanthanide ions could be used as shift reagents in aqueous solutions.[1371-1374] Since recent improvements in NMR spectroscopy have rendered it practical for the study of many biologically interesting compounds, aqueous LSR and relaxation probes are apt to see more use in the coming years. Examples of systems that have been studied with aqueous LSR include proteins, nucleotides and nucleic acids, carbohydrates, and membranes. Shift reagents have been developed for the study of biologically important metal cations. A limited number of chiral LSR for aqueous solutions have been described, and it would not be surprising if further developments are forthcoming in this area. Several review articles that discuss aqueous LSR have been published.[24,27,50,1375]

I. SHIFT REAGENT SELECTION

Most studies utilizing water-soluble LSR have employed the aquated 3+ ions. In 1972, Sanders et al. reported the shifts in the spectra of an arsine oxide, a phosphine oxide, nitrones, phosphates, phosphonates, and carboxylate compounds with $Eu(NO_3)_3$ and $Pr(NO_3)_3$.[68] Reuben has recommended the use of lanthanide(III) chlorides, as opposed to nitrates, because of reduced inner-sphere coordination of the chloride ion.[1376] The direction of the shifts with aqueous LSR are usually opposite to that observed with the tris chelates in organic solvents. Shift reagents with Eu(III) are therefore upfield reagents and those with Pr(III) are downfield reagents.

Chelates of ethylenediaminetetracetic acid(EDTA),[1377-1380] and other aminocarboxylate ligands[1378-1380] have been studied as LSR and in certain instances offer advantages over the aquated ions. The chelates of EDTA are soluble over a wider pH range as the aquated ions form insoluble hydroxides at pH values greater than 7.[1377] The multidentate EDTA ligand limits the coordination sphere of the lanthanide ion and favors the formation of 1:1 shift reagent-substrate complexes.[1377] The shifts in the ^2H and ^{17}O spectra of D_2O, and the ^{35}Cl spectrum of Cl^- have been compared with the aquated ions and lanthanide complexes of 1,4,7,10-tetraazacyclododecane-N,N',N'',N'''-tetraacetic acid(DOTA), 1,4,7-triazacyclononane-N,N',N''-triacetic acid(NOTA), and EDTA,[1379,1380] The complex with DOTA was rated as the most effective aqueous LSR. It was postulated that the complex with DOTA was the most sterically crowded of those studied and exhibited axial symmetry.[1379,1380] Further evidence for this conclusion came from a study that compared the effectiveness of lanthanide complexes with the ligands EDTA, diethylenetriaminepentaacetic acid(DTPA), pyridine-2,6-dicarboxylate(DPA), NOTA, nitrilotriacetic acid(NTA), DOTA, cyclohexanediaminetetraacetic acid(CyDTA), 1,4,8,11-tetraazacyclotetradecane-N,N',N'',N'''-tetraacetic acid-(TETA), and N-hydroxyethylethylenediaminetriacetic acid(HEDTA) as NMR shift reagents.[1378] Luminescence spectra of the europium complexes were used in an attempt to determine the symmetry and the magnitude of the crystal field splitting.[1378] The larger the crystal field splitting, the larger the magnitude of the dipolar shifts.[1378,1381] The lanthanide complexes with DOTA(I) were judged the most effective of the examples studied.[1378]

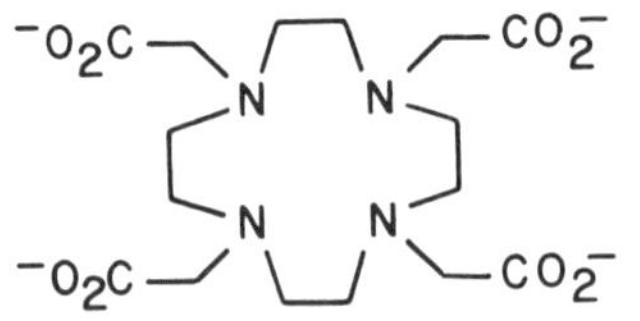

The shifts in the spectra of lysine and the dipeptide glycine-alanine with the Eu(III) complex with N-(pyridoxal)aspartic acid were measured, but were small in magnitude.[1382] Lanthanide complexes with the ligands pyridoxylaspartic acid, -asparagine, and -alanine have been evaluated as shift reagents for peptides.[1381] It was reported that the peptides associated with these shift reagents by hydrogen-bonding with the ligand, rather than coordinating to the metal. Once again, the shifts in the spectra of the substrates were small.

Horrocks et al. have evaluated lanthanide complexes with porphyrin ligands as aqueous shift reagents.[1383,1384] These had the formulas Ln(TPP) (β-dik) (TPP = meso-tetraphenylporphine and derivatives, β-dik = acac),[1383] and Ln(III)TPPS(OH) (imidazole)$_x$(x = 2) (TPPS = tetra-p-sulfonatophenylporphin).[1384] The Tm(III) complex of TPPS was found to be a particularly effective upfield shift reagent for neutral and cationic substances. Shifts with Er(III) and Yb(III) were also upfield, whereas shifts with Ho(III) were downfield. These complexes were useful over the pH range from 8 to 11. The sulfonate group was protonated below a pH of 8 causing precipitation of the complex. Above a pH of 11, the shifts in the spectrum of the substrate were smaller and the spectrum exhibited considerable line broadening.[1384]

A number of compounds including t-butanol,[1385,1386] dioxane,[1379,1387] acetone,[1388] 3-trimethylsilylpropionate,[1389] and 2,2-dimethyl-2-silapentane-5-sulfonate[1390] have been used as zero references with LSR in aqueous solutions. Horrocks and Hove have reported, however, that the spectra of dioxane and t-butanol were shifted in the presence of the lanthanide-porphyrin complexes.[1384] Bryden et al., on the other hand, have reported that the spectrum of dioxane was not shifted in the presence of lanthanide complexes with EDTA, NOTA, and DOTA.[1379] Horrocks and Hove could not find a suitable internal reference and recommended the use of the tetramethylammonium ion as an external reference.[1384] In this situation, the shift due to changes in the bulk susceptibility of the solution must be taken into account.

II. APPLICATIONS TO BIOLOGICAL SYSTEMS

A. Amino Acids, Peptides, and Proteins

The NMR spectra of amino acids including alanine,[1377,1391-1395] phenylalanine,[1393] histidine,[1396,1397] threonine,[1396] serine,[1396] tryptophan,[1393] proline,[1398,1399] hydroxy-L-proline,[1400] valine,[1399,1401] sarcosine,[1394,1402] glutamic acid,[1395] N-acetyl-L-aspartic acid,[1403] 3- and 4-aminobutyric acid,[1395,1401] and N-acetyl-L-3-nitrotyrosine ethyl ester[1404] have been recorded and analyzed using LSR.

Sherry et al. determined the stability constants for the complexes of alanine, histidine, threonine, and serine with Nd(III) by NMR and potentiometric methods and observed similar results.[1396] Several reports have assessed the structure of complexes of alanine with various lanthanide(III) ions.[1392-1394,1405] Sherry and Pascual analyzed the ^{1}H and ^{13}C spectra of alanine at a pH of 3 and reported that a structural change occurred between Tb(III) and Dy(III).[1392] The complexation of alanine with the larger ions(Pr through Tb) was reportedly monodentate, whereas bidentate complexation was indicated for the smaller ions(Dy through Yb). Contact shifts were reportedly observed in the spectrum of alanine and had to be separated from the pseudocontact shifts before an analysis of the geometry could be performed.[1392] Levine also reported that the complexation of alanine with Pr(III) and Tm(III) was different.[1393] A structural change was also noted for the complexes of phenylalanine and tryptophan with lanthanide ions.[1393] It was recommended that relaxation data obtained with Gd(III) be used to enhance conformation analyses of amino acids and peptides.[1393]

Elgavish and Reuben, however, analyzed shift and relaxation data for alanine and sarcosine and disputed the earlier findings of a structural change for the series of lanthanide complexes.[1394,1402,1405] These workers measured the relaxation times for the nuclei of the substrate with each lanthanide ion and calculated a set of internal ratios. Similarity of the internal ratios for all members of the series was the basis of their claim of isostructurality.[1402] The contact and pseudocontact shifts were determined by a method that assumed isostructurality, constance of the electronic nuclear hyperfine coupling constants, and axial symmetry for the series of lanthanide complexes.[1405] This method is similar to that previously described by Reilley et al.[1194] Nonaxial symmetry was reportedly the major departure from the model, and minimal contact contribution was found by these workers in the spectrum of alanine with the lanthanide ions.[1394]

Singh et al. subsequently analyzed relaxation data for L-proline at a pH of 3 with ten lanthanide ions, and the data were indicative of isostructural complexes.[1398] Problems were encountered, however, in applying Reilley's method for determining the extent of contact and pseudocontact shift mechanisms. It was postulated that the complexes had either nonaxial symmetry components or hyperfine coupling constants that were dependent on the lanthanide ion. It was also recognized that the previous conclusions for alanine might well be in error and that the nonaxial symmetry components might be different for the early and later lanthanides.[1398]

The solution structure of alanine at pH 10 has been analyzed with chelates of EDTA.[1377] It was reported that the alanine bonded in a bidentate manner through the carboxylate group and that a substantial contact shift mechanism was present. The exchange of alanine with the latter lanthanides was slow up to temperatures of 90°C, and the resulting broadening in the spectrum precluded any structural analysis for these complexes.[1377]

Conformational aspects of proline[1399] and valine[1399,1401] have been evaluated using paramagnetic lanthanide ions as probes. Lanthanide shift and relaxation data have been used to determine the conformation of hydroxy L-proline.[1400] The conformation that resulted from an analysis of the lanthanide data was compared to that obtained by considering the coupling constants. Similar, but not identical, results were achieved. Since the coupling constants did not change in the presence of the lanthanide ions, complexation did not alter the conformation of the substrate. The bonding of the alpha and gamma carboxylate groups of glutamic acid with lanthanide ions exhibited a pH dependence that was assessed through an analysis of the NMR spectrum.[1395] No binding occurred at the amine site.

The shifts in the ^{1}H spectrum of histidine with Eu(III), Pr(III), and La(III) have been measured.[1397] Significant variations in the internal ratios of the shifts with Eu(III) and Pr(III) were observed and were presented as evidence for a structural difference. The possibility of contact shifts or a nonaxial component to the dipolar shifts was not considered in this report.[1397] Lanthanide ions have been used to study 3-amino- and 4-aminobutyric acid.[1395,1401] The resonances of the diastereotopic methylene protons of the 3-amino- derivative were resolved in the presence of the shift reagent.[1401] The coupling constants for 3-aminobutyric acid changed on complexation with the shift reagent, which would indicate that the conformational population was altered.[1401]

The conformation of L-azetidine-2-carboxylic acid(I) was determined using shift and relaxation data with paramagnetic

[structure: azetidine ring with NH and COOH substituents]

lanthanide ions.[1406,1407] Bonding involved the carboxylate group, and the proton alpha to the carboxylate group exhibited a large contact shift. The shifts of all of the other proton resonances were reportedly pseudocontact in origin. Both 1:1 and 1:2 complexes were observed.[1407] The data with the lanthanide ions permitted an assessment of the degree of puckering of the four-member ring.[1406]

Lanthanide shift data for N-acetyl-L-aspartic acid and N-acetyl-L-aspartyl-L-glycyl-L-aspartylamide were used to probe aspects of the nature of Ca^{2+} binding with these substrates.[1403] Yb(III) was rated as the best shift reagent for these studies. The complex of Pr(III) with L-carnosine(β-alanyl-histidine) has been characterized by ^{1}H NMR spectroscopy and potentiometric titrations.[1408] It was reported that at low pH, the substrate coordinated only at the carboxylate site. At neutral pH, a stronger complex was noted and was believed to result from deprotonation and complexation of the imidazole ring.[1408] The complexes for a series of lanthanide ions with L-carnosine were reportedly isostructural and possessed effective axial symmetry.[1409] It was also reported, however, that the hyperfine coupling constant was not identical for all the lanthanides.[1409]

Lanthanide ions have been used as conformer probes for di- and tripeptides.[1393,1410] It was recommended that both shift data and relaxation data with Gd(III) be employed in conformation analyses such as these. Differences in the binding interaction with the lighter(bidentate) and heavier(monodentate) lanthanides were reportedly observed.[1393,1410] All substrates coordinated through the carboxyl group. The influence of temperature, ionic strength, and the anion on the conformation of these substrates was discussed.[1393,1410] The contact shifts for the methylene group alpha to the carboxylate group decreased in the order: amino acid > dipeptide > tripeptide > simple carboxylate > -O-.[1393] Examples of the peptides studied include Gly-Gly, Gly-Ala-CO_2^-, Gly-Phe-CO_2^-, Ala-Ala, Ala-His,[1393] Ala-Gly, Phe-Gly, Phe-Ala, and a series of alanine-glycine tripeptides.[1410] Shift data with $Pr(ClO_4)_3$ has been used to facilitate the assignment of the ^{13}C NMR spectra of glycyl-L-alanine, L-alanine-L-alanine, diglycine, triglycine, and alanine-glycine.[1411] Complexation of each substrate involved the terminal carboxylate group.

Paramagnetic lanthanide ions have been used to probe structural features of proteins. These studies often involve an analysis of calcium binding sites. The similarity in the size of Ca(II) and Ln(III) ions facilitates this type of analysis.

Parvalbumins have been the focus of several studies with LSR.[1412-1418] A general method for carrying out an analysis of a protein using LSR has been described by Lee and Sykes.[1414] Three resonances in the ^{1}H spectrum that could be assigned without the shift reagent were used to fix the geometry of the lanthanide-protein complex at the binding site.[1414] Once the geometry was set, it was possible to assign other resonances that were resolved in the shifted spectrum. Six methyl resonances about the first calcium binding site of carp parvalbumin were assigned in the shifted spectrum. Several -NH resonances were also resolved, but not assigned.[1414] In a subsequent study, shift data in the ^{1}H, ^{13}C, and ^{113}Cd spectra were used to improve the accuracy of the geometry determined for the lanthanide-protein complex.[1415] Cadmium is not found naturally in the protein, but was substituted into one of the binding sites. The cadmium resonance, as well as one ^{13}C resonance that could be assigned, when coupled with the three ^{1}H resonances that were assigned, provided a more accurate fit of the data.[1415] A further check on the geometry of the shift reagent-protein complex was obtained by monitoring the shifts in the ^{19}F spectrum of 4-fluorophenol-parvalbumin.[1418] Yb(III) was used in these studies to reduce the contact shift.[1414] Relaxation data with Gd(III) could not be collected because the complex was in the slow exchange region, and severe broadening occurred. Relaxation data with Yb(III) was collected, however, and was used to enhance the accuracy of the structural fit and assignment of resonances.[1416] One other problem was that the complexity of the initial spectrum precluded the determination of chemical shift values in the unshifted spectrum($δo$). A procedure to evaluate $δo$ values was described and involved monitoring the temperature dependence of the lanthanide-induced shifts. Extrapolation to infinite temperature(zero lanthanide shifts) yielded values of $δo$.[1414] These studies resulted in information about the structure of the binding site that was more detailed than structural information obtained from X-ray crystallographic studies.[1412]

It is possible to selectively fragment a protein with more than one calcium binding site into units that each contain one binding site.[1413] These can then be studied individually with lanthanide ions. Lee et al. have demonstrated this with parvalbumin and troponin C.[1413]

Other proteins that have been probed through the use of lanthanide ions include lysozyme,[1419-1425] bovine serum albumin, trypsinogen, trypsin, α-amylase,[1426] and phosphoglycerate kinase.[1427] Most of these reports employed the use of relaxation data with Gd(III) as a supplement to lanthanide shift data.[1419-1422,1427]

A series of reports by Campbell et al. described the application of lanthanide ions in the analysis of the NMR spectrum of lysozyme.[1422-1425] Two new methods of using lanthanide ions in the study of the NMR spectra of proteins were developed.[1422] The first involved recording and comparing the spectrum of the protein with and without La(III). Mathematical manipulation of the free induction decay was used to improve the resolution. The improved resolution permitted the determination of fine structural features of the protein. The second method was to record the difference spectrum of the protein with and without Gd(III). At low concentrations of Gd(III), it was possible to assign the resonances of nuclei close to the site of binding. Resonances of nuclei further removed from the binding site could be assigned at higher Gd(III) concentrations. These studies were used in conjunction with lanthanide shift data to assign 40 resonances in the proton NMR spectrum of hen egg-white lysozyme.[1423]

It is also possible to introduce binding sites for lanthanide ions into proteins, thereby providing sites at which to probe the structure. One example involves the derivatization of amine groups with diketene(II).[1428]

$$R-NH_2 + \begin{array}{c} CH_2-C-O \\ | \quad | \\ CH_2 C=O \end{array} \longrightarrow R-NH\overset{O}{\overset{\|}{C}}CH_2\overset{O}{\overset{\|}{C}}CH_3$$

II

The beta ketoamide group forms a chelate bond with lanthanide ions. A second is to nitrate tyrosine residues to prepare 3-nitrotyrosine groups.[1404,1429,1430] In this case, the nitro and phenol oxygen atoms of the 3-nitrotyrosine residue coordinate in a bidentate manner with lanthanide ions.[1404] The shifts in the spectrum of N-acetyl-L-3-nitrotyrosine ethyl ester with Eu(III), Pr(III), Gd(III), and La(III) have been analyzed.[1404] The line broadening with Yb(III) was too severe to permit its use. The information gained from this study was applied in the analysis of nitrated derivatives of bovine pancreatic trypsin inhibitor.[1429,1430] Bonding of the lanthanide ion occurred at the nitrotyrosine residue. Both the mononitro- (derivatized at tyr 10) and dinitro- derivatives(tyr 10 and

21) were studied.[1430] Analysis with the lanthanide ions facilitated the assignment of several -NH resonances and the determination of structural features about the binding site.

B. Nucleotides and Nucleic Acids

Nucleotides with adenine have been the focus of several studies with lanthanide shift reagents. Geraldes has studied the conformation of adenosine 5'-monophosphate(AMP) in D_2O/DMSO mixed solvents.[1431-1433] The effects of solvent composition, temperature, and the addition of chaotropic agents on the conformation of AMP were assessed. The conformation was assigned using lanthanide shift data, relaxation data with Gd(III), coupling constants, and nuclear Overhauser enhancements.[1433] The lanthanide ions bonded to the phosphate group and relaxation data for the series of ions indicated isostructural complexes.[1431]

In a study by Barry et al., the lanthanide complexes with AMP were also reported to be isostructural.[1434] Both ^1H and ^{13}C lanthanide shift data and relaxation data with Gd(III) were employed in this study. The conformation of AMP has been determined and compared in water and DMSO.[1435] The conformation of AMP has also been determined using shift data with lanthanide chelates of EDTA[1427] and pyridoxalidene aspartic acid.[1436]

The solution conformation of adenosine 2':3'-monophosphate was determined at a pH of 2.5 using lanthanide shift and relaxation data.[1437] Binding occurred at the phosphate group, and it was possible to determine the orientation of the base relative to the ribose unit. The conformations of 9-β-D-ribofuranosyladenine-5'-monophosphate and 1-β-D-deoxyribofuranosylthymine-5'-monophosphate were elucidated on the basis of lanthanide shift and relaxation data.[1438]

Cyclic-3',5'-adenosine monophosphate has been the focus of several studies with LSR.[1439-1441] Barry et al. determined the conformation at a pH of 2 and 5.5 using shift and relaxation data.[1441] Binding of the lanthanide ion at the phosphate group was confirmed by ^{31}P shift data. Lavallee and Zeltmann[1439] used lanthanide perchlorates as the shift reagent and observed similar results to those of Barry et al.[1441] Binding reportedly involved a bidentate interaction of both free oxygen atoms of the phosphate group. Both studies concluded that the shifts were pseudocontact in origin. Hayashi et al. used cyclic-3',5'-AMP as a substrate and compared the results obtained with two methods of conformational analysis.[1440] One involved the use of relaxation data with Gd(III). The second method was to monitor the effect of deuterium substitution on relaxation times to determine interproton distances. In these analyses, Pr(III) was added to resolve the resonances so that the relaxation times could be determined. The results of the two methods were judged comparable.[1440]

The conformation of adenosine diphosphate[1427] and adenosine triphosphate[1427,1436,1442] have been determined using LSR. Lanthanide shift data in the ^1H and ^{31}P spectra and relaxation data with Gd(III) were indicative of bonding at the beta and gamma phosphate groups of ATP.[1442] No bonding was indicated at the purine ring.

The conformation of nicotinamide mononucleotide has been determined through the fitting of lanthanide shift and relaxation data.[1443,1444] It was reported that the lanthanide ion complexed at the phosphate group and the conformation was not perturbed on bonding.[1443] The conformational analysis was facilitated by incorporating coupling constant data into the procedure.[1443] Assignment of the ^{13}C spectra of both the α and β anomer of ribose-5-phosphate,[1445] and nucleotides related to dihydronicotinamide adenine dinucleotide phosphate(NADPH) including NAD^+, NADH, and $NADP^+$[1446] was facilitated through the use of paramagnetic lanthanide ions. In these studies, the ^1H spectra were shifted and assigned. Through a series of selective decoupling experiments of the proton resonances, it was possible to assign the carbon resonances.[1445,1446] The conformations of NAD, NADH,[1436,1447,1448] NADP,[1436,1448] and NADPH[1448] have been analyzed through the use of lanthanide shift data. One report recommended the use of luminescence data with Eu(III) to improve the accuracy of the analysis.[1447] Another report concluded that the NMR data indicated lanthanide ion complexation at both the phosphate group and the nucleotide base.[1448]

The conformations of other mononucleotides including cytidine-5'-monophosphate(CMP),[1377,1449] and uridine-3'-monophosphate(UMP),[1444,1448,1450] have been analyzed using data collected with paramagnetic lanthanide ions. The coupling constants of UMP did not change in the presence of the lanthanide ions indicating that complexation did not alter the conformation of the substrate.[1450] Dobson et al. described a method to separate the contact and pseudocontact contributions to the shift of the ^{31}P resonance of CMP.[1449] The method assumed that the shifts in the proton spectrum were purely pseudocontact in origin and used them to fix the geometry of the shift reagent-CMP complex. The pseudocontact shift for the ^{31}P resonance could then be calculated. Relaxation data in the ^{1}H and ^{31}P spectrum of 3',5'-thymidine diphosphate bound in the active site of staphylococcal nuclease was used to determine distances in the substrate.[1451] These distances compared favorably with previous information available from X-ray crystallographic studies.

The conformations of dinucleoside phosphates including adenosyl-(3',5')-cytidylic acid, cytidyl-(3',5')-adenylic acid, guanyl(3',5')-cytidylic acid, citidyl-(3',5')-guanylic acid,[1452] and adenosyl-(3',5')adenylic acid[1389,1452] have been studied with the aid of lanthanide shift and relaxation data. One report found that the conformations obtained by fitting lanthanide shift data compared favorably to those obtained through evaluations of coupling constants, proton dimerization shifts, and *ab initio* molecular orbital calculations.[1389]

The spectrum of yeast tRNA has been recorded in the presence of Eu(III), Pr(III), and Dy(III).[1453] It was possible to assign certain of the -NH resonances and draw some conclusions about the location of bonding of the lanthanide ions.

C. Carbohydrates

Carbohydrates contain many sites for coordination with LSR. Most studies of carbohydrates with lanthanide ions have assessed the site of bonding. For compounds such as glyceric acid, gluconic acid, lactobionic acid, and glutaric acid, chelate bonding at the α-carboxy, β-hydroxy, and β-hydroxy oxygen atoms was indicated by the shift data.[1454] A particularly favorable arrangement of groups for tridentate complexation is an axial-equatorial-axial(ax-eq-ax) arrangement of three adjacent hydroxy groups.[1387,1455-1463]

The relative abilities of different arrangements of hydroxy groups to coordinate with lanthanide ions was determined by studying ten alditols(polyhydroxy substrates) with Pr(NO$_3$)$_3$.[1455,1456] The substrates included iditol, sorbitol, arabitol, galactitol, glycerol, threitol, xylitol, erythritol, mannitol, and ribitol. It was possible to assign all of the proton resonances on the basis of the lanthanide shift data.[1456] The greater rotational freedom of primary compared to secondary hydroxyl groups led to reduced association of primary groups with the lanthanide ion.[1455,1456] The relative ordering of complexation was reportedly: xylo(I) > threo(II) > arabino(lyxo) (III) > glycerol(IV) > erythro(V) > ribo(VI).[1456]

I II III IV V VI

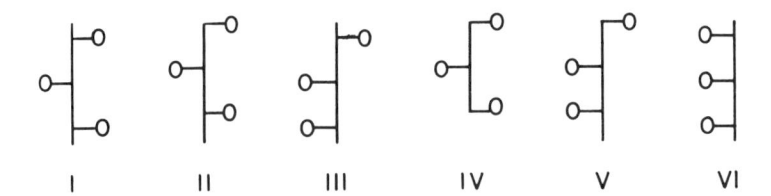

The conformations of the complexed substrates were also determined in this work.[1456]

The spectrum of epi-inositol, which has three adjacent hydroxy groups in an ax-eq-ax arrangement, exhibited large shifts in the presence of lanthanide ions.[1457,1459,1461] The spectra of cis-inositol, cyclohexane-1,2,3,4,5/O-pentol, and cyclohexane-1,2,3,4/5-pentol, which all have an ax-eq-ax arrangement of hydroxy groups, were also shifted by lanthanide ions.[1461] The spectra of myo-inositol, neo-inositol, cyclohexane-1,2,3,5/4-pentol, and cyclohexane cis-1,3,5-triol were not shifted in the presence of lanthanide ions.[1461] These substrates do not have an ax-eq-ax arrangement of three adjacent hydroxy groups.

The conformation of xylitol was determined through an analysis of shift data with Pr(III), Eu(III), and Nd(III).[1376] The coupling constants of xylitol were altered in the presence of the lanthanide ion. This indicates that the conformation changes on complexation. Large contact shifts were also noted for nuclei close to the coordination site.

Lanthanide ions have been used as models to study the binding of Ca(II) to sorbitol.[1387] The geometry of sorbitol in its complexed form was assigned. Sorbitol-1-d and -5-d were studied to facilitate assignment of the ^{13}C spectrum.

Methyl glycosides and 1,6-anhydrohexoses preferentially complex with lanthanide ions at adjacent ax-eq-ax oxygen atoms.[1457-1462] In addition to three adjacent hydroxy groups(VII), examples in which one of the bonding atoms was a methoxy oxygen(VIII), or an ether oxygen of the anhydro unit(IX) were observed.[1457]

VII VIII IX

The substrates employed in these studies included methyl α-D-gulopyranoside,[1457,1458] methyl β-D-hamamelopyranoside,[1457,1460,1462] methyl α-D-allopyranoside,[1462] methyl β-D-mannofuranoside,[1462] 1,6-anhydro-β-D-allopyranoside,[1457,1459,1462] 1,6-anhydro-β-D-mannopyranoside,[1457,1462] 1,6-anhydro-β-D-talopyranoside,[1457,1462] and 1,6-anhydro-β-D-glucopyranoside.[1462] All of these have an ax-eq-ax arrangement of oxygen atoms except 1,6-anhydro-β-D-glucopyranoside. In this case, the shifts were indicative of tridentate complexation of O-2, O-4, and the ring oxygen.[1462] Some of the resonances of these substrates exhibited "wrong way" shifts because of a sign change of the angle term of the dipolar shift equation.[1460,1462] Contact shifts were also noted in the spectra of these substrates.[1457,1459]

The conformation of sodium methyl α-D-galactopyranosiduronate(X) has been studied with the aid of lanthanide shift[1463,1464] and relaxation data.[1464] This compound

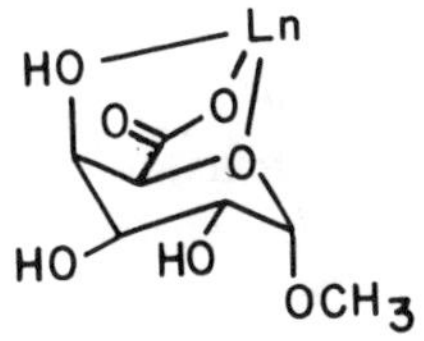

X

does not have an ax-eq-ax arrangement of oxygen atoms, and it was proposed that tridentate complexation occurred at O-4, O-5, and the carboxy oxygen atom.[1463,1464] The coupling constants did not change in the presence of the lanthanide ion, implying that no change in the conformation occurred upon complexation.[1463]

The structure of phosphonomannan(XI) was confirmed through an analysis of the NMR spectrum with Eu(III) and Pr(III).[1465] The ^{13}C spectra with Yb(III), Ho(III), Nd(III), and Gd(III) were too severely broadened to be used. The compounds α-D-mannose-1-phosphate and α-D-mannose-6-phosphate were employed as model substrates in this study.

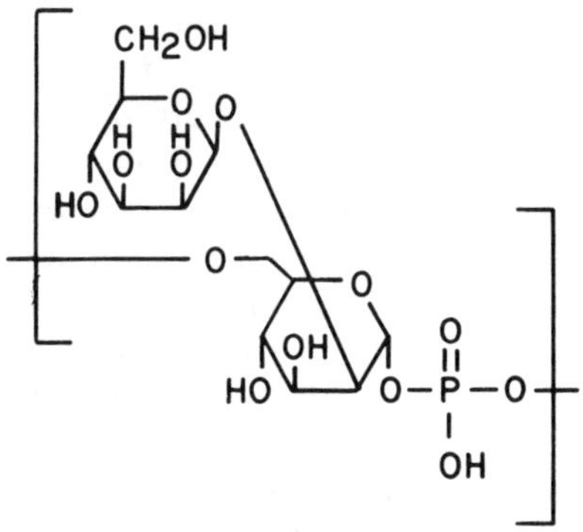

XI

D. Membranes

Paramagnetic lanthanide ions have been applied advantageously in the study of phospholipid membranes. It is possible to distinguish resonances of groups on the inner and outer surfaces provided the lanthanide ions do not migrate across the membrane. A method has been described for the preparation of bilayer vesicles that are impermeable toward lanthanide ions.[1466] This procedure was evaluated with L-α-distearoylphosphatidylcholine(DSL), L-α-dipalmitoylphosphatidylcholine(DPL), L-α-dimyristoylphosphatidylcholine, and mixed vesicles including DSL:DPL(1:1), and DPL with L-α-dipalmitoylphosphatidylethanolamine, cholesterol, and sphingomyelin. Most of the work on membranes has involved the study of phosphatidyl choline.[1467-1477] Phosphatidyl serine,[1467,1476] phosphatidyl inositol,[1467,1475] phosphatidyl glycerol,[1471,1474] and mixed systems[1467,1472,1476,1478] have also been studied. The lanthanide ions bond to the phosphate groups of the membranes.

The ^{13}C spectrum of egg phosphatidyl choline was recorded with Yb(III), Eu(III), Pr(III), and K$_3$Fe(CN)$_6$.[1468] A precipitate was observed with all of the shift reagents except Yb(III) before the inner and outer N-methyl carbon resonances were completely resolved. Gated decoupling was run on the Yb(III) sample to quantitate the signals.[1468] The spectrum of a sonicated lecithin dispersion was recorded with Eu(III) and Fe(CN)$_6^{3-}$, and less broadening was observed with the ferricyanide species.[1479] It has been recommended that relaxation data with Gd(III) be used in conjunction with shift data to facilitate structural analysis.[1472,1473,1477] A shifting ion such as Pr(III) was first added to resolve the resonances before relaxation data with Gd(III) was obtained.[1473]

A procedure has been described for the preparation of a vesicle that has one lanthanide ion inside and another outside.[1471,1475] The Eu(III) and Pr(III) ions were used in these studies. Since the shifts with Eu and Pr are in opposite directions, better resolution of the inner and outer signals was observed.[1471,1475] The resolution obtained in the spectra with the lanthanide ions facilitated the study of inside-outside distribution of species, intermembrane exchange, and transbilayer movement of phospholipids.[1471] The transmembrane asymmetry (charge difference between the inside and outside) of the liposome formed from phosphatidyl choline and phosphatidyl serine was also measured using Eu(III) and Pr(III).[1476]

The resonances of the outside glycerol and choline groups of egg yolk phosphatidyl choline were resolved from the inner groups in the presence of lanthanide ions.[1469] The ^{1}H resonance of the outer lipid acyl chains in phospholipid vesicular membranes were also shifted in the presence of lanthanide ions, and the ^{31}P resonances of the inner and outer phosphate groups were resolved.[1470,1472]

The shifts in the ^{1}H and ^{13}C spectra of mixed phospholipid membranes including phosphatidyl choline-phosphatidyl serine, phosphatidyl choline-phosphatidyl inositol, phosphatidyl choline-cholesterol, and phosphatidyl choline-sphingomyelin with lanthanide ions have been measured.[1472] The resolution achieved in the spectra with the lanthanide ions made it possible to study the effects of added materials, leakage of the vesicles, and intermembrane exchange of lipids.[1472] Shifts in the ^{1}H, ^{13}C, and ^{31}P spectra of liposomal material prepared from lecithins, sphingomyelins, and phosphatidyl inositols with Eu(III), Pr(III), and Nd(III) were used to assess vesicular size, features of the membrane surface, and the permeability of the membranes toward lanthanide ions.[1480]

The inner and outer -$N(CH_3)_3^+$ resonances of dipalmitoyllecithin vesicles were resolved in the presence of Eu(III), Nd(III), and UO_2^{2+}.[1481] The transport of Pr(III) across lecithin bilayer membranes in the presence of ionophores such as A23187[1482] and X-537A[1482,1483] was monitored by NMR spectroscopy. The loss of resolution of the inner and outer signals as the ionophore rendered the membrane permeable to Pr(III) was used to determine the transport rate. The influence of local anesthetics on the binding of Eu(III) and Pr(III) to lecithin liposomes was determined by monitoring the reduction of shifts of the outer resonance in the presence of the anesthetics.[1484]

E. Cations

The availability of multinuclear NMR spectrometers has made it a common practice to record the NMR spectra of metal atoms including biologically important cations such as $^{39}K^+$, $^{43}Ca^{2+}$, $^{23}Na^+$, and $^{25}Mg^{2+}$. Water-soluble NMR shift reagents have been developed for the study of these and other cations.

Several reports have evaluated the effectiveness of water-soluble shift reagents for cations.[1379,1380,1384,1485-1490] Among those studied are Ln(EDTA)$^-$ (EDTA = ethylenediaminetetraacetic acid),[1379,1380,1487,1491,1492] Ln(TTHA)$^{3-}$ (TTHA = triethylenetetraaminehexaacetic acid),[1488,1493] Ln(NTA)$_2^{3-}$ (NTA = nitrilotriacetic acid),[1487,1488,1494,1495] Ln(DPA)$_3^{3-}$ (DPA = dipicolinic acid),[1487,1488] Ln(CA)$_3^{6-}$ (CA = 4-hydroxypyridine-2,6-dicarboxylic acid),[1486] Ln(NOTA)[1379,1380] (NOTA = 1,4,7-triazacyclononane-N,N',N''-triacetic acid), Ln(DOTA)$^-$ (DOTA = 1,4,7,10-tetraazacyclododecane-N,N',N'',N'''-tetraacetic acid),[1379,1380] LnTPPS(OH) (imidazole)$_x$(x ≤ 2)[1384] (TPPS = tetra-p-sulfonatophenylporphin), Ln(NO$_3$)$_5^{2-}$,[1496,1497] LnCl$_6^{3+}$,[1497] Fe(CN)$_6^{3-}$,[1487] and Ln(PPP)$_2^{7-}$ (PPP = tripolyphosphoric acid).[1485,1488,1490,1493,1498-1500]

The shifts in the spectra of the *n*-butylammonium cation and Na⁺ with Ln(DOTA)⁻, Ln(EDTA)⁻, and Ln(NOTA) were compared and Ln(NOTA) was judged the most effective shift reagent of the three.[1379,1380] The spectrum of the *N*-methylpyridinium cation was measured with Ln(III)TPPS(OH) (imidazole)$_x$, and the Tm analog was found to be a particularly effective shift reagent.[1384] The shifts in the ^{23}Na spectrum of Na⁺ were recorded with Fe(CN)$_6^{3-}$, Pr(EDTA)⁻, Dy(EDTA)⁻, Dy(DPA)$_3^{3-}$, and Dy(NTA)$_2^{3-}$.[1487] No shift was observed with the ferricyanide species, and only a small shift was obtained with Pr(EDTA)⁻. The shift with Dy(EDTA)⁻ was larger, but the best results were obtained with Dy(DPA)$_3^{3-}$ and Dy(NTA)$_2^{3-}$.[1487] It was postulated that the latter two complexes were more effective because of the higher charge. The shift in the ^{7}Li spectrum of Li⁺ was recorded with Dy(DPA)$_3^{3-}$ and Dy(NTA)$_2^{3-}$ and was larger with Dy(DPA)$_3^{3-}$.[1487]

The effectiveness of Tm(CA)$_3^{6-}$ and Dy(CA)$_3^{6-}$ as shift reagents for cations provides further evidence for the influence of charge.[1486] In this report, the shift in the spectra of ^{25}Mg^{2+}, ^{39}K⁺, ^{23}Na⁺, ^{87}Rb⁺, and the ^{14}N spectrum of NH$_4^+$ were measured. The divalent magnesium ion associated most strongly with the shift reagent, and it was recommended that shift reagents of lower charge be used with divalent cations.[1486] The magnitude of the shifts was found to exhibit a pronounced dependence on pH.

The shifts in the spectra of ^{23}Na⁺, ^{39}K⁺, ^{87}Rb⁺, ^{25}Mg^{2+}, and the ^{14}N resonance of NH$_4^+$ with Dy(TTHA)$^{3-}$, Tm(TTHA)$^{3-}$, Dy(DPA)$_3^{3-}$, Dy(NTA)$_2^{3-}$, Dy(PPP)$_2^{7-}$, and Tm(PPP)$_2^{7-}$ have been measured and compared.[1488] The complexes with PPP, which were first introduced by Gupta and Gupta,[1490] caused the largest shifts. The shifts in the spectrum of ^{23}Na⁺ with Dy(PPP)$_2^{7-}$ were larger than those with the Dy(III) complexes with adenosine diphosphate, -pyrophosphate, and -tetrapolyphosphate.[1490] The shifts in the spectra of cations with complexes with PPP were highly dependent on pH and decreased in the presence of Ca^{2+} or Mg^{2+}.[1488] Complexes with TTHA were also susceptible to the influence of Ca^{2+} and Mg^{2+}, but were less so than those of PPP.[1488] The shifts in the spectra of cations with complexes of TTHA were also independent of pH over the range from 5.5 to 12. One other interesting observation was that, contrary to most complexes with Dy(III),[1486,1487] the shifts in the spectra of cations with Dy(TTHA)$^{3-}$ were downfield.[1488] It was postulated that the complexes with TTHA and PPP(I) might have specific cation binding sites that account for their particular effectiveness as shift reagents.[1488]

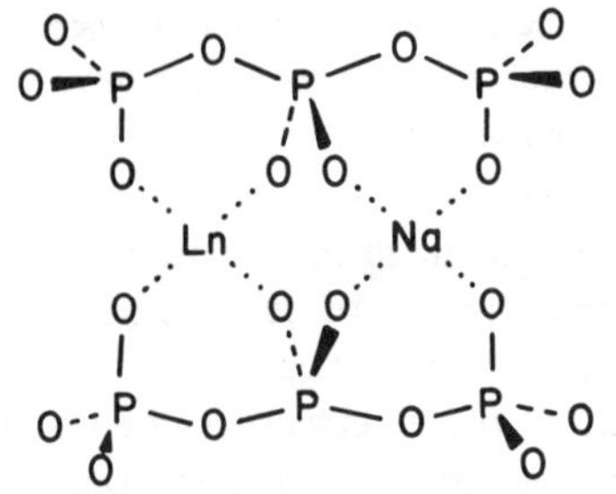

I

The bonding of lanthanide ions with PPP has been studied through an analysis of the ^{31}P and ^{17}O NMR spectra.[1489] Both the ^{31}P and ^{17}O data indicated a 1:2 metal:ligand complex. A proposed structure for the complex is shown in Figure 1.[1489] Slow exchange of the PPP ligands occurred for the heavier lanthanides including Dy(III), resulting in severely broadened ^{17}O NMR spectra. Faster ligand exchange was observed with the lighter lanthanides.[1489] Relaxation data for ^{23}Na⁺ with Tm(PPP)$_2^{7-}$ seemed to indicate

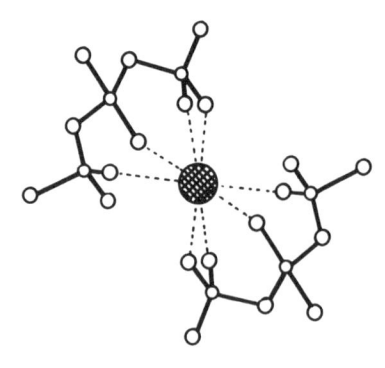

FIGURE 1. Proposed structure for Ln(PPP)$_2$$^{7-}$. (Reprinted with permission from Nieuwenhuizen, M. S., Peters, J. A., Sinnema, A., Kieboom, A. P. G., van Bekkum, H., J. Am. Chem. Soc., 107, 12, 1985. Copyright 1985, American Chemical Society.)

a larger distance than would be observed for structure I.[1489] The shifts in the spectrum of ^{39}K$^+$ with Dy(PPP)$_2$$^{7-}$, Tb(PPP)$_2$$^{7-}$, and Yb(PPP)$_2$$^{7-}$ have been compared.[1485] The Dy and Tb analogs were judged more effective than Yb.

The intra- and extracellular sodium in whole blood and washed human erythrocytes were distinguished using LSR.[1493] The shift reagent did not cross the membrane enabling the resolution of the intra- and extracellular cations. A similar distinction of intra- and extracellular potassium in human erythrocytes was achieved using Dy(PPP)$_2$$^{7-}$.[1500] The resonances of the sodium inside and outside the membrane of large unilamellar vesicles of egg lecithin were resolved with Dy(NTA)$_2$$^{3-}$.[1495] The transport properties of the membrane were studied by monitoring the relative signal intensities of the inner and outer species. Intra- and extracellular sodium resonances in RBC(red blood cells) and frog sartorius muscles were resolved through the use of Dy(PPP)$_2$$^{7-}$.[1490] Intra- and extracellular sodium, as well as transport properties, of the membranes of frog skin were monitored using Dy(PPP)$_2$$^{7-}$.[1499] Shift reagents have been used to simultaneously distinguish mucosal(urine). serosal(blood), and cytoplasmic sodium in a toad urinary bladder.[1488] A similar study was carried out on perfused rat hearts and shark rectal glands.[1488] The transport of sodium across the membranes of live *Saccharomyces cerevisiae* was investigated through the application of Dy(NTA)$_2$$^{3-}$.[1494] It is possible to deactivate these shift reagents by adding diamagnetic Lu(III) to the solution.[1487] A membrane can be prepared with shift reagent on both sides and then the extracellular shift reagent can be deactivated. This offers advantages in the study of some systems.[1487]

Shift reagents have been used in the study of the binding of calcium to the protein calmodulin in the presence and absence of calmodulin antagonist trifluoperazine.[1498] The resonances of free and bound calcium were resolved in the presence of Dy(PPP)$_2$$^{7-}$. This facilitated the determination of the effects of the antagonist on calcium binding.

The LSR described in this section can also be used with other cations including ammonium,[1492,1496,1497] amidinium,[1491] and phosphonium[1497] ions. The shifts in the spectra of pyrrolidinium and piperidinium ions with Yb(NO$_3$)$_5$$^{2-}$ and Er(NO$_3$)$_5$$^{2-}$ have been measured.[1496] The lanthanide species reportedly did not distinguish the stereochemical environment as well as tetraphenylborate.[1496] Shift data in the spectra of the N-alkylpyridinium, N,N,N-trimethylanilinium, and n-butyltriphenylphosphonium ions with LnCl$_6$$^{3-}$ and Ln(NO$_3$)$_5$$^{2-}$ have been compared to that with CoX$_4$$^{2-}$ and NiCl$_4$$^{2-}$.[1497] The lanthanide shifts were reportedly dipolar in nature. The shifts in the spectra of the (CH$_3$)$_n$NH$_{4-n}$$^+$, pyridinium, and methyl ethyl ammonium ions with

Ln(EDTA)⁻ have been measured.[1492] Lanthanide shift data with Eu(EDTA)⁻ and Pr(EDTA)⁻ were used to assign the -NH resonances of amidinium ions(II).[1491] The H_E and H_Z resonances

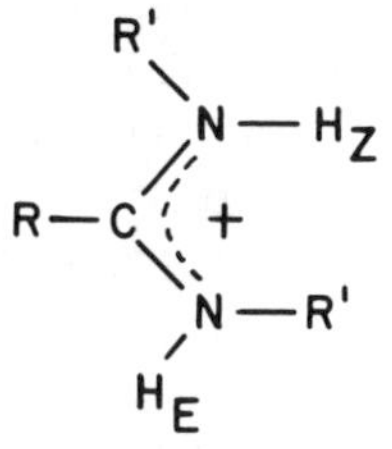

$$\begin{array}{c} R' \\ | \\ R-C \\ \diagdown \\ N-R' \\ | \\ H_E \end{array} \quad \begin{array}{c} N-H_Z \\ + \end{array}$$

II

of acetamidinium, benzamidinium, azoisobutyramidinium, and formamidine sulfinic acid were assigned on the basis of lanthanide shift data. A larger shift was observed for the H_Z resonance.

III. OTHER APPLICATIONS

A. Carboxylates

Carboxylate groups bond effectively to lanthanide ions in aqueous solutions. Acyclic monocarboxylate species including acetate,[1372,1396,1501-1503] methoxyacetate,[1405,1504] propanoate,[1388] butyrate,[1396] 3-hydroxybutyrate,[1501] 4-hydroxypentanoate,[1388] and n-valerate[1503] have been studied with LSR. The bonding of carboxylate compounds such as acetate[1502] and propanoate[1388] with the lanthanide ions is reportedly bidentate in nature. Shift data have been used to demonstrate that the acetate ion replaces water in the first coordination sphere of lanthanide ions.[1372,1502] The shifts in the ¹⁷O NMR spectrum of the acetate ion with Dy(III) have been reported.[1501] The stability constants for the complexes of acetate and butyrate with Nd(III) were determined using lanthanide shift data and potentiometric methods.[1396] The results of the two methods were judged comparable.

The influence of pH and ionic medium effects on the coordination of acetate and n-valerate with Ln(III) complexes with EDTA has been investigated.[1503] The spectrum of the acetate ion exhibited useful shifts over the pH range from 6 to 10. Below a pH of 6, acetic acid formed in appreciable quantities and did not effectively coordinate with the shift reagent. Lanthanide hydroxides formed at pH values above 10.[1503] Similar pH dependencies were observed for Pr(EDTA)⁻, Yb(EDTA)⁻, and Gd(EDTA)⁻. Evidence obtained from these experiments indicated that the complexes were isostructural. Larger shifts were observed with Pr(III) compared to Pr(EDTA)⁻, and it was recommended that aquated lanthanide ions be employed as shift reagents at pH values less than 7. The shift data were also used to analyze the structures of the lanthanide-EDTA complexes. The data were consistent with pentachelation of the EDTA ligand for the lighter lanthanides and hexachelation with the heavier members.[1503]

Other workers have analyzed the structures of lanthanide complexes with EDTA through the use of NMR spectral data.[1391,1505,1506] Sherry et al. also observed different structures for the lighter and heavier members of the series of lanthanide complexes with EDTA.[1391] The exchange of EDTA ligands was slow at low pH(4.5) and fast at alkaline pH values. An unsuccessful attempt was made to separate the contact and pseudocontact shifts based on different temperature dependencies. The method of Reilley et al.[1194] was applied with more success. It was recommended, however, that the lanthanide ions be divided into small subgroups when performing this analysis.[1391]

Two studies employed diamagnetic lanthanide ions.[1505,1506] Kostromina and Ternovaya reported that complexes of the formula Ln(EDTA) (HEDTA)$^{4-}$ were observed in solution for La(III) and Y(III).[1505] Complexes of diamagnetic lanthanide ions with EDTA,[1507] N-methylethylenediaminetriacetic acid,[1506] HEDTA,[1506] and diethylenetriaminepentaacetic acid[1506] have been studied through the analysis of ^{1}H NMR data.

The structures of lanthanide complexes with other multidentate ligands including 2,6-dipicolinate,[1386] ethylene glycol bis(aminoethyl)tetraacetic acid,[1421] and 1,4,7,10-tetraazacyclododecane-N,N',N'',N'''-tetraacetic acid(DOTA)[1508] have been studied through analyses of the effects of paramagnetic lanthanide ions on the spectra. Lanthanide shift data indicated that the dipicolinate ion coordinated in a tridentate manner and formed 3:1 complexes with lanthanide ions.[1386] Gd(III) and Eu(II) were used as relaxation probes in the study of complexes with ethylene glycol bis(aminoethyl)tetraacetic acid.[1421] The data was indicative of a 1:1 complex. The exchange properties and conformation of the DOTA ligand in complexes with Yb(III), Eu(III), and Pr(III) were determined through analysis of shift data in the ^{1}H and ^{13}C spectra.[1508] The metal was reportedly located under the macrocycle, and slow ligand exchange was observed at ambient probe temperatures.

The shifts in the ^{17}O NMR spectra of a variety of carboyxlate compounds including glycolate, lactate, malonate, malate, and citrate have been measured.[1501] These substrates exhibited fast exchange at 73°C, and it was reported that the shift mechanism was largely contact in origin. Oxydiacetate, (carboxymethoxy)succinate, and nitrilotriacetate exhibited slow exchange and the ^{17}O spectra were too broadened to observe resonances for the complexed substrate.[1501] The origin of the shifts in the ^{1}H NMR spectrum of methoxyacetate in the presence of lanthanide ions has been assessed.[1405,1504] The shift mechanism of the methyl group was reportedly dipolar in origin, whereas the shift of the methylene resonance could not be fit to a pure dipolar mechanism.

The bonding between oxaloacetates and Eu(III) has been assessed using shift data.[1509] Evidence indicated the formation of an α-oxocarboxylate complex with I. The bonding of II

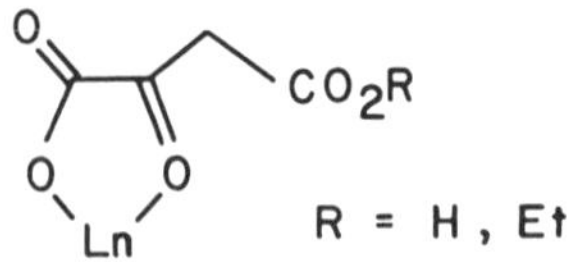

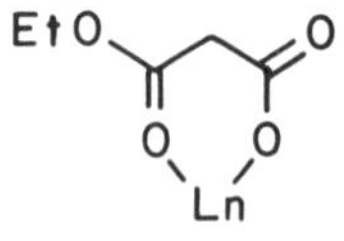

with Eu(III) was different than that of I. The association between pyruvate and trivalent lanthanide ions has been studied using NMR shift data.[1506] Pyruvate can exist in a keto(III), diol(IV), and dimer(V) form. The shifts in the spectra of IV and

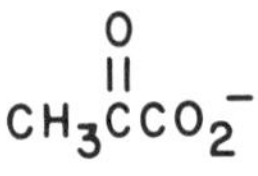

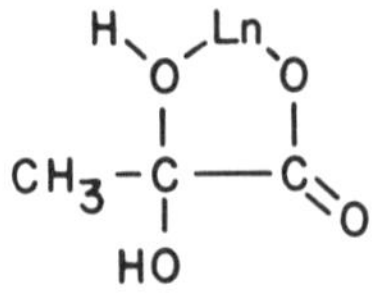

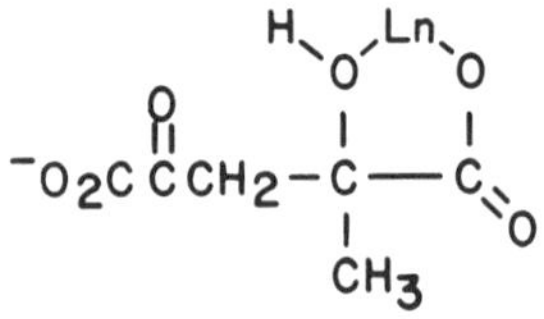

V were much larger than those in the spectrum of III. It was believed that IV and V formed a chelate bond with the lanthanide ions, whereas III coordinated in a monodentate manner.

Lanthanide shift and relaxation data have been recorded for the ^{1}H spectra of benzoate, o-toluate, and p-toluate.[1390] The shifts were reportedly pseudocontact in origin. It was concluded that the structures of the complexes with the lighter and heavier lanthanides were different. The structure and conformation of benzene-1,2-dioxydiacetate(VI) was determined by fitting lanthanide shift and relaxation data obtained from the ^{1}H and ^{13}C spectra.[1510] The shift data was indicative of tetradentate binding of the substrate.[1506,1510]

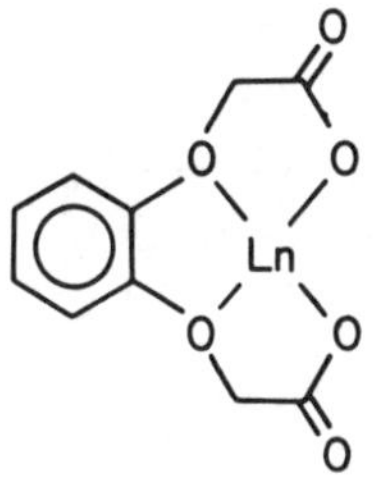

VI 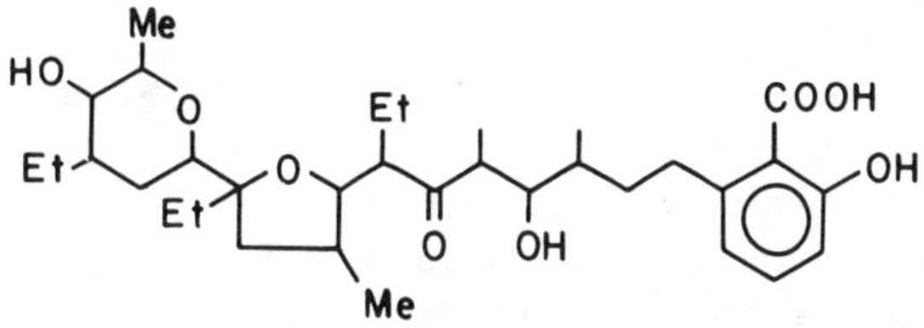 VII

A study of the lifetimes of complexes of antibiotic X-537A(lasalocid A) (VII) with metal cations(K^+, Ba^{2+}, Ca^{2+}, Sr^{2+}) was facilitated through the use of Pr(III).[1511] The Pr(III) was used as both a model cation and to shift the resonances of the "free" acid that was not bound to the cation under study. The methodology necessary to determine the lifetimes of such complexes in the presence of the lanthanide ion was described in this report.[1511]

The E,E and Z,E isomers of VIII were distinguished through the application of Eu(III) and Pr(III).[1512]

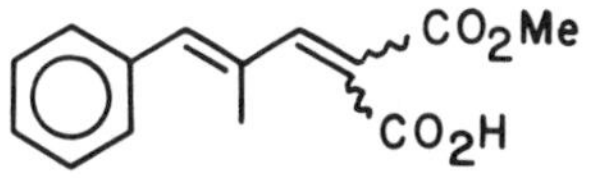

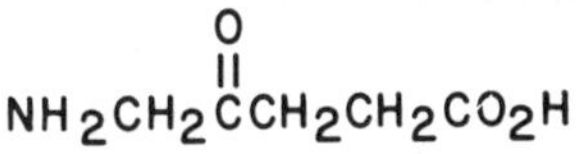

VIII IX

The lanthanide ions coordinated with the acid moiety of VIII. The location and extent of deuterium substitution in gamma-aminolevulinic acid(IX) was determined using Eu(III) and Pr(III).[1513] Complexation occurred at the carboxylate group.

Shift and relaxation data have been used to assess the bonding and structure of indol-3-acetate(X) with trivalent lanthanide ions.[1405,1514] The carboxylate group coordinated with the metal and

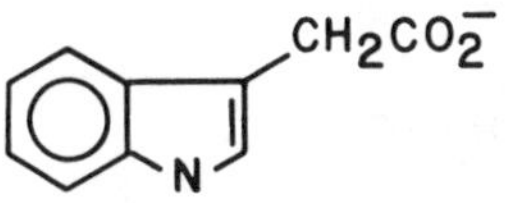

X

Levine et al. concluded that a structural change occurred at Tm(III) for the series of lanthanide ions.[1514] Reuben and Elgavish analyzed relaxation data for the entire series of ions and disputed the finding of a structural change.[1405] Reuben found that the internal ratios of the relaxation data were essentially identical for all of the lanthanide ions.

The structures of lanthanide complexes with *endo-cis*-bicyclo[2.2.1]hept-5-ene-2,3-dicarboxylic acid(XI) were assessed through an analysis of shift and relaxation data.[1515]

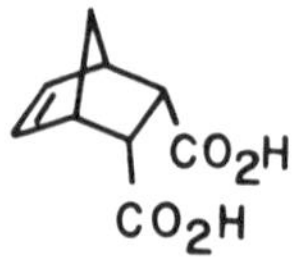

XI

Lanthanide complexes with EDTA, HEDTA, and CyDTA were used at a pH of 8 and aquated ions were used at a pH of 4.6. The relaxation data for the series of ions and HEDTA chelates was indicative of isostructural complexes.[1515] The shift data, however, were indicative of a structural change across the series. Since the relaxation data are considered to be more reliable for assessing structural differences, it was proposed that the anomalous shift data were the result of axial asymmetry of the susceptibility values. The nonaxial contribution to the shifts was largest for Tm(III) and Er(III). The smallest asymmetry was found for Pr(III) and Eu(III).[1515]

B. Phenols

The shifts in the spectra of a variety of phenols have been measured with lanthanide perchlorates.[1516-1518] The solvent employed in these studies was DMSO-d_6. The lanthanide shifts for *p*-substituted phenols, one example being *p*-nitrophenol, increased significantly in basic media when the substrate was converted into the phenolate ion.[1516-1518] The phenolate ions were obtained by adding diethylamine or piperidine to the solutions.[1516] *Ortho*-substituted phenols, such as *o*-nitrophenol, *o*-methoxyphenol, 2-hydroxyacetophenone(I), and 2-hydroxybenzaldehyde, coordinated in a chelate manner with the lanthanide ions.[1516,1517]

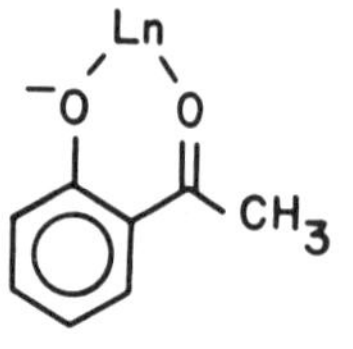

I 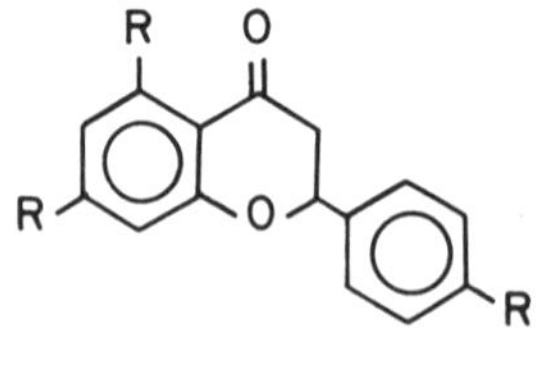 II 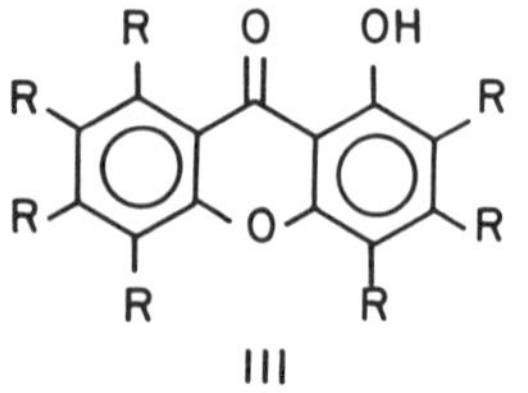 III

The *o*- and *p*-substituted phenols were used as model substrates for studies of polyphenols such as flavones(II) and xanthones(III).[1517] One or more of the R groups in II and III were -OH groups, and the shifts in the spectrum of each substrate were recorded with Eu(ClO$_4$)$_3$.[1517]

C. Oxides

Oxides such as 3-methylpyridine-1-oxide,[1519] triphenylphosphine oxide,[68] tri(*m*-tolyl)arsine oxide,[68] and I[68] bond effectively to lanthanide ions in aqueous solutions. The aromatic resonances of tri (m-tolyl)arsine oxide were fully resolved in the presence of Eu(III) and Pr(III).[68] The resonances of the α protons of 3-methylpyridine-1-oxide were resolved in complexes with lanthanide ions.[1519]

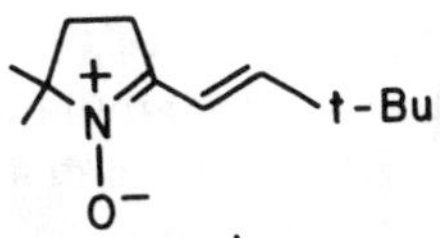

Shift data with Eu(III) were used to distinguish and assign the structures of 6-oxidodipyrido[2,1-b:2'3'-d]thiazolium(II) and 9-oxidodipyrido[2,1-b:3',2'-d]thiazolium(III).[1520] The lanthanide ion bonded to the oxide moiety, and it was impossible to fit the shift data for II to structure III.

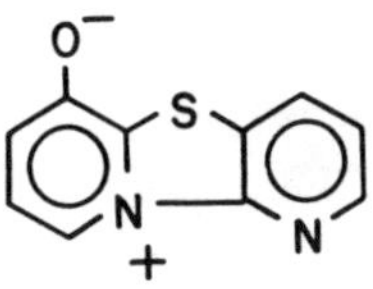

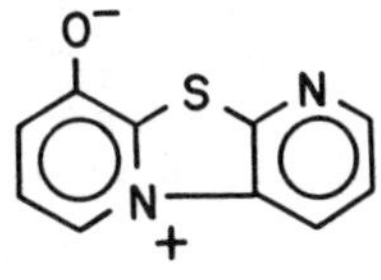

The binding of lanthanide ions with tetracycline(IV) has been studied using relaxation data with Gd(III).[1521] At a pH of 8.6 the

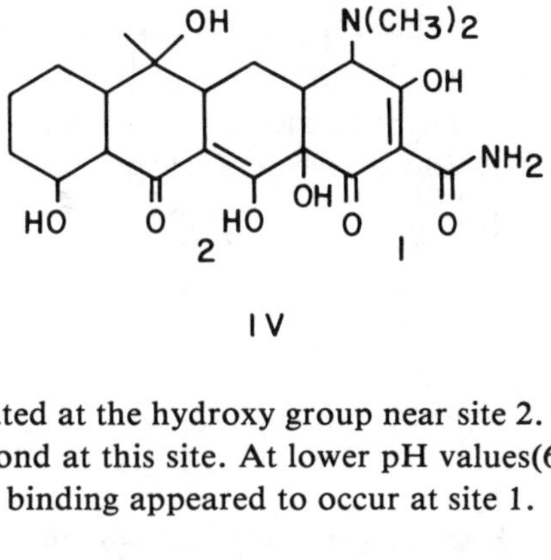

tetracycline is deprotonated at the hydroxy group near site 2. The relaxation data were indicative of a chelate bond at this site. At lower pH values(6.5 and 2.0), the hydroxy group is protonated and binding appeared to occur at site 1.

D. Amines

The structures of complexes of ethylenediamine with lanthanide perchlorates in acetonitrile were analyzed by fitting the NMR shift data.[1522] A tetrakis complex was indicated, and both contact and pseudocontact mechanisms were invoked to explain the shifts.

E. Nitrogen Heterocycles

The shifts in the spectra of 4-methylpyridine, 3,5-dimethylpyridine, 4-ethylpyridine, and 4-t-butylpyridine in acetonitrile have been measured with the nitrate and perchlorate salts of Pr(III) and Nd(III).[1523] The directions of the shifts were different for the nitrate and perchlorate salts. This was believed to result from geometric differences in the lanthanide-pyridine complexes. No lanthanide shifts were observed in the spectra

NMR Shift Reagents

of 2,6-dimethylpyridine and 2,4,6-trimethylpyridine.[1523] This is probably the result of the large steric hindrance about the nitrogen atom of these substrates. The shifts in the spectra of phenanthroline with trivalent lanthanide ions have been reported.[1385]

E. Substrates with a Sulfur Atom

The shifts in the spectrum of dimethylsulfoxide have been measured with $Eu(ClO_4)_3$ and $Pr(ClO_4)_3$.[1524] The products of a reaction of exo-2-norbornylbrosylate(I) were monitored by the ^{17}O NMR spectrum.[1525] A sulfonic acid was formed in the reaction and

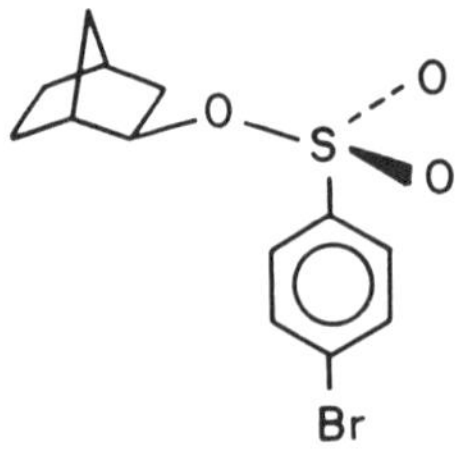

I

Pr(III) was added to the solution to shift the oxygen resonances of the acid away from the oxygen resonances of other products.

Lanthanide ions have been used to shift the spectrum of sodium dodecylsulfate dissolved as micelles in water.[1526] The shifts in the spectrum of acetone[1526,1527] and methanol[1526] solubilizates in these micelles were also measured.

F. Organometallics

The methyl resonances of I were resolved in the spectrum recorded with $Eu(ClO_4)_3$.[634] The shift data indicated that the Eu(III) bonded to the oxygen of the pendant acetate group.

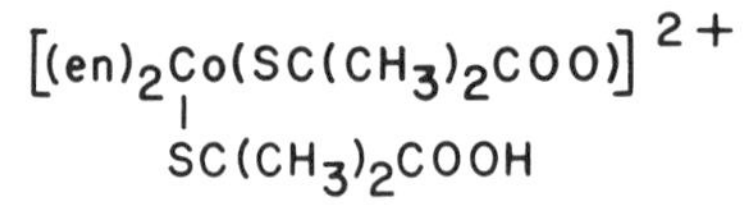

$$[(en)_2\underset{|}{Co}(SC(CH_3)_2COO)]^{2+}$$
$$SC(CH_3)_2COOH$$

I

IV. CHIRAL SHIFT REAGENTS

The first examples of chiral LSR for aqueous solutions were reported by Reuben.[1528,1529] These were formed by the addition of enantiomerically pure α-hydroxycarboxylates to solutions of lanthanide(III)chlorides. The optically pure ligand bonded to the lanthanide ion to create a chiral shift reagent that could then be used to enantiomerically resolve the spectra of chiral substrates. Restricted rotation of the carboxy group of the chiral additive was necessary to achieve enantiomeric resolution.[1528] Alpha-hydroxycarboxylates, by virtue of chelate bonding with the lanthanide ion, exhibit restricted rotation when complexed to the lanthanide ions.

The spectrum of lactate was enantiomerically resolved in the presence of lanthanide complexes of D-mandelate.[1529] The resonances of the enantiotopic methyl groups of α-hydroxyisobutyrate were resolved using lanthanide complexes with L-lactate[1528] and D-mandelate.[1529] The resonances of the enantiotopic methylene protons of glycolate($HOCH_2CO_2^-$) were resolved using $PrCl_3$ and L-lactate.[1529] Nonracemic mixtures of L- and D-lactate exhibited enantiomeric resolution in the presence of trivalent lanthanide ions.¹⁵²⁸ Phenomenological equations were derived to describe the optical resolution observed with these shift reagents.¹⁵²⁸ These equations demonstrated that the enantiomeric resolution was optimized with 3:1 ligand-lanthanide complexes.

Under certain circumstances, it was possible to assign the absolute configuration of the substrate.¹⁵²⁸ The chiral complex with L-lactate always caused a larger shift for the resonances of the d-enantiomer of α-hydroxycarboxylates. This trend was reversed for complexes with D-lactate. These trends were demonstrated with glycolate-d_1 and were used to assign the absolute configuration of citramalate(I).¹⁵²⁸

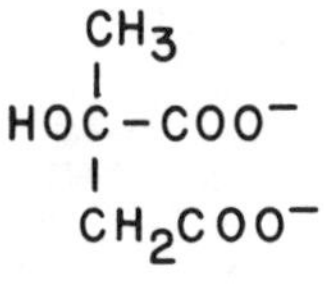

$$\begin{array}{c} CH_3 \\ | \\ HOC-COO^- \\ | \\ CH_2COO^- \end{array}$$

I

Peters et al. have employed the Eu(III) and Yb(III) complexes of (S)-carboxymethyloxysuccinic acid(II) as chiral shift reagents.¹⁵³⁰ These complexes are stable over the pH range

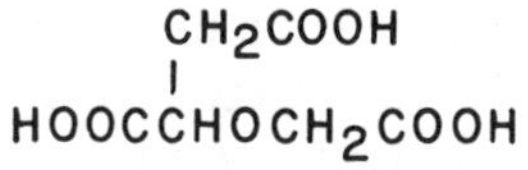

$$\begin{array}{c} CH_2COOH \\ | \\ HOOCCHOCH_2COOH \end{array}$$

II

from 3 to 10. Enantiomeric resolution was observed in the spectra of amino acids, (oxy)carboxylic acids, oxydilactate, alanine, and 3-hydroxyphenylalanine with these complexes.¹⁵³⁰ The methylene protons of oxydiacetate and nitrilotriacetate are enantiotopic by internal comparison and were resolved in the NMR spectrum recorded in the presence of these chiral reagents.

The Eu(III) complex of (R)-propylenediaminetetraacetic acid has been used as a chiral shift reagent.¹⁵³¹ Enantiomeric resolution was observed in the spectra of hydroxycarboxylic acids, amino acids, and carboxylic acids. The methyl groups of isobutyric acid are enantiotopic by internal comparison and were resolved in the presence of the chiral complex.¹⁵³¹

REFERENCES

1. Hinckley, C. C., Paramagnetic shifts in solutions of cholesterol and the dipyridine adduct of trisdipivalomethanatoeuropium(III). A shift reagent, *J. Am. Chem. Soc.*, 91, 5160, 1969.
2. Sanders, J. K. M. and Williams, D. H., A shift reagent for use in nuclear magnetic resonance spectroscopy. A first-order spectrum of n-hexanol, *J. Chem. Soc. Chem. Commun.*, p. 422, 1970.
3. Rondeau, R. E. and Sievers, R. E., New superior paramagnetic shift reagents for nuclear magnetic resonance spectral clarification, *J. Am. Chem. Soc.*, 93, 1522, 1971.
4. Whitesides, G. M. and Lewis, D. W., Tris[(3-tert-butylhydroxymethylene)-d-camphorato]europium(III). A reagent for determining enantiomeric purity, *J. Am. Chem. Soc.*, 92, 6979, 1970.
5. Goering, H. L., Eikenberry, J. N., and Koermer, G. S., Tris[3-(trifluoromethylhydroxymethylene)-d-camphorato]europium(III). A chiral shift reagent for direct enantiomeric compositions, *J. Am. Chem. Soc.*, 93, 5913, 1971.
6. Fraser, R. R., Petit, M. A., and Saunders, J. K., Determination of enantiomeric purity by an optically active nuclear magnetic resonance shift reagent of wide applicability, *J. Chem. Soc. Chem. Commun.*, p. 1450, 1971.
7. McCreary, M. D., Lewis, D. W., Wernick, D. L., and Whitesides, G. M., The determination of enantiomeric purity using chiral lanthanide shift reagents, *J. Am. Chem. Soc.*, 96, 1038, 1974.
8. Evans, D. F., Tucker, J. N., and de Villardi, G. C., Lanthanide shift reagents for alkenes, *J. Chem. Soc. Chem. Commun.*, p. 205, 1975.
9. Wenzel, T. J., Bettes, T. C., Sadlowski, J. E., and Sievers, R. E., New binuclear lanthanide NMR shift reagents effective for aromatic compounds, *J. Am. Chem. Soc.*, 102, 5903, 1980.
10. McConnell, H. M. and Robertson, R. E., Isotropic nuclear resonance shifts, *J. Chem. Phys.*, 29, 1361, 1958.
11. Briggs, J. M., Moss, G. P., Randall, E. W., and Sales, K. D., Pseudo-contact contributions to lanthanide-induced nuclear magnetic resonance shifts, *J. Chem. Soc. Chem. Commun.*, p. 1180, 1972.
12. Hofer, O., The lanthanide induced shift technique: applications in conformational analysis, *Top. Stereochem.*, 9, 111, 1976.
13. Flockhart, B. D., Lanthanide shift reagents in nuclear magnetic resonance spectroscopy, *CRC Crit. Rev. Anal. Chem.*, 6, 69, 1976.
14. Campbell, J. R., Lanthanide chemical shift reagents, *Aldrichchimica Acta*, 4, 55, 1971.
15. Begue, J. P., Deplacements paramagnetiques induits par les chelates de terres rares en resonance magnetique nucleaire: utilisation en chimie organique, *Bull. Soc. Chim. Fr.*, p. 2073, 1972.
16. Becker, E. D., Some recent developments in high resolution nuclear magnetic resonance, *Appl. Spectrosc.*, 26, 421, 1972.
17. von Ammon, R. and Fischer, R. D., Shift reagents in NMR spectroscopy, *Angew. Chem. Int. Ed.*, 11, 675, 1972.
18. Williams, D. H., Nuclear magnetic resonance shift reagents in organic chemistry, *Pure Appl. Chem.*, 40, 25, 1974.
19. Voronov, V. K., Paramagnetic reagents for the investigation of the structures of organic ligands, *Russ. Chem. Rev.*, 43, 432, 1974.
20. Slonim, I. Y. and Bulai, A. K., Paramagnetic shift reagents in nuclear magnetic resonance spectroscopy, *Russ. Chem. Rev.*, 42, 904, 1973.
21. Sinha, S. P., Applications of rare earth complexes as NMR shift reagents in elucidating the structure of organic molecules, *J. Mol. Struct.*, 19, 387, 1973.
22. Sanders, J. K. M. and Williams, D. H., Shift reagents in NMR spectroscopy, *Nature (London)*, 240, 385, 1972.
23. Cockerill, A. F., Davies, G. L. O., Harden, R. C., and Rackham, D. M., Lanthanide shift reagents for nuclear magnetic resonance spectroscopy, *Chem. Rev.*, 73, 553, 1973.
24. Inagaki, F. and Miyazawa, T., NMR analysis of molecular conformations and conformational equilibria with the lanthanide probe method, *Prog. Nucl. Magn. Reson. Spectrosc.*, 14, 67, 1981.
25. Reuben, J. and Elgavish, G. A., Shift reagents and nmr paramagnetic lanthanide complexes, in *Handbook on the Physics and Chemistry of Rare Earths*, Gschneidner, K. A., Jr. and Eyring, L., Eds., North-Holland, Amsterdam, 1979, 483.
26. Reuben, J., Paramagnetic lanthanide shift reagents in NMR spectroscopy: principles, methodology and applications, *Prog. Nucl. Magn. Reson. Spectrosc.*, 9, 1, 1975.
27. Reuben, J., The lanthanides as spectroscopic and magnetic resonance probes in biological systems, *Naturewissenschaften*, 62, 172, 1975.

28. Peterson, M. R., Jr. and Wahl. G. H., Jr., Lanthanide NMR shift reagents, *J. Chem. Educ.*, 49, 790, 1972.
29. Lefevre, F. and Martin, M. L., Utilisation des complexes de terres rares en resonance magnetique nucleaire, *Org. Magn. Reson.*, 4, 737, 1972.
30. Kime, K. A. and Sievers, R. E., A practical guide to uses of lanthanide NMR shift reagents, *Aldrichchimica Acta,* 10, 54, 1977.
31. Hinckley, C. C., Applications of lanthanide shift reagents, in *Modern Methods of Steroid Analysis,* Heftmann, E., Ed., Academic Press, New York, 1973, 265.
32. Grandjean, J., Utilisationsa des complexes paramagnetiques de lanthanides en RMN, *Ind. Chim. Belg.*, 37, 220, 1972.
33. Doerffel, K., Brunn, J., Hoebold, W., and Radeglia, R., Helpful reactions in NMR spectroscopy, *Z. Chem.*, 19, 129, 1979.
34. Danieli, B. and Palmisano, G., Shift reagents in NMR, *Relaz. Corso Teor. Prat. Risonanze Magn. Nucl.,* p. 231, 1973; *Chem. Abstr.*, 80, 53964a.
35. Shapiro, Y. E., Use of lanthanide shift reagents in the analysis of polymer microstructure by an NMR spectroscopic method, *Usp. Khim.*, 53, 1407, 1984.
36. Yamasaki, A., Basic knowledge of analytical reagents. NMR shift reagents, *Bunseki,* p. 916, 1983.
37. Jo, N. S., Lanthanide shift reagent for proton nuclear magnetic resonance, *Hwahak Kwa Kongop Ui Chinbo,* 22, 307, 1982; *Chem. Abstr.*, 98, 88412t.
38. Rebuffat, S., Davoust, D., Giraud, M., and Molho, D., Induced paramagnetic shifts by lanthanide chelates and salts in NMR. Principles, methods, applications in organic chemistry, *Bull. Mus. Natl. Hist. Natl. Sci. Phys. Chim.*, 8, 17, 1976; *Chem. Abstr.*, 87, 166938h.
39. Davidenko, N. K., Bidzilya, V. A., and Goryushko, A. G., Lanthanide shift reagents, *Probl. Koord. Khim.*, 13, 1977.
40. Hosoda, H., NMR shift reagents, *Farumashia,* 13, 122, 1977.
41. Voronov, V. K., Keiko, V. V., and Moskovskaya, T. E., Paramagnetic reagents for structure study of heteroatomic compounds according to NMR spectra, *Zh. Strukt. Khim.*, 18, 917, 1977.
42. Imamura, Y., NMR shift reagents, *Dojin Nyusu,* 13, 1, 1979; *Chem. Abstr.*, 94, 14621f.
43. Jo, N. S., Lanthanide shift reagent and proton nuclear magnetic resonance, *Hwahak Kwa Kongop Ui Chinbo,* 22, 219, 19; *Chem. Abstr.*, 97, 161874f.
44. Sohar, P., NMR as a shift reagent technique, *Kem. Kozl.*, 57, 315, 1982.
45. Nishio, M., Conformational analysis of mobile molecule by LIS simulation, *Kagaku No Ryoiki,* 36, 190, 1982; *Chem. Abstr.*, 97, 54887r.
46. Roth, K. and Rewicki, D., Methods for the use of nmr shift reagents, *Kontakte,* p. 9, 1978.
47. Hajek, M. and Mohyla, I., Shift reagents in the NMR spectroscopy of carbon-13, fluorine-19, phosphorus-31, nitrogen-14, and oxygen-17 nuclei, *Sb. Vys. Sk. Chem. -Technol. Praze Technol. Paliv.,* D39, 77, 1978; *Chem. Abstr.*, 93, 83706d.
48. Davidenko, N. K., Yatsimirskii, K. B., Bidzilya, V. A., and Golovkova, L. P., Use of lanthanide shift reagents for the proton NMR study of coordination properties of donor molecules, *Tezisy Dokl. Ukr. Resp. Konf. Fiz. Khim.,* 12th, 84, 1977; *Chem. Abstr.*, 92, 155186p.
49. Kornilov, M. Y. and Turov, A. V., Lanthanide shift reagents in the chemistry of heterocyclic compounds, *Khim. Geterotsikl. Soedin.,* p. 1299, 1979.
50. Glasel, J. A., Lanthanide ions as nuclear magnetic resonance chemical shift probes in biological systems, in *Current Research Topics in Bioinorganic Chemistry,* Lippard, S. J., Ed., John Wiley & Sons, New York, 1973, 383.
51. Kutal, C., Chiral shift reagents, in *Nuclear Magnetic Resonance Shift Reagents,* Sievers, R. E., Ed., Academic Press, New York, 1973, 87.
52. Fraser, R. R., Nuclear magnetic resonance analysis using chiral shift reagents, *Asymmetric Synth.,* 1, 173, 1983.
53. Sullivan, G. R., Chiral lanthanide shift reagents, *Top. Stereochem.*, 10, 287, 1978.
54. Weissman, S. I., On the action of europium shift reagents, *J. Am. Chem. Soc.*, 93, 4928, 1971.
55. Fischer, R. D., Lanthanide and actinide complexes, in *NMR of Paramagnetic Molecules Principles and Applications,* LaMar, G. N., Horrocks, W. D., Jr., Holm, R. H., Eds., Academic Press, New York, 1973, 521.
56. Cheng, H. N. and Gutowsky, H. S., The use of shift reagents in nuclear magnetic resonance studies of chemical exchange, *J. Am. Chem. Soc.*, 94, 5505, 1972.
57. Horrocks, W. D., Jr. and Sipe, J. P., III, Lanthanide complexes as nuclear magnetic resonance structural probes: paramagnetic anisotropy of shift reagent adducts, *Science,* 177, 994, 1972.
58. Kjosen, H. and Liaaen-Jensen, S., Application of the tris(dipivalomethanato)europium(III) nuclear magnetic resonance shift reagent to carotenoids, *Acta Chem. Scand.*, 26, 2185, 1972.

59. Servis, K. L., Bowler, D. J., and Ishii, C., Conformational analysis of 2-alkylcyclohexanone-lanthanide chelate complexes, *J. Am. Chem. Soc.*, 97, 74, 1975.
60. Kishi, M., Tori, K., and Komeno, T., Application of paramagnetic shift induced by tris(dipivalomethanato)europium(III) to configurational assignment of sulfinyl oxygen in 5α-cholestan-2α,5-episulfoxides. Examples of upfield shifts due to Eu(dpm)$_3$, *Tetrahedron Lett.*, 3525, 1971.
61. Shapiro, B. L., Hlubucek, J. R., Sullivan, G. R., and Johnson, L. F., Lanthanide-induced shifts in proton nuclear magnetic resonance spectra. I. Europium-induced shifts to higher fields, *J. Am. Chem. Soc.*, 93, 3281, 1971.
62. Iida, T., Kikuchi, M., Tamura, T., and Matsumoto, T., Proton magnetic resonance identification and discrimination of stereoisomers of C$_{27}$ steroids using lanthanide shift reagents, *J. Lipid Res.*, 20, 279, 1979.
63. Hlubucek, J. R. and Shapiro, B. L., Lanthanide-induced shifts in proton NMR spectra-IV. Praseodymium-induced shifts to lower applied fields and obscured chemical shifts from lanthanide-induced shift (LIS) data, *Org. Magn. Reson.*, 4, 825, 1972.
64. Hajek, M., Vodicka, L., Ksandr, Z., and Landa, S., NMR analysis of adamantane derivatives I. Eu(dpm)$_3$ as shift reagent for 2-alkyl-2-adamantanols, *Tetrahedron Lett.*, 4103, 1972.
65. Bhacca, N. S. and Wander, J. D., Aromatic proton magnetic resonances that are not shifted by Eu(dpm)$_3$, *J. Chem. Soc. Chem. Commun.*, p. 1505, 1971.
66. Yashimura, Y., Mori, Y., and Tori, K., Configurational assignment of α-phenyl-α,N-dimethylnitrones by a lanthanide shift reagent in proton magnetic resonance spectroscopy, *Chem. Lett.*, 181, 1972.
67. Siddall, T. H., III, An upfield H-1 nuclear magnetic resonance shift induced by tris(dipivalomethanato)europium, *J. Chem. Soc. Chem. Commun.*, p. 452, 1971.
68. Sanders, J. K. M., Hanson, S. W., and Williams, D. H., Paramagnetic shift reagents. The nature of the interactions, *J. Am. Chem. Soc.*, 94, 5325, 1972.
69. Mazzocchi, P. H., Tamburin, H. J., and Miller, G. R., Upfield and downfield shifts in the nuclear magnetic resonance spectrum of a tris(dipivalomethanato)europium(III) complex, *Tetrahedron Lett.*, p. 1819, 1976.
70. Lindoy, L. F. and Moody, W. E., Nuclear magnetic resonance studies of metal complexes using lanthanide shift reagents. Lanthanide-induced shifts in the H-1 (and C-13) spectra of diamagnetic metal complexes of quadridentate ligands incorporating oxygen-nitrogen donor atoms, *J. Am. Chem. Soc.*, 99, 5863, 1977.
71. Reuben, J. and Leigh, J. S., Jr., Effects of paramagnetic lanthanide shift reagents on the proton magnetic resonance spectra of quinoline and pyridine, *J. Am. Chem. Soc.*, 94, 2789, 1972.
72. Sievers, R. E., Brooks, J. J., Cunningham, J. A., and Rhine, W. E., Unusually volatile and soluble metal chelates: lanthanide NMR shift reagents, *Adv. Chem. Ser.*, p. 222, 1976.
73. Dyer, D. S., Cunningham, J. A., Brooks, J. J., Sievers, R. E., and Rondeau, R. E., Interactions of nucleophiles with lanthanide shift reagents, in *Nuclear Magnetic Resonance Shift Reagents*, Sievers, R. E., Ed., Academic Press, New York, 1973, 21.
74. Burgett, C. A. and Warner, P., A new lanthanide paramagnetic shift reagent containing fully fluorinated side chains, *J. Magn. Reson.*, 8, 87, 1972.
75. Burgett, C. A., Europium Complex of 1,1,1,2,2,6,6,7,7,7-decafluoro-3,5-heptanedione, U.S. Patent 3,867,418, 1975.
76. Morrill, T. C., Clark, R. A., Bilobran, D., and Youngs, D. S., New dimensions in lanthanide shift reagent-pmr analysis of organic compounds: Eu(tfn)$_3$, *Tetrahedron Lett.*, p. 397, 1975.
77. Lui, K. T., Application of ytterbium shift reagent to trifluoroacetate esters. Determination of deuterium distribution in exo-norbornyl-d trifluoroacetate, *Tetrahedron Lett.*, p. 1207, 1977.
78. Pickering, R. A. and Roling, P. V., NMR shift reagents: a stereoselective interaction of Eu(tfn)$_3$ with methyl ketones, *J. Magn. Reson.*, 22, 385, 1976.
79. Smith, W. B., Carbon-13 NMR spectroscopy of steroids, *Annu. Rep. NMR Spectrosc.*, 8, 199, 1978.
80. Shapiro, B. L., Johnston, M. D., Jr., Godwin, A. D., Proulx, T. W., and Shapiro, M. J., Concerning the relative shifting abilities of Eu(dpm)$_3$ and Eu(fod)$_3$, *Tetrahedron Lett.*, p. 3233, 1972.
81. Crump, D. R., Sanders, J. K. M., and Williams, D. H., Evaluation of some tris(dipivalomethanato)lanthanide complexes as paramagnetic shift reagents, *Tetrahedron Lett.*, p. 4419, 1970.
82. Ahmad, M., Bhacca, N. S., Selbin, J., and Wander, J. D., Nuclear magnetic resonance spectra of tris[2,2,6,6-tetramethyl-3,5-heptanedionato] complexes of the lanthanides. Temperature dependence of shift reagents, *J. Am. Chem. Soc.*, 93, 2564, 1971.
83. Wenzel, T. J., Ruggles, A. C., and Lalonde, D. R., Jr., Binuclear lanthanide(III)-silver(I) NMR shift reagents: investigations of new achiral and chiral analogs, *Magn. Reson. Chem.*, 23, 778, 1985.

84. Wei, H. H. and Hwang, J. M., The studies of lanthanide shift reagents in NMR spectroscopy, *J. Chinese Chem. Soc.*, 23, 101, 1976.
85. Smith, G. V., Boyd, W. A., and Hinckley, C. C., Isotope effects in nuclear magnetic resonance spectra modified by rare-earth shift reagents, *J. Am. Chem. Soc.*, 93, 6319, 1971.
86. Hinckley, C. C., Boyd, W. A., and Smith, G. V., Deuterium isotope effects observed in NMR shift reagents, *Tetrahedron Lett.*, p. 879, 1972.
87. Sanders, J. K. M. and Williams, D. H., Tris(dipivalomethanato)europium. A paramagnetic shift reagent for use in nuclear magnetic resonance spectroscopy, *J. Am. Chem. Soc.*, 93, 641, 1971.
88. Ohashi, M., Morishima, I., and Yonezawa, T., Application of proton NMR shift reagents to the stereochemical analysis of nicotine, *Bull. Chem. Soc. Jpn.*, 44, 576, 1971.
89. Johnson, B. F. G., Lewis, J., McArdle, P., and Norton, J. R., Applications of lanthanide shift reagents to C-13 and H-1 nuclear magnetic resonance spectroscopy, *J. Chem. Soc. Dalton Trans.*, 12, 1253, 1974.
90. Francis, H. E. and Wagner, W. F., Induced chemical shifts in organic molecules. Intermediate shift reagents, *Org. Magn. Reson.*, 4, 189, 1972.
91. Okigawa, M., Khan, N. U., and Kawano, N., Application of a lanthanide shift reagent, Eu(fod)$_3$ to the elucidation of the structures of flavones and related compounds, *J. Chem. Soc. Perkin Trans. 1*, p. 1563, 1975.
92. Potapov, V. M., Rukhadze, E. G., Il'ina, I. G., and Bakhmut-skaya, V. G., Europium and praseodymium thenoyltrifluoroacetonates as chemical shift reagents in NMR, *Z. Obsch. Khim.*, 44, 462, 1974.
93. Sinha, S. P. and Kong, S. C., Lanthanide shift reagents II: Pr(tta)$_3$ a "two-way" shift reagent for the ring protons of benzylic systems, *Spectrosc. Lett.*, 6, 423, 1973.
94. Beyer, K., Tris(4,4,4-trifluoro-1-(2-thienyl)-1,3-butanediono)-europium(III) als NMR verschiebungsreagens, *Org. Magn. Reson.*, 5, 471, 1973.
95. Davidenko, N. K., Bidzilya, V. A., Goryushko, A. G., Shokol, V. A., and Yatsimirskii, K. B., Investigation of the coordination properties of esters of phosphonecarboxylic acids by the pmr method, using lanthanide shift reagents, *Teor. Eksp. Khim.*, 10, 500, 1974.
96. Iida, T., Kikuchi, M., Tamura, T., and Matsummoto, T., NMR studies on natural products. V. On the evaluation of 13 lanthanide shift reagents, *Yakagaku*, 27, 390, 1978.
97. Joshi, K. C., Pathak, V. N., and Grover, V., Studies in fluorinated 1,3-diketones and related compounds. XIV. Search for new NMR shift reagents, *J. Indian Chem. Soc.*, 59, 1072, 1982.
98. Joshi, K. C., Pathak, V. N., and Grover, V., Studies in fluorinated 1,3-diketones and related compounds. XVI. Studies in lanthanide 1,3-diketonates as NMR shift reagents, *J. Indian Chem. Soc.*, 60, 802, 1983.
99. Gritsenko, T. V., Buikliskii, V. D., Panyushin, V. T., and Afanas'ev, Y. A., Study of the complexing of lanthanide shift reagents based on 1-trifluoromethoxy-1,1,2,2-tetrafluoro-5-phenyl-3,5-pentanedione with pyridine and picolines, *Koord. Khim.*, 9, 196, 1983.
100. Goryushko, A. G., Gruz, B. E., Davidenko, N. K., Kudryavtseva, L. S., Lozinskii, M. L., Fialkov, Y. A., Yagupol'skii, L. M., and Yatsimirskii, K. B., Coordination Compounds of Lanthanides with Fluorinated Beta-Diketones as Shift Reagents for NMR Spectroscopy, USSR Patent 555, 103, 1977.
101. Goryushko, A. G., Davidenko, N. K., Kudryavtseva, L. S., Lozinskii, M. O., Lugina, L. N., and Fialkov, Y. A., New coordination compounds of europium(III) with fluorinated beta-diketones and their use as lanthanide shift reagents, *Zh. Neorg. Khim.*, 25, 2400, 1980.
102. Potapov, V. M., Bakhmut-skaya, V. G., Il'ina, I. G., Rukhadze, E. G., and Vinnik, G. I., Lanthanide NMR chemical-shift reagents based on 2-(hydroxymethylene)cyclohexanones, *Zh. Obshch. Khim.*, 46, 2117, 1976.
103. Bakhmutskaya, V. G., Potapov, V. M., Il'ina, I. G., Vinnik, G. I., and Rukhadze, E. G., Hydroxymethylenecyclohexanone-based lanthanide shift reagents for NMR, *Tezisy Dokl. Vses. Chugaevskoe Soveshch. Khim. Kompleksn. Soedin.*, 12th, 3360, 1975; *Chem. Abstr.*, 86, 10409e.
104. Li, J. and Yang, R., Study of the synthesis and properties of solid complex compounds of rare earth elements with 1-phenyl-3-methyl-4-heptafluorobutyryl-5-pyrazolone, *Gaodeng Xuexiao Huaxue Xuebao*, 4, 145, 1983; *Chem. Abstr.*, 99, 186293v.
105. Briggs, J., Frost, G. H., Hart, F. A., Moss, G. P., and Staniforth, M. L., Lanthanide-induced shifts in nuclear magnetic resonance spectroscopy. Shifts to high field, *J. Chem. Soc. Chem. Commun.*, p. 749, 1970.
106. Sayeed, M. and Ahmad, N., Mixed ligand complexes of trivalent lanthanide ions with beta-diketones and heterocyclic amines and their use as possible shift reagents, *J. Inorg. Nucl. Chem.*, 43, 3197, 1981.
107. Iftikhar, K., Sayeed, M., and Ahmad, N., Lanthanoid shift reagents, Synthesis and spectral studies, *Bull. Chem. Soc. Jpn.*, 55, 2258, 1982.

108. Horrocks, W. D., Jr. and Wong, C. P., Lanthanide porphyrin complexes. Evaluation of nuclear magnetic resonance dipolar probe and shift reagent capabilities, *J. Am. Chem. Soc.*, 98, 7157, 1976.
109. Green, R. D. and Sinha, S. P., Lanthanide shift reagents: acetylacetone complex of Yb(III) ion, *Spectrosc. Lett.*, 4, 411, 1971.
110. Horrocks, W. D., Jr. and Sipe, J. P., III, Lanthanide shift reagents. A survey, *J. Am. Chem. Soc.*, 93, 6800, 1971.
111. Horrocks, W. D., Jr., Sipe, J. P., III, and Sudnick, O., Magnetic anisotropy and dipolar shifts in shift reagent systems, in *Nuclear Magnetic Resonance Shift Reagents*, Sievers, R. E., Ed., Academic Press, New York, 1973, 53.
112. Rondeau, R. E. and Sievers, R. E., New nuclear magnetic resonance shift reagents, *Anal. Chem.*, 45, 2145, 1973.
113. Mohyla, I., Ksandr, Z., Hajek, M., and Vodicka, L., Application of shift reagents in NMR analysis of 1-adamantanol, *Collect. Czech. Chem. Commun.*, 39, 2935, 1974.
114. Demarco, P. V., Elzey, T. K., Lewis, R. B., and Wenkert, E., Paramagnetic induced shifts in the proton magnetic resonance spectra of alcohols using tris(dipivalomethanato)europium(III), *J. Am. Chem. Soc.*, 92, 5734, 1970.
115. Raber, D. J., Johnston, M. D., Jr., Janks, C. M., Perry, J. W., and Jackson, G. F., III, Structure elucidation with lanthanide induced shifts. 6. Solvent effects on bound shifts and association constants, *Org. Magn. Reson.*, 14, 32, 1980.
116. Susaki, Y., Kawaki, H., and Okazaki, Y., Studies on the proton magnetic resonance spectra in aliphatic systems. VI. Tris(dipivalomethanato)europium induced paramagnetic shifts of aliphatic amines and alcohols: steric, electronic and solvent effects, *Chem. Pharm. Bull.*, 23, 1899, 1975.
117. Walters, D. B., A new nmr method for the analysis of fatty acid methyl ester mixtures with Eu(dpm)$_3$ in carbon disulfide, *Anal. Chim. Acta*, 66, 134, 1973.
118. Walters, D. B., Carbon disulfide as a solvent for the application of Eu(dpm)$_3$ to nmr spectroscopy, *Anal. Chim. Acta*, 60, 421, 1972.
119. Yanagawa, H., Kato, T., and Kitahara, Y., Effect of tris(dipivalomethanato)europium(III) on the NMR spectra of sulfhydryl compounds, *Tetrahedron Lett.*, p. 2137, 1973.
120. Marks, T. J., Kristoff, J. S., Alich, A., and Shriver, D. F., Rare earth shift reagents as chemical structural probes for organometallic compounds, *J. Organomet. Chem.*, 33, C35, 1971.
121. Bulkowski, J. E. and Van Dyke, C. H., Application of a lanthanide shift reagent to ethylfluorogermane, *Inorg. Nucl. Chem. Lett.*, 11, 749, 1975.
122. Angerman, N. S., Danyluk, S. S., and Victor, T. A., A direct determination of the spatial geometry of molecules in solution. I. Conformation of chloroquine, an antimalarial, *J. Am. Chem. Soc.*, 94, 7137, 1972.
123. Fujiwara, H., Bose, A. K., Manhas, M. S., and van der Veen, J. M., C-13 nuclear magnetic resonance studies on the conformation of substituted hydantoins, *J. Chem. Soc. Perkin Trans. 2*, 1573, 1980.
124. Schwendiman, D. and Zink, J. I., Solvent effects on the isotropic shifts and magnetic susceptibility of the shift reagent tris(dipivaloylmethanato)europium(III), *Inorg. Chem.*, 11, 3051, 1972.
125. Sasaki, Y., Fujiwara, H., Kawaki, H., and Okazaki, Y., Studies on the proton magnetic resonance spectra of aliphatic systems. VIII. Complex shift and equilibrium constant of aliphatic alcohol-tris(dipivalomethanato)europium complex in solution, *Chem. Pharm. Bull.*, 26, 1066, 1978.
126. Bouquant, J. and Chuche, J., NMR lanthanide shift reagents. II. Solvent effects, *Tetrahedron Lett.*, p. 493, 1973.
127. Armitage, I. and Hall, L. D., Novel chemical shift changes of carbohydrates induced by lanthanide shift reagents: some experimental optimizations, *Can. J. Chem.*, 49, 2770, 1971.
128. Arduini, A., Armitage, I. M., Hall, L. D., and Marshall, A. G., Evaluation of the binding constants, bound chemical shifts, and equilibrium stoichiometry for the association of lanthanide shift reagents with carbohydrate derivatives, *Carbohydr. Res.*, 31, 255, 1973.
129. Lewis, R. B. and Wenkert, E., Structure elucidation of natural products, in *Nuclear Magnetic Resonance Shift Reagents*, Sievers, R. E., Ed., Academic Press, New York, 1973, 99.
130. Karhan, J., Hajek, M., and Ksandr, Z., Effect of water on NMR spectra measured with shift reagents, *Vysoke. Skoly. Chem. Tech. Praze Anal. Chem.*, 11, 213, 1976.
131. Haines, A. H., Harris, R. K., and Rao, R. C., Silicon-29 and carbon-13 NMR studies of organosilicon chemistry. VIII. Assignment of the Si-29 and related H-1 resonances of trimethylsilylated sugars by deuteration and shift reagent studies, *Org. Magn. Reson.*, 9, 432, 1977.
132. Evans, D. F. and Wyatt, M., Nuclear magnetic resonance studies of lanthanide complexes. I. Solvation numbers and kinetics of substrate exchange in lanthanide shift reagent systems, *J. Chem. Soc. Dalton Trans.*, 6, 765, 1974.

133. Beattie, J. K., Lindoy, L. F., and Moody, W. E., Dynamic hydrogen-1 nuclear magnetic resonance line broadening in adducts formed between transition metal complexes and lanthanide shift reagents, *Inorg. Chem.*, 15, 3170, 1976.
134. Ammon, H. L., Mazzocchi, P. H., and Colicelli, E. J., An investigation of the utility of using NMR lanthanide shift reagents for C-13 assignments in selected aromatic ethers and ketones. A detailed description of the lanthanide induced shift ratio method for lanthanide shift reagent-substrate geometry analysis, *Org. Magn. Reson.*, 11, 1, 1978.
135. de Boer, J. W. M., Hilbers, C. W., and de Boer, E., Lanthanide shift reagents. I. Complexation of mono- and bifunctional ethers, *J. Magn. Reson.*, 25, 437, 1977.
136. Shapiro, B. L. and Johnston, M. D., Jr., Lanthanide-induced shifts in proton nuclear magnetic resonance spectra. III. Lanthanide shift reagent-substrate equilibria, *J. Am. Chem. Soc.*, 94, 8185, 1972.
137. Eisentraut, K. J. and Sievers, R. E., Volatile rare earth chelates, *J. Am. Chem. Soc.*, 87, 5254, 1965.
138. Springer, C. S., Jr., Meek, D. W., and Sievers, R. E., Rare earth chelates of 1,1,1,2,2,3,3-heptafluoro-7,7-dimethyl-4,6-octanedione, *Inorg. Chem.*, 6, 1105, 1967.
139. Springer, C. S., Jr., Bruder, A. H., Tanny, S. R., Pickering, M., and Rockefeller, H. A., Ln(fod)$_3$ complexes as NMR shift reagents: states of hydration; self association; solution adduct formation and changes of the NMR time scale, in *Nuclear Magnetic Resonance Shift Reagents*, Sievers, R. E., Ed., Academic Press, New York, 1973, 283.
140. Graham, L. L., Vanderkooi, G., and Getz, J. A., Conformational analysis of N,N-diisopropylamides by combined use of NMR lanthanide-induced shifts and conformational energy calculations, *Org. Magn. Reson.*, 9, 80, 1977.
141. Peters, J. A., Schuyl, P. J. W., Bovee, W. M. M. J., Alberts, J. H., and van Bekkum, H., Potential causes of erroneous results of analysis of lanthanide-induced shifts: contamination of Ln(fod)$_3$ NMR shift reagents with Ln(fod)$_3$ M(fod) and self-association of Ln(fod)$_3$, *J. Org. Chem.*, 46, 2784, 1971.
142. Lippard, S. J., A volatile inorganic salt, Cs[Y(CF$_3$COCHCOCF$_3$)$_4$], *J. Am. Chem. Soc.*, 88, 4300, 1966.
143. Buckley, D. G., Green, G. H., Ritchie, E., and Taylor, W. C., Application of the tris(dipivalomethanato)europium nuclear magnetic resonance shift reagent to the study of triterpenoids, *Chem. Ind.*, p. 298, 1971.
144. Porter, R., Marks, T. J., and Shriver, D. F., Delineation of shift reagent-substrate equilibria, *J. Am. Chem. Soc.*, 95, 3548, 1973.
145. Schwarberg, J. E., Gere, D. R., Sievers, R. E., and Eisentraut, K. J., Clarification of discrepancies in the characterization of lanthanum series complexes of 2,2,6,6-tetramethyl-3,5-heptanedione, *Inorg. Chem.*, 6, 1933, 1967.
146. Ghotra, J. S., Hart, F. A., Moss, G. P., and Staniforth, M. L., Nature of the species present in solutions of lanthanide nuclear magnetic resonance shift reagents, *J. Chem. Soc. Chem. Commun.*, p. 113, 1973.
147. Johnston, M. D., Jr., Shapiro, B. L., Shaprio, M. J., Proulx, T. W., Godwin, A. D., and Pearce, H. L., Lanthanide induced shifts in protein nuclear magnetic resonance spectra. XI. Equilibrium constants and bound shifts for cyclohexanones and cyclohexanols, *J. Am. Chem. Soc.*, 97, 542, 1975.
148. Briggs, J. M., Hart, F. A., Moss, G. P., Randall, E. W., Sales, K. D., and Staniforth, M. L., Studies of lanthanide shift reagents at Queen Mary College, in *Nuclear Magnetic Resonance Shift Reagents*, Sievers, R. E., Ed., Academic Press, New York, 1973, 197.
149. Archer, M. K., Fell, D. S., and Jotham, R. W., Self-induced pseudocontact shifts in tris(2,2,6,6-tetramethyl-3,5-heptanedionato) praseodymium dimers, Pr$_2$(thd)$_6$, and its europium homologue, *Inorg. Nucl. Chem. Lett.*, 7, 1135, 1971.
150. Shapiro, B. L., Johnston, M. D., Jr., and Shapiro, M. J., Lanthanide induced shifts in proton NMR spectra of cyclohexanones, *Org. Magn. Reson.*, 5, 21, 1973.
151. Shapiro, B. L., Shapiro, M. J., Godwin, A. D., and Johnston, M. D., Jr., Lanthanide-induced effects in proton NMR spectra. VIII. "Scavenging" effects — a problem and a solution, *J. Magn. Reson.*, 8, 402, 1972.
152. Wenzel, T. J., unpublished data, 1984.
153. Juneau, G. P., Removal of europium shift reagents from nuclear magnetic resonance samples, *Anal. Chem.*, 49, 2375, 1977.
154. Leibfritz, D. and Roberts, J. D., Nuclear magnetic resonance spectroscopy. Carbon-13 spectra of cholic acids and hydrocarbons included in sodium deoxycholate solutions, *J. Am. Chem. Soc.*, 95, 4996, 1973.
155. Stolzenberg, G. E., Zaylskie, R. G., and Olson, P. A., Nuclear magnetic resonance identification of o,p-isomers in an ethoxylated alkylphenol nonionic surfactant as tris(2,2,6,6-tetramethylheptane-3,5-dione)europium(III) complexes, *Anal. Chem.*, 43, 908, 1971.

156. Desreux, J. F., Fox, L. E., and Reilley, C. N., Aggregation studies of some nuclear magnetic resonance shift reagents by vapor phase osmometry, *Anal. Chem.*, 44, 2217, 1972.
157. Bulsing, J. M., Sanders, J. K. M., and Hall, L. D., Spin-echo methods for resolution control of lanthanide-shifted NMR spectra, *J. Chem. Soc. Chem. Commun.*, p. 1201, 1981.
158. Boudreux, G. J., Bailey, A. V., and Tripp, V. W., Induced chemical shifts in the NMR spectrum of methyl petroselinate, *J. Am. Oil Chem. Soc.*, 49, 200, 1972.
159. Tomic, L., Majerski, Z., Tomic, M., and Sunko, D. E., Tris(dipivalomethanato)holium induced NMR shifts, *Croat. Chem. Acta*, 43, 267, 1971.
160. Smentowski, F. J. and Stipanovic, R. D., Lanthanide shift reagents as an aid in the NMR analysis of the normal alcohols C_6 to C_{11}, *J. Am. Oil Chem. Soc.*, 49, 48, 1972.
161. Rabenstein, D. L., Applications of paramagnetic shift reagents in proton magnetic resonance spectrometry, *Anal. Chem.*, 43, 1599, 1971.
162. Ernst, L., Paramagnatische lanthaniden-komplexe als, "Verschiebungsreagantien" in der NMR-spektroskopie, *Nachr. Chem. Tech.*, 18, 439, 1970.
163. Kawaki, H., Okazaki, Y., Fujiwara, H., and Sasaki, Y., Nuclear magnetic resonance studies of acid-base association in solution. I. Thermodynamic parameters of the association between Lewis bases and tris(dipivalomethanato)europium, *Chem. Pharm. Bull.*, 28, 871, 1980.
164. Sasaki, Y., Fujiwara, H., Kawaki, H., and Okazaki, Y., Studies on the proton magnetic resonance spectra of aliphatic systems. VIII. Complex shift and equilibrium constant of aliphatic alcohol-tris(dipivalomethanato)europium complex in solution, *Chem. Pharm. Bull.*, 26, 1066, 1978.
165. Chadwick, D. J. and Williams, D. H., Lanthanide-induced shifts in the carbon-13 nuclear magnetic resonance spectra of some ketones, alcohols, and amines. An analysis of contact, pseudo-contact, and complex-formation contributions to the observed shifts, *J. Chem. Soc. Perkin Trans. 2*, p. 1202, 1974.
166. Ikeda, N. and Fukuzumi, K., Study on quantitative analyses of hydroperoxides and alcohols by NMR shift reagent, *J. Am. Oil Chem. Soc.*, 51, 340, 1974.
167. Williamson, K. L., Clutter, D. R., Emch, R., Alexander, M., Burroughs, A. E., Chua, C., and Bogel, M. E., Conformation analysis by nuclear magnetic resonance. Shift reagent studies on acyclic alcohols. H-1 and C-13 spectra of the six-carbon aliphatic alcohols, *J. Am. Chem. Soc.*, 96, 1471, 1974.
168. Vitullo, V. P., Cashen, M. L., Marx, J. N., Caudle, L. J., and Fritz, J. R., Cyclohexadienyl cations. VII. Kinetics and mechanism of the acid-catalyzed dienol-benzene rearrangement, *J. Am. Chem. Soc.*, 100, 1205, 1978.
169. Sasaki, Y., Kawaki, H., and Okazaki, Y., Studies on the proton magnetic resonance spectra in aliphatic systems. V. Tris(dipivalomethanato)europium induced shift parameters of aliphatic amines and alcohols, *Chem. Pharm. Bull.*, 21, 2488, 1973.
170. Mariano, P. S. and McElroy, R., Lanthanide induced shift difference of diastereotopic protons, *Tetrahedron Lett.*, p. 5305, 1972.
171. Kawaki, H., Fujiwara, H., and Sasaki, Y., Studies on the proton magnetic resonance spectra of aliphatic systems. IX. Complex shift equilibrium constant of association between excess amount of aliphatic alcohol and tris(dipivalomethanato)europium, *Chem. Pharm. Bull.*, 26, 2694, 1978.
172. Cawley, J. J. and Petrocine, D. V., Comparison and evaluation of some available methods for the analysis of lanthanide induced shift data, *Org. Magn. Reson.*, 6, 544, 1974.
173. Bell, H. M., Lanthanide-induced shifts of diastereotopic groups, *Org. Magn. Reson.*, 7, 240, 1975.
174. Armitage, I., Campbell, J. R., and Hall, L. D., Applications of lanthanide shift reagents to the identification of C-13 resonances, *Can. J. Chem.*, 50, 2139, 1972.
175. Alvarez Ibarra, C., Arias Perez, M. S., Garcia Romo, M. T., and Quiroga, M. L., Assignment of the relative configurations to diastereomeric carbinols with lanthanide shift reagent [Eu(fod)$_3$]. Application to 1-mesityl-2-methyl-3-phenyl-1-propanol, *Tetrahedron*, 37, 1249, 1981.
176. Spassov, S. L. and Stefanova, R., Conformational studies of some 2-phenylpropyl derivatives by NMR spectroscopy and use of lanthanide shift reagents, *J. Mol. Struct.*, 42, 109, 1077.
177. Bangov, I. P., Spassov, S. L., and Stefanova, R., Application of computer-simulated lanthanide-induced shifts for conformational analysis of 2-phenylpropyl derivatives, *C. R. Acad. Bulg. Sci.*, 31, 305, 1978.
178. Rajeswari, K., Dubey, R., and Ranganayakulu, K., Use of a lanthanide shift reagent in the analysis of isomer ratio, *Indian J. Chem.*, 17B, 504, 1979.
179. Kunieda, N., Endo, H., Hirota, M., Kodama, Y., and Nishio, M., The conformational analysis of 1-p-tolyl-2-phenyl-1-propanols, 1-p-totyl-2-phenylethanol, and 1-p-tolyl-2-phenyl-1-propanone by means of NMR spectroscopy, *Bull. Chem. Soc. Jpn.*, 56, 3110, 1983.
180. Alvarez Ibarra, C., Quiroga Feijoo, M. L., Arias Perez, M. S., and de la Orden Parra, J., Assignment of relative configurations to acyclic diastereomeric carbinols with the lanthanide shift reagent Eu(fod)$_3$ by H-1 and C-13 NMR, *Org. Magn. Reson.*, 21, 520, 1983.

181. Lasperas, M., Perez-Ossorio, R., Perez-Rubalcaba, A., and Quiraga Feijoo, M. L., Stereochemistry of addition to a carbonyl group. XVII. Effect of solvent nature on the asymmetric induction of reactions of methylmagnesium bromide and (±)-1-aryl-2,3-diphenyl-2-methylpropanones, *An. Quim. Ser. C,* 77, 112, 1981.
182. Alvarez Ibarra, C., Arias Perez, M. S., Fernandez Dominguez, M. J., and Moya Molina, E., Conformational study of diastereoisomers. VIII. Assignment of relative configurations to diastereoisomeric acyclic carbinols using Eu(fod)$_3$. Application to 4,4-dimethyl-4-phenyl-2-pentanol, *An. Quim. Ser. C,* 78, 244, 1982.
183. Alvarez Ibarra, C., Garcia Gomez, R., Garcia Romo, M. T., and Arias Perez, M. S., Conformational study of diastereomers. IX. Assignment of the relative configurations of acyclic diastereomeric carbinols with the lanthanide shift reagent tris(6,6,7,7,8,8,8-heptafluoro-2,2-dimethyl-3,5-octanediónato)europium [Eu(fod)$_3$]. Application to (2R4R,2S4S)- and (2R4S,2S4R)-2,4-diphenyl-5,5-dimethyl-2-hexanol. *An. Quim. Ser. C,* 79, 122, 1983.
184. Brooks, J. J. and Sievers, R. E., Gas chromatographic studies of interactions between selected nucleophiles and the NMR shift reagent, tris(1,1,1,2,2,3,3-heptafluoro-7,7-dimethyl-3,5-octanediónato)-europium(III), *J. Chromatogr. Sci.,* 11, 303, 1973.
185. Raber, D. J., Johnston, M. D., Jr., Campbell, C. M., Guida, A., Jackson, G. F., III, Janks, C. M., Perry, J. W., Propeck, G. J., Raber, N. K., Schwalke, M. A., and Sutton, P. M., Structure elucidation with lanthanide-induced shifts. V. Evaluation of the binding ability of various functional groups, *Monatsch. Chem.,* 111, 43, 1980.
186. Wolkowski, Z. W., Beaute, C., and Jantzen, R., Shifts induced by tris(dipivalomethanato)-lanthanides in F-19 nuclear magnetic resonance spectroscopy, *J. Chem. Soc. Chem. Commun.,* p. 619, 1972.
187. Skacel, F., Hajek, M., Liska, F., Trska, P., and Ksandr, Z., Application of shift reagents in H-1 and F-19 NMR spectroscopy. Interactions with chlorofluoroalkanols C$_3$, C$_4$, *Collect. Czech. Chem. Commun.,* 44, 1440, 1979.
188. Khetrapal, C. L. and Kunwar, A. C., PMR spectra of methyl alcohol dissolved in a nematic solvent with and without a lanthanide shift reagent, *J. Magn. Reson.,* 15, 389, 1974.
189. Kintzinger, J. P. and Nguyen, T. T. T., Selection of shift reagents for 0-17 NMR. Application to line assignments, *Org. Magn. Reson.,* 13, 464, 1980.
190. MacMillan, J. H. and Washburne, S. S., Lanthanide chemical shift reagents as tools for determining isomer distributions in 2,4-hexadienoates and related compounds, *Org. Magn. Reson.,* 6, 250, 1974.
191. Richey, H. G. and Von Rein, F. W., Determinations of the configurations (cis or trans) of alkenols by the effects of paramagnetic shift reagents on their spectra, *Tetrahedron Lett.,* p. 3781, 1971.
192. Martin, G. J., Naulet, N., Lefevre, M. L., and Martin, M. L., Determination rapide de la configuration cis trans des alcenes > CH−CH=CH− utilisation des complexes de terres rares, *Org. Magn. Reson.,* 4, 121, 1972.
193. de Haan, J. W. and van der Ven, L. J. M., Z-E conformational isomerism of nerol, geraniol and their acetates, *Tetrahedron Lett.,* p. 2703, 1971.
194. Heldman, D. A. and Gilde, H. G., Use of a lanthanide NMR shift reagent in the analysis of the trans-retinol spectrum, *J. Chem. Educ.,* 57, 390, 1980.
195. Najib, L. M. and Khalil, S. M., Proton magnetic resonance studies of citronellol and nerol with lanthanide shift reagents, *Iraqi J. Sci.,* 21, 563, 1980.
196. Yoshimoto, M., Hiraoka, T., Kuwano, H., and Kishida, Y., Use of a shift reagent in first-order analysis of cyclopropane derivatives in NMR spectroscopy, *Chem. Pharm. Bull.,* 19, 849, 1971.
197. Crombie, L., Findley, D. A. R., and Whiting, D. A., Lanthanide induced chemical shifts in natural cyclopropanes: stereochemistry of chrysanthemyl and presqualene alcohols and esters, *Tetrahedron Lett.,* p. 4027, 1972.
198. Altman, L. J., Kowerski, R. C., and Laungani, D. R., Studies in terpene biosynthesis. Synthesis and resolution of presqualene and prephytoene alcohols, *J. Am. Chem. Soc.,* 100, 6174, 1978.
199. Altman, L. J., Kowerski, R. C., and Rilling, H. C., Synthesis and conversion of presqualene alcohol to squalene, *J. Am. Chem. Soc.,* 93, 1782, 1971.
200. Tomic, L., Majerski, Z., Tomic, M., and Sunko, D. E., Temperature and concentration dependence of the paramagnetic induced shifts in proton magnetic spectroscopy, *J. Chem. Soc. Chem. Commun.,* p. 719, 1971.
201. Richey, H. G., Jr. and Bension, R. M., Stereochemistry of addition of allylic Grignard reagents to 3-(hydroxymethyl)cyclopropenes, *J. Org. Chem.,* 45, 5036, 1980.
202. Vincens, M., Dumont, C., and Vidal, M., Identification des configurations Z it E en serie cyclopropylidenique. Stereoselectivite de la magration exocyclique de la double liaison cyclopropenique en milieu basique, *Tetrahedron,* 15, 2683, 1981.

203. Patrick, T. B. and Patrick, P. H., Interactions of benzocycloalkenols with tris(dipivalomethanato)europium, *J. Am. Chem. Soc.,* 95, 5192, 1973.
204. Patrick, T. B. and Patrick, P. H., Reliability of coupling constants obtained from tris(dipivalomethanato)europium shifted proton magnetic resonance spectra, *J. Am. Chem. Soc.,* 94, 6230, 1972.
205. Lewis, F. D. and Hirsch, R. H., Photochemical addition of stilbene to dienes, *Tetrahedron Lett.,* p. 4947, 1973.
206. Brooke, G. M. and Matthews, R. S., The use of a lanthanide shift reagent in F-19 NMR spectroscopy. The orientation of fluorine atoms in a trifluoroindanol, *Tetrahedron Lett.,* p. 3469, 1973.
207. Takakis, I. M. and Rhodes, Y. E., Resolution of cyclopentanol methylene proton signals with Eu(fod)$_3$, *J. Magn. Reson.,* 35, 13, 1979.
208. Grosse, M., Roth, K., and Rewicki, D., Zur anwendung von NMR-verschiebungreagenzien: Die domplexbildungskonstante als strukturparamater, *Org. Magn. Reson.,* 10, 115, 1977.
209. Cockerill, A. F. and Rackham, D. M., Quantitation of the chemical shifts induced by tris(dipivalomethanato)europium III in the PMR spectra of hydroxyadamantanes and cyclopentanol, *Tetrahedron Lett.,* p. 5149, 1970.
210. Christl, M., Reich, H. J., and Roberts, J. D., Nuclear magnetic resonance spectroscopy. Carbon-13 chemical shifts of methylcyclopentanes, cyclopentanols, and cyclopentyl acetates, *J. Am. Chem. Soc.,* 93, 3463, 1971.
211. Shapiro, B. L., Johnston, M. D., Jr., and Proulx, T. W., 3-(α-naphthyl)-5,5-dimethylcyclohexanone and derived alcohols. Synthesis and stereochemical studies by means of lanthanide-induced proton nuclear magnetic resonance shifts, *J. Am. Chem. Soc.,* 95, 520, 1973.
212. Yamada, K., Ishihara, S., and Iida, H., Effects of paramagnetic shift reagents on the rotation of an isopropyl group, *Chem. Lett.,* p. 549, 1973.
213. Takagi, Y., Teratani, S., and Uzawa, J., Application of nuclear magnetic resonance shift reagents to kinetic studies on catalytic deuteration of 4-t-butylcyclohexanone, *J. Chem. Soc. Chem. Commun.,* p. 280, 1972.
214. Konavalov, E. V., Lavrenyuk, T. Y., and Saenko, E. P., F-19 investigation of the diastereomeric behavior of the trifluoromethyl groups in molecules of the cyclohexane series using a shift reagent, *Teor. Eksp. Khim.,* 10, 687, 1974.
215. Iida, T., Ibaraki, S., and Kikuchi, M., Application of Eu(dpm)$_3$ to the assignments of the methyl resonances in some monocylic monoterpene isomers, *Agric. Biol. Chem.,* p. 41, 2471, 1977.
216. Higgs, M. D., Vanderah, D. J., and Faulkner, D. J., Polyhalogenated monoterpenes from *Plocamium cartilagineum* from the British coast, *Tetrahedron,* 33, 2775, 1977.
217. Granger, P., Claudon, M. M., and Guinet, J. F., Utilisation du tris(dipivalomethanato)europium III en etudes conformationnelles cas des α-monobenzyl-cyclohexanols cis et trans, *Tetrahedron Lett.,* p. 4167, 1971.
218. Demarco, P. V., Cerimele, B. J., Crane, R. W., and Thakkar, A. L., Time-averaged solution geometries for ligand-europium complexes: the significance of functional group rotation, basicity, and steric environment, *Tetrahedron Lett.,* p. 3539, 1972.
219. Belanger, P., Freppel, C., Tizane, D., and Richer, J. C., Deplacement induit par un lanthanide en resonance magnetique nucleaire. Une application a un probleme stereochimique, *Can. J. Chem.,* 49, 1985, 1971.
220. Bouquant, J., Wuilmet, M., Maujean, A., and Chuche, J., Nuclear magnetic resonance lanthanide-shift reagents. Conformation analysis of 1-methylcyclohexanol, *J. Chem. Soc. Chem. Commun.,* p. 778, 1974.
221. Crabbe, P., Dollat, J. M., Gallina, J., Luche, J. L., Velarde, E., Maddox, M. L., and Tokes, L., Steric course of cross coupling of organocopper reagents with allylic actates, *J. Chem. Soc. Perkin Trans. 1,* p. 730, 1978.
222. Fischer, A., Henderson, G. N., Smyth, T. A., Einstein, F. W. B., and Cobbledick, R. E., Ipso nitration. XXII. The stereochemistry of 1,4-dimethyl-4-nitrocyclohexa-2,5-dienols and their acetates and methyl esters, *Can. J. Chem.,* 59, 584, 1981.
223. Dixon, J. R., McIntyre, P. S., Morris, I. G., and Williams, D. L., Effect of lanthanide shift reagents on conformational equilibria, *Org. Magn. Reson.,* 15, 273, 1981.
224. Bohlmann, F., Zeisberg, R., and Klein, E., C-13 NMR-spektren von monoterpenen, *Org. Magn. Reson.,* 7, 426, 1975.
225. Glukhovtsev, M. N., Methods for calculating organic molecule conformations using lanthanide paramagnetic shift reagents, *Izv. Sev. Kavk. Nauchn. Tsentra Vyssh. Shk. Estestv. Nauki,* 6, 59, 1978; *Chem. Abstr.,* 90, 203396g.
226. Christl, M. and Roberts, J. D., Carbon-13 nuclear magnetic resonance spectroscopy. Conformational analysis of methyl-substituted cycloheptanes, cycloheptanols, and cycloheptanones, *J. Org. Chem.,* 37, 3443, 1972.

227. Grandjean, J., Un example de l'utilisation des complexes paramagnetiques dans les etudes conformationnelles, *Bull. Soc. Chim. Belges,* 81, 513, 1972.
228. Drake, J. A. G. and Jones, D. W., High-resolution NMR spectroscopy of tricyclic non-alterant systems containing seven-membered rings. II. H-1 and C-13 solvent and lanthanide-shift-reagent studies of 5-keto- and 5-hydroxy-substituted 5H-dibenzo[a,d]cycloheptene, *Spectrochim. Acta,* 37A, 77, 1981.
229. Foldesi, P. and Hofer, O., Model calculations of LIS. III. A simple force field model for Ln(III)-shift-reagent- substrate complexes. I. Carbinols, *Monatsch. Chem.,* 111, 351, 1980.
230. Paasivirta, J., Chemical shift reagents in the study of polycyclic alcohols. I. PMR spectra of 2-norbornanols and dehydronorborneols, *Suom. Kemistil. B,* p. 131, 1971.
231. Paasivirta, J., Chemical shift reagents in the study of polycyclic alcohols. II. PMR spectra of 3-nortricyclanol and the stereoisomeric 1-methyl-3-nortricyclanols, *Suom. Kemistil. B,* p. 135, 1971.
232. Paasivirta, J. and Malkonen, P. J., Chemical shift reagents in the study of polycyclic alcohols. III. PMR spectra of methyl-substituted 5-norbornen-2-ols, *Suom. Kemistil. B,* p. 230, 1971.
233. Liu, K-T., Nuclear magnetic resonance studies. IV. Proton chemical shifts and coupling constants of 2-norbornanols and methyl-substituted 2-norbornanols, *J. Chinese Chem. Soc.,* 23, 1, 1976.
234. Liu, K-T., Proton magnetic resonance study of 2-endo-hydroxymethyl-5-norbornene. The effect of tris(dipivalomethanato)europium on coupling constants, *Tetrahedron Lett.,* p. 5039, 1972.
235. Liu, K-T., Nuclear magnetic resonance studies. II. The absence of a significant influence of tris(dipivalomethanato)europium on coupling constants of norbornanols and 7,7-dimethylnorbornanols, *Tetrahedron Lett.,* p. 2747, 1973.
236. Laihia, K. and Kantolahti, E., Lanthanide shift reagent studies of three unsaturated norbornanols and their catalytically deuterated products, *Finn. Chem. Lett.,* p. 10, 1975.
238. Korvola, J., The effects of a chemical shift reagent on the PMR spectra of stereoisomeric 2-methylfenchols, *Suom. Kemistil. B,* 46, 265, 1973.
236. Hogeveen, H., Roobeek, C. F., and Volger, H. C., Inversion of configuration in a fused cyclopropane ring opening by hydrochloric acid, *Tetrahedron Lett.,* p. 221, 1972.
239. Gansow, O. A., Willcott, M. R., and Lenkinski, R. E., Carbon magnetic resonance. Signal assignment by alternately pulsed nuclear magnetic resonance and lanthanide-induced chemical shifts, *J. Am. Chem. Soc.,* 93, 4295, 1971.
240. Briggs, J., Hart, F. A., Moss, G. P., and Randall, E. W., A ready method of assignment for C-13 nuclear magnetic resonance spectra: the complete assignment of the C-13 spectrum of borneol, *J. Chem. Soc. Chem. Commun.,* p. 364, 1971.
241. Briggs, J., Hart, F. A., and Moss, G. P., The application of lanthanide-induced shifts to the complete analysis of the borneol nuclear magnetic resonance spectrum, *J. Chem. Soc. Chem. Commun.,* p. 1506, 1970.
242. Abraham, R. J., Coppell, S. M., and Ramage, R., The H-1 NMR spectra of some norbornene derivatives, a LIS study, *Org. Magn. Reson.,* 6, 658, 1974.
243. Schneider, H. J. and Weigand, E. F., Lanthanide induced H-1 and C-13 NMR shifts and their use for geometry analysis with alicyclic compounds, *Tetrahedron,* 31, 2125, 1975.
244. Wing, R. M., Uebel, J. J., and Anderson, K. K., Lanthanide induced nuclear magnetic resonance shifts. A structural and computational study, *J. Am. Chem. Soc.,* 95, 6046, 1973.
245. Tori, K., Yoshimura, Y., and Muneyuki, R., Application of paramagnetic induced shifts in PMR spectra to the determination of the positions of deuterium substitution in bornanes derived from α-pinene using tris(dipivalomethanato)europium(III), *Tetrahedron Lett.,* p. 333, 1971.
246. Vedejs, E. and Salomon, M. F., Borohydride reduction of sigma-bonded organopalladium complexes in the norbornenyl-nortricyclenyl system. Evidence against a radical mechanism, *J. Am. Chem. Soc.,* 92, 6965, 1970.
247. Duddeck, H. and Dietrich, W., Lanthanide induced shifts on the carbon-13 chemical shifts of 2-adamantanol and 2-adamantanethiol, *Tetrahedron Lett.,* p. 2925, 1975.
248. Doerffel, V. K., Ehrig, R., Hauthal, H. G., Kasper, H., and Zimmermann, G., Korrelationsbeziehungen bei paramagnetischen verschiebungen mittels tris-(dipivalomethanato)-europium(III), *J. Prakt. Chem.,* 314, 385, 1972.
249. Hajek, M., Trska, P., Vodicka, L., and Hlavaty, J., Use of lanthanide shift reagents for the determination of the alkyl conformation in 2-alkyl-2-adamantanols, *Org. Magn. Reson.,* 10, 52, 1977.
250. Wahl, G. H., Jr. and Peterson, M. R. Jr., On the mechanism of deshielding of the tris(dipivalomethanato)europium nuclear magnetic resonance shift reagent, *J. Chem. Soc. Chem. Commun.,* p. 1167, 1970.
251. Nordlander, J. E. and Haky, J. E., Complete substitution stereochemistry of solvolysis of 1-methyl-2-adamantyl tosylate and 4-methyl-exo- and 4-methyl-endo-4-protoadamantyl 3,5-dinitrobenzoate, *J. Am. Chem. Soc.,* 103, 1518, 1981.

252. Volka, K., Suchanek, M., Karhan, J., and Hajek, M., Infrared and proton magnetic resonance spectroscopic study of the shift reagent-1-adamantanol complexes, *J. Mol. Struct.,* 46, 329, 1978.
253. Norin, T., Stromberg, S., and Weber, M., Lanthanide-induced shifts in proton magnetic resonance; studies on the conformations of thujane derivatives, *Acta Chem. Scand.,* 27, 1579, 1973.
254. Durr, H. and Bujnoch, W., Stereochemische studien an bicyclen. Einfluss von verschiebungreagenzien auf H-1-NMR-spektren, *Justus Liebigs Ann. Chem.,* 10, 1691, 1973.
255. Crumrine, D. S. and Yen, H. H. B., Europium shift reagents. The assignment of aryl stereochemistry in 6,6-diarylbicyclo[3.1.0]hexan-3-exo-ols, *J. Org. Chem.,* 41, 1273, 1976.
256. Reuvers, A. J. M., Sinnema, A., and van Bekkum, H., 2-bromo-4,4,7-trimethyltricyclo[2.2.1.0^{2,6}]heptan-3-one. Configurational analysis of 4,7,7-trimethyltricyclo[2.2.1.0^{2,6}]heptan-3-ols using Eu(dpm)$_3$, *Tetrahedron,* 28, 4353, 1972.
257. Perraud, R. and Pierre, J. L., Etudes conformationnelles en serie bicyclo[4.2.1]heptane. Utilisation des deplacements induits par Eu(dpm)$_3$, *Bull. Soc. Chim. Fr.,* 11, 2615, 1974.
258. Nishino, C. and Takayanagi, H., Synthesis of verbenols and related alcohols and their PMR spectra with shift reagent, *Agric. Biol. Chem.,* 43, 1967, 1979.
259. Nishino, C. and Takayanagi, H., Conformational analysis of verbenols and related alcohols by PMR spectra with a chemical shift reagent, *Agric. Biol. Chem.,* 43, 2323, 1979.
260. Coxon, J. M., Hydes, G. J., and Steel, P. J., Carbon-13 nuclear magnetic resonance spectra of pinane monoterpenoids, *J. Chem. Soc. Perkin Trans. 2,* p. 1351, 1984.
261. Hinckley, C. C. and Brumley, W. C., Errors in analyses of lanthanide-induced shifts. Cis- and trans-pinocarveol, *J. Magn. Reson.,* 24, 239, 1976.
262. Velez, H. T., New evaluation of chemical shifts induced in trans-pinocarveol by rare earth chelates, *Rev. Cienc. Quim.,* 14, 51, 1983.
263. Ogawa, Y., Matsusaki, H., Hanaoka, K., Ohkata, K., and Hanafusa, T., Stereochemical studies on 3,4-benzobicyclo[4.1.0]hept-3-en-2-ol systems and solvolytic studies on tis p-nitrobenzoates, *J. Org. Chem.,* 43, 849, 1978.
264. Schnieder, H. J., Buchheit, U., and Agrawal, P. K., Topological, stereochemical and NMR line assignments in alcohols and ketones by normalized NMR shifts induced by ytterbium, *Tetrahedron,* 40, 1017, 1984.
265. Caubere, P. and Brunet, J. J., Condensation en nilieu aprotique des enolates de cetones sur le chloro-1-cyclohexene en presence de bases-III, *Tetrahedron,* 28, 4859, 1972.
266. Hart, H. and Kuzuya, M., Circumambulation, bridge shifts, and cyclopropylcarbinyl rearrangements in bicyclo[3.2.1]octadienyl carbocations, *J. Am. Chem. Soc.,* 98, 1551, 1976.
267. Roff, A. A. M., van Wageningen, A., Kruk, C., and Cerfontain, H., Photochemistry of non-conjugated dienones. II. Electrocyclic reactions of some 1,2-dimethylenecyclohexanes, *Tetrahedron Lett.,* p. 367, 1972.
268. Schneider, H. J., Lonsdorfer, M., and Weigand, E. F., C-13 NMR-spectroskopische und stereochemische untersuchungen. XI. Konformationen von bycyclo[3.3.1]nonanen und ihre unterscheidbarkeit durch lanthanideninduzierte verschiebungen, *Org. Magn. Reson.,* 8, 363, 1976.
269. Vegar, M. R. and Wells, R. J., The conformational analysis of some 3-substituted bicyclo[3.3.1]nonanes by means of Eu(dpm)$_3$ induced NMR shifts, *Tetrahedron Lett.,* p. 2847, 1971.
270. Cherr, C. J., Rosen, W., and Uebel, J. J., Lanthanide induced shift NMR studies of some bicyclo-[6.1.0]-nona-2,4,6-triene derivatives, *Tetrahedron Lett.,* p. 4045, 1974.
271. Berson, J. A., Luibrand, R. T., Kundu, N. G., and Morris, D. G., Carbonium ion rearrangements of bicyclo[2.2.]oct-2-ylcarbinyl derivatives, *J. Am. Chem. Soc.,* 93, 3075, 1971.
272. Maier, G., Fritschi, B., and Hoppe, B., Chlorsubstituierte cyclobutadiene [1,2], *Tetrahedron Lett.,* p. 1463, 1971.
273. Willcott, M. R., Oth, J. F. M., Thio, J., Plinke, G., and Schroder, G., Rapid solution of stereochemical problems with europium-tris(tetramethylheptanedione), Eu(thd)$_3$, *Tetrahedron Lett.,* p. 1579, 1971.
274. Miyashi, T., Hazato, A., and Mukai, T., Deep-seated rearrangement in the anionic oxy-cope system. Extremely facile epimerically unfavorable anionic oxy-cope rearrangement of anti-bisallylic 1,5,7-triene alkoxides, *J. Am. Chem. Soc.,* 100, 1008, 1978.
275. Mehta, G. and Pandey, P. N., PMR spectral studies on pentacyclo[5.3.0.0^{2,5}.0^{3,9}.0^{3,9}.0^{4,8}]decan-6-one and derivatives with shift reagent Eu(fod)$_3$, *Indian J. Chem.,* 13, 1351, 1975.
276. Talvitie, A. and Borg-Karlson, A. K., Chemical shift reagent study of cubenol and epicubenol, two sesquiterpene alcohols from the wood of *Anthrotaxis selaginoides* Don, *Finn. Chem. Lett.,* p. 93, 1979.
277. Andersen, N. H., Uh, H. S., Smith, S. E., and Wuts, P. G. M., Stereochemical course of the cyclization of olefinic aldehydes, *J. Chem. Soc. Chem. Commun.,* p. 956, 1972.

278. Ito, S. and Itoh, I., Cycloaddition reaction of tropylium ion and cyclopentadiene, *Tetrahedron Lett.*, p. 2968, 1971.
279. Carey, F. A., Application of europium(III) chelate induced chemical shifts to stereochemical assignments of isomeric perhydrophenalenols, *J. Org. Chem.*, 36, 2199, 1971.
280. Mason, T. J., The configuration and conformation of the tricyclo[4.4.1.1^{2,5}]dodecan-11-ols determined by H-1 NMR spectroscopy using the shift reagent Eu(fod)$_3$, *Org. Magn. Reson.*, 15, 321, 1981.
281. McKinney, J. D., Keith, L. H., Alford, A., and Fletcher, C. E., The proton magnetic resonance spectra of some chlorinated polycyclodiene pesticide metabolites. Rapid assessment of stereochemistry, *Can. J. Chem.*, 49, 1993, 1971.
282. ApSimon, J. W., Beierbeck, H., and Saunders, J. K., Lanthanide shift reagents in C-13 nuclear magnetic resonance: quantitative determination of pseudocontact shifts and assignment of C-13 chemical shifts of steroids, *Can. J. Chem.*, 51, 3874, 1973.
283. Wenkert, E. and Buckwalter, B. L., Carbon-13 nuclear magnetic resonance spectroscopy of naturally occurring substances. X. Pimaradienes, *J. Am. Chem. Soc.*, 94, 4367, 1972.
284. Tsuda, M. and Schroepfer, G. J., Jr., Carbon-13 nuclear magnetic resonance studies of C$_{27}$ sterol precursors of cholesterol, *J. Org. Chem.*, 44, 1290, 1979.
285. Smith, W. B. and Deavenport, D. L., The effect of Eu(dpm)$_3$ on the C-13 NMR spectrum of cholesterol, *J. Magn. Reson.*, 6, 256, 1972.
286. Smith, W. B., Deavenport, D. L., Swanzy, J. A., and Pate, G. A., Steroid C-13 chemical shifts. Assignments via shift reagents, *J. Magn. Reson.*, 12, 15, 1973.
287. Chadwick, D. J. and Williams, D. H., The full assignment of the carbon-13 nuclear magnetic resonance spectrum of 5α-cholestan-3β-ol with the aid of the lanthanide shift reagent Yb(dpm)$_3$, *J. Chem. Soc. Perkin Trans. 2*, 15, 1903, 1974.
288. Berman, E., Luz, Z., Mazur, Y., and Sheves, M., Conformational analysis of vitamin D analogues. C-13 and H-1 nuclear magnetic resonance study, *J. Org. Chem.*, 42, 3325, 1977.
289. Coxon, J. M., Hoskins, P. R., and Ridley, T. K., The assignment of the carbon-13 NMR spectra of Westphalen-rearranged steroid olefins with the aid of the lanthanide shift reagent Eu(fod)$_3$, *Aust. J. Chem.*, 30, 1735, 1977.
290. Okamura, W. H., Hammond, M. L., Rego, A., Norman, A. W., and Wing, R. M., Studies on vitamin D(calciferol) and its analogues. XII. Structural and synthetic studies of 5,6-trans-vitamin D$_3$ and the stereoisomers of 10,19-dihydrovitamin D$_3$ including dihydrotachysterol$_3$, *J. Org. Chem.*, 42, 2284, 1977.
291. La Mar, G. N. and Budd, D. L., Elucidation of the solution conformation of the A ring in vitamin D using proton coupling constants and a shift reagent, *J. Am. Chem. Soc.*, 96, 7317, 1974.
292. Demarco, P. V., Elzey, T. K., Lewis, R. B., and Wenkert, E., Tris(dipivalomethanato)europium(III). A shift reagent for use in the proton magnetic resonance analysis of steroids and terpenoids, *J. Am. Chem. Soc.*, 92, 5737, 1970.
293. Achmatowicz, O., Jr., Ejchart, A., Jurczak, J., Kozerski, L., and St. Pyrek, J., Confirmation of the structure of a new diterpene trachyloban-19-ol, by tris(dipivalomethanato)europium-shifted nuclear magnetic resonance spectroscopy, *J. Chem. Soc. Chem. Commun.*, p. 98, 1971.
294. Wing, R. M., Okamura, W. H., Pirio, M. R., Sine, S. M., and Norman, A. W., Vitamin D in solution: conformations of vitamin D$_3$, 1α-25-dihydroxyvitamin D$_3$, and dihydrotachysterol$_3$, *Science*, 186, 939, 1974.
295. Romeo, G., Giannetto, P., and Aversa, M. C., Constituents of *Satureia calamintha*. Application of Eu(fod)$_3$ to the assignments of the methyl resonances of triterpenes related to 12-ursene, *Org. Magn. Reson.*, 9, 29, 1977.
296. Rahier, A., Cattel, L., and Benveniste, P., Mechanism of the enzymatic cleavage of the 9β,19-cyclopropane ring of cycloeucalenol, *Phytochemistry*, 16, 1187, 1977.
297. Corbett, R. E. and Wilkins, A. L., Lichens and fungi. XII. Dehydration and isomerization of stictane triterpenoids, *J. Chem. Soc. Perkin Trans. 1*, p. 857, 1976.
298. Romeo, G., Giannetto, P., and Aversa, M. C., Costituenti della *Satureia calamintha*. Nota II. Applicazione dello -shift reagent- Eu(fod)$_3$ all'assegnazione delle risonanze dei metili nella serie del 12-ursene, *Chim. Ind.*, 58, 448, 1976.
299. Iida, T., Kikuchi, M., Tamura, T., and Matsumoto, T., Lanthanide- and aromatic solvent-induced shift effects on proton resonances in C-4-methylated steroids and tetracyclic triterpenoids, *Chem. Phys. Lipids*, 20, 157, 1977.
300. Iida, T., Proton NMR spectra of isomeric sterols. I. On the identification of ring structures, *Nihon Daiguku Kogakubu Kiyo Bunrui A*, 21, 227, 1980; *Chem. Abstr.*, 94, 47605g.
301. Iida, T., Proton NMR spectra of isomeric sterols. II. On the identification of side chain structures, *Nihon Daiguku Kogakubu Kiyo Bunrui A*, 21, 233, 1980; *Chem. Abstr.*, 94, 47606h.

302. Iida, T., Tamura, T., and Matsumoto, T., Proton nuclear magnetic resonance identification and discrimation of side chain isomers of phytosterols using a lanthanide shift reagent, *J. Lipid Res.*, 21, 326, 1980.
303. Wittstruck, T. A., Analysis of steroid nuclear magnetic resonance spectra using paramagnetic shift reagents, *J. Am. Chem. Soc.*, 94, 5130, 1972.
304. Barry, C. D., Dobson, C. M., Swiegart, D. A., Ford, L. E., and Williams, R. J. P., The structure of a cholesterol-shift reagent complex in solution, in *Nuclear Magnetic Resonance Shift Reagents*, Sievers, R. E., Ed., Academic Press, New York, 1973, 173.
305. ApSimon, J. W., Beirbeck, H., and Fruchier, A., Automatic sorting of signals in shift reagent spectra, *Can. J. Chem.*, 50, 2905, 1972.
306. Matsuo, M. and Matsumoto, S., C-13 nuclear magnetic resonance studies on phenols; effects of a lanthanide shift reagent and triethylamine, *Chem. Pharm. Bull.*, 25, 1399, 1977.
307. Schneider, H. J. and Agrawal, P. K., Normalized ytterbium induced C-13-NMR shifts as a simple aid for structural and C-13 signal assignments in multifunctional and natural compounds, *Tetrahedron*, 40, 1025, 1984.
308. Mashimo, K., Hayashibe, T., and Wainai, T., Determination of monoalkyl phenol isomers using shift reagent, *Bunseki Kagaku*, 26, 672, 1977.
309. Werstler, D. D. and Suman, P. T., Application of paramagnetic shift reagents to phenolic systems in proton nuclear magnetic resonance spectrometry, *Anal. Chem.*, 47, 144, 1975.
310. Shoffner, J. P., Use of tris-(6,6,7,7,8,8,8-heptafluoro-2,2-dimethyl-3,5-octanedionato)europium(III) for the structure determination and quantitative analysis of phenols, *Anal. Chem.*, 47, 341, 1975.
311. Shoffner, J. P., On the use of tris(6,6,7,7,8,8,8-heptafluoro-2,2-dimethyl-3,5-octanedionato)europium(III) as a shift reagent for carboyxlic acids and phenols, *J. Am. Chem. Soc.*, 96, 1599, 1974.
312. Pilkington, J. W. and Waring, A. J., Cyclohexadienones. Use of the dienone-phenol rearrangement in measuring migratory aptitudes of alkyl groups, *J. Chem. Soc. Perkin Trans. 2*, p. 1349, 1976.
313. Liu, K. T., Hsu, M. F., and Chen, J. S., Nuclear magnetic resonance studies. III. Application of tris(1,1,1,2,2,3,3-heptafluoro-7,7-dimethyl-4,6-octanedionato)europium as a shift reagent for phenols, *Tetrahedron Lett.*, p. 2179, 1974.
314. Chadwick, D. J., The problem of a unique ketal position in lanthanide shift reagent-ketone complexes in solution, *Tetrahedron Lett.*, p. 1375, 1974.
315. Finocchiaro, P., Recca, A., Maravigna, P., and Montaudo, G., A topological approach to the interaction of lanthanide shift reagents with the carbonyl group, *Tetrahedron*, 30, 4159, 1974.
316. Newman, R. H., The importance of non-axial symmetry in the interpretation of lanthanide-induced shifts for ketones, *Tetrahedron*, 30, 969, 1974.
317. Lienard, B. H. S. and Thomson, A. J., Evidence for two binding sites of lanthanoid shift reagents at the carbonyl groups of camphor and canthaxanthin, *J. Chem. Soc. Perkin Trans. 2*, p. 1390, 1977.
318. Talvitie, A., Paasivirta, J., and Widen, K. G., Structure proof of valeranone with NMR shift reagent, *Finn. Chem. Lett.*, p. 197, 1977.
319. Abraham, R. J., Chadwick, D. J., and Sancassan, F., Refinement of the solution structure of bicyclo[3.1.0]hexan-3-one by H-1 and C-13 lanthanide induced shift (LIS) analysis, *Tetrahedron Lett.*, p. 265, 1979.
320. Servis, K. L. and Shue, F. F., Conformational analysis of 2-butanone and 3-methyl-2-butanone using lanthanide shift reagents, *J. Magn. Reson.*, 40, 293, 1980.
321. Raber, D. J., Janks, C. M., Johnston, M. D., Jr., and Raber, N. K., Structure elucidation with lanthanide-induced shifts. 8. Geometry of europium-ketone complexes, *J. Am. Chem. Soc.*, 102, 6591, 1980.
322. Hofer, O., Conformational analysis by LIS. II. 2- And 3-alkyl substituted 1-indanones, *Monatsch. Chem.*, 110, 979, 1979.
323. Foldesi, P. and Hofer, O., Concerning the one site resp. two site coordination model for ketones in LIS calculations, *Tetrahedron Lett.*, 21, 2137, 1980.
324. Filippova, T. M., Bekker, A. R., and Lavrukhin, B. D., Application of paramagnetic shift reagents to the conformational analysis of isomeric dienones, *Org. Magn. Reson.*, 14, 337, 1980.
325. Abraham, R. J., Chadwick, D. J., and Sancassan, F., Conformational analysis. II. A lanthanide-induced shift NMR and theoretical study of bicyclo[3.1.0]hexan-3-one and adamantanone, *Tetrahedron*, 37, 1081, 1981.
326. Abraham, R. J., Bovill, M. J., Chadwick, D. J., Griffiths, L., and Sancassan, F., A lanthanide induced shift (LIS) investigation of the conformation of cyclohexanones in solution, *Tetrahedron*, 36, 279, 1980.
327. Abraham, R. J., Chadwick, D. J., Griffiths, L., and Sancassan, F., Direct lanthanide-induced-shift NMR determination of conformer populations in substituted cyclohexanones, *J. Am. Chem. Soc.*, 102, 5128, 1980.

328. Abraham, R. J., Chadwick, D. J., Griffiths, L., and Sancassan, F., On the importance of a correct choice of model for lanthanide ion binding in the interpretation of lanthanide-induced NMR shifts. The adamantanone problem, *Tetrahedron Lett.*, p. 4691, 1979.
329. Abraham, R. J., Bergen, H. A., and Chadwick, D. J., Conformational analysis. IV. A lanthanide induced shift (LIS) NMR investigation of the conformation and relative conformer energies of trans and cis 2-decalone, *Tetrahedron*, 38, 3271, 1982.
330. Chadwick, D. J. and Williams, D. H., Lanthanide-induced shifts in the carbon-13 nuclear magnetic resonance spectra of ketones, *J. Chem. Soc. Chem. Commun.*, p. 128, 1974.
331. Kessler, H. and Molter, M., The contact contribution in the lanthanoid-induced shift of C=O NMR signals, *Angew. Chem. Int. Ed.*, 13, 537, 1974.
332. Platzer, N., Basselier, J. J., and Demerseman, R., Etude des deplacements des signaux RMN induits par les chelates de lanthanides dans le cas de molecules en echange conformationnel. Cas de cetones aromatiques, *Bull. Soc. Chim. Fr.*, p. 1717, 1973.
333. Wolkowski, Z. W., Tris(dipivalomethanato)ytterbium — induced shifts in PMR of ketones and aldehydes, *Tetrahedron Lett.*, p. 821, 1971.
334. Thoai, N. and Chau, T. M., Etude conformationnelle des alcenyl phenylcetones par spectrographics infrarouge, ultraviolette et par l'utilisation de l'effet Overhauser nucleaire et de complexe de lanthanide Eu(dpm)$_3$ en resonance megnetique nucleaire, *Can. J. Chem.*, 52, 1331, 1974.
335. Grimaud, M. and Pfister-Guillouzo, G., Influence de la substitution sur la structure de la benzophenone X-Etude conformationnelle de benzaldehydes, acetophenones et benzophenones substitues par complexation avec les lanthanides, *Org. Magn. Reson.*, 7, 386, 1975.
336. Gryko, J. and Wielogorski, Z. A., The interaction of lanthanide shift reagents with ketones, *Rocz. Chem.*, 49, 2033, 1975.
337. Galera, E., Walkowiak, U., Roszak, S., and Zabza, A., A lanthanide-induced shift investigation of the conformation of pseudoionone and its derivatives in solution, *J. Mol. Struct.*, 101, 287, 1983.
338. Lenkinski, R. E. and Reuben, J., Line broadenings induced by lanthanide shift reagents. Concentration, frequency, and temperature effects, *J. Magn. Reson.*, 21, 47, 1976.
339. Zakharova, N. I., Filippova, T. M., Bekker, A. R., Miropol'skaya, M. A., and Samokhalov, G. I., Synthetic studies on polyene compounds. XLIII. Synthesis and study of α,β-isomerism of conjugated α,β-cis-dienones by proton NMR spectroscopy using Eu(fod)$_3$, *Zh. Org. Khim.*, 14, 1413, 1978.
340. Potapov, V. N., Dem'yanovich, V. M., and Zaitsev, V. P., Stereochemical studies. XLVII. Chiraloptical properties of (+)-2-methylbutyrophenones, *Zh. Org. Khim.*, 14, 91, 1978.
341. Kawaki, H., Takagi, T., Fujiwara, H., and Sasaki, Y., Nuclear magnetic resonance studies of acid-base association in solution. II. Association between ketones and tris(dipivalomethanato)europium, *Chem. Pharm. Bull.*, 29, 2397, 1981.
342. Rackham, D. M., Lanthanide shift reagents. XII. Equilibrium binding constants for aromatic aldehydes and ketones with tris-(tetramethylheptanedionato)europium, Eu(thd)$_3$, *Org. Magn. Reson.*, 12, 388, 1979.
343. Abraham, R. J., Bergen, H. A., Chadwick, D. J., and Sancassan, F., Lanthanum-induced C-13 NMR shifts: a novel probe of pi-electron delocalisation, *J. Chem. Soc. Chem. Commun.*, p. 998, 1982.
344. Abraham, R. J., Chadwick, D. J., and Sancassan, F. A. E. G., Conformational analysis. V. A lanthanide induced shift (LIS) NMR investigation of conformational isomerism in aromatic aldehydes and ketones, *Tetrahedron*, 38, 3245, 1982.
345. Abraham, R. J., Bergen, H. A., and Chadwick, D. J., Conformational analysis. VI. A lanthanide-induced shift nuclear magnetic resonance investigation of steric effects in mesitaldehyde and 2,4,6-trimethylacetophenone, *J. Chem. Soc. Perkin Trans. 2*, p. 1161, 1983.
346. Belanger, P., Freppel, C., Tizane, D., and Richer, J. C., Lanthanide-induced shifts in nuclear magnetic resonance spectroscopy: application to ketones, *J. Chem. Soc. Chem. Commun.*, p. 266, 1971.
347. Borgen, G., The conformation of 4,4,7,7-tetramethylcyclononanone; low-temperature NMR-spectroscopy in conjunction with the shift reagent Eu(dpm)$_3$, *Acta Chem. Scand.*, 26, 1740, 1972.
348. Casey, C. P. and Boggs, R. A., Stereospecific addition of lithium dipropenylcuprate to cyclohexenone, *Tetrahedron Lett.*, p. 2455, 1971.
349. Cory, R. M. and Hassner, A., Proton magnetic resonance assignments in dichloroketene-olefin adducts by lanthanide-induced shifts, *Tetrahedron Lett.*, p. 1245, 1972.
350. Kristiansen, P. and Ledaal, T., Tris(dipivalomethanato)europium(III) induced PMR shifts of cyclic ketones, *Tetrahedron Lett.*, p. 2817, 1971.
351. Kristiansen, P. and Ledaal, T., Lanthanide induced PMR shifts of cyclic ketones. A comparison of Pr(dpm)$_3$ and Eu(dpm)$_3$ as shift reagents, *Tetrahedron Lett.*, p. 4457, 1971.
352. Servis, K. and Bowler, D. J., Conformational analysis of 2-alkylcyclohexanone-lanthanide chelate complexes, *J. Am. Chem. Soc.*, 95, 3392, 1973.

353. Shapiro, B. L., Johnston, M. D., Jr., and Towns, R. L. R., Lanthanide-induced changes in proton spin-spin coupling constants, *J. Am. Chem. Soc.*, 94, 4381, 1972.
354. Andrieu, C. G., Lemarie, B., and Paquer, D., Analyse conformationnelle de cetones cyclopropaniques par RMN a l'aide des deplacements paramagnetiques induits par Eu(dpm)$_3$, *Org. Magn. Reson.*, 6, 479, 1974.
355. Enriquez, R., Taboada, J., Salazar, I., and Diaz, E., NMR stereotopic induced shifts in epimeric mixtures, *Org. Magn. Reson.*, 5, 291, 1973.
356. Fabre, J. M., Torreilles, E., Giral, L., and Mousserson, M., Application a des determinations de configurations des deplacements chimiques induits par les sels d'europium III, *C. R. Acad. Sci. Paris*, 278, 479, 1974.
357. Metzger, P. and Casadevall, E., Utilisation des deplacements induits par un complexe de L'europium sur les signaux des protons voisins d'un carbonyle pour la determination des vitesses d'echange H-D de chacun des quatre protons en alpha du C=O de la decalone-2 trans, *Tetrahedron Lett.*, p. 3341, 1973.
358. Servis, K. L. and Bowler, D. J., Conformational analysis using lanthanide shift reagents. Determination of alkyl group conformations in 2-alkyl-4-tert-butylcyclohexanones, *J. Am. Chem. Soc.*, 97, 80, 1975.
359. Hart, H. and Love, G. M., A new degenerate cyclopropylcarbinyl cation?, *J. Am. Chem. Soc.*, 93, 6264, 1971.
360. Hart, H. and Love, G. M., A novel intramolecular ketene cycloaddition. Functionalized tetracyclo[3.3.0.0^{2,8}.0^{3,6}]octanes, *J. Am. Chem. Soc.*, 93, 6266, 1971.
361. Hinckley, C. C., Applications of rare earth nuclear magnetic resonance shift reagents. II. The assignment of the methyl proton magnetic resonance of d-camphor, *J. Org. Chem.*, 35, 2834, 1970.
362. Leitereg, T. J., Synthesis of sesquiterpenoids related to nootkatone — structure determination by NMR using tris(dipivalomethanato)europium, *Tetrahedron Lett.*, p. 2617, 1972.
363. Jones, R. A., Application of lanthanide shift reagents to the analysis of NMR spectra of flavour constituents — a configurational and conformational analysis of nootkatone, *Flavour Ind.*, May/June, 125, 1974.
364. Krymskaya, E. B., Moskvichev, V. I., Gavrilova, T. F., Antonova, N. D., Aul'chenko, I. S., and Kheifits, L. A., Determination of the structure of methyl-substituted bicyclo[2.2.1]heptan-2-ones by PMR spectroscopy using a paramagnetic-shift reagent, *Zh. Prikl. Spektros.*, 22, 865, 1975.
365. Mazzocchi, P. H., Ammon, H. L., and Jameson, C. W., Lanthanide shift reagents. III. Errors resulting from the neglect of angle dependence, *Tetrahedron Lett.*, p. 573, 1973.
366. Hassner, A., Cory, R. M., and Sartoris, N., The stereochemistry of cycloadditions of ketenes to unsymmetrical alkenes. Evidence for nonparallel transition states, *J. Am. Chem. Soc.*, 98, 7698, 1976.
367. Weissberger, E. and Page, G., Enantiomeric recognition during cyclopentanone formation with iron(0), *J. Am. Chem. Soc.*, 99, 147, 1977.
368. Haag, R., Wirz, J., and Wagner, P. J., The photoenolization of 2-methylacetophenone and related compounds, *Helv. Chim. Acta*, 60, 2595, 1977.
369. McGuirk, P. R., Marfat, A., and Helquist, P., Determination of the configurations of trisubstituted olefins obtained from addition of alkylcopper complexes to acetylenes. Use of a lanthanide NMR shift reagent, *Tetrahedron Lett.*, p. 2973, 1978.
370. Tavares, R. F., and Katten, E., Total synthesis of 1-acetyl-7,10-ethano-4,4,7-trimethyl-1(9)-octalin, *Tetrahedron Lett.*, p. 1713, 1977.
371. Raber, D. J., Janks, C. M., Johnston, M. D., Jr., and Raber, N. K., Structure elucidation with lanthanide induced shifts. IX. Bicyclo[3.3.1]nonan-9-one, *Tetrahedron Lett.*, 21, 677, 1980.
372. Salaun, J. R. and Conia, J. M., The stereospecific thermal ring enlargement of 1-vinyl cyclopropanols into cyclobutanone derivatives, *Tetrahedron Lett.*, p. 2849, 1972.
373. Gibb, V. G., Armitage, I. M., Hall, L. D., and Marshall, A. G., Variation of stoichiometry with solvent in a lanthanide nuclear magnetic resonance shift reagent complex, *J. Am. Chem. Soc.*, 94, 8919, 1972.
374. Dougherty, D. A., Mislow, K., Huffman, J. W., and Jacobus, J., Study of the A-ring conformation in 4,4-dimethyl-3-keto steroids, *J. Org. Chem.*, 44, 1585, 1979.
375. Zushi, S., Kodama, Y., Nishihata, K., Umemura, K., Nishio, M., Uzawa, J., and Hirota, M., The conformations of 2-phenylpropionaldehyde and some aliphatic ketones. The possible importance of the CH/pi and CH/n interactions in determining the molecular geometry of a mobile system, *Bull. Chem. Soc. Jpn.*, 53, 3631, 1980.
376. Wooten, J., Savitsky, G. B., and Jacobus, J., H-2 spin-lattice relaxation in the presence of paramagnetic shift reagents, *J. Am. Chem. Soc.*, 97, 5027, 1975.
377. Barlet, R., Conformational study of ketones in the small ring series. IV. Case of methylcyclopropyl ketones gem-dihalogenated on the ring, *Bull. Soc. Chim. Fr.*, p. 543, 1977.

378. Cho, N. S. and Yun, S. S., Study on NMR induced chemical shift by lanthanide chemical shift reagent. II. Effect on lanthanide shift reagent-substrate equilibria and bound shifts of phenylcyclopropane derivatives, *Rep. Res. Inst. Chem. Spectrosc. Chungnam Natl. Univ.*, 4, 13, 1983; *Chem. Abstr.*, 100, 131191u.
379. Rackham, D. M., Lanthanide shift reagents paper 19. Equilibrium binding constants for alicyclic ketones, ethers and epoxides, *Spectrosc. Lett.*, 14, 117, 1981.
380. Shakirov, M. M., Detsina, A. N., and Koptyug, V. A., Determination of the structure of polymethylated cyclohexadienones by a PMR method using a shift reagent, *Izv. Sib. Otd. Akad. Nauk SSSR Ser. Khim. Nauk*, p. 94, 1977; *Chem. Abstr.*, 87, 38964b.
381. Abraham, R. J., Bergen, H. A., and Chadwick, D. J., The conformation of 4-phenylcyclohexanone in solution by H-1 and C-13 lanthanide induced shift (LIS) analysis: evidence in favour of phenyl over t-butyl as a conformational lock, *Tetrahedron Lett.*, 22, 2807, 1981.
382. Pons, A. and Chapat, J. P., Effets de substituants en serie dialkyl-1,2 cyclohexanique. II. Modifications conformationnelles liees a la symetrie des substituants, *Tetrahedron*, 36, 2297, 1980.
383. Abraham, R. J., Bergen, H. A., and Chadwick, D. J., The effect of locking groups on the geometry of the cyclohexanone ring: a comparative lanthanide-induced shift NMR study, *J. Chem. Res.(S)*, p. 118, 1983.
384. Abraham, R. J., Conformational and structural studies by lanthanide-induced shifts, *Anal. Proc. (London)*, 18, 364, 1981.
385. Yates, P. and Cong, D. D., Substituent parameters v. lanthanide-induced shifts in the interpretation of the C-13 nuclear magnetic resonance spectrum of 5,5-dimethyl-2-norbornanone, *Org. Magn. Reson.*, 20, 199, 1982.
386. Sasaki, T., Hayakawa, K., Manabe, T., and Nishida, S., Molecular design by cycloaddition reactions. 37. Peri-, stereo-, and regioselectivities in cycloaddition reactions of 7-isopropylidenebenzonorbornadiene, *J. Am. Chem. Soc.*, 103, 565, 1981.
387. Struchkova, M. I., Margaryan, A. K., and Serebryakova, E. P., Carbon-13 NMR spectroscopic study of substituted bicyclo[4.2.0]oct-7-en-2-ones, *Izv. Akad. Nauk SSSR Ser. Khim.*, p. 56, 1980; *Chem. Abstr.*, 92, 214422n.
388. Nitta, M., Omata, A., and Sugiyama, H., The solvomercuration, bromination, and related reactions of 1,5-dimethyl-6-methylenetricyclo[3.2.1.0^{2,7}]oct-3-en-8-one and its related compounds, *Bull. Chem. Soc. Jpn.*, 55, 569, 1982.
389. Fray, G. I. and Saxton, R. G., Reactions of cyclo-octatetraene and its derivatives. VII. Reactions of cyclo-octatetraene tetramer and of the dimer heptacyclo[8.6.0.0^{2,6}.0^{3,9}.0^{7,15}.0^{8,12}.0^{11,16}]-hexadeca-4,13-diene, *Tetrahedron*, 34, 2663, 1978.
390. Drake, J. A. G. and Jones, D. W., High-resolution NMR spectra of fluorene and its derivatives. V. H-1 and C-13 solvent and lanthanide-shift-reagent studies of fluorene-9-one, *Spectrochim. Acta*, 36A, 23, 1980.
391. Akporiaye, D. E., Farrant, R. D., and Kirk, D. N., Deuterium incorporation in the backbone rearrangement of cholest-5-ene, *J. Chem. Res. (S)*, p. 210, 1981.
392. Kutner, A. and Jaworska, R., Application of Eu(fod)$_3$ as a paramagnetic shift reagent in the proton magnetic resonance analysis of testosterone and its methyl derivatives, *Chem. Anal. (Warsaw)*, 23, 653, 1978.
393. Sugiyama, T., Kobayashi, A., and Yamashita, K., Paramagnetic induced shifts in the proton magnetic resonance spectra of cyclopropanecarboxylic acid esters using Eu(fod)$_3$, *Agric. Biol. Chem.*, 37, 1497, 1973.
394. Sakamoto, K. and Oki, M., Determination of conformations of esters with the use of lanthanide shift reagents, *Bull. Chem. Soc. Jpn.*, 47, 2623, 1974.
395. Wineburg, J. P. and Swern, D., NMR chemical shift reagents in structural determination of lipid derivatives. II. Methyl petroselinate and methyl oleate, *J. Am. Oil Chem. Soc.*, 49, 267, 1972.
396. Walters, D. B. and Horvat, R. J., Eu(dpm)$_3$ for the determination of the cis-trans composition of methyl elaidate-oleate by NMR spectroscopy, *Anal. Chim. Acta*, 65, 198, 1973.
397. Pfeffer, P. E., Foglia, T. A., Barr, P. A., and Obenauf, R. H., Deuterium nuclear magnetic resonance. Evaluation of the positional distribution of low levels of deuterium in the presence of Eu(fod)$_3$, *J. Org. Chem.*, 43, 3429, 1978.
398. Casey, C. P. and Cyr, C. R., Hydroformylation of 3-methyl-1-hexene-3-d$_1$. Evidence against direct formylation of a methyl group in the "oxo" reaction, *J. Am. Chem. Soc.*, 93, 1280, 1971.
399. Lunazzi, L., Placucci, G., Grossi, L., and Strocchi, A., Lanthanide induced shift as a tool for isomer assignment in esters of unsaturated fatty acids, *Chem. Phys. Lipids*, 30, 347, 1982.
400. Barton, F. E., II, Himmelsbach, D. S., and Walters, D. B., C-13 nuclear magnetic resonance of homologous methyl esters of saturated acids and the effect of chemical shift reagents, *J. Am. Oil Chem. Soc.*, 55, 574, 1978.

401. Noda, M. and Takahashi, T., NMR analysis of saturated and unsaturated fatty acid methyl esters by using shift reagents, *Yukagaku,* 28, 411, 1979.
402. Noda, M. and Miyake, K., Synthesis and H-1 NMR analysis of 2,2-dideuterio fatty acid methyl esters, *Yukagaku,* 31, 154, 1982.
403. Koller, H. L. and Dorn, H. C., Acid-catalyzed reactions of 2,2,2-trifluorodiazoethane for analysis of functional groups by F-19 nuclear magnetic resonance spectroscopy, *Anal. Chem.,* 54, 529, 1982.
404. Spassov, S. L., Stefanova, R., and Ladd, J. A., NMR studies of 3-phenylbutyric acid and its methyl ester. A possible approach for conformational analysis using lanthanide shift reagents, *J. Mol. Struct.,* 36, 93, 1977.
405. Rebuffat, S., Davoust, D., Giraud, M., and Molho, D., Application des deplacements paramagnetiques induits par Pr(fod)$_3$ a l'etude conformationnelle des esters methyliques des acides methyl-4 phenyl-5 pentadiene-2 cis, 4 trans et -2 trans, 4 trans oiques, *Bull. Soc. Chim. Fr.,* 2892, 1974.
406. Tsukida, K., Ito, M., and Ikeda, F., Application of a shift reagent in nuclear magnetic resonance spectroscopy. IV. A simple method for stereochemical assignment and simultaneous determination of cis-trans isomeric trisubstituted allylic alcohols, *Chem. Pharm. Bull.,* 21, 248, 1973.
407. Tsukida, K., Ito, M., and Ikedo, F., Application of a shift reagent in nuclear magnetic resonance spectroscopy of some fat-soluble vitamin acetates, *J. Vitaminol.,* 18, 24, 1972.
408. Tsukida, K., Ito, M., and Ikeda, F., Application of a shift reagent in nuclear magnetic resonance spectroscopy. III. A simple identification and simultaneous determination of geometrical isomers of vitamin A, *Int. J. Vitam. Nutr. Res.,* 42, 91, 1972.
409. Tsukida, K. and Ito, M., Application of a shift reagent in nuclear magnetic resonance spectroscopy of esters. An approach for simple identification and simultaneous determination of tocopherols, *Experientia,* 27, 1004, 1971.
410. Ranganayakulu, K., Dubey, R., and Rajeswari, K., A novel method for correlation of substituent constants in some aromatic esters using the shift reagent Eu(fod)$_3$, *Tetrahedron Lett.,* 22, 4359, 1981.
411. Radeglia, R., Application of NMR shift reagents for substance mixtures, *Z. Chem.,* 14, 72, 1974.
412. Anderson, R., Haines, A. H., and Stark, B. P., Paramagnetic shifts induced by europium and praseodymium complexes in the NMR spectra of some substituted phenols, *Angew. Makromol. Chem.,* 27, 151, 1972.
413. Ogorodnikov, V. D., Tishkina, O. A., and Maksyutin, Y. K., Use of lanthanide shift reagents for the analysis of heteroatomic petroleum concentrates, *Soversh. Metodov Anal. Neftei,* p. 148, 1983; *Chem. Abstr.,* 101, 194716k.
414. Cho, N. S., Yun, S. S., Kho, K. H., and Park, D. S., Study on NMR induced chemical shift by lanthanide chemical shift reagents. I. Effect on structure and basicity of phenylcyclopropane compounds on lanthanide induced shift, *Rep. Res. Inst. Chem. Spectrosc. Chungnam Natl. Univ.,* 3, 13, 1982; *Chem. Abstr.,* 98, 208909j.
415. Taylor, R. C. and Walters, D. B., Europium-induced shifts in the H-1 and P-31 NMR of "mixed" nitrogen-phosphorus bidentate ligands: evidence for a Eu(dpm)$_3$-P(III) interaction, *Tetrahedron Lett.,* p. 63, 1972.
416. Mironov, V. A., Luk'yanov, V. T., Yankovskii, S. A., and Gorshkova, L. V., Cyclic unsaturated compounds. LXIX. Analysis of compositions of mixtures of esters of the cyclopentadiene series by proton NMR spectroscopy using a paramagnetic shift reagent, *Izv. Vyssh. Uchebn. Zaved. Khim. Khim. Tekhnol.,* 24, 688, 1981; *Chem. Abstr.,* 95, 149420f.
417. Roth, K., Analysis of lanthanide induced shift data of mixtures, *Anal. Chem.,* 48, 2277, 1976.
418. Blanch, C., Camps, F., and Coll, J., Configurational analysis of the methyl esters of p-menthan-7-oic, p-menth-2-en-7-oic and p-mentha-2-5-dien-7-oic acids by the action of NMR lanthanide shift reagents, *Afinidad,* 35, 111, 1978.
419. Chervov, P. P., Bazyl'chik, V. V., and Samitov, Y. Y., Study of menthane series compounds. XV. Configuration and conformations of 2- and 4-methyl-1-(ethoxycarbonyl)cyclohexanes, their dynamic stereochemistry, and Eu(fod)$_3$ shift reagent NMR effects, *Zh. Org. Khim.,* 18, 1222, 1982.
420. Whitesides, G. M. and San Filippo, J., Jr., The mechanism of reduction of alkylmercuric halides by metal hydrides, *J. Am. Chem. Soc.,* 92, 6611, 1970.
421. Pizzala, L. and Bodot, H., Interactions intramoleculaires. Etude RMN de la configuration et de la conformation des deux trimethyl-1.3.6 methoxycarbonyl-3 bicyclo(4.1.0) heptanes; utilisation de faibles concentrations d'un reactif de deplacement, *C. R. Acad. Sci. Paris,* 276, 1267, 1973.
422. Harding, K. E. and Trotter, J. W., Synthesis via chloroketene adducts. Synthesis of demethylsesquicarene, *J. Org. Chem.,* 42, 4157, 1977.
423. Johnson, L. F., Chakravarty, J., Dasgupta, R., and Ghatak, U. R., Application in the use of a shift reagent in proton magnetic resonance spectroscopy for the elucidation of structures and stereochemistry of epimeric methyl 4,9-dimethyl-7,8-benzobicyclo(3.3.1)non-7-ene-4-carboxylates, *Tetrahedron Lett.,* p. 1703, 1971.

424. Talvitie, A., Norin, T., Stroemberg, S., and Weber, M., LIS (lanthanide-induced shift) NMR studies on the conformation of hinokiic acid, *Finn. Chem. Lett.*, p. 149, 1979.
425. Tsukida, K., Akutsu, K., and Ito, M., Simultaneous determination of eight vitamin D_2 isomers by proton magnetic resonance spectroscopy, *J. Nutr. Sci. Vitam.*, 22, 7, 1976.
426. Iida, T., Ishikawa, T., Tamura, T., and Matsumoto, T., H-1 NMR lanthanide-induced shifts of sterol acetates by Ho(fod)$_3$, *Yukagaku*, 29, 683, 1980.
427. Rae, I. D., Lanthanide shifts in the H-1 NMR spectra of thioamides, selenoamides and 1,2-dithioles, *Aust. J. Chem.*, 28, 2527, 1975.
428. Manni, P. E., Howie, G. A., Katz, B., and Cassady, J. M., Simplification of epoxide and lactone proton magnetic resonance spectra using tris(dipivalomethanato)europium shift reagent, *J. Org. Chem.*, 37, 2769, 1972.
429. De Lima, R. A., Monache, F. D., Marletti, F., and Marini-Bettolo, G., Lanthanide induced shifts of some coumarins substituted at the heterocyclic ring, *Gazz. Chim. Ital.*, 107, 427, 1977.
430. Carroll, F. I. and Blackwell, J. T., Structure and conformation of cis and trans-3,5-dimethylvalerolactones, *Tetrahedron Lett.*, p. 4173, 1970.
431. Karpf, M. and Dreiding, A. S., Anwendung von ringerweiterungen zur herstellung von rac-muscon und exalton, *Helv. Chim. Acta*, 58, 2409, 1975.
432. Tori, K., Horibe, I., Tamura, Y., and Tada, H., Simultaneous application of the nuclear Overhauser effect and an NMR shift reagent. Conformations of costunolide and dihydrocostunolide in solution, *J. Chem. Soc. Chem. Commun.*, p. 620, 1973.
433. Beaute, C., Wolkowski, Z. W., Merda, J. P., and Lelandais, D., Structural assignment of chloro- and bromo-vinyl aldehydes by Yb(dpm)$_3$-assisted proton nuclear magnetic resonance, *Tetrahedron Lett.*, p. 2473, 1971.
434. Abraham, R. J., Chadwick, D. J., and Sancassan, F., On the lack of evidence for contact contributions to lanthanide-induced NMR shifts (LIS's) in aromatic aldehydes: a rigorous error analysis and the isolation of diamagnetic(complexation) shifts, *Tetrahedron Lett.*, 22, 2139, 1981.
435. Abraham, R. J., Chadwick, D. J., and Sancassan, F., Conformational analysis. III. A lanthanide induced shift (LIS) NMR investigation of benzaldehyde, and thiophen- and furan-2-aldehyde, *Tetrahedron*, 38, 1485, 1982.
436. Abraham, R. J., Chadwick, D. J., and Sancassan, F., Conformational analysis. VII. A lanthanide induced shift (LIS) nuclear magnetic resonance investigation of conformational isomerism in ortho- and meta-substituted benzaldehydes, *J. Chem. Soc. Perkin Trans. 2*, p. 1037, 1984.
437. Hajek, M., Vodicka, L., and Ksandr, Z., Lanthanide induced shifts of carboxylic acids with Eu(dpm)$_3$ and Eu(fod)$_3$ as shift reagents, *Sb. Vys. SK. Chem. Technol. Praze Technol. Paliv.*, D32, 73, 1976.
438. Mizugaki, M., Hoshino, T., Ito, Y., Sakamoto, T., Shiraishi, T., and Yamanaka, H., Studies on the metabolism of unsaturated fatty acids. III. Structural determination of cis- and trans-isomers of 3- and 2-alkenoic acids by nuclear magnetic resonance spectroscopy using a chemical shift reagent, *Chem. Pharm. Bull.*, 28, 2347, 1980.
439. Cohen-Addad, C., Dipropylacetic, diethylacetic and dimethylacetic acid conformations in chloroform solutions as observed from lanthanide-induced NMR shifts. Comparison with primary amide conformations, *Spectrochim. Acta*, 36A, 587, 1980.
440. Mashimo, K., Matsukawa, M., and Wainai, T., Analysis of aromatic carboxylic acids using shift reagents, *Bunseki Kagaku*, 28, 147, 1979.
441. Craig, R. E. R., Craig, A. C., Larsen, R. D., and Caughlan, C. N., Molecular geometry studies. The crystal and molecular structure of a 7-spirocyclopentylbicyclo[2.2.1]heptene anhydride, *J. Org. Chem.*, 42, 3188, 1977.
442. Beaute, C., Wolkowski, Z. W., and Thoai, N., Tris(dipivalomethanato)ytterbium — induced shifts in PMR amines, *Tetrahedron Lett.*, p. 817, 1971.
443. Serve, P., Rondeau, R. E., and Rosenberg, H. M., Application of Eu(fod)$_3$ in elucidation of structures of bicyclic ethers, *J. Heterocycl. Chem.*, 9, 721, 1972.
444. Peters, J. A., Cranenburgh, P. E. J. P. V., Van Der Toorn, J. M., Wortel, T. M., and Van Bekkum, H., Conformational analysis of 7-alkyl-3-oxabicyclo[3.3.1]nonanes and complexes with lanthanide shift reagents, *Tetrahedron*, 34, 2217, 1978.
445. Harris, M. M. and Patel, P. K., Optically active nine-membered rings incorporating the 8- and 8'-positions of 1,1'-binaphthyl. II. Oxygen, sulphur, and selenium heterocycles. Synthesis, H-1 nuclear magnetic resonance spectra, and absolute configuration, *J. Chem. Soc. Perkin Trans. 2*, p. 304, 1977.
446. Ribo, J. M. and Serra, X., Uber die anwendung von lanthaniden-verschiebungs-reagenzien (LSR) bei der strukturbestimmung flexibler molekule, am beispiel des 2-methyltetrahydropyrans und des cis-4-methyl-2-(2'-methyl-1'-propenyl)-tetrahydropyrans, *Org. Magn. Reson.*, 12, 467, 1979.

447. Wineburg, J. P. and Swern, D., NMR chemical shift reagents in structural determination of lipid derivatives. IV. Methyl cis- and trans-9,10-epoxystearate and methyl erythro- and threo-9,10-dihydroxystearate, *J. Am. Oil Chem. Soc.,* 51, 528, 1974.
448. Vidal, J. P., Girard, J. P., Rossi, J. C., Chapat, J. P., and Granger, R., Epoxydes cyclopentaniques conformationnellement homogenex. II. Determination des configurations au moyen des deplacements chimiques induits par complexation a l'europium, *Org. Magn. Reson.,* 6, 522, 1974.
449. Samitov, Y. Y. and Bikeev, S. S., Dispersion of the pseudocontact shifts in NMR spectra, *Dokl. Akad. Nauk SSSR,* 218, 145, 1974.
450. Keith, L. H., Eu(dpm)$_3$ transannularly induced paramagnetic chemical shifts in the PMR spectra of endrin, dieldrin, and photodieldrin, *Tetrahedron Lett.,* p. 3, 1971.
451. Marcos, M., Melendez, E., Serrano, J. L., Camps, P., Figueredo, M., and Jamie, C., A study on the solid-solid phase transitions of (E)- and (Z)-9-(bicyclo[4.2.1]nonan-9-ylidene)bicyclo[4.2.1]nonane and some related compounds. Assignment of the configuration to the product of monoepoxidation of (E)-9-(bicyclo[4.2.1]non-3-en-9-ylidene)bicyclo[4.2.1]non-3-ene by H-1 nuclear magnetic resonance spectroscopy using the shift reagent Eu(fod)$_3$, *J. Chem. Soc. Perkin Trans. 2,* p. 7, 1984.
452. Willcott, M. R., III and Davis, R. E., Configurational assessment of conformationally mobile molecules by the LIS experiment, in *Nuclear Magnetic Resonance Shift Reagent,* Sievers, R. E., Ed., Academic Press, New York, 1973, 159.
453. Samitov, Y. Y. and Bikeev, S. S., Influence of chemical exchange on the temperature dependence of Eu(fod)$_3$-induced shifts, *Org. Magn. Reson.,* 7, 467, 1975.
454. Caple, R. and Kuo, S. C., Angular dependency of lanthanide-induced shifts in proton nuclear magnetic resonance spectra. Analysis of rigid bicyclic ethers, *Tetrahedron Lett.,* p. 4413, 1971.
455. Caple, R., Harriss, D. K., and Kuo, S. C., Dipolar nature of lanthanide-induced shifts. Detection of the angular dependency factor, *J. Org. Chem.,* 38, 381, 1973.
456. Sohar, P., Vajna, Z. M., and Bernath, G., Determination of the stereostructure of geometrical isomers by proton NMR, using a shift reagent, *Kem. Kozl.,* 46, 486, 1976.
457. Goldstein, M. J. and Kline, S. A., Synthesis of bicyclo[4.3.2]undeca-2,4,8,10-tetraen-7-one. II. Thermodynamic and kinetic control of intramolecular cycloaddition, *Tetrahedron Lett.,* p. 1089, 1973.
458. Shue, F. F. and Yen, T. F., Application of lanthanide NMR shift reagents for functional group studies of shale bitumen asphaltenes, *Fuel,* 62, 127, 1982.
459. Heut, J., Fabre, O., and Zimmermann, D., Etude de la structure et de le reactivite d'ethers d'enol par utilisation d'un reactif de deplacements chimiques et jesures d'effets Overhauser nucleaire en RMN H-1, *Tetrahedron,* 37, 3739, 1981.
460. Yastrebov, V. V. and Chernyshev, A. I., Dynamic models of associates of paramagnetic reagents with flexible-structure substrates for interpretation of paramagnetic proton NMR shifts as illustrated by Eu(fod)$_3$-1-methoxy-4-bromo-2-butyne, *Zh. Strukt. Khim.,* 17, 1013, 1976.
461. Ernst, L. and Mannschreck, A., Eu(dpm)$_3$-induced shifts in substituted anilines. The importance of basicity and steric effects, *Tetrahedron Lett.,* p. 3023, 1971.
462. Burzynska, H., Dabrowski, J., and Krowczynski, A., Europium- and praseodymium-induced shifts in the PMR spectra of amines, *Bull. Acad. Pol. Sci.,* 19, 587, 1971.
463. Hooper, D. L. and Kardos, A., Lanthanide induced shifts in the proton magnetic resonance spectra of cyclic amines, *Can. J. Chem.,* 51, 4080, 1973.
464. Hirayama, M. and Ishida, N., The correlation of steric effects and equilibrium constants evaluated from lanthanoid-induced H-1 shifts in aniline derivative-Eu(fod)$_3$ systems, *Bull. Chem. Soc. Jpn.,* 50, 779, 1977.
465. Witanowski, M., Stefaniak, L., and Januszewski, H., Structure differentiation by means of lanthanide-induced nitrogen-14 chemical shifts, *J. Cryst. Mol. Struct.,* 5, 141, 1975.
466. Witanowski, M., Stefaniak, L., and Januszewski, H., Effect of Yb(dpm)$_3$ and Eu(dpm)$_3$ on nitrogen-14 magnetic resonance, *Tetrahedron Lett.,* p. 1653, 1971.
467. Sasaki, Y., Kawaki, H., and Okazaki, Y., Shift parameter induced by tris(dipivalomethanato)europium and steric strain energies of aliphatic amines, *Chem. Pharm. Bull.,* 21, 917, 1973.
468. Rackham, D. M., Lanthanide shift reagents paper 16. Binding constants for europium shift reagent with sterically hindered amine donors, *Spectrosc. Lett.,* 13, 509, 1980.
469. Ajisaka, K., Kainosho, M., Shigemoto, H., Tori, K., Wolkowski, Z. W., and Yoshimura, Y., Conspicuous effects due to complex formation and contact term upon lanthanide-induced shifts in F-19 NMR of some fluoroaromatic compounds, *Chem. Lett.,* p. 1205, 1973.
470. Hirayama, M. and Owada, M., The correlation of basicity and equilibrium constants evaluated from lanthanoid-induced shifts in the Eu(fod)$_3$-p-substituted aniline systems, *Bull. Chem. Soc. Jpn.,* 52, 1786, 1979.

471. Marzin, C., Leibfritz, D., Hawkes, G. E., and Roberts, J. D., Nuclear magnetic resonance shift reagents: abnormal C-13 shifts produced by complexation of lanthanide chelates with saturated amines and n-butyl isocyanide, *Proc. Natl. Acad. Sci. U.S.A.*, 70, 562, 1973.
472. Hawkes, G. E., Marzin, C., Johns, S. R., and Roberts, J. D., Nuclear magnetic resonance shift reagents. Quantitative estimates of contact contributions to lanthanide-induced chemical-shift changes for exo-norbornylamine, *J. Am. Chem. Soc.*, 95, 1661, 1973.
473. Cushley, R. J., Anderson, D. R., and Lipsky, S. R., An upfield Carbon-13 shift induced by tris(dipivalomethanato)europium, *J. Chem. Soc. Chem. Commun.*, 636, 1972.
474. Hirayama, M. and Sato, M., Quantitative estimates of pi-contact term contributions to the lanthanide-induced H-1 and C-13 shifts in aniline and p-toluidine, *Chem. Lett.*, p. 725, 1974.
475. Hirayama, M., Sato, M., Takeuchi, M., and Saito, M., Quantitative estimates of pi-contact-term contributions to the lanthanide-induced paramagnetic H-1 shifts in aniline and m-toluidine, *Bull. Chem. Soc. Jpn.*, 48, 2690, 1975.
476. Johnson, B. F. G., Lewis, J., McArdle, P., and Norton, J. R., Contact effects in nuclear magnetic resonance spectra with lanthanide shift reagents, *J. Chem. Soc. Chem. Commun.*, p. 535, 1972.
477. Ajisaka, K. and Kainosho, M., Diastereomeric interaction of partially resolved amines facilitated by lanthanide chelates. Evidence for dynamic equilibrium between seven-coordinate and eight-coordinate alkylamine-lanthanide chelate adducts, *J. Am. Chem. Soc.*, 97, 1761, 1975.
478. Wright, G. E., NMR assignment of diastereotopic protons in amphetamine by the use of Eu(fod)$_3$, *Tetrahedron Lett.*, p. 1097, 1973.
479. Alvarez Ibarra, C., Arjona Lorague, O., Perez-Rubalcaba, A., Quiroga Feijoo, M. L., and Valdes, F., Conformational study of diastereomers. XI. Assignment of the relative configuration of diastereomeric acyclic amines with the lanthanide shift reagent Eu(fod)$_3$ by proton NMR. Isomers of N-(2-phenyl-1-methyl)propyl-N-phenylamine, *An. Quim. Ser. C*, 80, 7, 1984.
480. Stock, L. M. and Wasielewski, M. R., Pseudocontact and contact shifts for 6-aminobenzobicyclo[2.2.2]octene and 2-aminotriptycene. The sign of B_0^1, *J. Am. Chem. Soc.*, 95, 2743, 1973.
481. Jolidon, S. and Hansen, H. J., Untersuchungen uber aromatische amino-Claisen-umlagerungen, *Helv. Chim. Acta*, 60, 978, 1977.
482. Torocheshnikov, V. N. and Sergeev, N. M., Effect of paramagnetic shift reagents on the time of carbon-13 spin-lattice relaxation and on constants of carbon-13/proton spin-spin interaction, *Zh. Struckt. Khim.*, 24, 121, 1983.
483. Ernst, V. L., Zur kernresonanz-spektroskopie mit lanthaniden-komplexes: konfigurationsbestimmung der isomeren 1,3-dimethylcyclohexylamine-(2) mit hilfe von tris(dipivalomethanato)europium(III), *Chem. Z.*, 95, 325, 1971.
484. Remy, D. C. and Van Saun, W. A., Jr., Determination of the stereochemical configuration of 3-substituted N,N-dimethyl-5H-dibenzo[a,d]cycloheptene-delta5,gamma-propylamine derivatives using tris-(dipivalomethanato)europium as a shift reagent, *Tetrahedron Lett.*, p. 2463, 1971.
485. Barraclough, D., Oakland, J. S., and Sheinmann, F., Studies in the norbornene series. I. Elucidation of the structure of norbornene derivatives by use of the nuclear magnetic resonance shift reagent trisdipivaloylmethanatoeuropium(III), the nuclear Overhauser effect, and mass spectrometry, *J. Chem. Soc. Perkin Trans. 1*, p. 1500, 1972.
486. Sen, E. C. and Jones, R. A., The chemistry of terpenes. IV. The configuration and conformation of the isomeric 3-aminopinanes, *Tetrahedron*, 28, 2871, 1972.
487. Anderson, A. G., Jr. and Wade, P. C., Synthesis of 1-azatricyclo[5.2.1.0^{4,10}]decane, *J. Org. Chem.*, 43, 54, 1978.
488. Chalmers, A. A. and Pachler, K. G. R., Contact and pseudo-contact lanthanide-induced shifts in the nuclear magnetic resonance spectrum of quinoline, *J. Chem. Soc. Perkin Trans. 2*, p. 748, 1974.
489. Beech, G. and Morgan, R. J., Anomalous substituent effects observed for lanthanide-induced-shifts in substituted pyridines, *Tetrahedron Lett.*, p. 973, 1974.
490. Chalmers, A. A. and Pachler, K. G. R., C-13 and H-1 nuclear magnetic resonance spectra of quinoline in the presence of lanthanide shift reagents, *Tetrahedron Lett.*, p. 4033, 1972.
491. Hirayama, M. and Hanyu, Y., The contact-term contributions to lanthanide-induced C-13 paramagnetic shifts in acridine, quinoline, and isoquinoline, *Bull. Chem. Soc. Jpn.*, 46, 2687, 1973.
492. Hirayama, M., Edagawa, E., and Hanyu, Y., Contact term contribution to lanthanide-induced C-13 nuclear magnetic resonance shifts in pyridine and beta-picoline, *J. Chem. Soc. Chem. Commun.*, p. 1343, 1972.
493. Kordova, I., Fomichev, A. A., Zvolinskii, V. P., Gusarov, A. N., Pleshakov, V. G., and Prostakov, N. S., Calculations of lanthanide induced chemical NMR shifts. Paradox of a pseudocontact model for europium reagents during coordination on a pyridinium nitrogen atom, *Zh. Strukt. Khim.*, 22, 27, 1981.

494. Armitage, I. M., Burnell, E. E., Dunn, M. B., Hall, L. D., and Malcolm, R. B., The effect of lanthanide shift reagents on the NMR spectra of molecules partially oriented in a nematic phase: pyridine, *J. Magn. Reson.*, 13, 167, 1974.
495. Armarego, W. L. F., Batterham, T. J., and Kershaw, J. R., Tris(dipivalomethanato)europium induced shifts of proton resonances in pi-deficient nitrogen heterocycles, *Org. Magn. Reson.*, 3, 575, 1971.
496. Fletton, R. A., Green, G. F. H., and Page, J. E., Lanthanide-induced displacements in the NMR spectra of simple heterocyclic N-oxides, *Chem. Ind.*, 167, 1972.
497. Huber, H. and Seelig, J., On the complex formation of tris-(dipivalomethanato)europium with pyridine, *Helv. Chim. Acta*, 55, 135, 1972.
498. Witanowski, M., Stefaniak, L., and Januszewski, H., Selection of dipivalomethanate chelates as shift reagents for nitrogen-14 nuclear magnetic resonance, *J. Chem. Soc. Chem. Commun.*, p. 1573, 1971.
499. Stensio, K. E. and Ahlin, U., Synthesis and europium shifted NMR spectra of trans- and cis 4[β(1-naphthyl)vinyl]pyridine, *Tetrahedron Lett.*, p. 4729, 1971.
500. Nagawa, M., Ono, M., Hirota, M., Hamada, Y., and Takeuchi, I., Lanthanide induced shifts of the PMR spectra of polyaza-aromatic compounds, *Bull. Chem. Soc. Jpn.*, 49, 1322, 1976.
501. Heigl, T. and Mucklow, G. K., Position of lanthanide ion in NMR shift reagent — ligand chelates, *Tetrahedron Lett.*, p. 649, 1973.
502. Rackham, D. M., Lanthanide shift reagents. Paper 18. Equilibrium binding constants for europium shift reagent with nitrogen heterocycles, *Spectrosc. Lett.*, 13, 517, 1980.
503. Keiko, V. V. and Voronov, V. K., Determination of the mutual orientation of the ligand molecule and the coordinating ion in paramagnetic complexes, *Izv. Sib. Otd. Akad. Nauk SSSR Ser. Khim. Nauk*, p. 125, 1976; *Chem. Abstr.*, 86, 113345k.
504. Huber, H. and Pascual, C., Uber den einfluss des kontaktterms auf die lanthaniden-verschiebung der NMR-signale aromatischer stickstoffheterocyclen, *Helv. Chim. Acta*, 54, 913, 1971.
505. Rosen, B. I. and Weber, W. P., Electrocyclic synthesis of 5,6- and 7,8-dihydroquinolines and 5,6- and 7,8-dihydroisoquinolines, *J. Org. Chem.*, 42, 47, 1977.
506. Atkins, R. L., Moore, D. W., and Henry, R. A., H-1 nuclear magnetic resonance structure elucidation of substituted isoquinolines by means of Eu(fod)$_3$-induced paramagnetic shifts, *J. Org. Chem.*, 38, 400, 1973.
507. Young, J. A., Grasselli, J. G., and Ritchey, W. M., Use of paramagnetic shift reagents for simplification of nuclear magnetic resonance spectra of organic nitriles, *Anal. Chem.*, 45, 1410, 1973.
508. Davis, R. E., Willcott, M. R., III, Lenkinski, R. E., Doering, W. E., and Birladeanu, L., Interpretation of the pseudocontact model for nuclear magnetic resonance shift reagents. V. Collinearity in the structural elucidation of nitriles, *J. Am. Chem. Soc.*, 95, 6846, 1973.
509. Raber, D. J., Johnston, M. D., Jr., Perry, J. W., and Jackson, G. F., III, Structure elucidation with lanthanide-induced shifts. III. Acyclic aliphatic nitriles, *J. Org. Chem.*, 43, 229, 1978.
510. Raber, D. J., Johnston, M. D., Jr., and Schwalke, M. A., Structure elucidation with lanthanide-induced shifts. 2. Conformational analysis of cyclohexanecarbonitrile, *J. Am. Chem. Soc.*, 99, 7671, 1977.
511. Raber, D. J., Beaumont, W. E., and Johnston, M. D., Jr., Structure elucidation with lanthanide induced shifts. 14. Structural effects on equilibrium between nitriles and Eu(fod)$_3$, *Spectrosc. Lett.*, 15, 329, 1982.
512. Beaute, C., Wolkowski, Z. W., and Thoai, N., Tris(dipivalomethanato)ytterbium-induced shifts in the H-1 nuclear magnetic resonance spectra of some nitrogen compounds, *J. Chem. Soc. Chem. Commun.*, p. 700, 1971.
513. Seux, R., Morel, G., and Foucaud, A., Decyanuration des succinonitriles trisubstitues. Configurations Z et E des cinnamonitriles α,β-dialcoyles obtenus, *Tetrahedron Lett.*, p. 1003, 1972.
514. Anderson, S. J. and Norbury, A. H., Lanthanide-induced shifts in inorganic and organometallic compounds: a method for distinguishing between N- and S-thiocyanato-compounds, *J. Chem. Soc. Chem. Commun.*, p. 48, 1975.
515. Raber, D. J., Caines, G. H., Johnston, M. D., Jr., and Raber, N. K., Structure elucidation with lanthanide-induced shifts. 11. Analysis of alkyl-substituted benzonitriles, *J. Magn. Reson.*, 47, 38, 1982.
516. Raber, D. J., Janks, C. M., Johnston, M. D., Jr., and Raber, N. K., Structure elucidation with lanthanide induced shifts. 7. Development of a reliable method for structure evaluation and the application to organic nitriles, *Org. Magn. Reson.*, 15, 57, 1981.
517. Klarner, F. G., Yaslak, S., and Wette, M., Die synthese optisch aktiver [norcaradien = cycloheptatrien]-derivate, *Chem. Ber.*, 110, 107, 1977.
518. Klarner, F. G. and Wette, M., Eine entartete umlagerung des cis-bicyclo[6.1.0]nona-2,4,6-triens, *Chem. Ber.*, 111, 282, 1978.

519. Khalilov, L. M., Sadykov, R. A., Fatykhov, A. A., and Panasenko, A. A., Calculation of the geometry of a complex Eu(fod)₃-2-cyano(trimethylsilyl)bicyclo[2.2.1]heptane and stereochemical assignments in the proton NMR spectra, *Zh. Strukt. Khim.*, 22, 163, 1981.
520. Ward, T. M., Allcox, I. L., and Wahl, G. H., Jr., Lanthanide induced shifts due to coordination at phosphoryl and amide sites, *Tetrahedron Lett.*, p. 4421, 1971.
521. Kleinpeter, V. E., Widera, R., and Muhlstadt, M., N-substituierte carbamid- und thiocarbamidsaure-O-athylester, *J. Prak. Chem.*, 319, 133, 1977.
522. Montaudo, G. and Finocchiaro, P., Determination of the molecular geometry of Eu(fod)₃ complexes with amides and diamides and its conformational significance, *J. Org. Chem.*, 37, 3434, 1972.
523. Graham, L. L., Primary and secondary binding sites for lanthanide shift reagents with amides. Differential behavior of cis vs. trans amide isomers with LSR, *Org. Magn. Reson.*, 14, 40, 1980.
524. Gutierrez, M. A., Martins, M. A. P., and Rittner, R., NMR chemical shift substituent effects 3; alpha-monosubstituted N,N-diethylacetamides; bifunctional compounds in LSR experiments, *Org. Magn. Reson.*, 20, 20, 1982.
525. Rackham, D. M. and Chitty, C. J., Lanthanide shift reagents. Paper 20. Lanthanide binding constants for amides and quaternary salts, *Spectrosc. Lett.*, 14, 249, 1981.
526. Radeglia, R., Differentiation of contact and pseudocontact interaction contributions with tris(dipivalomethanato)europium(III)-induced H-1 and carbon 13 NMR chemical shifts of vinylogous N,N-dimethylformamides (merocyanines), *Z. Chem.*, 14, 417, 1974.
527. Fletton, R. A., Green, G. F. H., and Page, J. E., Effect of lanthanide shift reagents on the nuclear magnetic resonance spectra of secondary amides, *J. Chem. Soc. Chem. Commun.*, p. 1134, 1972.
528. Montaudo, G., Maravigna, P., Caccamese, S., and Librando, V., Structural analysis by lanthanide-induced shifts. V. Influence of steric and conjugative effects on the barriers to rotation in N,N-dimethylamides, *J. Org. Chem.*, 39, 2806, 1974.
529. Lewin, A. H., Restricted rotation in amides. IV. Resonance assignments in tertiary amides and thioamides utilizing NMR shift reagents, *Tetrahedron Lett.*, p. 3583, 1971.
530. Isbrandt, L. R. and Rogers, M. T., Lanthanide-induced shifts in H-1 nuclear magnetic resonance spectroscopy: resonance assignments in tertiary amides, *J. Chem. Soc. Chem. Commun.*, p. 1378, 1971.
531. Cohen-Addad, C. and Cohen-Addad, J. P., Dipropylacetamide and diethylacetamide conformations in chloroform solutions as observed from lanthanide induced NMR shifts, *Spectrochim. Acta*, 33A, 821, 1977.
532. Cohen-Addad, C. and Cohen-Addad, J. P., Crystal structure of isobutyramide and comparison with its conformation in chloroform solution as observed from lanthanide-induced nuclear magnetic resonance shifts, *J. Chem. Soc. Perkin Trans. 2*, p. 168, 1978.
533. Lovy, J., Doskocilova, D., Schmidt, P., and Schneider, B., Conformational structure of N-methylpropionamide and N-methylisobutyroamide. NMR shift reagent and IR study, *J. Mol. Struct.*, 50, 81, 1978.
534. Graham, L. L., Alteration of the cis:trans isomer ratio in N-methylformamide using lanthanide shift reagents. Determination of the equilibrium constants for complexation, *J. Chem. Soc. Perkin Trans. 2*, p. 1481, 1981.
535. Cheng, H. N. and Gutowsky, H. S., Lanthanide shift reagents and their use in dynamic NMR studies, *J. Phys. Chem.*, 84, 1039, 1980.
536. Turov, A. V. and Kornilov, M. Y., Study of stereoisomerism of tertiary amides using lanthanide shift reagents, *Dopov. Akad. Nauk Ukr RSR Ser. B: Geol Khim-Biol Nauki*, p. 817, 1976; *Chem. Abstr.*, 86, 16229t.
537. Montaudo, G., Recca, A., Verhoeven, J., and Kruk, C., IX/lanthanide-induced contact shifts in the carbon-13 spectra of amides, *Gazz. Chim. Ital.*, 105, 443, 1975.
538. Furst, G. T., Wachsman, M. A., Pieroni, J., White, J. G., and Moriconi, E. J., Concerted cycloaddition of chlorosulfonyl isocyanate to alpha-pinene stepwise rearrangement of the beta-lactam cycloadduct to a gamma-lactam, *Tetrahedron*, 29, 1675, 1973.
539. Montaudo, G., Caccamese, S., Librando, V., and Recca, A., Structural analysis by lanthanide shift reagents. IV. Conformation of the amide group as a function of the ring size in lactams, *J. Mol. Struct.*, 27, 303, 1975.
540. Krow, G. R. and Ramey, K. C., The Cope rearrangement: substituent effects on equilibria of bridged homotropilidenes, *Tetrahedron Lett.*, 3141, 1971.
541. Bose, A. K., Manhas, M. S., Srinivasan, P. R., Chawla, H. P. S., Dayal, B., and Foley, D. A., NMR spectral studies. XIII. Titanium tetrachloride-induced shifts in the proton NMR spectra of beta-lactams, *Org. Magn. Reson.*, 8, 151, 1976.
542. Sasaki, T., Hayakawa, K., Manabe, T., and Nishida, S., Molecular design by cycloaddition reactions. 37. Peri-, stereo-, and regioselectivities in cycloaddition reactions of 7-isopropylidenebenzonorbornadiene, *J. Am. Chem. Soc.*, 103, 565, 1981.

543. Virmani, V., Srivastava, B. B. P., and Jain, P. C., A quantitative study of Eu(fod)$_3$-induced PMR shift in 1-substituted-3-benzylidene-2-pyrrolidones/piperidones and 1-substituted-2-methylenepyrrolidines/piperidines — stereochemical assignments, *Indian J. Chem. Sect. B*, 15B, 981, 1977.
544. Verma, S. M. and Prasad, R., NMR signal assignment of camphorimide using tris(dipivalomethanato)-Eu(III) as a shift reagent, *Indian J. Chem. Sect. B*, 15B, 742, 1977.
545. Bruder, A. H., Tanny, S. R., Rockefeller, H. A., and Springer, C. S., Jr., Complexes of nucleophiles with rare earth chelates. II. Self-association and adduct formation of the lanthanide tris(1,1,1,2,2,3,3-heptafluoro-7,7-dimethyl-4,6-octanedionate) chelates Pr(fod)$_3$ and Eu(fod)$_3$, *Inorg. Chem.*, 13, 880, 1974.
546. Tanny, S. R., Pickering, M., and Springer, C. S., Jr., Increasing the time resolution of dynamic nuclear magnetic resonance spectroscopy through the use of lanthanide shift reagents, *J. Am. Chem. Soc.*, 95, 6227, 1973.
547. Kornberg, N. and Kost, D., Torsional barriers in substituted N,N-dimethylcarbamates. A probe for perturbational molecular orbital analyses of amide rotation, *J. Chem. Soc. Perkin Trans. 2*, p. 1661, 1979.
548. Kleinpeter, E., Kretschmer, M., Borsdorf, R., Widera, R., and Muehlstaedt, M., Dynamic NMR study of hindered rotation at the N-C(X) bonding fragment. XII. Effect of substitutents on the carbon-nitrogen rotation barrier to variously N-substituted O-ethyl carbamates, *J. Prakt. Chem.*, 322, 793, 1980.
549. Tori, K., Yoshimura, Y., Kainosho, M., and Ajisaka, K., Evidence for the presence of contact term contribution to lanthanide induced shifts in H-1 and C-13 NMR spectra of pyridine N-oxides, *Tetrahedron Lett.*, p. 1573, 1973.
550. Sakamoto, T., Niitsuma, S., Mizugaki, M., and Yamanaka, H., Studies on pyrimidine derivatives. XIV. On the structural determination of pyrimidine N-oxides, *Chem. Pharm. Bull.*, 27, 2653, 1979.
551. Rackham, D. M., Lanthanide shift reagents. Paper 17. Equilibrium binding constants for europium shift reagent with NO, SO and PO functions, *Spectrosc. Lett.*, 13, 513, 1980.
552. ApSimon, J. W. and Cooney, J. D., Structure and nuclear magnetic resonance spectrum of N-nitrosocamphidine, *Can. J. Chem.*, 49, 2377, 1971.
553. Fraser, R. R. and Wigfield, Y. Y., Effects of stereochemistry on the stability of alpha-nitrosamino carbanions, *Tetrahedron Lett.*, p. 2515, 1971.
554. Forrest, T. P., Hopper, D. L., and Ray, S., Conformational analysis by lanthanide induced shifts. N-nitrosopiperidines, *J. Am. Chem. Soc.*, 96, 4286, 1974.
555. Perry, R. A. and Chow, Y. L., Association of a europium shift reagent with dialkylnitrosamines, *Can. J. Chem.*, 52, 315, 1974.
556. Hoesch, L. and Weber, H. P., Crystal structure of cis-(4′-methyl-N,N,O)-azoxybenzene, *Helv. Chim. Acta*, 60, 3015, 1977.
557. Snyder, J. P., Bandurco, V. T., Darack, F., and Olsen, H., Cis-azoxyalkanes. IV. Preparation and nuclear magnetic resonance spectra, *J. Am. Chem. Soc.*, 96, 5158, 1974.
558. Bearden, W. H., Davis, R., Willcott, M. R., III, and Snyder, J. P., cis-Alkoxyalkenes. XI. Lanthanide induced nuclear magnetic resonance shifts; metal complexation at nitrogen, *J. Org. Chem.*, 44, 1974, 1979.
559. Masson, M. A., Bouchy, A., Roussy, G., Serratrice, G., and Delpuech, J. J., Microwave spectroscopy and nuclear magnetic resonance study of the molecular geometry of 2-fluorophenylisocyanate, *J. Mol. Struct.*, 68, 307, 1980.
560. Berlin, K. D. and Rengaraju, S., A study of syn/anti oxime ratios from the paramagnetic-induced shifts in the proton magnetic resonance spectra using tris(dipivalomethanato)europium(III), *J. Org. Chem.*, 36, 2912, 1971.
561. Wolkowski, Z. W., Cassan, J., Elegant, L., and Azzaro, M., Etude par resonance magnetique du proton de quelques oximes terpeniques, *C. R. Acad. Sci. Paris*, 272, 1244, 1971.
562. Wolkowski, Z. W., Tris (dipivalomethanato) europium-induced shifts in PMR oximes, *Tetrahedron Lett.*, p. 825, 1971.
563. Singh, A. K. and Verma, S. M., Stereochemical assignment of camphoroxime by PMR spectroscopy using tris(dipivalomethanato)europium(III), *Indian J. Chem.*, 208, 33, 1981.
564. Fraser, R. R., Capoor, R., Bovenkamp, J. W., Lacroix, B. V., and Pagotto, J., The use of shift reagents and C-13 nuclear magnetic resonance for assignment of stereochemistry to oximes, *Can. J. Chem.*, 61, 2616, 1983.
565. Wilson, S. R. and Turner, R. B., Pyrazoline stereochemistry: use of nuclear magnetic resonance shift reagents with azo-compounds, *J. Chem. Soc. Chem. Commun.*, p. 557, 1973.
566. Vitt, G., Hadicke, E., and Quinkert, G., Die vier stereoisomeren 3,8-diphenyl-1,2-diaza-1-cyclooctene, *Chem. Ber.*, 109, 578, 1976.
567. Trost, B. M., Scudder, P. H., Cory, R. M., Turro, N. J., Ramamurthy, V., and Katz, T. J., 1,2-diaza-2,4,6-cyclooctatetraene, *J. Org. Chem.*, 44, 1264, 1979.

568. Franck-Neumann, M. and Sedrati, M., Deplacements chimiques induits par complexation a l'europium en serie delta-pyrazoline, *Org. Magn. Reson.*, 5, 217, 1973.
569. Saleem, L. M. N., trans-cis Isomerization of Schiff's bases (N-benzylideneanilines) on addition of lanthanide shift reagents, *Org. Magn. Reson.*, 19, 176, 1982.
570. Nielsen, K. and Kjaer, A., Lanthanide-induced shifts in proton magnetic resonance spectra of some simple di-, tri-, and tetracoordinate sulfur compounds, *Acta Chem. Scand.*, 26, 852, 1972.
571. Kodama, Y., Nishihata, K., and Nishio, M., Conformation of some aliphatic sulphoxides as evidenced by computer simulation of lanthanide induced shifts, *J. Chem. Res. (S)*, p. 102, 1977.
572. Fraser, R. R. and Wigfield, Y. Y., Assignment of configuration to sulphoxides by use of the "shift reagent", *J. Chem. Soc. Chem. Commun.*, p. 1471, 1970.
573. Andersen, K. K. and Uebel, J. J., Use of trisdipivalomethanatoeuropium(III) as a nuclear magnetic resonance chemical shift reagent in the analysis of the spectra of some sulfoxides, *Tetrahedron Lett.*, p. 5253, 1970.
574. Caccamese, S., Finocchiaro, P., Maravigna, P., Montaudo, G., and Recca, A., Steric effects in the interaction of lanthanide shift reagents with bidentate substrates. A stochastic model for the simulation of lanthanide induced shifts in molecules bearing the sulphonyl group, *Gazz. Chim. Ital.*, 107, 163, 1977.
575. Smith, D. J. H., Finlay, J. D., Hall, C. R., and Uebel, J. J., Conformational study of 3-substituted thietane 1-oxides. Lanthanide shift reagent approach, *J. Org. Chem.*, 44, 4757, 1979.
576. Kodama, Y., Zushi, S., Nishihata, K., Nishio, M., and Uzawa, J., A general preference for gauche alkyl-phenyl interactions. The use of lanthanide shift reagents in determining the preferred conformations of some alkyl 1-phenylethyl sulphoxides, *J. Chem. Soc. Perkin Trans. 2*, p. 1306, 1980.
577. Konovalov, E. V., Lavrenyuk, T. Y., Egorov, Y. P., Gaidamaka, S. N., and Aleksandrov, A. M., Study of compounds containing asymmetric phosphorus and sulfur atoms by an NMR method using tris(1,1,1,2,2,3,3-heptafluoro-7,7-dimethyloctane-4,6-dionato)-europium, *Teor. Eksp. Khim.*, 13, 407, 1977.
578. Lavrenyuk, T. Y., Turov, A. V., Kornilov, M. Y., Boldeskul, I. E., and Bezmenova, T. E., Study of substituted 2-thiolene and 3-thiolene 1,1-dioxides by proton NMR using a lanthanide shift reagent, *Khim. Geterotsikl. Soedin.*, p. 314, 1984.
579. Cazaux, L., Chassaing, G., and Maroni, P., Deplacements chimiques induits en RMN par un reactif lanthanideque I-application a l'analyse conformationnelle de sulfites cycliques, *Org. Magn. Reson.*, 8, 461, 1976.
580. Wood, G., Buchanan, G. W., and Miskow, M. H., Conformational studies of substituted trimethylene sulfites by proton magnetic resonance, *Can. J. Chem.*, 50, 521, 1972.
581. Albriktsen, P. and Thorstenson, T., NMR experiments on cyclic sulfites. VII. Lanthanide induced chemical shifts in trimethylene sulfites with respect to the orientation of the S = O bond, *Acta Chem. Scand.*, A30, 763, 1976.
582. Dale, A. J., Lanthanide induced chemical shifts in 5,5-dimethyl-1,3,2-dioxaphosphorinan-2-ones with respect to the conformational preference of the 2-substituent, *Acta Chem. Scand.*, 26, 298, 1972.
583. Dale, A. J., Lanthanide-induced chemical shifts in proton NMR spectra of 5,5-dimethyl-2-oxo-1,3,2-dioxaphosphorinanes, *Acta Chem. Scand.*, B30, 255, 1976.
584. Tangerman, A. and Zwanenburg, B., The effect of shift reagent on the conformational equilibrium of 3,3'-disubstituted diphenylsulfines, *Tetrahedron Lett.*, p. 5195, 1973.
585. Tangerman, A. and Zwanenburg, B., Assignment of configuration to sulfines by means of shift reagent and ASIS, *Tetrahedron Lett.*, p. 79, 1973.
586. Legler, L. E., Jindal, S. L., and Murray, R. W., NMR spectra of thiolsulfinates containing heterosteric groups. Use of a chemical shift reagent, *Tetrahedron Lett.*, p. 3907, 1972.
587. Gavezzotti, E., Tempesti, E., Marchello, G., Airoldi, G., Petrillo, V., and Giuffre, L., XXX/6-Hexadecyl-1,2-oxathiane 2,2-dioxide: structural assignment using a lanthanide shift reagent, *Gazz. Chim. Ital.*, 106, 1111, 1976.
588. Hashimoto, S. and Nagai, T., The use of NMR shift reagents for analysis of surface-active agents containing sulfonic groups. II. Determination of sodium alpha-olefin sulfonate by NMR., *Bunseki Kagaku*, 26, 10, 1977.
589. Hashimoto, S., Asano, K., and Nagai, T., The use of NMR shift reagents for the analysis of surface-active agents containing sulfonic groups. I. Determination of sodium alkanesulfonate by an NMR method, *Bunseki Kagaku*, 26, 5, 1977.
590. Lowe, G. and Salamone, S. J., Application of a lanthanide shift reagent in 0-17 NMR spectroscopy to determine the stereochemical course of oxidation of cyclic sulphite diesters to cyclic sulphate diesters with ruthenium tetroxide, *J. Chem. Soc. Chem. Commun.*, p. 1392, 1983.
591. Tisnes, P., Maroni, P., and Cazaux, L., Deplacements chimiques induits en RMN par un reactif lanthanidique. II. Analyse conformationnelle de sulfinamates cycliques: oxo-2-oxathiazannes-1,2,3 et oxo-2 benzo-5,6 dihydro-3,4 oxathiazines-1,2,3, *Org. Magn. Reson.*, 12, 490, 1979.

592. Bauman, R. A., Differentiation of cis and trans isomers of thionocarbamate esters by use of tris(dipivalomethanato)europium(III), *Tetrahedron Lett.,* p. 419, 1971.
593. Morrill, T. C., Opitz, R. J., and Mozzer, R., Lanthanide shift reagent effects on NMR spectra of organosulfur compounds-HSAB theory, *Tetrahedron Lett.,* p. 3715, 1973.
594. Walter, W., Becker, R. F., and Thiem, J., Configurational assignment of thioamides using pseudo contact shifts induced by Eu(dpm)$_3$, *Tetrahedron Lett.,* p. 1971, 1971.
595. Fritz, H., Hug, P., Lawesson, S. O., Logemann, E., Pedersen, B. S., Sauter, H., Scheibye, S., and Winkler, T., Studies on organophosphorus compounds. XXVI. Synthesis and C-13 NMR spectra of N,N-dialkyl thioamides, *Bull. Soc. Chim. Belg.,* 87, 525, 1978.
596. Mandel, F. S., Cox, R. H., and Taylor, R. C., Interaction of lanthanide shift reagents with organophorphorus substrates, *J. Magn. Reson.,* 14, 235, 1974.
597. Gerken, T. A. and Ritchey, W. M., Lanthanide-induced proton, carbon, and phosphorus NMR shifts for a series of organophosphorus compounds, *J. Magn. Reson.,* 24, 155, 1976.
598. Borowitz, I. J., Yee, K. C., and Crouch, R. K., Determination of stereochemistry in vinyl phosphorylated species by nuclear magnetic resonance shift reagents. Revised mechanistic pathways for the Perkow reaction, *J. Org. Chem.,* 38, 1713, 1973.
599. Galakhov, I. V., Verenikin, G. V., and Knunyants, I. L., Reaction of lanthanide shift reagents with organophosphorus compounds, *Zh. Vses. Khim. O-va,* 23, 234, 1978; *Chem. Abstr.,* 89, 59364n.
600. Yee, K. C. and Bentrude, W. G., Six-membered ring phosphorus heterocycles. Use of europium shift reagent in the analysis of the PMR spectrum of trans-2-methyl-5-t-butyl-2-oxo-1,3,2-dioxaphosphorinane, *Tetrahedron Lett.,* p. 2775, 1971.
601. Bentrude, W. G., Tan, H. W., and Yee, K. C., Conformations of saturated phosphorus heterocycles. Effects of europium dipivaloylmethane and europium heptafluorodimethyloctanedione on conformational equilibria of 2-substituted 5-tert-butyl-2-oxo-1,3,2-dioxaphosphorinanes, *J. Am. Chem. Soc.,* 94, 3264, 1972.
602. Finocchiaro, P., Recca, A., Bentrude, W. G., Tan, H. W., and Yee, K. C., Conformational analysis by lanthanide shift reagents. Applications of a topological approach to the study of the conformational equilibria of 2-substituted 5-tert-butyl-2-oxo-1,3,2-dioxaphosphorinanes, *J. Am. Chem. Soc.,* 98, 3537, 1976.
603. Kashman, Y. and Awerbouch, O., Complexing of phosphoryl-containing compounds by tris(dipivalomethano)europium, *Tetrahedron,* 27, 5593, 1971.
604. Berkova, G. A., Vafina, G. S., Zakharov, V. I., Mashlyakovskii, L. N., and Ionin, B. I., NMR study of organophosphorus compounds using lanthanide shift reagents. Stereochemistry of 1,3-alkadienyl phosphonates, *Zh. Obshch. Khim.,* 51, 745, 1981.
605. Berkova, G. A., Zakharov, V. I., Ionin, B. I., and Petrov, A. A., NMR study of organophosphorus compounds using lanthanide shift reagents. Rotational isomerism of alkenephosphonic acid derivatives, *Zh. Obshch. Khim.,* 48, 66, 1978.
606. Berkova, G. A., Shekhade, A. M., Zakharov, V. I., Ionin, B. I., and Petrov, A. A., Analysis of rotational isomerism of 1,3-alkadienephosphonates by NMR method using lanthanide shift reagents, *Zh. Obshch. Khim.,* 47, 957, 1977.
607. Berkova, G. A., Zakharov, V. I., Smirnov, S. A., Markovin, N. V., and Ionin, B. I., Contact contribution to proton and phosphorus-31 lanthanide shifts in NMR spectra of phosphonates, *Zh. Obshch. Khim.,* 47, 1431, 1977.
608. Wetzel, R. B. and Kenyon, G. L., 1-Phosphabicyclo[2.2.1]heptane 1-oxide and 1-phosphabicyclo[2.2.2]octane 1-oxide. Syntheses and some properties relative to their monocyclic and acyclic analogs, *J. Am. Chem. Soc.,* 96, 5189, 1974.
609. Grayson, J. I., Norrish, H. K., and Warren, S., Synthesis of phosphindoline oxides, tetrahydrophosphinoline oxides, and related compounds by cyclisation of allyl- and vinyl-phosphine oxides, *J. Chem. Soc. Perkin Trans. 1,* p. 2556, 1976.
610. Cuddy, B. D., Treon, K., and Walker, B. J., The use of lanthanide shift reagents in structural studies of phosphine oxides, *Tetrahedron Lett.,* p. 4433, 1971.
611. Corfield, J. R. and Trippett, S., Assignment of configuration to 2,2,3,4,4-pentamethylphosphetan oxides using tris(dipivalomethanato)europium(III), *J. Chem. Soc. Chem. Commun.,* p. 721, 1971.
612. Quaegebeur, J. P., Belald, S., Chachaty, C., and Le Ball, H., Tertiary phosphine oxide complexes with lanthanide beta-diketonates. NMR-relaxation and pseudocontact-shift studies, *J. Phys. Chem.,* 85, 417, 1981.
613. Houalla, D., Sanchez, M., Wolf, R., Bois, M., Gagnaire, D., and Robert, J. B., Stéréochimie en pentacoordination. Apport des réactifs inducteurs de déplacement chimique à l'étude de spirophosphoranes à liaison P-H, *Org. Magn. Reson.,* 6, 340, 1974.
614. Balitskii, Y. V., Kornilov, M. Y., and Gololobov, Y. G., Synthesis and PMR study of 1,3,2-oxazaphospholine 2-oxides using lanthanide shift reagents, *Zh. Obshch. Khim.,* 47, 227, 1977.

615. Turov, A. V., Povolotskii, M. I., Balitskii, Y. V., Kornilov, M. Y., and Boldeskul, I. E., Study of the reaction of 2-methyl-2-oxo-3,5-di-tert-butyl-delta-1,3,2-oxazapholine with lanthanide shift reagents, *Zh. Obshch. Khim.*, 53, 2199, 1983.
616. Yatsimirskii, K. B., Bidzilya, V. A., Davidenko, N. K., and Golovkova, L. P., Estimation of limiting shifts and stability constants of adducts of lanthanide shift reagents with hexamethylphosphoramide, *Teor. Eksp. Khim.*, 10, 115, 1974.
617. Bidzilya, V. A., Davidenko, N. K., Golovkova, L. P., and Yatsimirskii, K. B., Composition and stability of adducts of the lanthanide shift reagent $Pr(fod)_3$ with hexamethylphosphoramide, *Teor. Eksp. Khim.*, 11, 388, 1975.
618. Bidzilya, V. A., Davidenko, N. K., and Golovokova, L. P., The use of paramagnetic shifts, induced by lanthanide shift reagents, to study the kinetics of ligand exchange, *Teor. Eksp. Khim.*, 11, 687, 1975.
619. Bidzilya, V. A., Davidenko, N. K., Golovokova, L. P., and Yatsimirskii, K. B., Study of the kinetics of ligand exchange by the paramagnetic shifts induced by lanthanide shift reagents, *Dokl. Akad. Nauk SSSR*, 225, 842, 1975.
620. San Filippo, J., Jr., Nuzzo, R. G., and Romano, L. J., Application of lanthanide shift reagents to alkyl fluorides, *J. Am. Chem. Soc.*, 97, 2546, 1975.
621. Graves, R. E. and Rose, P. I., Application of lanthanide induced shift reagents to organic cations by outer sphere complexation, *J. Chem. Soc. Chem. Commun.*, p. 630, 1973.
622. Lipkowitz, K. B., Chevaliar, T., Mundy, B. P., and Theodore, J. J., Organic salts as powerful lanthanide shift donors, *Tetrahedron Lett.*, 21, 1297, 1980.
623. Seeman, J. I. and Bassfield, R. L., Quaternary ammonium halides as powerful lanthanide shift donors, *J. Org. Chem.*, 42, 2337, 1977.
624. Montaudo, G., Kruk, G., and Verhoeven, J. W., Conformational dependence of solvent and lanthanide induced shifts in ortho-substituted N-benzylpyridinium ions, *Tetrahedron Lett.*, p. 1845, 1974.
625. Balaban, A. T., Lanthanide-induced shifts of pyridines, pyridinium and pyrylium salts, *Tetrahedron Lett.*, p. 5055, 1978.
626. Bassfield, R. L., The interaction of lanthanide shift reagents with cationic sites: A H-1 and C-13 NMR study of the solution geometry of nicotine N-methiodide, *J. Am. Chem. Soc.*, 105, 4168, 1983.
627. Caret, R. L. and Vennos, A. N., Lanthanide-induced chemical shifts of sulfonium salts, *J. Org. Chem.*, 45, 361, 1980.
628. Caret, R. L., Vennos, A., Zapf, M., and Uebel, J. J., Sulfonium salts as lanthanide shift donors in PMR and CMR spectroscopy, *Tetrahedron Lett.*, 22, 2085, 1981.
629. Lefevre, F., Rabiller, C., Mannschreck, A., and Martin, G. J., Cation-anion association in iminium salts. Use of europium complexes as H-1 NMR auxiliary compounds, *J. Chem. Soc. Chem. Commun.*, p. 942, 1979.
630. Wenzel, T. J. and Zaia, J., Lanthanide tetrakis(beta-diketonates) as effective NMR shift reagents for organic salts, *J. Org. Chem.*, 50, 1322, 1985.
631. Lindoy, L. F. and Moody, W. E., Nuclear magnetic resonance studies of metal complexes using lanthanide shift reagents. Lanthanide induced shifts in the spectra of oxygen-donor ligands coordinated to nickel(II), *J. Am. Chem. Soc.*, 97, 2275, 1975.
632. Lindoy, L. F., Lip, H. C., Louie, H. W., Drew, M. G. B., and Hudson, M. J., Interaction of lanthanide shift reagents with co-ordination complexes; direct observation of nuclear magnetic resonance signals for free and complexed tris(pentane-2,4-dionato)cobalt(III) at ambient temperatures, and X-ray crystal and molecular structure of the 1:1 adduct formed, *J. Chem. Soc. Chem. Commun.*, p. 778, 1977.
633. Lindoy, L. F. and Louie, H. W., Nuclear magnetic resonance study of the interaction of lanthanide shift reagent with tris(beta-diketonato)cobalt(III) complexes. Kinetics of adduct formation involving slow chemical exchange at ambient temperature, *J. Am. Chem. Soc.*, 101, 841, 1979.
634. Lindoy, L. F., Nuclear magnetic resonance studies of coordination metal complexes using lanthanide shift reagents, *Coord. Chem. Rev.*, 48, 83, 1983.
635. Hirayama, M. and Kitami, K., Observation of Al-27 NMR paramagnetic shifts (large contact shift contributions) induced by lanthanoid shift-reagents in tris(acetylacetonato)aluminum(III), *J. Chem. Soc. Chem. Commun.*, p. 1030, 1980.
636. Hirayama, M., Kawamata, Y., Fuji, Y., and Nakano, Y., A Co-59 nuclear magnetic resonance study of the interaction of lanthanoid shift reagents with tris(beta-diketonato)cobalt(III) complexes, *Bull. Chem. Soc. Jpn.*, 55, 1798, 1982.
637. Hirayama, M. and Kawamata, Y., Lanthanoid-induced Co-59 NMR shifts of the adducts of tris(acetylacetonato)Co(III) with $Ln(fod)_3$ and $Ln(dpm)_3$ in solutions, *Chem. Lett.*, p. 1295, 1980.

638. Hirayama, M. and Sasaki, Y., Observation of lanthanoid-induced Pt-195 shifts for the adduct of bis(acetylacetonato)Pt(II) with tris(8,8,8,7,7,6,6-heptafluoro-2,2-dimethyloctane-3,5-dionato)Pr(III) in CDCl₃ solution, *Chem. Lett.*, p. 195, 1982.
639. Graddon, D. P., Muir, L., Lindoy, L. F., and Louie, H. W., Calorimetric titration and hydrogen-1 nuclear magnetic resonance studies of adduct formation between cobalt(III) beta-diketonates and a lanthanide shift reagent, *J. Chem. Soc. Dalton Trans.*, p. 2596, 1981.
640. Lindoy, L. F. and Louie, H. W., Dynamic H-1 nuclear magnetic resonance line-broadening study of adduct formation between azidocobalt(III) complexes containing organic ligands and lanthanide shift reagent, *Inorg. Chem.*, 20, 4186, 1981.
641. Paul, J., Schlogl, K., and Silhan, W., Die anwendung von tris(dipivalomethanato)europium in der NMR-spektroskopie von metallocenen, *Monatsch. Chem.*, 103, 243, 1972.
642. Foreman, M. I. and Leppard, D. G., NMR spectra of organometallic compounds: use of a lanthanide shift reagent, *J. Organomet. Chem.*, 31, C31, 1971.
643. Dickson, R. S., Johnson, S. H., and Rae, I. D., The lanthanide-shifted NMR spectra of some cyclopentadienone-cobalt complexes, *Aust. J. Chem.*, 28, 1681, 1975.
644. Willcott, M. R., Bearden, W. H., Davis, R. E., and Pettit, R., An NMR study of the conformation of certain alcohol derivatives of tricarbonyl(diene)iron compounds: a critique, *Org. Magn. Reson.*, 7, 557, 1975.
645. Sakurai, H. and Hayashi, J., Stereochemical assignment of a silanol complex by use of a lanthanide shift reagent, *J. Organomet. Chem.*, 63, C7, 1973.
646. Faller, J. W. and Johnson, B. V., Organometallic conformational equilibria. XIX. Lanthanide shift reagent studies of rotational barriers and preferred orientations in phosphine derivatives of cyclopentadienyl complexes of iron and nickel, *J. Organomet. Chem.*, 96, 99, 1975.
647. Marks, T. J., Porter, R., Kristoff, J. S., and Shriver, D. F., Organometallic aspects of shift reagent chemistry, in *Nuclear Magnetic Resonance Shift Reagents*, Sievers, R. E., Ed., Academic Press, New York, 1973, 247.
648. Schloegl, K. and Schoelm, R., Stereochemistry of metallocenes. XXXXV. Conformational analysis of tricarbonylchromium complexes of diphenic acid and their derivatives by LIS method, *J. Organomet. Chem.*, 194, 69, 1980.
649. Voronov, V. K., Characteristics of employing lanthanide chelates in NMR spectroscopy of weak Lewis bases, *Izv. Akad. Nauk SSSR Ser. Khim.*, p. 2110, 1976.
650. Lycka, A., Snobl, D., and Vencl, J., The effect of tris(dipivalomethane)europium chelate on the H-1 NMR spectra of tertiary silanols, *J. Organomet. Chem.*, 149, C37, 1978.
651. Yastrebov, V. V., Krylov, A. V., Yashtulov, N. A., and Nikishina, I. S., Proton NMR shifts of diorganyldimethoxysilanes with paramagnetic shift reagents and the possible structure of associates, *Tr. Mosk. In-ta Tonk. Khim. Tekhnol.*, 9, 19, 1979; *Chem. Abstr.*, 92, 110157p.
652. Yastrebov, V. V., Krylov, A. V., and Yashtulov, N. A., Effect of lanthanide shift reagents on the proton NMR spectra of vinyl derivatives of silicon, *Koord. Khim.*, 6, 1398, 1980.
653. Yastrebov, V. V., Krylov, A. V., Nikishina, I. S., Zueva, G. Y., and Bystrov, L. V., Reaction of phenyltrimethoxysilanes with lanthanide shift reagents, *Izv. Akad. Nauk SSSR Ser. Khim.*, p. 1429, 1980.
654. Yastrebov, V. V., Yashtulov, N. A., Zueva, G. Y., and Lipatova, G. V., NMR paramagnetic shifts in dimethyldialkoxygermanes in the presence of lanthanide shift reagents, *Izv. Akad. Nauk SSSR Ser. Khim.*, p. 119, 1983.
655. Sidorkin, V. F., Keiko, V. V., Pestunovich, V. A., Kalinina, N. A., and Voronkov, M. G., Structure and nature of induced shifts in NMR of complexes of silatranes with paramagnetic reagents, *Zh. Obshch. Khim.*, 53, 581, 1983.
656. Yastrebov, V. V., Yashtulov, N. A., Sheludyakov, V. D., and Lakhtin, V. G., Characteristics of the interaction of beta-chlorovinyltrimethoxysilanes with lanthanide shift reagents, *Koord. Khim.*, 7, 1659, 1981.
657. Yastrebov, V. V. and Nikishina, I. S., Study of NMR in anisoles using shifting reagents, *Zh. Vses. Khim. O-va*, 27, 96, 1982; *Chem. Abstr.*, 96, 198954n.
658. Lycka, A., Snobl, D., and Vencl, J., Determination of hydroxyl groups bonded to silicon atoms in siloxane preparations, *Czech. CS*, 191, 118, 1981.
659. Nikishina, I. S., Krylov, A. V., and Mikishina, I. S., Simple method of determining the characteristics of complexes formed with paramagnetic shift reagents, *Zh. Vses. Khim. O-va*, 27, 97, 1982; *Chem. Abstr.*, 96, 173167v.
660. Yastrebov, V. V., Nikishina, I. S., and Krylov, A. V., Characteristics of the paramagnetic effect of lanthanum shift reagents on the paramagnetic resonance in monoalkoxymethylphenylsilanes, *Zh. Obshch. Khim.*, 51, 1336, 1981.
661. Voronov, V. K., Keiko, V. V., Baryshok, V. P., D'yakov, V. M., and Voronkov, M. G., Coordination of 1-methylsilatrane with lanthanide shift reagents, *Dokl. Akad. Nauk SSSR*, 236, 147, 1977.

662. Yastrebov, V. V., Chernyshev, A. I., and Zhavoronkov, I. P., Study of the characteristics of associates of trimethylmethoxy derivatives of silicon, germanium, and carbon with the paramagnetic shift reagent Eu(fod)$_3$ by a proton NMR method, *Koord. Khim.*, 4, 697, 1978.

663. Yastrebov, V. V., Zhavoronkov, I. P., and Tyurikov, V. A., Preferred internal conformation of dimethyl(chloromethyl)methoxysilane in associates with lanthanide shift reagents, *Zh. Strukt. Khim.*, 20, 143, 1979.

664. Gielen, M., Goffin, N., and Topart, J., Organometallic compounds. XXXIII. Influence of tris(dipivalomethanato)europium(III) on the PMR spectra of, and evidence for intramolecular complexation in functionally-substituted organotin compounds, *J. Organomet. Chem.*, 32, C38, 1971.

665. McArdle, P., Wood, J. O., and Lee, E. E., Application of lanthanide shift reagents to H-1 and C-13 NMR spectra of some carbohydrate derivatives, *Carbohydr. Res.*, 69, 39, 1979.

666. Lala, A. K. and Kulkarni, A. B., [Eu(dpm)$_3$]-induced shifts in steroids. I. A study of bifunctional molecules, *Indian J. Chem.*, 12, 926, 1974.

667. Rodewald, W. J., Wielogorski, Z. A., and Borodziewicz, W. J., Computer-assisted analysis of lanthanide-induced shifts for bifunctional steroids. Ring B conformation of 3β-hydroxy-β-homocholestan-7,8α-lactone, *Biol. Soc. Quim. Peru*, 47, 64, 1981.

668. Farid, S. and Scholz, K. H., Ring expansion of hydroxyoxetanes to dihydrofurans, *J. Org. Chem.*, 37, 481, 1972.

669. Farid, S., Ateya, A., and Maggio, M., Lanthanide-induced shifts in H-1 nuclear magnetic resonance: the significance of the angle term in the geometric factor of pseudocontact shift, *J. Chem. Soc. Chem. Commun.*, p. 1285, 1971.

670. Chizhov, O. S., Shashkov, A. S., Usov, A. I., and Shienok, A. I., Use of paramagnetic shifting reagents to study some 3,6-anhydropyranosides, *Izv. Akad. Nauk SSSR Ser. Khim.*, p. 2591, 1975.

671. Chappell, G. S., Grabowski, B. F., Sandmann, R. A., and Yourtee, D. M., Simplified NMR spectra of bifunctional tropanes induced by the paramagnetic shift reagent tris(dipivalomethanato)europium(III), *J. Pharm. Sci.*, 62, 414, 1973.

672. Wiegrebe, W., Fricke, J., Budzikiewicz, H., and Pohl, L., Synthese eines 3-phenylisochromans, *Tetrahedron*, 28, 2849, 1972.

673. Aversa, M. C. and Giannetto, P., Synthesis and stereochemical characterization of 1,2,7,11b-tetrahydropyrrolo[1,2-d][1,4]benzodiazepine-3,6-(5H)-diones, obtained via Raney nickel hydrogenation of tetrahydroisoxazolo[2,3-d][1,4]benzodiazepinones, *J. Chem. Soc. Perkin Trans. 2*, p. 81, 1984.

674. Kingsbury, C. A., Asymmetric synthesis of diastereomeric hydroxy sulfides, sulfoxides, and sulfones by condensation and oxidation reactions, *J. Org. Chem.*, 37, 102, 1972.

675. Porter, G. B. and Simpson, J., Effect of shift reagents on NMR coupling constants, *Angew. Chem.*, 90, 51, 1978.

676. Richer, J. C., Beljean, M., and Pays, M., Etudes d'alkylidenehydrazono-2 methyl-3 dihydro-2,3 benzothiazoles par RMN H-1 en presence d'Eu(fod)$_3$, *Org. Magn. Reson.*, 10, 226, 1977.

677. Lafuma, F. and Quivoron, C., Etude des spectres de resonance magnetique nucleaire de quelques acetal-alcools derives du tetrahydropyranne, en presence de tri-(dipivalomethanato)d'europium, *C. R. Acad. Sci. Paris*, 276, 359, 1973.

678. Armitage, I. and Hall, L. D., Some observation on the chemical shift changes induced by europium tris(dipivalomethane), *Chem. Ind.*, 28, 1537, 1970.

679. McArdle, P., O'Reilly, J. P., Simmie, J., and Lee, E. E., Quantitative treatment of lanthanide-induced shifts for some carbohydrate systems where chelation is observed, *Carbohydr. Res.*, 90, 165, 1981.

680. Meyer, A. L. and Turner, R. B., An interesting synthesis of 3-methoxy-2,6-dimethylphenethyl alcohol, *Tetrahedron*, 27, 2609, 1971.

681. Kearney, P. C., Plimmer, J. R., Williams, V. P., Klingebiel, U. I., Isensee, A. R., Laanio, T. L., Stolzenberg, G. E., and Zaylskie, R. G., Soil persistence and metabolism of N-sec-butyl-4-tert-butyl-2,6-dinitroaniline, *J. Agric. Food Chem.*, 22, 856, 1974.

682. Mitchenko, Y. I., Fenin, V. A., and Khar'kov, S. N., Study of heptyl alcohol and monoheptyl glycol ether conormations by NMR with paramagnetic shift reagents, *Zh. Struckt. Khim.*, 18, 808, 1977.

683. Ribeiro, A. A. and Dennis, E. A., A carbon-13 and proton nuclear magnetic resonance study on the structure and mobility of nonionic alkyl polyoxyethylene ether micelles, *J. Phys. Chem.*, 81, 957, 1977.

684. Wineburg, J. P. and Swern, D., NMR chemical shift reagents in structural determination of lipid derivatives. III. Methyl ricinoleate and methyl 12-hydroxystearate, *J. Am. Oil Chem. Soc.*, 50, 142, 1973.

685. Garcia, G. A., Diaz, E., and Crabbe, P., Lanthanide-induced shifts in proton nuclear magnetic resonance spectra of prostaglandins, *Chem. Ind.*, 16, 585, 1973.

686. Strain, H. H., Svec, W. A., Aitzetmuller, K., Grandolfo, M. C., Katz, J. J., Kjosen, H., Norgard, S., Liaaen-Jensen, S., Haxo, F. T., Wegfahrt, P., and Rapoport, H., The structure of peridinin, the characteristic dinoflagellate carotenoid, *J. Am. Chem. Soc.*, 93, 1823, 1971.
687. Smith, J. C., Russ, P., Cooperman, B. S., and Chance, B., Synthesis, structure determination, spectral properties, and energy-linked spectral responses of the extrinsic probe oxonol V in membranes, *Biochemistry*, 15, 5094, 1976.
688. Dambska, A., Janowski, A., and Razniewska, G., Use of lanthanide shift reagents for compounds containing intramolecular hydrogen bonds, *Mater. Ogolnopol. Semin. "Magn. Rezon Jad. Jego Zastosow."*, 13th, 203, 1980; *Chem. Abstr.*, 96, 162017v.
689. Diaz, E., Guzman, A., Cruz, M., Mares, J., Rameirz, D. J., and Joseph-Nathan, P., Lanthanide induced shifts as a conformational probe for o,o,o',o'-tetrasubstituted diphenyls, *Org. Magn. Reson.*, 13, 180, 1980.
690. Hofer, O., Griengl, H., and Nowak, P., Ein modell aur beschreibung der lanthaniden-induzierten verschiebungen bei zweizahnigen liganded-1,3-diole, *Monatsch. Chem.*, 109, 21, 1978.
691. Gualtieri, F., Melchiorre, C., Giannella, M., and Pigini, M., Synthesis and identification by shift reagents of isomeric 2-methyl-2-n-propylcyclopentane-1,3-diols, *J. Org. Chem.*, 40, 2241, 1975.
692. Burgot, J. L., Masson, J., and Vialle, J., Homo-1,4-addition of Grignard reagents to β-oxothiocarbonyl compounds cis cyclopropane ring closure, *Tetrahedron Lett.*, p. 4775, 1976.
693. Brown, E., Dhal, R., and Casals, P. F., Stereochimie des dialcoyl-2,6 piperidinols-3, *Tetrahedron*, 28, 5607, 1972.
694. Padwa, A., Au, A., Lee, G. A., and Owens, W., Carbonyl group photochemistry via the enol form. Photoisomerization of 4-substituted 3-chromanones, *J. Am. Chem. Soc.*, 98, 3555, 1976.
695. Okutani, T., Morimoto, A., Kaneko, T., and Masuda, K., Studies on azetidine derivatives. III. Configurational assignment with a shift reagent, *Tetrahedron Lett.*, p. 1115, 1971.
696. Sugimoto, Y., Sakita, T., Moriyama, Y., Murae, T., Tsuyuki, T., and Takahashi, T., Structure of nigakialcohol, a new ionone derivative from *Picrasma ailanthoides*, planchon, *Tetrahedron Lett.*, p. 4285, 1978.
697. Rebuffat, S., Davoust, D., Giraud, M., and Molho, D., Configuration and spectral study of hydroperoxide diastereoisomers, *Bull. Mus. Natl. Hist. Nat. Sci. Phys. Chim.*, 16, 85, 1977; *Chem. Abstr.*, 89, 128982b.
698. Dombi, G., Pelczer, I., Szabo, J. A., Gondos, G., and Bernath, G., Stereochemical studies. XXXVI. Studies on cyclic 2-hydroxycarboxylic acids. VII. Proton NMR investigations of ethyl cis- and trans-2-hydroxy-1-cyclopentane-, -hexane-, -heptane-, and octanecarboxylates and their trichloroacetyl carboxamido derivatives by Eu(fod)$_3$ shift reagent, *Acta Chim. Acad. Sci. Hung.*, 104, 287, 1980.
699. Hofer, O., Conformational analysis by means of lanthanide induced shifts: the aryl-methoxy bond, *Tetrahedron Lett.*, p. 3415, 1975.
700. Shingu, T., Hayashi, T., and Inouye, H., Zur stereochemie des catalponols, ein beispiel der anwendung des verschiebungs-reagenzes Zur konfugurationsaufklarung, *Tetrahedron Lett.*, p. 3619, 1971.
701. Paquette, L. A., Lang, S. A., Jr., Porter, S. K., and Clardy, J., Hydration of hexamethyl (Dewar benzene) oxide. Stereochemical aspects of the resulting rearrangement, *Tetrahedron Lett.*, p. 3137, 1972.
702. Clark, R. D. and Untch, K. G., [2 + 2] Cycloaddition of ethyl propiolate and silyl enol ethers, *J. Org. Chem.*, 44, 248, 1979.
703. Fleming, I., Hanson, S. W., and Sanders, J. K. M., The effect of Eu(dpm)$_3$ on the NMR spectra of bifunctional compounds, *Tetrahedron Lett.*, p. 3733, 1971.
704. Butkus, E., Kadziauskas, P., and Kornilov, M. Y., Determination of the ratio of conformers of 2,6-derivatives of bicyclo[3.3.1]nonane by proton NMR spectroscopy in the presence of Eu(dpm)$_3$, *Zh. Org. Khim.*, 16, 2446, 1980.
705. Peters, J. A., Remijnse, J. D., van der Wiele, A., and van Bekkum, H., Synthesis and (non-chair) conformation of some 3α,7α-disubstituted bicyclo[3.3.1]nonanes, *Tetrahedron Lett.*, p. 3065, 1971.
706. Santilli, A. A. and Scotese, A. C., Synthesis of thiopyrano[2,3-d]pyrimidines and thieno[2,3-d]pyrimidines, *J. Heterocycl. Chem.*, 14, 361, 1977.
707. Norin, T., Sundin, S., and Theander, O., The constituents of conifer needles. VII. The configuration of dehydropinifolic acid, a diterpene acid from the needles of *Pinus silvestris* L., *Acta Chem. Scand.*, B34, 301, 1980.
708. Jankowski, K., Israeli, J., and Rabczenko, A., Two-donor atom complexes with the shift reagents, *Bull. Acad. Pol. Sci.*, 24, 453, 1976.
709. Jankowski, K., Application of shift reagents to partially deuterated alkaloid, *Bull. Acad. Pol. Sci.*, 21, 741, 1973.
710. Coxon, D. T., Price, K. R., Howard, B., and Curtis, R. F., Metabolites from microbially infected potato. I. Structure of phytuberin, *J. Chem. Soc. Perkin Trans. 1*, p. 53, 1977.

711. Bradshaw, A. P. W., Hanson, J. R., and Siverns, M., Biosynthesis of illudin sesquiterpenoids from [1,2-$^{13}C_2$]acetate, *J. Chem. Soc. Chem. Commun.*, p. 303, 1978.
712. Kleinpeter, E., Kunk, H., and Muhlstadt, M., NMR-untersuchungen an dicyclopentadienderivaten. I. H-1-NMR: anwendung von paramagnetischen verschiebungreagenzien (Eu(fod)$_3$) zur vereinfachung der spektren exo/endo-isomerer diole, *Org. Magn. Reson.*, 8, 261, 1976.
713. Duggan, J. C., Urry, W. H., and Schaefer, J., Carbon-13 pseudo-contact shifts in structure determination: the hemiketal dimer from 2-hydroxy-2-methylcyclobutanone, *Tetrahedron Lett.*, p. 4197, 1971.
714. Trost, B. M., Weber, L., Strege, P. E., Fullerton, T. J., and Dietsche, T. J., Allylic alkylation: nucleophilic attack on pi-allylpalladium complexes, *J. Am. Chem. Soc.*, 100, 3416, 1978.
715. Mompon, B. and Toubiana, R., Configuration du subsluteolide; nouveau guaianolide isole du vernonia sublutea scott elliot(composees), *Tetrahedron*, 33, 2199, 1977.
716. Gray, A. I., Waigh, R. D., and Waterman, P. G., Interactions of coumarins with a lanthanide shift reagent: determination of substitution patterns, *J. Chem. Soc. Perkin Trans. 2*, p. 391, 1978.
717. Tada, M., Moriyama, Y., Tanahashi, Y., Takahashi, T., Fukuyama, M., and Sato, K., New furanosesquiterpenes from Ligularia japonica less. Furanoeremophilane-6β,10β-diol, 10β-hydroxy-6β-methoxyfuranoeremophilane and 10β-hydroxyfuranoeremophilan-6β-yl 2'-methylbutanoate, *Tetrahedron Lett.*, p. 4010, 1971.
718. Hajek, M., Janku, J., Burkhard, J., and Vodicka, L., NMR study of thia derivatives of adamantane with shift reagents, *Coll. Czech. Chem. Commun.*, 41, 2533, 1976.
719. Hajek, M., Vodicka, L., and Hlavaty, J., Evaluation of limiting-induced shifts of disubstituted adamantane derivatives on the basis of shift additivity, *Org. Magn. Reson.*, 7, 529, 1975.
720. Cockerill, A. F. and Rackham, D. M., Elucidation of the PMR spectrum of 2-hydroxy-1-(2-hydroxyethyl)adamantane in presence of tris(dipivalomethanato)europium III, *Tetrahedron Lett.*, p. 5153, 1970.
721. Vodicka, L., Hajek, M., Ksandr, Z., and Hlavaty, J., Application of shift reagents in the study of disubstituted derivatives of adamantane by NMR spectroscopy, *Coll. Czech. Chem. Commun.*, 40, 293, 1975.
722. Duddeck, H. and Wiskamp, V., Carbon-13 nuclear magnetic resonance spectra. XVI. Lanthanide induced signal shifts of diastereotopic C-13 nuclei as a probe for a conformational equilibrium, *Org. Magn. Reson.*, 15, 361, 1981.
723. Perel'son, M. E., Kir'yanov, A. A., Sklyar, Y. E., and Vandyshev, V. V., Use of the paramagnetic shift agent Eu(dpm)$_3$ for studying the structure and stereochemistry of conferol and conferone, *Khim. Prirod. Soedin.*, p. 726, 1973.
724. Kleinpeter, V. E., Kuhn, H., and Muhlstadt, M., Endo-substituierte 7-oxatetracyclo(6,3,0,0^{2,6},0^{3,10})- und 3-oxatetracyclo(5,4,0,0^{2,9},0^{4,8})undecane, *J. Prak. Chem.*, 319, 732, 1977.
725. Stoll, M. S., Elder, G. H., Games, D. E., O'Hanlon, P., Millington, D. S., and Jackson, A. H., Isocoproporphyrin: nuclear-magnetic-resonance- and mass-spectral methods for the determination of porphyrin structure, *Biochem. J.*, 131, 429, 1973.
726. Williams, T. H., Naphthomycin, a novel ansa macrocyclic antimetabolite. Proton NMR spectra and structure elucidation using lanthanide shift reagent, *J. Antibiotics*, 28, 85, 1975.
727. Gall, R. E. and Taylor, J., Acetolytic cleavage of 2β,19-epoxy-5α-cholestane involving a 1,2-hydride shift to C2, *Aust. J. Chem.*, 30, 2249, 1977.
728. Crossley, N. S., Seco-steroids. I. 15,16-secoprogesterone, *J. Chem. Soc. C*, p. 2491, 1971.
729. Ekong, D. E. U., Okogun, J. I., and Shok, M., The meliacins (limonoids). Tris(dipivalomethanato)europium-induced upfield and downfield shifts in the nuclear magnetic resonance spectra of the meliacins, *J. Chem. Soc. Perkin Trans. 1*, p. 653, 1972.
730. Huntoon, S., Fourcans, B., Lutsky, B. N., Parish, E. J., Emery, H., Knapp, F. F., Jr., and Schroepfer, G. J., Jr., Sterol synthesis. Chemical syntheses, spectral properties, and metabolism of 5α-cholest-8(14)-en-3β,15β-diol and 5α-cholest-8(14)-en-3β,15α-diol, *J. Biol. Chem.*, 253, 775, 1978.
731. Wachter, M. P., Adams, R. E., Cotter, M. L., and Settepani, J. A., Lumi-mestranol and epi-lumi-mestranol, *Steroids*, 33, 287, 1979.
732. Cheung, H. T. A., Coombe, R. G., Sidwell, W. T. L., and Watson, T. R., Afroside, a 15β-hydroxycardenolide, *J. Chem. Soc. Perkin Trans. 1*, p. 64, 1981.
733. Noam, M., Tamir, I., Breuer, E., and Mechoulam, R., Conversion of ruscogenin into 1α- and 1β-hydroxycholesterol derivatives. Structure elucidation by computer assisted analysis of lanthanide-induced NMR shifts, *Tetrahedron*, 37, 597, 1981.
734. Lukacs, G., Lusinchi, X., Girard, P., and Kagan, H., Alcaloides steroidiques. CXVIII. Attribution en RMN des protons d'un systeme AB par complexation avec Eu(dpm)$_3$, *Bull. Soc. Chim. Fr.*, p. 3200, 1971.

735. Wang, Q. W., Chen, Z. H., and Chen, Y. Q., LSR study of the configuration of the 16α-hydroxyl-16β-methyl-delta$^{4,17(20)}$-pregnadien-3-one — interpretation of the pseudocontact model or LIS, *Hua Hsueh Hsueh Pao,* 39, 203, 1981; *Chem. Abstr.,* 95, 204268q.
736. Ius, A., Vecchio, G., and Carrea, G., The use of Eu(dpm$_3$) with bifunctional molecules. Addivity of the induced chemical shift changes in the NMR spectrum, *Tetrahedron Lett.,* p. 1543, 1972.
737. Hinckley, C. C., Klotz, M. R., and Patil, F., Applications of rare earth nuclear magnetic resonance shift reagents. III. Graphical analysis of paramagnetic shifts for systems having two coordination sites. Testosterone and 17α-methyltestosterone, *J. Am. Chem. Soc.,* 93, 2417, 1971.
738. Girard, P., Kagan, H., and David, S., Spectres de RMN de quelques derives de sucres en presence de chelates du terres rares, *Tetrahedron,* 27, 5911, 1971.
739. Gero, S. D., Horton, D., Sepulchre, A. M., and Wander, J. D., A simple method for determining the configurations of tertiary alcoholic centers in branched-chain carbohydrate derivatives by use of europium(III)-induced shifts in the H-1 nuclear magnetic resonance spectrum, *J. Org. Chem.,* 40, 1061, 1975.
740. Nieuwenhuis, J. J. and Jordaan, J. H., Configurational assignments of C-nitromethyl-C-hydroxy branched-chain carbohydrate derivatives by use of a europium shift-reagent in H-1 NMR spectroscopy, *Carbohydr. Res.,* 51, 207, 1976.
741. Girard, P., Kagan, H., and David, S., Spectroscopie RMN dans la serie des glucides en presence de chelates de terres rares, *Bull. Soc. Chim. Fr.,* p. 4515, 1970.
742. Armitage, I. and Hall, L. D., Evaluation of the binding of lanthanide shift reagents with carbohydrate derivatives, *Carbohydr. Res.,* 24, 221, 1972.
743. Horton, D. and Thomson, J. K., Application of a lanthanide shift-reagent for conformational and configurational assignment in the carbohydrate field, *J. Chem. Soc. Chem. Commun.,* p. 1389, 1971.
744. Canas-Rodriguez, A., Martinez Tobed, A., Gomez Sanchez, A., and Martin Madero, C., Synthesis of 6-azido- and 6-amino-2,3,6-trideoxy-D-erythro-hexose, *Carbohydr. Res.,* 56, 289, 1977.
745. Crump, D. R., Sanders, J. K. M., and Williams, D. H., Some applications of paramagnetic shift reagents in organic chemistry, *Tetrahedron Lett.,* p. 4949, 1970.
746. Hosoda, H., Yamashita, K., Ikegawa, S., and Nambara, T., Blockage of coordination with shift reagent by tert-butyldimethylsilylation in nuclear magnetic resonance spectroscopy and its applications to deuterium-labeled steroids, *Chem. Pharm. Bull.,* 25, 2545, 1977.
747. Hosoda, H., Yamashita, K., and Nambara, T., Steroid tert-butyl-dimethyl-silyl ether: selective blockage of coordination with lanthanide shift reagent in proton magnetic resonance spectroscopy, *Chem. Ind.,* p. 650, 1976.
748. Raber, D. J. and Propeck, G. J., Structure elucidation with lanthanide-induced shifts. 15. Blocking groups for polyfunctional compounds, *J. Org. Chem.,* 47, 3324, 1982.
749. Kashman, Y. and Awerbouch, O., The cycloaddition reaction of 3,4-dimethyl-1-thio-1-phenylphosphole with tropone, *Tetrahedron,* 29, 191, 1973.
750. Remy, D. C., Van Saun, W. A., Jr., Engelhardt, E. L., and Arison, B. H., Amitriptyline metabolites. The synthesis of (RS)-(Z) and (RS)-(E)-N-methyl-(10,11-dihydro-10-hydroxy-5H-dibenzo[a,d]cycloheptene-delta^5-propylamine, *J. Org. Chem.,* 38, 700, 1973.
751. Malherbe, R. and Dahn, H., Pi-participation in diazoketone hydrolysis. II. Exo-endo cyclization ratio in the hydrolyses of 7-syn- and 5-endo-diazoacetyl-2-norbornene, *Helv. Chim. Acta,* 60, 2539, 1977.
752. Cameron, D. W., Feutrill, G. I., and Patti, A. F., Lanthanide shifts in the H-1 NMR spectra of 1,4-naphthoquinones, *Aust. J. Chem.,* 32, 575, 1979.
753. Wilbur, D. S., Structural determination of some chloroazepine-2,5-diones using a lanthanide shift reagent, *J. Heterocycl. Chem.,* 21, 801, 1984.
754. Platzer, N., Lang, C., Basselier, J. J., and Demerseman, P., Etude de l'association d'un chelate de lanthanide Eu(fod)$_3$ avec des molecules possedant deux sites basiques proches dans l'espace. Cas de sites ethers, cetones et esters aromatiques, *Bull. Soc. Chim. Fr.,* p. 227, 1975.
755. Dunkelblum, E. and Hart, H., Conformational changes induced by europium shift reagent in medium-ring 3-methoxycycloalkanones, *J. Org. Chem.,* 42, 3958, 1977.
756. Dennis, N., Katritzky, A. R., and Rittner, R., 1,3-dipolar character of six-membered aromatic rings. XXV. 5-aryl-1-methyl-3-oxidopyridiniums, *J. Chem. Soc. Perkin Trans.,* p. 2329, 1976.
757. Cameron, D. W., Feutrill, G. I., and Patti, A. F., Lanthanide shifts in the H-1 NMR spectra of perialkoxy- and -alkylamino-quinones, *Aust. J. Chem.,* 30, 1255, 1977.
758. Cairns, H. and Hunter, D., Further studies on the structural determination of esters of 4-oxo-4H-1-benzopyran-2-carboxylic acids using the paramagnetic shift reagent Eu(fod)$_3$, *J. Heterocycl. Chem.,* 14, 245, 1977.
759. Okigawa, M., Khan, N. U., Kawano, N., and Rahman, W., Application of shift reagent, Eu(fod)$_3$, for the structural determination of isoflavone and xanthone derivatives, *Chem. Ind.,* 20, 575, 1974.

760. Okigawa, M., Kawano, N., Agil, M., and Rahman, W., The structure of ochnaflavone, a new type of biflavone and the synthesis of its pentamethyl ether, *Tetrahedron Lett.,* p. 2003, 1973.
761. Joseph-Nathan, P., Mares, J., and Ramirez, D. J., Computer assisted shift reagent PMR study of methoxyflavones, *J. Magn. Reson.,* 34, 57, 1979.
762. Tjan, S. B. and Visser, F. R., PMR shift to high field induced by tris(dipivalomethanato)europium, *Tetrahedron Lett.,* p. 2833, 1971.
763. Camps, P. and Jaime, C., A one step synthesis of 2,4-dialkoxybicyclo[3.2.1]octan-8-ones. Stereochemical assignments using the lanthanide NMR shift reagent, Eu(fod)$_3$, *Tetrahedron,* 36, 393, 1980.
764. Joseph-Nathan, P., Abramo-Bruno, D., and Torres, M. A., Structural elucidation of polymethoxyflavones from shift reagent proton NMR measurements, *Phytochemistry,* 20, 313, 1981.
765. Joseph-Nathan, P. and Santillan, R. L., H-1 NMR study of naphthoflavones with shift reagents, *Org. Magn. Reson.,* 22, 129, 1984.
766. Dittmer, D. C., Parker, E. J., and Bodwell, J. R., Large lanthanide induced chemical shifts in cobalt complexes of enethials (α,β-unsaturated thioaldehydes, *Org. Magn. Reson.,* 22, 609, 1984.
767. Camps, P. and Jaime, C., The conformation of (1R,2S,4R,5S)-2,4-dimethoxybicyclo[3.3.1]nonan-9-one and some related compounds, *Org. Magn. Reson.,* 14, 177, 1980.
768. Kessler, H. and Rosenthal, D., Z,E-isomerie aromatischer diazoketone, *Tetrahedron Lett.,* p. 393, 1973.
769. Hirota, M., Yoshida, S., Nagawa, Y., Ono, M., Endo, H., and Satonaka, H., The lanthanoid-induced shifts of proton NMR spectra and the conformations of β-(2-thienyl)acrylic esters and related compounds, *Bull. Chem. Soc. Jpn.,* 49, 3200, 1976.
770. Hart, H. and Love, G. M., Coordination sites and the use of chemical shift reagents in polyfunctional molecules, *Tetrahedron Lett.,* p. 625, 1971.
771. Gacel, G., Fournie-Zaluski, M. C., and Roques, B. P., Utilisation des deplacements chemiques induits par les lanthanides dans l'analyse conformationnelle d'heterocycles aromatiques β-carbonyles, *Org. Magn. Reson.,* 8, 525, 1976.
772. Caccamese, S., Montaudo, G., Recca, A., Fringuelli, F., and Taticchi, A., Conformational preferences and electronic effects in selenophene and tellurophene carbonyl derivatives investigated by lanthanide induced shifts, *Tetrahedron,* 30, 4129, 1974.
773. Montaudo, G., Caccamese, S., Librando, V., and Maravigna, P., Simulation of the lanthanide induced shifts. Description of a computer method and its applications to conformational equilibria of simple systems, *Tetrahedron,* 29, 3915, 1973.
774. Balaban, A. T., Gheorghiu, M. D., and Draghici, C., H-1 NMR conformational study of 2-acyl-3,5-dialkylfurans using lanthanide shift reagents, *Isr. J. Chem.,* 20, 168, 1980.
775. Gheorghiu, M. D., Draghici, C., Stanescu, L., and Avram, M., 2-Cyclobuten-1-one derivatives from acetylenes and t-butylcyanoketene, *Tetrahedron Lett.,* p. 9, 1973.
776. Gheorghiu, M. D., Ciobanu, O. B., and Elian, M., Solvent- and lanthanide-induced shifts in the exact solution of three-proton systems: case study of 2-t-butyl-2-cyano-3-phenylcyclobutanone, *J. Magn. Reson.,* 44, 330, 1981.
777. Balaban, A. T. and Gheorghiu, M. D., Solvent- and lanthanide-induced shifts in the H-1 NMR spectra of 4-acetonylidene-2,6-dimethyl-[4H]pyran and its tautomeric protonated pyrylium cations, *Rev. Roum. Chim.,* 23, 1065, 1978.
778. Bohlmann, F. and Jacob, J., Uber den einfluss von Eu(fod)$_3$ als shift-reagenz auf die H-1 NMR spektren vonzimtsaurederivaten, *Chem. Ber.,* 107, 2578, 1974.
779. Grandjean, J., Structure of averufin: a metabolite of aspergillus versicolor(vuill.) tiraboschi, *J. Chem. Soc. D,* p. 1060, 1971.
780. Okigawa, M., Kawano, N., Rahman, W., and Dhar, M. M., Paramagnetic induced shifts in the NMR spectra of flavone compounds by Eu(fod)$_3$, *Tetrahedron Lett.,* p. 4125, 1972.
781. Ghozland, F., Maroni-Bernaud, Y., and Maroni, P., Fixation en 1-4 de l'enolate chloromagnesien de la (+) pulegone sur la (+) pulegone. Configuration par RMN, en presence de sel de lanthanide, des deux dicetones diastereoisomeres obtenues, *Bull. Soc. Chim. Fr.,* p. 978, 1976.
782. Gornilov, M. Y., Turov, A. V., Fedotov, K. V., and Romanov, N. N., Mesoionic compounds with a bridging nitrogen atom. 8. Proton NMR study of the structure of condensed thiazole derivatives, *Khim. Geterostikl. Soedin.,* p. 619, 1983.
783. Dennis, N., Katritzky, A. R., Parton, S. K., Nomura, Y., Takahashi, Y., and Takeuchi, Y., 1,3-Dipolar character of six-membered aromatic rings. XVIII. Adducts from 3-oxide-1-phenylpyridinium and their quaternisation and conversion into tropone derivatives, *J. Chem. Soc. Perkin Trans. 1,* p. 2289, 1976.
784. Camps, P., Font, J., and Marques, J. M., Additivity of the lanthanide induced chemical shifts in rigid compounds with two identical functional groups, *Tetrahedron,* 31, 2581, 1975.

785. Aversa, M. C., Cum, G., Giannetto, P. D., Romeo, G., and Uccella, N., 1,2-Oxazolidines. V. Spiro and condensed biheterocyclic isomers from nitrones and 1,3-oxazolidin-2-ones, *J. Chem. Soc. Perkin Trans. 1*, p. 209, 1974.
786. Falk, H., Gergely, S., and Hofer, O., Beitrage zur chemie der pyrrolpigmente. 5. Mitt. Die N-H tautomeric von substituierten pyrromethenen: konformationsanalytische studien mit hilfe der lanthaniden-verschiebungstechnik, *Monatsch. Chem.*, 105, 1004, 1974.
787. Cavalcante, S. H., Giesbrecht, A. M., Gottlieb, O. R., Mourao, J. C., and Yoshida, M., The chemistry of Brazilian Lauraceae. L. Lanthanide shift reagents in neolignan analysis: revision of structure of canellin-B, *Phytochemistry*, 17, 983, 1978.
788. Gottlieb, O. R., Mourao, J. C., Yoshida, M., Mascarenhas, Y. P., Rodrigues, M., Rosenstein, R. D., and Tomita, K., Absolute configuration of the benzofuranoid neolignans, *Phytochemistry*, 16, 1003, 1977.
789. Camps, P., Ortuno, R. M., and Serratosa, F., Lanthanide NMR shift reagents and stereochemical assignments. Stereochemistry in the reduction of 6-dicyanomethylidene and 6-(1-cyanoethylidene) derivatives of the cis-8α-methyl-1-decalone, *Tetrahedron Lett.*, 32, 2583, 1976.
790. van Bruijnsvoort, A., Kruk, C., de Waard, E. R., and Huisman, H. O., Elucidation of relative configuration of thiadecalones with Eu(dpm)$_3$; preferred site of complexation, *Tetrahedron Lett.*, p. 1737, 1972.
791. Luskus, L. J. and Houk, K. N., Pyrazolotropones from the cycloaddition of diazomethane to tropone, *Tetrahedron Lett.*, p. 1925, 1972.
792. Triepel, J. and Otto, H. H., Struktur von 7,11-diphenyl-2,4-diazaspiro[5,5]undecan-1,3,5,9-tetraonen durch LIS-NMR. Reaktionen von 1,4-pentadien-3-onen, 14. Mitt., *Monatsch. Chem.*, 108, 1085, 1977.
793. Levine, S. G. and Hicks, R. E., The conformation of griseofulvin. Application of an NMR shift reagent, *Tetrahedron Lett.*, p. 311, 1971.
794. Eckroth, D. R., Secondary orbital effects vs. steric effects in some Diels-Alder additions, *J. Org. Chem.*, 41, 394, 1976.
795. Dennis, N., Ibrahim, B. E. D., and Katritzky, A. R., A facile synthesis of (2RS, 4αRS, 9SR, 9αSR)-1,2,4,4α,9,9α-hexahydro-1-aryl-2,9-methanoindeno-[2,1-b]pyridin-3-ones. A new ring system, *Synthesis*, p. 105, 1976.
796. Casals, P. F. and Boccaccio, G., Photodimers de le dimethyl-4,4 cyclohexene-2 one synthese de bis a enones de la serie du tricyclo (6,4,0,0^{2,7}) dodecane, *Tetrahedron Lett.*, p. 1647, 1972.
797. Bohlmann, F. and Zdero, C., Zwei neue sesquiterpen-lactone aus libdbeckia pectinata berg und pentzia elegans DC, *Tetrahedron Lett.*, p. 621, 1972.
798. Duddeck, H., Steric repulsion and intramolecular interaction of substituents in 4-substituted adamantan-2,6-diones, *Tetrahedron*, 39, 1365, 1983.
799. Bull, J. R. and Hodgkinson, A. J., Steroidal analogues of unnatural configuration. VI. A-ring transformations of 4,4,14α-trimethyl-19(10-9β)abeo-10α-pregn-5-ene-3,11,20-trione, *Tetrahedron*, 28, 3969, 1972.
800. Cooper, P. S., Culshaw, C. M., Gall, R. E., and Nemorin, J. E., 19-Nor and aromatic steroids. III. The conversion of 3β-acetoxy-5α-lanost-8-en-7-one into steroids with A and B rings aromatic, *Aust. J. Chem.*, 32, 179, 1979.
801. Kim, D. J., Colebrook, L. D., and Adley, T. J., C-13 nuclear magnetic resonance spectra of steroids: the assignments of chemical shifts for C-15 and C-17 in 17β-acetyl steroids, *Can. J. Chem.*, 54, 3766, 1976.
802. Wazeer, M. I. M., Gunatilaka, A. A. L., and Nanayakkara, N. P. D., Studies on terpenoids and steroids. 8. Proton nuclear magnetic resonance lanthanide-induced shift studies of some D:A-friedooleanones, *Arabian J. Sci. Eng.*, 9, 7, 1984.
803. Trifunac, A. D. and Katz, J. J., Structure of chlorophyll a dimers in solution from proton magnetic resonance and visible absorption spectroscopy, *J. Am. Chem. Soc.*, 96, 5233, 1974.
804. Bohlmann, F. and Jacob, J., Uber den einfluss von Eu(fod)$_3$ als verschiebungsreagenz auf die H-1 NMR spektren von zimtsaurederivaten. II, *Chem. Ber.*, 108, 2809, 1975.
805. Butterworth, R. F., Pernet, A. G., and Hanessian, S., Utility of shift reagents in nuclear magnetic resonance studies of polyfunctional compounds: N- and O-acetylated carbohydrates and nucleosides, *Can. J. Chem.*, 49, 981, 1971.
806. Izumi, K., NMR spectra of peracetylated methyl D-hexopyranosides and their 2-acetamido-2-deoxy derivatives in the presence of an europium shift reagent, *J. Biochem.*, 76, 535, 1974.
807. Rondeau, R. E., Berwick, M. A., Steppel, R. N., and Serve, M. P., Central linkage influence upon mesomorphic and electrooptical behavior of diaryl nematics. A general proton magnetic resonance method employing a lanthanide shift reagent for analysis of isomeric azoxybenzenes, *J. Am. Chem. Soc.*, 94, 1096, 1972.

808. Yatsimirskii, K. B., Davidenko, N. K., Bidzilya, V. A., and Golovkova, L. P., Analysis of the local bonding constant of a lanthanoid shift reagent with bifunctional ligands, *Koord. Khim.*, 4, 1306, 1978.
809. Bose, A. K., Dayal, B., Chawla, H. P. S., and Manhas, M. S., Unexpected effect of shift reagents on diastereotopic protons in some β-lactams, *Tetrahedron Lett.*, p. 3599, 1972.
810. Alexander, R. G. and Southgate, R., 1-Oxadethiapenicillins: synthesis and stereochemical assignments using lanthanide induced shifts, *J. Chem. Soc. Chem. Commun.*, p. 405, 1977.
811. Servis, K. L. and Patel, D. J., Aspects of the valinomycin backbone conformation in chloroform solution using lanthanide shift reagents. Evaluation of phi(D-val) and psi(D-HyIv) rotation angles, *Tetrahedron*, 31, 1359, 1975.
812. Schiemenz, G. P., Zur wirkung von verschiebungsreagentien auf diastereotope protonen in β-lactamen, *Tetrahedron Lett.*, p. 4267, 1972.
813. Martinelli, L. C., Honigberg, I. L., and Sternson, L. A., Conformational studies of peptides in the presence of lanthanide shift reagent, Eu(dpm)$_3$, *Tetrahedron*, 29, 1671, 1973.
814. Kessler, H. and Molter, M., Conformation of protected amino acids. V. Application of lanthanide-shift reagents for conformational studies of tert-butyoxycarbonyl alpha-amino acid esters: equilibrium changes and kinetics of the isomerization, *J. Am. Chem. Soc.*, 98, 5969, 1976.
815. Kessler, H. and Molter, M., Use of the Eu(fod)$_3$-induced shift for determination of the complexation site and for signal assignment in the C-13 NMR spectra of protected dipeptides, *Angew. Chem. Int. Ed.*, 13, 538, 1974.
816. Kessler, H. and Molter, M., Dynamic NMR measurements of boc-amino esters with application of lanthanoid shift reagents, *Angew. Chem. Int. Ed.*, 12, 1011, 1973.
817. Ochiai, M., Mizuta, E., Aki, O., Morimota, A., and Okada, T., The determination of stereochemistry of 3-methylenecepham derivatives by means of lanthanide-induced shifts, *Tetrahedron Lett.*, p. 3245, 1972.
818. Kricheldorf, H. R. and Hull, W. E., N-15 NMR spectroscopy. 22. Influence of shift reagents and paramagnetic ions on the spectra of peptides and polyamides, *Makromol. Chem.*, 181, 507, 1980.
819. Salmon, M., Structure of azomesobilirubin isomers, *J. Heterocycl. Chem.*, 14, 1101, 1977.
820. Berson, J. A., Dervan, P. B., Malherbe, R., and Jenkins, J. A., Formation and thermal rearrangements of some dimers of butadiene and piperylene. Tests of the validity of thermochemical-kinetic arguments for identification of common biradical intermediates, *J. Am. Chem. Soc.*, 98, 5937, 1976.
821. Kato, A. and Numata, M., Brugierol and isobrugierol, trans- and cis-1,2-dithiolane-1-oxide, from brugiera conjugata, *Tetrahedron Lett.*, p. 203, 1972.
822. Ellis, G. P. and Jones, R. T., One-step synthesis and spectral study of some 1-methylbenzimidazoles including use of a lanthanide shift reagent, *J. Chem. Soc. Perkin Trans. 1*, p. 903, 1974.
823. Elguero, J., Navarro, P., and Rodriguez Franco, M. I., Synthesis of new macrocyclic polyether di- or tetraester ligands containing pyrazole units, *Chem. Lett.*, p. 425, 1984.
824. Schuttler, R. and Hoffman, R. W., Strukturzuordnung Z/E-isomerer norbornadien-derivate anhand der LIS-sattigungswerte, *Tetrahedron Lett.*, p. 5109, 1973.
825. Izumi, K., Proton NMR spectra of peracetylated derivatives of α,α-trehalose and sucrose in the presence of a europium shift reagent, *Agric. Biol. Chem.*, 43, 2615, 1979.
826. Izumi, K., Proton NMR spectra of peracetylated derivatives of methyl L-arabinofuranosides in the presence of a europium shift-reagent, *Carbohydr. Res.*, 77, 218, 1979.
827. Pfeffer, P. E. and Rothbart, H. L., Effects of a europium-shift reagent upon the PMR spectrum of some triglycerides, *Acta Chem. Scand.*, 27, 3131, 1973.
828. Pfeffer, P. E. and Rothbart, H. L., PMR spectra of triglycerides: discrimination of isomers with the aid of a chemical shift reagent, *Tetrahedron Lett.*, p. 2533, 1972.
829. Wedmid, Y. and Litchfield, C., Positional analysis of isovaleroyl triglycerides using proton magnetic resonance with Eu(fod)$_3$ and Pr(fod)$_3$ shift reagents. I. Model compounds, *Lipids*, 10, 145, 1975.
830. Ceder, O. and Beijer, B., The effect of Eu(fod)$_3$ on the NMR spectra of methyl esters of some cis and trans α,β-unsaturated mono and dicarboxylic acids, *Acta Chem. Scand.*, 26, 2977, 1972.
831. Knippel, E., Knippel, M., Michalik, M., Kelling, H., and Kristen, H., On the geometric isomers of ethyl 2-nitro-3-ethoxyacrylate, ethyl 2-nitro-3-phenylaminoacrylate and ethyl 2-cyano-3-phenylaminoacrylate, *Tetrahedron*, 33, 231, 1977.
832. Nulu, J. R. and Bell, E. A., Configuration of L-gamma-ethylideneglutamic acid from Guilandina crista, *Phytochemistry*, 11, 2573, 1972.
833. Kuhnz, W. and Rembold, H., Application of lanthanide induced shifts in proton magnetic resonance of spectroscopy of juvenile hormones, *Z. Naturforsch.*, 32, 563, 1977.
834. Brunn, J. and Grossman, M., Der einfluss von benzol und Eu(fod)$_3$ auf die protonen-signale des dimeren und trimeren acrylsauremethylesters im h-H-1 NMR spectrum, *Z. Chem.*, 15, 107, 1975.

835. Abbott, P. J., Acheson, R. M., Kornilov, M. Y., and Stubbs, J. K., Addition reactions of heterocyclic compounds. LXII. A new rearrangement in the quinazoline series, *J. Chem. Soc. Perkin Trans. 1*, p. 2322, 1975.
836. Yanagawa, H., Kato, T., and Kitahara, Y., Asparagusic acid S-oxides. New plant growth regulators in etiolated young asparagus shoots, *Tetrahedron Lett.*, p. 1073, 1973.
837. Bucci, P., Ghidichimo, G., Liquori, A., Menniti, G., and Uccella, N., H-1 NMR separation of epimeric mixtures by lanthanide shift reagents: analysis of some isoxazolidine derivatives, *Org. Magn. Reson.*, 22, 399, 1984.
838. Cavagna, F. and Groebel, A., PMR configurational analysis of pinitol with the aid of lanthanide shift reagents and the homonuclear INDOR technique, *Org. Magn. Reson.*, 5, 133, 1973.
839. Achmatowicz, O., Jr., Ejchart, A., Jurczak, J., Kozerski, L., St. Pyrek, J., and Zamojski, A., The NMR spectra and conformations of dihydropyran derivatives. III. Application of Eu(dpm)$_3$ in studies of the PMR spectra of 2-methoxy-5,6-dihydro-α-pyran and 2-methoxytetrahydropyran derivatives, *Rocz. Chem.*, 46, 903, 1972.
840. Amin, H. B., Al-Showiman, S. S., and Al-Najjar, I. M., Effects of lanthanide shift reagents on proton NMR spectra of 1-(X-benzo[b]thienyl)ethyl acetate derivatives, *J. Chem. Soc. Pak.*, 5, 207, 1983.
841. Amin, H. B., Al-Showiman, S. S., and Al-Najjar, I. M., Determination of the binding constants, bound chemical shifts, and stoichiometry of lanthanide-substrate complexes, *J. Chem. Soc. Pak.*, 5, 201, 1983.
842. Ishii, H. and Murakami, Y., Fischer indolization and its related compounds. XIII. Measurement of the nuclear magnetic resonance spectra of ethyl indole-2-carboxylate derivatives using the shift reagent and its application, *Yakugaku Zasshi*, 99, 413, 1979.
843. Franzus, B., Wu, S., Baird, W. C., Jr., and Scheinbaum, M. L., A nuclear magnetic resonance technique for distinguishing isomers of 3,5-disubstituted nortricyclenes, *J. Org. Chem.*, 37, 2759, 1972.
844. Covey, D. F. and Nickon, A., Difunctionalized brendanes via thallium triacetate cleavage of the cyclopropyl ring of triaxane, *J. Org. Chem.*, 42, 794, 1977.
845. Paquette, L. A., Beckley, R. S., Truesdell, D., and Clardy, J., Substituent control of stereochemistry in the rearrangements of 1,8-bishomocubanes catalyzed by silver(I) ion, *Tetrahedron Lett.*, p. 4913, 1972.
846. Jankowski, K., Pelletier, O., and Tower, R., NMR shift reagent application to reserpine, *Bull. Acad. Pol. Sci.*, 22, 867, 1974.
847. Wright, G. E. and Tang Wei, T. Y., Lanthanide induced shifts in the NMR spectra of methyl methoxybenzoates and methoxybenzenes, *Tetrahedron*, 29, 3775, 1973.
848. Sakamoto, K. and Oki, M., Hetera-p-carbophane. VIII. The change in binding sites of NMR shift reagents in polyoxadioxo[n]paracyclophanes and their open-chain analogs, *Bull. Chem. Soc. Jpn.*, 49, 3159, 1976.
849. Izumi, K., Proton NMR spectra of peracetylated D-galactopyranose derivatives in the presence of lanthanide shift reagents, *J. Biochem.*, 81, 1605, 1977.
850. Boren, H. B., Garegg, P. J., Pilotti, A., and Swahn, C. G., NMR spectra of some glycoside acetates in the presence of tris(dipivalomethanato)europium, *Acta Chem. Scand.*, 26, 3261, 1972.
851. Matsui, M. and Okada, M., The application of paramagnetic shift reagent Eu(fod)$_3$ in the nuclear magnetic resonance studies of acetylated aryl glycopyranosides, *Chem. Pharm. Bull.*, 20, 1033, 1972.
852. Izumi, K., Proton NMR spectra of some acetylated sugars in the presence of a europium shift reagent, *Mem. Fac. Ind. Arts, Kyoto Tech. Univ. Sci. Technol.*, 29, 24, 1980; *Chem. Abstr.*, 95, 88585q.
853. McArdle, P., O'Reilly, J. P., Lee, E., and Simmie, J., Lanthanide-induced shift and relaxation studies on methyl 4,6-O-benzylidene-2,3-di-O-methyl-α-D-glucopyranoside, *Proc. R. Ir. Acad. Sect. B*, 83B, 115, 1983.
854. Izumi, K., Proton NMR spectra of some 3,6-anhydro-D-galactose derivatives in the presence of a europium shift-reagent, *Carbohydr. Res.*, 62, 368, 1978.
855. Izumi, K., Proton NMR spectra of peracetylated derivatives of some pentopyranosides in the presence of a europium shift reagent, *Agric. Biol. Chem.*, 43, 95, 1979.
856. Liska, K. J., Fentiman, A. F., Jr., and Foltz, R. L., Use of tris-(dipivalomethanato)europium as a shift reagent in the identification of 3-H-pyrano[3.2-f]quinolin-3-one, *Tetrahedron Lett.*, p. 4657, 1970.
857. Joseph-Nathan, P., Dominguez, M., and Ortega, D. A., Shift reagent H-1 NMR study of methoxycoumarins, *J. Heterocycl. Chem.*, 21, 1141, 1984.
858. Gray, A. I., Waigh, R. D., and Waterman, P. G., A method for the determination of substitution pattern in coumarins using a lanthanide shift reagent, *J. Chem. Soc. Chem. Commun.*, p. 632, 1974.

859. Kojima, Y. and Kato, N., Syntheses and conformational analyses of the perhydrofuro[2,3-b]furan compounds by using lanthanide shift reagent and empirical force-field calculation, *Tetrahedron Lett.*, p. 4667, 1979.
860. Lewis, F. D., Hirsch, R. H., Roach, P. M., and Johnson, D. E., Photochemical cycloaddition of singlet and triplet diphenylvinylene carbonate with vinyl ethers, *J. Am. Chem. Soc.*, 98, 8438, 1976.
861. Kretschmer, M., Kleinpeter, E., Pulst, M., and Borsdorf, R., Kombinierte d-NMR- und LIS-untersuchungen zum konformativen verhalten von formylmethylenthiopyranen, *Monatsch. Chem.*, 114, 289, 1983.
862. Sheinker, V. N., Lifintseva, T. V., Perel'son, M. E., Vasil'eva, I. A., Garnovskii, A. D., and Osipov, O. A., Study of the structure and properties of heterocyclic compounds and their complexes. XXXVIII. Study of the conformations of carbonyl derivatives of azoles by a PMR spectroscopic method using shift reagents, *Zh. Org. Khim.*, 13, 1067, 1977.
863. Kornilov, Y. M., Turov, A. V., and Kutrov, G. P., Study of the conformations of formyl derivatives of indolizine using lanthanide shift reagents, *Ukr. Khim. Zh.*, 48, 758, 1982; *Chem. Abstr.*, 97, 143974j.
864. Radeglia, V. R. and Weber, A., Eu(dpm)$_3$-induzierte verschiebungen in den H-1 NMR spektren von vinylogen N,N-dimethylformamiden, *J. Prak. Chem.*, 314, 884, 1972.
865. Radeglia, V. R., C-13 NMR spektroskopische untersuchungen der elektronenstruktur von einfachen polymethinen. C-13 Chemische verschiebungen, C-13-H kopplungskonstanten und Eu(dpm)$_3$-verschiebungseffekte, *J. Prak. Chem.*, 315, 1121, 1973.
866. Radeglia, R., Complex application of shift reagents in NMR spectroscopy, *Z. Chem.*, 13, 73, 1973.
867. Neville, G. A., Characterization of di- and trimethoxybenzaldehydes by analysis of deshielding gradients obtained with Eu(fod)$_3$ shift reagent, *Org. Magn. Reson.*, 4, 633, 1972.
868. Piegsa, U., Radeglia, R., and Doerffel, K., Interaction between p-substituted benzaldehydes and NMR shift reagents, *J. Prakt. Chem.*, 322, 742, 1980.
869. Young, J. A., Grasselli, J. G., and Ritchey, W. M., A carbon-13 study of complexes of organonitriles with shift reagents, *J. Magn. Reson.*, 14, 194, 1974.
870. McCullough, J. J., Miller, R. C., and Wu, W. S., Photoreactions of 1- and 2-naphthonitriles with tetramethylethylene in various solvents, *Can. J. Chem.*, 55, 2909, 1977.
871. Herz, J. E., Rodriguez, V. M., and Joseph-Nathan, P., Stereospecific interactions of ketals with tris(dipivalomethanato)europium(III), *Tetrahedron Lett.*, p. 2949, 1971.
872. de Wit, J. and Wynberg, H., Configuration of a naphtho[2,3-b]thiophene Diels-Alder adduct from shift reagent data, *Spectrosc. Lett.*, 5, 119, 1972.
873. Pirkle, W. H., Rinaldi, P. L., and Simmons, K. A., Binding of lanthanide shift reagents to oxaziridines, *J. Magn. Reson.*, 34, 251, 1979.
874. Vilsmaier, E., Klein, C. M., and Adam, R., Assignment of configuration of larger bicyclo[n.1.0]alkylamines. The use of a lanthanide shift reagent, *J. Chem. Soc. Perkin Trans. 2*, p. 23, 1984.
875. Latypova, F. N., Zorin, V. V., Krasutskii, P. A., Karakhanova, N. A., Apjok, J., and Bartok, M., PMR study of the complexation of 1,3-dihetero analogs of cycloalkanes with europium, *Acta Phys. Chem.*, 29, 73, 1983.
876. Grotens, A. M., Smid, J., and de Boer, E., Radical anions and lanthanide complexes as NMR shift reagents, *J. Magn. Reson.*, 6, 612, 1972.
877. Grotens, A. M., Smid, J., and de Boer, E., Lanthanide-induced contact shifts in polyglycoldimethylethers, *Tetrahedron Lett.*, p. 4863, 1971.
878. Grotens, A. M., Backus, J. J. M., Pijpers, F. W., and de Boer, E., Lanthanide-induced shifts in polyglycoldimethylethers. IV. Chemical exchange, *Tetrahedron Lett.*, p. 1467, 1973.
879. Grotens, A. M., de Boer, E., and Smid, J., Lanthanide-induced contact shifts in polyglycoldimethylethers. V. Additivity of the shifts, *Tetrahedron Lett.*, p. 1471, 1973.
880. Grotens, A. M., Backus, J. J. M., and de Boer, E., Lanthanide-induced contact shifts in polyglycoldimethylethers. VI. Calculation of Fermi contact and dipolar contributions from the temperature dependence of the shifts, *Tetrahedron Lett.*, p. 4343, 1973.
881. Lockhart, J. C., Atkinson, B., Marshall, G., and Davies, B., Relative basicity of oxygen atoms in some benzo-crown ethers as displayed towards shift reagents, *J. Chem. Res. (S)*, p. 32, 1979.
882. Newkome, G. R. and Kawato, T., Metal ion site complexation of polyfunctional ligands. Nicotinamides with NMR shift reagents, *Tetrahedron Lett.*, p. 4643, 1978.
883. Kaifer, A., Echegoyen, L., and Gokel, G. W., Evidence of side-arm involvement in lariat ether complexes: a lanthanide shift reagent study, *J. Org. Chem.*, 49, 3029, 1984.
884. Braz Fo, R., Mourao, J. C., Gottlieb, O. R., and Maia, J. G. S., Lanthanide induced shifts as an aid in the structural determination of eusiderins, *Tetrahedron Lett.*, p. 1157, 1976.
885. Smith, R. V., Erhardt, P. W., Rusterholz, D. B., and Bardknecht, C. F., NMR study of amphetamines using europium shift reagents, *J. Pharm. Sci.*, 65, 412, 1976.

886. Smith, R. V. and Stocklinski, A. W., Synthesis of 10,11-dimethoxyaporphine-d$_8$. Use of Eu(fod)$_3$ reagent in structure determination, *Tetrahedron Lett.,* p. 1819, 1973.
887. Streefkerk, D. G. and Stephen, A. M., PMR studies on fully methylated disaccharides using lanthanide shift reagents: assignments of the methoxyl signals, *Carbohydr. Res.,* 57, 25, 1977.
888. Eliel, E. L. and Juaristi, E., Conformational analysis. 37. Gauche-repulsive interactions in 5-methoxy- and 5-methylthio-1,3-dithianes, *J. Am. Chem. Soc.,* 100, 6114, 1978.
889. Bidzilya, V. A. and Golovkova, L. P., Proton NMR study of the complexing of dithiotetraoxo-18-crown-6 with a lanthanide shift reagent, *Teor. Eksp. Khim.,* 16, 258, 1980.
890. Bidzilya, V. A. and Golovkova, L. P., Study of an adduct of a lanthanide shift reagent with an 18-crown-6-macrocyclic ligand, *Teor. Eksp. Khim.,* 16, 261, 1980.
891. Davidenko, N. K., Bidzilya, V. A., Golovkova, L. P., Davydova, S. L., and Yatsimirskii, K. B., Reaction of a macrocyclic polyether of dibenzo-18-crown-6 with the lanthanide shift reagent tris(1,1,1,2,2,3,3-heptafluoro-7,7-dimethyloctane-4,6-dionato)-europium, *Teor. Eksp. Khim.,* 13, 404, 1977.
892. Shen, L., Gao, Y., Xiao, Y., and Ni, J., NMR study on the equilibrium between the lanthanide shift reagent and crown ether, *Zhongguo Kexueyan Changchun Yingyong Huaxue Yanjiuso Jikan,* 20, 1, 1983; *Chem. Abstr.,* 101, 98650p.
893. Erk, C., Effect of Eu(fod)$_3$ on the carbon-13 NMR spectra of macrocyclic ethers and polyoxa lactones, *Fresenius Z. Anal. Chem.,* 316, 477, 1983.
894. Lafuma, F. and Quivoron, C., Etudes des spectres de resonance magnetique nucleaire de quelques acetals, en presence de tri-(dipivalomethanate) d'europium, *C. R. Acad. Sci. Paris,* 272, 2020, 1971.
895. Couturier, D., Fargeau, M. C., and Maitte, P., Existence de zones d'anisotropie dans les molecules complexees par (dpm)$_3$Eu. Applications a l'etude conformationnelle d'alcoxy-2 tetrahydropyranne ou chromanne, *C. R. Acad. Sci. Paris,* 274, 1853, 1972.
896. Mundy, B. P., Dirks, G. W., Larter, R. M., Craig, A. C., Lipkowitz, K. B., and Carter, J., On structural determination of C-7-substituted 6,8-dioxabicylo[3.2.1]octanes. A reevaluation, *J. Org. Chem.,* 46, 4005, 1981.
897. Warrener, R. N., Pitt, I. G., and Russell, R. A., Photodimers of isobenzofuran: a novel application of lanthanide induced shift spectroscopy to determine stereochemistry, *J. Chem. Soc. Chem. Commun.,* p. 1195, 1982.
898. Joseph-Nathan, P., Herz, J. E., and Rodriguez, V. M., Doubling proton magnetic resonance signals by use of tris(dipivalomethanato)europium(III), *Can. J. Chem.,* 50, 2788, 1972.
899. Chandler, J. F. and Stark, B. P., Paramagnetic shifts induced by europium and praseodymium chelates in the NMR spectrum of bisphenol A diglycidyl ether, *Die Angew. Makromol. Chem.,* 27, 159, 1972.
900. Lycka, A. and Snobl, D., Effect of lanthanide shift reagents on H-1 NMR spectra of aminopyridines, *Coll. Czech. Chem. Commun.,* 44, 908, 1979.
901. Deswarte, S., Bellec, C., and Souchay, P., Etude spectroscopique de quelques β-cyano-enamines et enamines N-substitue, *Bull. Soc. Chim. Belg.,* 84, 321, 1975.
902. Cockerill, A. F., Gutteridge, N. J. A., Rackham, D. M., and Smith, C. W., Formation of bis(3-amino-2-naphthalene)trisulphide by fusion of 1-naphthylamine and sulphur, *Tetrahedron Lett.,* p. 3059, 1972.
903. van Brederode, H. and Huysmans, W. G. B., Analysis of diamine stereoisomers from proton paramagnetic shifts, *Tetrahedron Lett.,* p. 1695, 1971.
904. Skala, V., Kuthan, J., Dedina, J., and Schraml, J., Reaction of some dicyanopyridines with methylmagnesium iodide. Observation of the nuclear Overhauser effect in the presence of NMR shift reagent, *Coll. Czech. Chem. Commun.,* 39, 834, 1974.
905. Rengaraju, S. and Berlin, K. D., A case of slow nitrogen inversion due to intramolecular hydrogen bonding. Study of slow nitrogen inversion in diethyl 2-aziridinylphosphonate from the paramagnetic induced shifts in the proton magnetic resonance spectra using tris(dipivalomethanato)europium(III), and solvent shifts, *J. Org. Chem.,* 37, 3304, 1972.
906. Wamhoff, H., Materne, C., and Knoll, F., Die synthese von furo[2.3-e]1.4-diazepinin und deren NMR- analyse unter verwendung des tris(dipivalo-methanato)-europium-komplexes, *Chem. Ber.,* 105, 753, 1972.
907. Moroi, R., Tomita, K., and Sano, M., The application of lanthanide shift reagent. I. Conformational analysis of 1-[2-(1,3-dimethyl-2-butylidene)hydrazino]phthalazine by lanthanide induced shifts, *Chem. Pharm. Bull.,* 24, 2541, 1976.
908. Hogberg, T., Determination of the Z/E-configuration of 3-phenyl-3-pyridylallylamines related to zimelidine by means of lanthanide-induced shifts in proton magnetic resonance, *Acta Chem. Scand.,* B34, 629, 1980.
909. Morgan, L. W. and Bourlas, M. C., Eu(fod)$_3$ induced shifts in a heteronuclear bicyclic amine, *Tetrahedron Lett.,* p. 2631, 1972.

910. Sega, A., Moimas, F., Decorte, E., Toso, R., and Sunjic, V., LIS-NMR study of 2[([5-(dimethylamino)methyl-2-furanyl]thio)ethyl]amino-2-methylamino-1-nitroethene, *Gazz. Chim. Ital.*, 112, 421, 1982.
911. Tesse, J., Glacet, C., and Couturier, D., Etude conformationnelle par RMN d'α-aminotetrahydropyrannes, *C. R. Acad. Sci. Paris*, 280, 1525, 1975.
912. Bhakuni, D. S., Silva, M., Matlin, S. A., and Sammes, P. G., Aristoteline and aristotelone, unusual indole alkaloids from Aristotelia chilensis, *Phytochemistry*, 15, 574, 1976.
913. Ricci, A., Danieli, R., Phillips, R. A., and Ridd, J. H., The effects of lanthanide shift reagents on the NMR spectra of diaza- and dithiacyclophanes, *J. Heterocycl. Chem.*, 11, 551, 1974.
914. Wrobel, J. T., Ruszkowska, J., and Kabzinska, K., The use of $Eu(fod)_3$-induced shifts of furan protons for the elucidation of the structure of natural nuphar S-oxides, *Collect. Lect. Int. Symp. Furan Chem.*, 3rd, 161, 1979; *Chem. Abstr.*, 92, 164143z.
915. Barciszewski, J., Rafalski, A. J., and Wiewiorowski, M., Europium shifts in the NMR spectra of lactams. I. Stereochemical assignment of lupanine, *Bull. Acad. Pol. Sci.*, 19, 545, 1971.
916. Romeo, G., Aversa, M. C., Gianetto, P., Vigorita, M. G., and Ficarra, P., Nuclear magnetic resonance of 1,4-benzodiazepines. II. Stereochemistry of 1,3-dihydro-2H-1,4-benzodiazepin-2-ones by lanthanide shift reagents, *Org. Magn. Reson.*, 12, 593, 1979.
917. Esaki, T., Study on C-13 NMR lanthanoid-induced shifts of antipyrine by tris(1,1,1,2,2,3,3-heptafluoro-7,7-dimethyl-4,6-octane-dionato)praseodymium, *Nippon Kagaku Kaishi*, p. 765, 1978.
918. Geissler, G., Menz, I., Koeppel, H., Kretschmer, M., and Kulpe, S., LIS and X-ray studies of the structure of 1,5-diazabicyclo[3.1.0]hexan-2-ones obtained by irradiation of pyrazolidin-3-one azomethinimines, *J. Prakt. Chem.*, 325, 995, 1983.
919. Claramunt, R. M., Elguiro, J., and Jacquier, R., Etudes RMN en serie heterocyclique. VIII. Effet des dipivalomethanates de nickel(II), cobalt(II), europium(III) et praseodyme(III) sur les deplacements chimiques de quelques derives du pyrazoles, *Org. Magn. Reson.*, 3, 595, 1971.
920. Kleinpeter, E. and Borsdorf, R., NMR-untersuchungen an polymethinfarbstoffen und farbstoffzwisehenprodukten; $Eu(fod)_3$-induzierte verschiebungen an anilen, *Z. Chem.*, 13, 183, 1973.
921. Romeo, G., Aversa, M. C., Giannetto, P., Ficarra, P., and Vigorita, M. G., Nuclear magnetic resonance of psychotherapeutic agents. IV. Conformational analysis of 2,3-dihydro-1H-1,4-benzodiazepines, *Org. Magn. Reson.*, 15, 33, 1981.
922. Rackham, D. M., Cockerill, A. F., and Harrison, R. G., Lanthanide shift reagents. IX. Proton and carbon magnetic resonance study of specific contact interactions of a bifunctional oxazole amide with fluorinated lanthanide shift reagents, *Org. Magn. Reson.*, 11, 424, 1978.
923. Newkome, G. R. and Kowato, T., Conformational studies of N,N-disubstituted nicotinamides, NMR peak assignments and utilization of shift reagents with 2,6-dichloronicotinamides, *J. Org. Chem.*, 45, 629, 1980.
924. Cazaux, L., Vidal, C., and Pasdeloup, M., NMR of some benzodiazepine drugs: structure elucidation with lanthanide-induced shifts of N-1 substituted benzodiazepinones, *Org. Magn. Reson.*, 21, 190, 1983.
925. Paul, H. H., Sapper, H., Lohmann, W., and Kalinowski, H. O., Analysis and applications of C-13 NMR lanthanide induced shifts of 1,4-benzodiazepines, *Org. Magn. Reson.*, 19, 49, 1982.
926. Sakamoto, T., Niitsuma, S., Mizugaki, M., and Yamanaka, H., On the structural determination of pyrimidine N-oxides, *Heterocycles*, 8, 257, 1977.
927. Rondeau, R. E., Steppel, R. N., Rosenberg, H. M., and Knaak, L. E., Structural differentiation of isomeric N-oxides, *J. Heterocycl. Chem.*, 10, 495, 1973.
928. Dakterneiks, D. R., Enthalpies of adduct formation of some tris(dipivaloylmethanato)lanthanide complexes with 1,8-naphthyridine and bipyridyl, *J. Inorg. Nucl. Chem.*, 38, 141, 1976.
929. Bhacca, N. S., Selbin, J., and Wander, J. D., Nuclear magnetic resonance spectra of 1:1 adducts of 1,10-phenanthroline and α,α'-bipyridyl with tris[2,2,6,6-tetramethylheptane-3,5-dionato] complexes of the lanthanides, *J. Am. Chem. Soc.*, 94, 8719, 1972.
930. Hart, F. A., Newberry, J. E., and Shaw, D., Lanthanide complexes. X. PMR studies of alkyl-substituted bipyridine complexes of lanthanides: paramagnetic shifts and reaction kinetics, *J. Inorg. Nucl. Chem.*, 32, 3585, 1970.
931. Glover, E. E. and Pointer, D. J., Use of lanthanide shift reagents in a reaction product analysis: methylation of 2-methyl-4(5)-phenylimidazole, *Chem. Ind.*, p. 412, 1976.
932. Elguero, J., Marzin, C., and Roberts, J., Carbon-13 magnetic resonance studies of azoles. Tautomerism, shift reagent effects, and solvent effects, *J. Org. Chem.*, 39, 357, 1974.
933. Scahill, T. A. and Smith, S L., Nitrogen-15 NMR studies of 1,4-benzodiazepines. 2. The triazolobenzodiazepines, *Org. Magn. Reson.*, 21, 662, 1983.
934. Bramwell, A. F., Riezebos, G., and Wells, R. D., Correlations of NMR spectral parameters with structure: benzylic coupling in substituted pyrazines, *Tetrahedron Lett.*, p. 2489, 1971.

935. Bramwell, A. F. and Wells, R. D., The nuclear magnetic resonance spectra of pyrazines. Benzylic coupling in substituted methylpyrazines, *Tetrahedron*, 28, 4155, 1972.

936. Primc, J., Stanovnik, B., and Tisler, M., Heterocycles. CLII. Pyridazines. LXXXV. Tris(6,6,7,7,8,8,8-heptafluoro-2,2-dimethyl-3,5-octane-dionato)europium- and -praseodymium-induced pseudocontact shifts of proton resonances in azolopyridazines with bridgehead nitrogen, *Vestn. Slov. Kem. Drus.*, 23, 45, 1976.

937. Dusemund, J. and Roth, K., Uber die anwendung von NMR-verschiebungreagenzien zur strukturaufklarung von chinazolinderivaten, *Z. Naturforsch.*, 316, 509, 1976.

938. Gacel, G., Fournie-Zaluski, M. C., and Roques, B. P., Utilisation des deplacements chimiques induits par les lanthanides a la determination de la structure de benzothienopyridines, *J. Heterocycl. Chem.*, 12, 623, 1975.

939. Zvolinskii, V. P., Galiullin, M. A., Zakharov, V. F., Pleshakov, V. G., and Prostakov, N. S., Proton NMR study of configuration isomerism of N-(4-aza-9-fluorenylidene)arylamines using the paramagnetic shift reagent europium tris(dipivalomethanate), Deposited Doc. 1977, VINTI 3624-77; *Chem. Abstr.*, 90, 120574v.

940. Galiullin, M. A., Evolinskii, V. P., Fomichev, A. A., Pleshakov, V. G., Gusarov, A. N., Zakharov, V. F., Grigor'ev, G. V., Obukhov, A. E., and Prostakov, N. S., Proton NMR study of the isomerism of N-(4-aza-9-fluorenylidene)aryl(cyclohexyl)amines using the lanthanide shift reagent Eu(dpm)$_3$, *Koord. Khim.*, 7, 523, 1981.

941. Saraswathi, T. V. and Srinivasan, V. R., A one-step synthesis of 3,6-bis-substituted imidazo[1,2-b]-as-triazines, a set of highly fluorescent heterocycles, *Indian J. Chem.*, 15B, 607, 1977.

942. Hoegel, J. and Schmidpeter, A., Phosphazenes. LXVI. Phosphorus-31 NMR studies. XI. Preparation of cyclotri(phosphazene)derivatives via the corresponding lithium compound and unfolding of a phosphorus-31 NMR spectrum by a shift reagent, *Z. Anorg. Allg. Chem.*, 458, 168, 1979.

943. Bidzilya, V. A., Davidenko, N. K., Golovkova, L. P., Drach, B. S., and Sviridov, E. P., Use of lanthanide shift reagents for detecting diastereotopic groups in α-aminophosphonic acid esters, *Ukr. Khim. Zh.*, 42, 1150, 1976.

944. Otazo, E., Gva, R., and Macias, A., Study of the properties of furoylthioureas. II. Effect of the paramagnetic chelates of europium on the chemical shift of 1-furoyl-3-(p-tolyl)thiourea, *Rev. CENIC Cienc. Fis.*, 10, 331, 1979; *Chem. Abstr.*, 94, 174165p.

945. Gra, R. and Suarez, M., Conformational changes to furancarboxanilides by (Eu(dpm)$_3$ and Eu(fod)$_3$, *Rev. CENIC Cienc. Fis.*, 7, 301, 1976; *Chem. Abstr.*, 89, 128965y.

946. Montaudo, G., Finocchiaro, P., and Overberger, C. G., Application of the lanthanide shift reagents to the study of the conformation of structurally rigid polyamides, *J. Polym. Sci.*, 11, 619, 1973.

947. Wenkert, E., Cochran, D. W., Hagaman, E. W., Lewis, R. B., and Schell, F. M., Carbon-13 nuclear magnetic resonance spectroscopy with the aid of a paramagnetic shift agent, *J. Am. Chem. Soc.*, 93, 6271, 1971.

948. Lal, B., Singh, P., and Bhaduri, A. P., Stereochemistry of 5,6-dimethoxyisatylidenes, *Indian J. Chem.*, 12, 906, 1974.

949. Buti, M., Vigne, C., Montginoul, C., Torreilles, E., and Giral, L., Lanthanide reagent effects on 3-(2',4'-dimethylphenyl)-1H-3H-quinazoline-2,4-dione, *Org. Magn. Reson.*, 8, 505, 1976.

950. Neville, G. A., Use of Eu(fod)$_3$ shift reagent and solvent effects in structural elucidation of novel isomeric pyrimidines and model methoxylated pyrimidines, *Can. J. Chem.*, 50, 1253, 1972.

951. Samitov, Y. Y., Goncharova, I. N., Ramzaeva, N. P., Mishnev, A. F., and Bleidelis, J., Synthesis and structure of 2'-substituted 1-(1,3-dioxan-5-yl)uracils. Positive role of NMR shift reagent Eu(fod)$_3$, *Khim. Geterosikl. Soedin.*, p. 1523, 1981.

952. Ammon, H. L., Mazzocchi, P. H., and Liu, L., Conformational analysis with lanthanide shift reagents. Differences in the solution and crystal structure conformations of a compound, *Chem. Lett.*, p. 897, 1980.

953. Ammon, H. L., Mazzocchi, P. H., Colicelli, E., Jameson, C. W., and Liu, L., A convenient method for mixing H-1 and C-13 lanthanide induced shift (LIS) calculations. A technique for facilitating C-13 assignments, *Tetrahedron Lett.*, p. 1745, 1976.

954. Ammon, H. L., Mazzocchi, P. H., Kopecky, W. J., Jr., Tamburin, H. J., and Watts, P. H., Jr., Lanthanide shift reagents. II. (a) Photochemical ring expansion of a β-lactam and product identification using LSR NMR shifts and X-ray crystallography. (b) Probable structure of an LSR-substrate complex in solution. (c) Conformational analysis using LSR NMR data, *J. Am. Chem. Soc.*, 95, 1968, 1973.

955. Caputo, J. F. and Martin, A. R., Stereochemical elucidation of isomeric tricyclic 1,4-benzodioxans by the use of the NMR shift reagent, Eu(od)$_3$, *Tetrahedron Lett.*, p. 4547, 1971.

956. Reisberg, P., Brenner, I. A., and Bodin, J. I., NMR assay of diastereoisomers of 7-chloro-3,3α-dihydro-2-methyl-2H,9H-isoxazolo-(3,2-b)(1,3)benzoxazin-9-one using deuterated tris(1,1,1,2,2,3,3-heptafluoro-7,7-dimethyl-4,6-octane-dionato)europium III shift reagent, *J. Pharm. Sci.*, 63, 1586, 1974.
957. Skolik, J., Barciszewski, J., Rafalski, A. J., and Wiewiorowski, M., Europium shifts in the lactam series. II. Infrared spectroscopic study of shift reagent influence on substrate conformation, *Bull. Acad. Pol. Sci.*, 19, 599, 1971.
958. Chimirri, A., Grasso, S., and Monforte, P., Stereochemistry of phthalazino[2,3-b]phthalazine-5,12-(14H,7H)dione by lanthanide induced nuclear magnetic resonance shifts, *J. Heterocycl. Chem.*, 17, 1509, 1980.
959. Young, P. E., Madison, V., and Blout, E. R., Cyclic peptides. VI. Europium-assisted nuclear magnetic resonance study of the solution conformations of cyclo(L-pro-L-pro) and cyclo(L-pro-D-pro), *J. Am. Chem. Soc.*, 95, 6142, 1973.
960. Wasylishen, R. and Schaefer, T., Proton magnetic resonance spectra, conformational preferences, and approximate molecular orbital calculations for the syn and anti-2-furanaldoximes, *Can. J. Chem.*, 50, 274, 1972.
961. Tronchet, J. M. J., Barbalat-Rey, F., and Le-Hong, N., Equilibres conformationnels de glucides au niveau de liaisons sigma sp^2-sp^3 C-C. III. Derives d'oximes d'aldehydo-sucres utilisation de tris-dipivaloylmethanato-europium, *Helv. Chem. Acta*, 54, 2615, 1971.
962. Hansen, R. S. and Trahanovsky, W. S., Preferential complexation of one of the diastereomers of 1,2-diazido-1,2-di-tert-butylethane with an europium nuclear magnetic resonance shift reagent, *J. Org. Chem.*, 39, 570, 1974.
963. Greene, J. L., Jr. and Shevlin, P. B., Meso- and (+)-bis(phenylsulphinyl)methane. Characterization using nuclear magnetic resonance chemical shift reagents, *J. Chem. Soc. Chem. Commun.*, p. 1092, 1971.
964. Selling, H. A., Assessment of configuration of four isomeric α,β-unsaturated sulfones using a lanthanide shift reagent, *Tetrahedron*, 31, 2543, 1975.
965. Yatsimirskii, K. B., Bidzilya, V. A., Golovkova, L. P., and Shtepanek, A. S., Proton NMR study of phosphonyl-containing macrocyclic polyesters using lanthanide reagents, *Dokl. Akad. Nauk SSSR*, 244, 1142, 1979.
966. Bulai, A. K., Gruznov, A. G., Urman, Y. G., Romanov, L. M., and Slonim, I. Y., The effect of paramagnetic shift reagents on the NMR spectra of oligomers containing ether linkages, *Vysokomol. Soyed.*, A16, 2203, 1974.
967. Okada, T., H-1 NMR spectra of poly(ethylene glycols) in the presence of shift reagent, *J. Polym. Sci.*, 17, 155, 1979.
968. Ho, F. F. L., Application of the Eu(dpm)$_3$ chemical shift reagent to the determination of the molecular weight of poly(propylene glycol) by NMR, *Polym. Lett.*, 9, 491, 1971.
969. Katritzky, A. R. and Smith, A., Application of europium and praseodymium shift reagents to the NMR spectra of polymers, *Br. Polym. J.*, 4, 199, 1979.
970. Fleischer, D. and Schulz, R. C., Sequenzanalyse bei trioxan/dioxolan-copolymeren durch NMR-spektroskopie unter zusatz von lanthaniden-komplexen, *Die. Makromol. Chem.*, 152, 311, 1972.
971. Fleischer, D. and Schulz, R. C., Sequenzanalyse bei trioxocan/dioxolan-copolymeren durch NMR-spektroskopie unter zusatz von lanthaniden-komplexen, *Die. Makromol. Chem.*, 162, 103, 1972.
972. Perrett, B. S. and Stenhouse, I. A., The Application of Tri(dipivalomethanato)europium(III) as a Shift Reagent in Nuclear Magnetic Resonance Spectroscopy, Res. Group Rep., AERE-R 7042, U.K. Atomic Energy Authority, Winfrith, U. K., 1972.
973. Yuki, H., Hatada, K., Hasegawa, T., Terawaki, Y., and Okuda, H., PMR spectra of poly(methyl vinyl ether) in the presence of shift reagent, *Polym. J.*, 3, 645, 1972.
974. Guillet, J. E., Peat, I. R., and Reynolds, W. F., The effect of lanthanide shift reagents on the NMR spectrum of stereoregular poly(methyl methacrylate), *Tetrahedron Lett.*, p. 3493, 1971.
975. Inoue, Y. and Konno, T., NMR studies of conformation and molecular motion of poly(methyl methacrylate) in solution, *Makromol. Chem.*, 179, 1311, 1978.
976. Katritzky, A. R. and Smith, A., Application of contact shift reagents to the NMR spectra of polymers, *Tetrahedron Lett.*, p. 1765, 1971.
977. Arrington, K. and Taylor, R. C., The Effect of Lanthanide Shift Reagents on the Proton NMR Spectra of Vinyl Polymers, *U.S. NTIS AD Rep.*, 77, 207, 1977; *Chem. Abstr.*, 87, 118227d.
978. Amiya, S., Ando, I., Watanabe, S., and Chujo, R., Paramagnetically shifted NMR spectra and conformation of isotactic and syndiotactic poly(methyl methacrylate)s, *Polym. J.*, 6, 194, 1974.
979. Amiya, S., Ando, I., and Chujo, R., Paramagnetically shifted high-resolution proton NMR spectra of poly(methyl methacrylate) by the use of tris(dipivalomethanato)europium(III), *Polym. J.*, 4, 385, 1973.

980. Natansohn, A., Maxim, S., and Feldman, D., Methoxy signal resolution in the proton NMR spectra of methyl acrylate-styrene copolymers, *Bul. Inst. Politeh. Iasi Sect. 2: Chim. Ing. Chim.*, 26, 83, 1980.

981. Okada, T., Izuhara, M., and Hashimoto, T., A study of chloroprene and methylmethacrylate radical copolymers by H-1 NMR spectra shifted by the use of tris(dipivalomethanato)europium(III), *Polym. J.*, 7, 1, 1975.

982. Okada, T., Hashimoto, K., and Ikushige, T., Sequence distributions of vinyl chloride-vinyl acetate and vinyl chloride-vinyl propionate copolymers, *J. Polym. Sci.*, 19, 1821, 1981.

983. Okada, T., Microstructure analysis of vinyl chloride copolymers by C-13 NMR with lanthanide shift reagents, *J. Sci. Hiroshima Univ. Ser. A*, 45, 215, 1981.

984. Okada, T. and Ikushige, T., H-1 NMR spectra of ethylene-vinyl acetate copolymers in the presence of shift reagent, *Polym. J.*, 9, 121, 1977.

985. Delsarte, J. and Weill, G., Application de la methode des deplacements chimiques induits par les chelates de terre rare a la determination RMN de la conformation D'un polymere helicoidal, *J. Phys. C*, 34, C8, 1973.

986. Delsarte, J. and Weill, G., Application of the lanthanide-induced shifts to the nuclear magnetic resonance determination of the helical conformation of poly(β-hydroxybutyrate) in solution, *Macromolecules*, 7, 343, 1974.

987. Chujo, R., Koyama, K., Ando, I., and Inoue, Y., Paramagnetically shifted high-resolution NMR spectra of copolyesters by the use of tris(dipivalomethanato)europium(III), *Polym. J.*, 3, 394, 1972.

988. Matlengiewicz, M. and Turska, E., Application of lanthanide shift reagents to the determination of sequences in an aromatic copolyterephthalate by proton NMR spectroscopy, *Polym. Bull. (Berlin)*, 6, 603, 1982.

989. Bulai, A. K., Kalinina, V. S., Arshava, B. M., Urman, Y. G., Barshtein, R. S., and Slonim, I. Y., Study of equilibrium during alcoholysis of polyester plasticizers by proton and carbon-13 NMR methods using lanthanide shift reagents, *Vysokomol. Soedin. Ser. A*, 18, 2472, 1976; *Chem. Abstr.*, 86, 73177y.

990. Iida, M., Hayase, S., and Araki, T., C-13 NMR spectroscopy of poly(β-substituted β-propiolactone)s. Tacticity recognition in 1,5-disubstituted polymer system and stereospecific contact of shift reagent, *Macromolecules*, 11, 490, 1978.

991. Shapiro, Y. E., Dozorova, N. P., Turov, B. S., and Shvetsov, O. K., Analysis of the microtacticity of copolymers by NMR spectroscopy with the use of shift reagents. Microtacticity of isoprene-nitrile rubber, *Zh. Anal. Khim.*, 33, 393, 1978.

992. Dozorova, N. P., Shapiro, Y. E., and Zakharov, N. D., Study of the microstructure of butadiene-nitrile rubbers by proton NMR spectroscopy, *Izv. Vyssh. Uchebn. Zaved. Khim. Khim. Tekhnol.*, 20, 428, 1977; *Chem. Abstr.*, 86, 191096.

993. Vorontsov, E. D., Rusak, A. F., Gusev, V. V., Filippova, E. E., Nikolaev, N. N., and Ekdakov, V. P., Study of the structure of atactic poly(4-vinylpyridine) using paramagnetic shift reagents, *Vysokomol. Soedin. Ser. A*, 21, 1415, 1979; *Chem. Abstr.*, 91, 75001p.

994. Roth, K., Sign determination of spin-spin coupling constants with paramagnetic shift reagents, *J. Chem. Res. (S)*, p. 270, 1977.

995. Schiemenz, G. P. and Rast, H., Lanthanide shift reagents as aids in diastereotopic problems, *Tetrahedron Lett.*, p. 4685, 1971.

996. Schiemenz, G. P. and Rast, H., Ions. IV. Isochronous heterotopic nuclei, *Tetrahedron Lett.*, p. 1697, 1972.

997. Sanders, J. K. M. and Williams, D. H., Secondary deuterium isotope effect on Lewis basicity: tris(dipivaloylmethanato)europium as a simple and effective probe, *J. Chem. Soc. Chem. Commun.*, p. 436, 1972.

998. Grotens, A. M., Hilbers, C. W., de Boer, E., and Smid, J., Isotope effects in nuclear magnetic resonance spectra of polyglycoldimethylethers, complexes with rare-earth shift reagents, *Tetrahedron Lett.*, p. 2067, 1972.

999. Hinkley, C. C., Boyd, W. A., Smith, G. V., and Behbahany, F., Chemistry of lanthanide shift reagents: secondary deuterium isotope effects, in *Nuclear Magnetic Resonance Shift Reagents*, Sievers, R. E., Ed., Academic Press, New York, 1973, 1.

1000. Balaban, A. T., Stanoiu, I. I., and Chiraleu, F., Steric origin of isotope effects in nuclear magnetic resonance shifts induced by lanthanide shift reagents; effect of deuteration of methyl substituted pyridines, *J. Chem. Soc. Chem. Commun.*, p. 984, 1976.

1001. Balaban, A. T., Stanoiu, I. I., and Chiraleu, F., Isotope effects in lanthanide-induced shifts of H-1 NMR spectra of selectively side-chain deuterated pyridines, *Rev. Roum. Chim.*, 23, 187, 1978.

1002. Bargon, J., Chemically induced dynamic nuclear polarization in the presence of paramagnetic shift reagents, *J. Am. Chem. Soc.*, 95, 941, 1973.

1003. Bargon, J., Chemically induced dynamic nuclear polarization in the presence of paramagnetic shift reagents, in *Nuclear Magnetic Resonance Shift Reagents,* Sievers, R. E., Ed., Academic Press, New York, 1973, 265.
1004. Whitesides, G. M. and Lewis, D. W., The determination of enantiomeric purity using chiral lanthanide shift reagents, *J. Am. Chem. Soc.,* 93, 5914, 1971.
1005. Goering, H. L., Backus, A. C., Chang, C. S., and Masilamani, D., Preparation and determination of absolute rotations and configurations of 6,7-dimethoxy-1,2-dimethyl-exo-2-benzonorbornenyl derivatives, *J. Org. Chem.,* 40, 1533, 1975.
1006. Goering, H. L., Eikenberry, J. N., Koermer, G. S., and Lattimer, C. J., Direct determination of enantiomeric compositions with optically active nuclear magnetic resonance lanthanide shift reagents, *J. Am. Chem. Soc.,* 96, 1493, 1974.
1007. Sumitomo Chem. Co., Analysis of optical isomer ratio, *Jpn. Kokai Tokkyo Koho,* JP 57, 168, 147; *Chem Abstr.,* 98, 178598e.
1008. Fraser, R. R., Stothers, J. B., and Tan, C. T., Determination of optical purity with chiral shift reagents and C-13 magnetic resonance, *J. Magn. Reson.,* 10, 95, 1973.
1009. Seebach, D., Ehrig, V., and Teschner, M., Erzeugung und reaktionen des chiralen lithiumenolats von (+)-(S)-3-methyl-2-pentanon, *Justus Liebigs Ann. Chem.,* p. 1357, 1976.
1010. Kawa, H., Yamaguchi, F., and Ishikawa, N., Optically active di(perfluoro-2-propoxypropionyl)methane: a novel ligand for NMR shift reagent, *Chem. Lett.,* p. 153, 1982.
1011. Schurig, V., Chiral shift reagents for NMR-spectroscopy. A simple and improved access to lanthanide-tris-chelates of d-3-tfa-camphor, *Tetrahedron Lett.,* p. 3297, 1972.
1012. Crout, D. H. G. and Whitehouse, D., Absolute configuration of 2,3-dihydroxy-3-methylpentanoic acid, an intermediate in the biosynthesis of isoleucine, and its identity with the esterifying acid of the pyrrolizidine alkaloid strigosine, *J. Chem. Soc. Perkin Trans. 1,* p. 544, 1977.
1013. Pasto, D. J. and Borchardt, J. K., Partial resolution and correlation of the absolute configuration of (−)-(R)-2-phenylisobutenylidencyclopropane, *Tetrahedron Lett.,* p. 2517, 1973.
1014. Dewar, G. H., Kwakye, J. K., Parfitt, R. T., and Sibson, R., Optical purity determination by NMR: use of chiral lanthanide shift reagents and a base line technique, *J. Pharm. Sci.,* 71, 802, 1982.
1015. Nakanishi, K. and Dillon, J., A simple method for determining the chirality of cyclic alpha-glycols with Pr(dpm)$_3$ and Eu(dpm)$_3$, *J. Am. Chem. Soc.,* 93, 4058, 1971.
1016. Nakanishi, K., Schooley, D. A., Koreeda, M., and Dillon, J., Absolute configuration of the C_{18} juvenile hormone: application of a new circular dichroism method using tris(dipivalomethanato)praseodymium, *J. Chem. Soc. Chem. Commun.,* p. 1235, 1971.
1017. Langenfeld, N. and Welzel, P., D-moenuronsaure (4-C-methyl-D-glucuronsaure), ein neuer baustein des antibiotikums moenomycin A, *Tetrahedron Lett.,* p. 1833, 1978.
1018. Dillon, J. and Nakanishi, K., Absolute configuration studies of vicinal glycols and amino alcohols. II. With Pr(dpm)$_3$, *J. Am. Chem. Soc.,* 97, 5417, 1975.
1019. Mitchell, G. N. and Carroll, F. I., Use of tris(dipivalomethanato)praseodymium(III) for the determination of the chirality of simple amines and cyclic 1,2-amino alcohols, *J. Am. Chem. Soc.,* 95, 7912, 1973.
1020. Dale, J. A. and Mosher, H. S., Nuclear magnetic resonance enantiomer reagents. Configurational correlations via nuclear magnetic resonance chemical shifts of diastereomeric mandelate. O-methyl-mandelate, and α-methoxy-α-trifluoromethylphenylacetate(MTPA) esters, *J. Am. Chem. Soc.,* 95, 512, 1973.
1021. Dale, J. A., Dull, D. L., and Mosher, H. S., α-Methoxy-α-trifluoromethylphenylacetic acid, a versatile reagent for the determination of enantiomeric composition of alcohols and amines, *J. Org. Chem.,* 34, 2543, 1969.
1022. Yasuhara, F. and Yamaguchi, S., Use of shift reagent with MTPA derivatives in H-1 NMR spectroscopy. III. Determination of absolute configuration and enantiomeric purity of primary carbinols with chiral center at the C-2 position, *Tetrahedron Lett.,* p. 4085, 1977.
1023. Reich, C. J., Sullivan, G. R., and Mosher, H. S., Chiral 1-deuterio alcohols. Synthesis and determination of enantiomeric purity by chiral lanthanide NMR shift reagents, *Tetrahedron Lett.,* p. 1505, 1973.
1024. Yamaguchi, S. and Mosher, H. S., Asymmetric reductions with chiral reagents from lithium aluminum hydride and (+)-(2S,3R)-4-dimethylamino-3-methyl-1,2-diphenyl-2-butanol, *J. Org. Chem.,* 38, 1870, 1973.
1025. Yasuhara, F. and Yamaguchi, S., Determination of absolute configuration and enantiomeric purity of 2- and 3-hydroxycarboxylic acid esters, *Tetrahedron Lett.,* p. 2827, 1980.
1026. Yamaguchi, S., Yasuhara, F., and Kabuto, K., Use of shift reagent with diastereomeric MTPA esters for determination of configuration and enantiomeric purity of secondary carbinols in H-1 NMR spectroscopy, *Tetrahedron,* p. 1363, 1976.

1027. Yamaguchi, S. and Yasuhara, F., Use of shift reagent with MTPA derivatives in H-1 NMR spectroscopy. II. Determination of absolute configuration and diastereomeric composition of secondary carbinols in epimeric mixture, *Tetrahedron Lett.*, p. 89, 1977.

1028. Vanhoeck, L., Bossaerts, J., Dommisse, R. A., Lepoivre, J. A., and Alderweireldt, F. C., The use of lanthanide-induced shifts for conformational and structural studies of (R)-MTPA (α-methoxy-α-trifluoromethyl-α-phenylacetic acid) esters, *Org. Magn. Reson.*, 22, 24, 1984.

1029. Vanhoeck, L., Merckx, E. M., Bossaerts, J., Lepoivre, J. A., and Alderweireldt, F. C., A model for the complex of the (R)-MTPA ester of trans-4-tert-butylcyclohexanol with Eu(fod)$_3$, based on lanthanide induced shifts, *Org. Magn. Reson.*, 21, 214, 1983.

1030. Merckx, E. M., Lepoivre, J. A., Lemiere, G. L., and Alderweireldt, F. C., Use of shift reagent with MTPA derivatives in F-19 NMR spectroscopy, *Org. Magn. Reson.*, 21, 380, 1983.

1031. Merckx, E. M., Van de Wal, A. J., Lepoivre, J. A., and Alderweireldt, F. C., Use of shift reagent with MTPA derivatives in F-19 NMR spectroscopy: determination of absolute configuration of stereomeric 2- and 3-substituted cyclohexanols, *Bull. Soc. Chim. Belg.*, 87, 21, 1978.

1032. Kalyanam, N. and Lightner, D. A., A convenient method for the determination of absolute configuration and enantiomeric excess of bicyclic secondary carbinols, *Tetrahedron Lett.*, p. 415, 1979.

1033. Murai, A., Sasamori, H., and Masamune, T., The configuration of 17-ethyl-3β-hydroxyetiojerva-5,12,17(20)-trien-11-one and related compounds, *Chem. Lett.*, p. 669, 1977.

1034. Lightner, D. A., Bouman, T. D., Gawronski, J. K., Gawronska, K., Chappuis, J. L., Crist, B. V., and Hansen, A. E., Dissymmetric chromophores. 7. On the optical activity of conjugated cisoid dienes: an experimental-theoretical study of 5-alkyl-1,3-cyclohexadienes, *J. Am. Chem. Soc.*, 103, 5314, 1981.

1035. Van de Wal, A. J., Merckx, E. M., Lemiere, G. L., Van Osselaer, T. A., Lepoivre, J. A., and Alderweireldt, F. C., Use of shift reagent with MTPA derivatives in F-19 NMR spectroscopy. III. Determination of absolute configuration and enantiomeric purity of secondary alcohols, *Bull. Soc. Chim. Belg.*, 87, 545, 1978.

1036. Yasuhara, F., Kabuto, K., and Yamaguchi, S., Use of shift reagent with MTPA derivatives in H-1 NMR spectroscopy. IV. Determination of absolute configuration and enantiomeric purity of amino acid derivatives, *Tetrahedron Lett.*, p. 4289, 1978.

1037. Kabuto, K., Yasuhara, F., and Yamaguchi, S., Use of shift reagent with MTPA (α-methoxy-α-trifluoromethylphenylacetic acid) derivatives. V. Determination of absolute configuration and enantiomeric purity of axially chiral compounds, *Koen Yoshishu — Hibenzenkei Hokozoku Kagaku Toronkai[oyobi] Kozo Yuki Kagaku Toronkai*, 12th, 217, 1979; *Chem. Abstr.*, 92, 180570s.

1038. Kabuto, K., Yasuhara, F., and Yamaguchi, S., A revised absolute configuration of 2-hydroxy-1,1'-binaphthyl, *Bull. Chem. Soc. Jpn.*, 56, 1263, 1983.

1039. Kabuto, K., Yasuhara, F., and Yamaguchi, S., Determination of enantiomeric purity of axially chiral biaryls, *Tetrahedron Lett.*, 21, 307, 1980.

1040. Kabuto, K., Yasuhara, F., and Yamaguchi, S., Determination of absolute configuration of axially chiral biaryls, *Tetrahedron Lett.*, 22, 659, 1981.

1041. Van Os, C. P. A., Vente, M., and Vliegenthart, J. F. G., A NMR method for determination of the enantiomeric composition of hydroperoxides formed by lipoxygenase, *Biochim. Biophys. Acta*, 574, 103, 1979.

1042. Jurczak, J. and Konowal, A., Use of Eu(fod)$_3$ with diastereomeric (−)-w-camphanic esters for determination of enantiomeric purity of alcohols by H-1 NMR spectroscopy. A convenient method for monitoring racemate resolution, *Pol. J. Chem.*, 52, 1967, 1978.

1043. Elsenbaumer, R. L., Mosher, H. S., Morrison, J. D., and Tomaszewski, J. E., Stereochemical course of the "mixed hydride" (AlD$_3$ and AlCl$_2$H) reduction of optically active styrene-2,2-d$_2$ oxide, *J. Org. Chem.*, 46, 4034, 1981.

1044. Gerlach, H. and Zagalak, B., Determination of the enantiomeric purity and absolute configuration of α-deuterated primary alcohols, *J. Chem. Soc. Chem. Commun.*, p. 274, 1973.

1045. Caspi, E. and Eck, C. R., Preparative scale synthesis of (1R) [1-^{2}H$_1$] or [1-^{3}H$_1$] primary alcohols of high optical purity, *J. Org. Chem.*, 42, 767, 1977.

1046. Jurczak, J., Konowal, A., Krawczyk, Z., and Ejchart, A., Application of H-1 NMR — Eu(fod)$_3$-shifted spectra for the determination of the enantiomeric composition and absolute configuration of secondary alcohols, using (−)-w-camphanic esters, *Org. Magn. Reson.*, 17, 50, 1981.

1047. De Munari, S., Marazzi, G., and Forgione, A., Determination of absolute configuration of the derivative from 2-[4-(1-oxo-2-isoindolinyl)-phenyl]-propionic acid and R-(+)-1-phenylethylamine by H-1 NMR spectroscopy, *Tetrahedron Lett.*, 21, 2273, 1980.

1048. Zaitsev, V. P., Potopov, V. M., Dem'yanovich, V. M., and Solov'eva, L. D., Stereochemical investigations. XLIII. Certain features of the use of paramagnetic shift reagents for determination of optical purity, *Z. Org. Khim.*, 12, 2326, 1976.

1049. Furukawa, J., Iwasaki, S., and Okuda, S., Method for determination of enantiomeric composition and absolute configuration of 2,3-deuterated 3-alkylpropanols, *Tetrahedron Lett.*, 24, 5257, 1983.
1050. Furukawa, J., Iwasaki, S., and Okuda, S., Eu(fod)$_3$-shifted ^{2}H-NMR as a probe of chirality due to deuterium substitution: stereospecific deuterium incorporation into 2-n-hexyl-5-n-propylresorcinol. A polyketide produced by *Pseudomonus* sp. B-9004, *Tetrahedron Lett.*, 24, 5261, 1983.
1051. Pirkle, W. H. and Sikkenga, D. L., Use of achiral shift reagents to indicate relative stabilities of diastereomeric solvates, *J. Org. Chem.*, 40, 3430, 1975.
1052. Pirkle, W. H. and Sikkenga, D. L., The use of chiral solvating agents for nuclear magnetic resonance determination of enantiomeric purity and absolute configuration of lactones. Consequences of three-point interactions, *J. Org. Chem.*, 42, 1370, 1977.
1053. Jennison, C. P. R. and Mackay, D., Lanthanide induced enhancement of enantiomeric shifts in chiral solvents and its use in the determination of optical purity, *Can. J. Chem.*, 51, 3726, 1973.
1054. Bus, J. and Lok, C. M., Stereospecific H-1 NMR analysis of trimethylsilyl derivatives of glycerides with a chiral shift reagent, *Chem. Phys. Lipids*, 21, 253, 1978.
1055. Diaz, E., Rojas-Davila, E., Guzman, A., and Joseph-Nathan, P., The use of chiral shift reagents in an NMR study of 2,2',6,6'-tetrasubstituted biphenyls, *Org. Magn. Reson.*, 14, 439, 1980.
1056. Damiano, J. C., Luche, J. L., and Crabbe, P., Stereochemistry of allenic compounds. Absolute configuration of cycloallenols and cycloallenones, *Tetrahedron*, 34, 3137, 1978.
1057. Gallina, C., Lucente, G., and Pinnen, F., Asymmetric hydrogenation of β-methylcinnamates of phenylalkylmethanols. Catalyst-phenyl groups interactions in the preferred adsorption conformation, *Tetrahedron*, 34, 2361, 1978.
1058. Hofer, E., Keuper, R., and Renken, H., Chiral relaxation reagents (CRR) in C-13 NMR spectroscopy, *Tetrahedron Lett.*, 25, 1141, 1984.
1059. Haller, R. and Bruer, H. J., Zur bestimmung det enantiomerenreinheit bei phenyl-alkyl-carbinolen mittels chiraler lanthaniden-verschiebungsreagenzien, *Arch. Pharm.*, 309, 367, 1976.
1060. Capillon, J. and Lacombe, L., Effets des substituants dans les spectres de rmn de benzhydrols substitues en presence de Eu(dcm)$_3$, *Can. J. Chem.*, 57, 1446, 1979.
1061. Nonaka, T., Ota, T., and Fuchigami, T., Stereochemical studies of the electrolytic reactions of organic compounds. IV. Electrolytic reduction of optically-active 1-pyridylalkanols to the corresponding substituted-alkyl pyridines, *Bull. Chem. Soc. Jpn.*, 50, 2965, 1977.
1062. Lau, K. S. Y., Wong, P. K., and Stille, J. K., Oxidative addition of benzyl halides to zero-valent palladium complexes. Inversion of configuration at carbon, *J. Am. Chem. Soc.*, 98, 5832, 1976.
1063. Morrison, J. D., Tomaszewski, J. E., Mosher, H. S., Dale, J., Miller, D., and Elsenbaumer, R. L., Hydrogen vs. deuterium transfer in asymmetric reductions: reduction of phenyl trifluoromethyl ketone by the chiral Grignard reagent from (S)-2-phenyl-1-bromoethane-1,1,2-d$_3$, *J. Am. Chem. Soc.*, 99, 3167, 1977.
1064. Dongala, E. B., Solladie-Cavallo, A., and Solladie, G., Determination de la purete enantiomerique de β-hydroxyesters partiellement actifs par RMN. Utilisation du tris (trifluoromethylhydroxymethylene)-3-camphorato-d europium III, *Tetrahedron Lett.*, p. 4233, 1972.
1065. Korver, O. and van Gorkom, M., Optically active 2-methyl substituted acids and esters: chiroptical properties, conformational equilibria and NMR with optically active shift reagents, *Tetrahedron*, 30, 4041, 1974.
1066. Liu, J. H. and Tsay, J. T., Use of chiral lanthanide shift reagents for the nuclear magnetic resonance spectrometric determination of amphetamine enantiomers, *Analyst (London)*, 107, 544, 1982.
1067. Jakovac, I. J. and Jones, J. B., Determination of enantiomeric purity of chiral lactones. A general method using nuclear magnetic resonance, *J. Org. Chem.*, 44, 2165, 1979.
1068. Bus, J., Lok, C. M., and Groenewegen, A., Determination of enantiomeric purity of glycerides with a chiral PMR shift reagent, *Chem. Phys. Lipids*, 16, 123, 1976.
1069. Harrison, D. M. and Quinn, P., The stereospecific synthesis of (S)-2-[^{2}H$_3$]methyl-2-methylbutanol. Characterisation of the (R) and (S) enantiomers of the recemic [^{2}H$_3$] alcohol by H-2 NMR in the presence of a chiral shift reagent, *Tetrahedron Lett.*, 24, 831, 1983.
1070. Wilson, W. K., Scallen, T. J., and Morrow, C. J., Determination of the enantiomeric purity of mevalonolactone via NMR using a chiral lanthanide shift reagent, *J. Lipid Res.*, 23, 645, 1982.
1071. Midland, M. M., McDowell, D. C., Hatch, R. L., and Tramontano, A., Reduction of α,β-acetylenic ketones with β-3-pinanyl-9-borabicyclo[3.3.1]nonane. High asymmetric induction in aliphatic systems, *J. Am. Chem. Soc.*, 102, 867, 1980.
1072. Claesson, A., Olsson, L. I., Sullivan, G. R., and Mosher, H. S., Synthesis of chiral allenic alcohols and nuclear magnetic resonance determination of their enantiomeric purities using a chiral lanthanide shift reagent, *J. Am. Chem. Soc.*, 97, 2919, 1975.
1073. Karasawa, D., Studies on the optical purity on terpinen-4-ol. Determination of optical purity using chiral shift reagent for NMR spectroscopy, *Shinshu Daigaku Nogakubu Kiyo*, 14, 119, 1977; *Chem. Abstr.*, 89, 24543t.

1074. Mikami, Y., Watanabe, E., Fukunaga, Y., and Kisaki, T., Formation of 2S-hydroxy-β-ionone and 4E-hydroxy-β-ionone by microbial hydroxylation of β-ionone, *Agric. Biol. Chem.*, 42, 1075, 1978.
1075. Takeya, K. and Itokawa, H., Stereochemistry in oxidation of allylic alcohols by cell-free system of callus induced from *Cannabis sativa* L., *Chem. Pharm. Bull.*, 25, 1947, 1977.
1076. Meyer, H. and Seebach, D., Synthese einiger pilzmetabolite mit 4-methoxy-5,6-dihydro-2-pyronstruktur, *Justus Liebigs Ann. Chem.*, p. 2261, 1975.
1077. He, X. C., Wen, Y. C., Yi, D. N., and Xu, G. Y., Application of NMR chiral shift reagent Eu(facam)$_3$, *Yu Chi Hua Hsueh*, 2, 115, 1981; *Chem. Abstr.*, 95, 96358h.
1078. Hoffmann, R. W., Gerlach, R., and Goldmann, S., Stereochemistry of a [2,3]sigmatropic allylsulfoxide/allyl sulfenate rearrangement, *Tetrahedron Lett.*, p. 2599, 1978.
1079. Mori, K., Synthesis of the both enantiomers of grandisol, the boll weevil pheromone, *Tetrahedron*, 34, 915, 1978.
1080. Connor, A. H. and Rowe, J. W., Differentiating manool and 13-epimanool with NMR chiral shift reagents, *Phytochemistry*, 15, 1949, 1976.
1081. Hellwinkel, D. and Krapp, W., Polycyclische triaryldioxyphosphorane extremer stabilitat, *Chem. Ber.*, 111, 13, 1978.
1082. Jankowski, K., Israeli, J., Chiasson, J. B., and Rabczenko, A., Configuration of some alkaloids by a chiral NMR shift reagent, *Heterocycles*, 19, 1215, 1982.
1083. McKinney, J. D., Matthews, H. B., and Wilson, N. K., Determination of optical purity and prochirality of chlorinated polycyclodiene pesticide metabolites, *Tetrahedron Lett.*, p. 1895, 1973.
1084. Hikino, H., Mohri, K., Hikino, Y., Arihara, S., Takemoto, T., Mori, H., and Shibata, K., Inokosterone, an insect metamorphosing substance from *Achyranthes fauriei*. Absolute configuration and synthesis, *Tetrahedron*, 32, 3015, 1976.
1085. Behnam, B. A., Hall, D. M., and Modarai, B., Detection of enantiomers in dissymmetric biaryls with chiral shift reagents, *Tetrahedron Lett.*, p. 2619, 1979.
1086. Mannschreck, A., Jonas, V., Bodecker, H. O., Elbe, H. L., and Kobrich, G., Diastereomeric association complexes of chiral butadienes, *Tetrahedron Lett.*, p. 2153, 1974.
1087. Becher, G., Burgemeister, T., Henschel, H. H., and Mannschreck, A. A., Applications of NMR spectroscopy of chiral association complexes. 5. Stereodynamics and diastereotopism in chiral 1,3-dienes HOCMe$_2$-(CCl=CCl)$_2$-X, *Org. Magn. Reson.*, 11, 481, 1978.
1088. Fraser, R. R., Petit, M. A., and Miskow, W. M., Separation of nuclear magnetic resonance signals of internally enantiotropic protons using a chiral shift reagent. The deuterium isotope effect on geminal proton-proton coupling constants, *J. Am. Chem. Soc.*, 94, 3253, 1972.
1089. Kainosho, M., Ajisaka, K., Pirkle, W. H., and Beare, S. D., The use of chiral solvents or lanthanide shift reagents to distinguish meso from d or l diastereomers, *J. Am. Chem. Soc.*, 94, 5924, 1972.
1090. Bhole, S. I. and Gogte, V. N., Optical induction. II. Determination of enantiomeric excess in optically active β-phenylmercaptoethyl aryl/alkyl ketones, *Indian J. Chem.*, 20B, 222, 1981.
1091. Sadler, D. E., Hildenbrand, K., and Schaffner, K., Concerning the stereochemical course and mechanism of the photochemical 1,3-acetyl shift in a β-unsaturated ketone, *Helv. Chem. Acta*, 65, 2071, 1982.
1092. Duggan, P. G. and Murphy, W. S., Intramolecular alkylation of phenols. III. Asymmetric induction by a chiral leaving group, *J. Chem. Soc. Perkin Trans. 1*, p. 634, 1976.
1093. Grubbs, R. H. and Pancoast, T. A., The mechanism of the oxidative decomposition of cyclobutadienyliron tricarbonyl complexes: intramolecular trapping, *J. Am. Chem. Soc.*, 99, 2382, 1977.
1094. Sinnema, A., van Rantwijk, F., de Koning, A. J., van Wijk, A. M., and van Bekkum, H., Thermally induced skeletal inversion and ring opening of hexamethylbicyclo[2.2.0]hexanes, *Tetrahedron*, 32, 2269, 1976.
1095. Anastassiou, A. G. and Hasan, M., Evidence for skeletal enantiomerism in benzo[9]annulenone, *Tetrahedron Lett.*, 24, 4279, 1983.
1096. Anastassiou, A. G. and Hasan, M., The existence of helical chirality in a sterically congested naphtho[9]annulenone, *Helv. Chim. Acta*, 65, 2526, 1982.
1097. Agranat, I., Tapuhi, Y., and Lallemand, J. Y., Manifestation of chirality in dynamic stereoisomers lacking asymmetric centers by the LIS technique. The chirality of bianthrones, *Nouv. J. Chim.*, 3, 59, 1979.
1098. Cameron, D. W., Edmonds, J. S., Feutrill, G. I., and Hoy, A. E., The use of chiral NMR shift reagents in a study of the stereochemistry of bianthrones, *Aust. J. Chem.*, 29, 2257, 1976.
1099. Rahman, W., Ilyas, M., Okigawa, M., and Kawano, N., Atropisomerism of biflavones. Confirmation of the enantiomeric purity of WB1 by using a chiral nuclear magnetic resonance shift reagent, *Chem. Pharm. Bull.*, 30, 1491, 1982.
1100. Holik, M. and Mannschreck, A., Applications of NMR spectroscopy of chiral association complexes. 6. Rotation about the C(sp^2)-C(aryl) bond in 2,6-disubstituted pivalophenones, *Org. Magn. Reson.*, 12, 28, 1979.

1101. Domagala, J. M., Bach, R. D., and Wemple, J., Rearrangement of optically active ethyl (E)-3-methyl-3-phenylglycidate. Evidence for concerted carbethoxy migration, *J. Am. Chem. Soc.*, 98, 1975, 1976.

1102. Hayashi, T., Konishi, M., Fukushima, M., Kanehira, K., Hioki, T., and Kumada, M., Chiral (β-aminoalkyl)phosphines. Highly efficient phosphine ligands for catalytic asymmetric Grignard cross-coupling, *J. Org. Chem.*, 48, 2195, 1983.

1103. Schmid, M. and Barner, R., Totalsynthese von naturlichem a-tocopherol. 2. Mitteilung. Aufbau der seitenkette aus (−)-(S)-3-methyl- -butyrolacton, *Helv. Chim. Acta*, 62, 464, 1979.

1104. Valentine, D., Jr., Chan, K. K., Scott, C. G., Johnson, K. K., Toth, K., and Saucy, G., Direct determination of R/S enantiomer ratios of citronellic acid and related substances by nuclear magnetic resonance spectroscopy and high pressure liquid chromatography, *J. Org. Chem.*, 41, 62, 1976.

1105. Valentine, D., Jr., Johnson, K. K., Priester, W., Sun, R. C., Toth, K., and Saucy, G., Rhodium chiral monophosphine complex catalyzed hydrogenations of terpenic and α-(acylamino)-substituted acrylic acids, *J. Org. Chem.*, 45, 3698, 1980.

1106. Schneider, W. P., Bundy, G. L., Lincoln, F. H., Daniels, E. G., and Pike, J. E., Isolation and chemical conversions of prostaglandins from *Plexaura homomalla:* preparation of prostaglandin E_2, prostaglandin $F_{2\alpha}$, and their 5,6-trans isomers, *J. Am. Chem. Soc.*, 99, 1222, 1977.

1107. Rackham, D. M., Lanthanide shift reagents. XV. Quantitation of D,L-amino acid enantiomers using chiral lanthanide shift reagents, *Spectrosc. Lett.*, 13, 321, 1980.

1108. Wilson, M. E. and Whitesides, G. M., Conversion of a protein to a homogeneous asymmetric hydrogenation catalyst by site-specific modification with a diphosphinerhodium(I) moiety, *J. Am. Chem. Soc.*, 100, 306, 1978.

1109. Ajisaka, K., Kamisaku, M., and Kainosho, M., Enantiomeric shift difference induced by the lanthanide shift reagent: its correlation with absolute configuration of α-amino acids, *Chem. Lett.*, p. 857, 1972.

1110. Lang, R. W. and Hansen, H. J., H-1 NMR spektroskopische bestimmung der enantiomerenreinheit von allencarbonsaureestern mit optisch aktiven europium-verschiebungsreagenzien, *Helv. Chim. Acta*, 62, 1025, 1979.

1111. Coe, G. L., Photochemical addition of dimethyl maleate to 2,3-dimethyl-2-butene. Use of a chiral shift reagent, *J. Org. Chem.*, 38, 4285, 1973.

1112. Trost, B. M. and Strege, P. E., Asymmetric induction in catalytic allylic alkylation, *J. Am. Chem. Soc.*, 99, 1649, 1977.

1113. Jankowski, K. and Rabczenko, A., Application of an optically active nuclear magnetic resonance shift reagent to configurational problems, *J. Org. Chem.*, 40, 960, 1975.

1114. Morton, D. R. and Morge, R. A., Total synthesis of 3-oxa-4,5,6-trinor-3,7-inter-m-phenylene prostaglandins. I. Photochemical approach, *J. Org. Chem.*, 43, 2093, 1978.

1115. Goering, H. L. and Kantner, S. S., Absolute configurations and rotations of exo- and endo-2-methylbicyclo[3.2.1]oct-3-ene, *J. Org. Chem.*, 46, 4605, 1981.

1116. Janusz, J. M. and Berson, J. A., Trisection of reaction pathways in automerization of 6-methylenebicyclo[3.2.1]oct-2-ene, *J. Am. Chem. Soc.*, 100, 2237, 1978.

1117. Kroll, J. A., Determination of the enantiomorphic composition of cocaine using the chiral lanthanide shift reagent europium tris-d-trifluroacetylcamphorate, *J. Forensic Sci.*, 24, 303, 1979.

1118. Nakazaki, M., Naemura, K., and Arashiba, N., Synthesis and absolute configuration of (−)-D_{2d}-bisnoradamantan-2-one (tricyclo[3.3.0.0^{3,7}]octan-2-one), *J. Org. Chem.*, 43, 888, 1978.

1119. Nakazaki, M., Naemura, K., and Arashiba, N., Syntheses and chiroptical properties of (−)-ditwistbrendane and (+)-D_3-trishomocubane, *J. Org. Chem.*, 43, 689, 1978.

1120. Hasaka, N., Okigawa, M., Kauno, I., and Kawano, N., Optical resolution of 2,2′,6,6′-tetrafluorobiphenyl-3,3′-dicarboxylic acid, *Bull. Chem. Soc. Jpn.*, 55, 3828, 1982.

1121. Kawano, N., Okigawa, M., Hasaka, N., Kouno, I., Kawahara, Y., and Fujita, Y., Atropisomerism of biphenyl compounds. An important role of ortho-substituted methoxy groups and fluorine atoms, *J. Org. Chem.*, 46, 389, 1981.

1122. Stridsberg, B. and Allenmark, S., A stereoselective Pummerer rearrangement of an optically active sulfoxide, *Acta Chem. Scand. B*, 30, 219, 1976.

1123. Koch, T. H., Olesen, J. A., and DeNiro, J., Photochemical reactivity of amino lactones. Photoreduction and photoelimination, *J. Org. Chem.*, 40, 14, 1975.

1124. Matsushita, H., Noguchi, M., and Yoshikawa, S., Optical activation of racemic α-substituted carbonyl compounds using optically active amines, *Bull. Chem. Soc. Jpn.*, 49, 1928, 1976.

1125. Mannschreck, A., Magnetische kernresonanz in gegenwart optisch-aktiver hilfsverbindungen, *Nachr. Chem. Techn.*, 23, 295, 1975.

1126. Colonna, S., Fornasier, R., and Pfeiffer, U., Asymmetric induction in the Darzens reaction by means of chiral phase-transfer in a two-phase system. The effect of binding the catalyst to solid polymeric support, *J. Chem. Soc. Perkin Trans. 1*, p. 8, 1978.

1127. Pluim, H. and Wynberg, H., Catalytic asymmetric induction in oxidation reactions. Synthesis of optically active epoxynaphthoquinones, *J. Org. Chem.*, 45, 2498, 1980.
1128. Baldwin, J. E. and Broline, B. M., Kinetics of the stereomutations of (+)-2-deuterio-3,7-dimethyl-7-methoxymethylcyclohepta-1,3,5-triene: one-centered epimerization at C(7) and [1,5] carbon migration with retention of configuration in a norcaradiene, *J. Am. Chem. Soc.*, 100, 4599, 1978.
1129. Wainer, I. W., Schneider, L. C., and Weber, J. D., Application of chiral lanthanide nuclear magnetic resonance shift reagents to pharmaceutical analyses. II. Determination of dextro- and levoamphetamine mixtures, *J. Assoc. Off. Anal. Chem.*, 64, 848, 1981.
1130. Ali, A. R. E. N. O., Application of chiral lanthanide shift reagents for the determination of enantiomeric and diastereoisomeric ratios of chiral mixtures of amphetamine and ephedrines, *Indian J. Chem. Sect. B*, 22, 762, 1983.
1131. Ossman, A. R. E. N. and Mathieson, D. W., Determination of the enantiomeric ratios of chiral mixtures of amphetamine by NMR, *J. Pharm. Belg.*, 36, 348, 1981.
1132. Buyuktimkin, N., Enantiomeric purity of ephedrine samples with chiral NMR shift reagent Eu(hfc)$_3$, *Sci. Pharm.*, 52, 158, 1984.
1133. Buyuktimkin, N., Determination of the enantiomeric purity of norephedrine samples by using the chiral lanthanide shift reagent Eu(hfc)$_3$, *Eczacilik Bul.*, 25, 65, 1983; *Chem. Abstr.*, 100, 126981x.
1134. Buyuktimkin, N., Determination of the enantiomeric purity of methadone using the chiral NMR shift reagent Eu(hfc)$_3$, *Arch. Pharm.*, 317, 653, 1984.
1135. Cockerill, A. F., Davies, G. L. O., Harrison, R. G., and Rackham, D. M., NMR determination of the enantiomer composition of penicillamine using an optically active europium shift reagent, *Org. Magn. Reson.*, 6, 669, 1974.
1136. Gogte, V. N., Nanda, R. K., Natu, A. A., Pandit, V. S., and Sastry, M. K., Determination of enantiomeric excess by C-13 NMR spectroscopy, *Org. Magn. Reson.*, 22, 624, 1984.
1137. Shaath, N. A. and Soine, T. O., Determination of the enantiomeric purity of isoquinoline alkaloids by the use of chiral lanthanide nuclear magnetic resonance shift reagents, *J. Org. Chem.*, 40, 1987, 1975.
1138. Wainer, I. W., Tischler, M. A., and Sheinin, E. B., Determination of dextro- and levomethorphan mixtures using chiral lanthanide NMR shift reagents, *J. Pharm. Sci.*, 69, 459, 1980.
1139. Tatone, D., Dich, T. C., Nacco, R., and Botteghi, C., Optically active heteroaromatic compounds. VII. Synthesis of the three optically active sec-butylpyridines, *J. Org. Chem.*, 40, 2987, 1975.
1140. Buyuktimkin, N. and Schunack, W., Determination of the enantiomeric purity of levamisole and dexamisole using a lanthanide shift reagent, *Arch. Pharm.*, 316, 1042, 1983.
1141. Holik, M. and Mannschreck, A., Application of NMR spectroscopy of chiral association complexes 8. Rotation of the C(sp^2)-C(aryl) bond in 2,6-disubstituted benzamides, *Org. Magn. Reson.*, 12, 223, 1979.
1142. Avolio, J. and Rothchild, R., Optical purity determination and H-1 NMR spectral simplification with lanthanide shift reagents. IV. Miphobarbital, 1-methyl-5-ethyl-5-phenylbarbituric acid, *J. Magn. Reson.*, 58, 328, 1984.
1143. Knabe, J. and Gradmann, V., Bestimmung der enantiomeren reinheit chiraler barbiturate mit lanthaniden-shiftreagenzien, *Arch. Pharm.*, 310, 468, 1977.
1144. Rothchild, R. and Simons, P., Optical purity determination and H-1 NMR spectral simplification with lanthanide shift reagents. VI. Methohexital, α-dl-5-allyl-1-methyl-5-(1-methyl-2-pentynyl) barbituric acid, *Spectrochim. Acta*, 40A, 881, 1984.
1145. Eberhart, S. and Rothchild, R., Optical purity determination and proton NMR spectral simplification with lanthanide shift reagents. II. Thiamylal, 5-allyl-5-(1-methylbutyl)-2-thiobarbituric acid, *Appl. Spectrosc.*, 38, 74, 1984.
1146. Eberhart, S. and Rothchild, R., Optical purity determination and proton NMR spectral simplification with lanthanide shift reagents: glutethimide, 3-ethyl-3-phenyl-2,6-piperidinedione, *Appl. Spectrosc.*, 37, 292, 1983.
1147. Avolio, J. and Rothchild, R., Optical purity determination and proton NMR spectral simplification with lanthanide shift reagents. III. Ethotoin, 3-ethyl-5-phenyl-2,4-imidazolidinedione, *Appl. Spectrosc.*, 38, 734, 1984.
1148. Fourrey, J. L. and Moron, J., Thiocarbonyl photochemistry. VII. Photochemistry of 1,3-dimethyl 4-thiouracil in presence of triethylamine, *Tetrahedron Lett.*, p. 301, 1976.
1149. Yasumoto, M., Ueda, S., Yamashita, J., and Hashimoto, S., Studies on tetrahydrofuranyl-5-fluorouracils. II. NMR studies of 1-(tetrahydro-2-furanyl)-, 3-(tetrahydro-2-furanyl)- and 1,3-bis(tetrahydro-2-furanyl)-5-fluoruracils, *Nucl. Acids Symp. Ser.*, 6, 585, 1979; *Chem. Abstr.*, 93, 94330c.

1150. Reisberg, P., Brenner, I. A., and Bodin, J. I., NMR determination of enantiomers of 7-chloro-3,3α-dihydro-2-methyl-2H,9H-isoxazolo-[3,2-b][1,3]benzoxazin-9-one using chiral shift reagent, tris[3-(heptafluorobutyryl)-d-camphorato]europium(III), *J. Pharm. Sci.,* 65, 592, 1976.

1151. Seitz, G. and Kromeke, G., Notiz sur reaktion von all-cis cyclopentantetracarbonsaure-dianhydrid mit methylamin, *Arch. Pharm.,* 309, 930, 1976.

1152. Lefevre, F., Burgemeister, T., and Mannschreck, A., H-1 NMR lineshape kinetics of intramolecular motions by means of an optically active lanthanide complex, *Tetrahedron Lett.,* p. 1125, 1977.

1153. Mannschreck, A., Jonas, V., and Kolb, B., Determination of rate of internal rotation by H-1 NMR spectroscopy in the presence of optically active auxiliary compounds, *Angew. Chem. Int. Ed.,* 12, 909, 1973.

1154. Uncuta, C., Bally, I., Draghici, C., Chiraleu, F., and Balaban, A. T., Heterocyclic organoboron compounds. XVII. Axial chirality of boron chelates evidenced by H-1 NMR spectroscopy in the presence of chiral additives, *Rev. Roum. Chim.,* 29, 121, 1984.

1155. Annunziata, R., Cinquini, M., and Cozzi, F., Enantiomeric excess determination of new classes of chiral sulphur compounds by the use of europium shift reagents, *Org. Magn. Reson.,* 21, 183, 1983.

1156. Deshmukh, M., Dunach, E., Juge, S., and Kagan, H. B., A convenient family of chiral shift reagents for measurement of enantiomeric excesses of sulfoxides, *Tetrahedron Lett.,* 25, 3467, 1984.

1157. Hoffman, R. W. and Goldmann, S., Intramolekulare substitution von sulfensaure-allylestern, *Chem. Ber.,* 111, 2716, 1978.

1158. Nozaki, H., Yoshino, K., Oshima, K., and Yamamoto, Y., The determination of the enantiomeric purity of methyl p-tolyl sulfoxide by means of an NMR shift reagent, *Bull. Chem. Soc. Jpn.,* 45, 3495, 1972.

1159. Tangerman, A. and Zwanenburg, B., Recognition of atropisomerism and thermally labile prochirality, and measurement of barriers to rotation in sulfines (thione oxides) by the use of chiral shift reagents, *J. R. Neth. Chem. Soc.,* 96, 196, 1977.

1160. Lanzilotta, R. P., Bradley, D. G., and Maddox, M. L., Stereospecific sulfur oxidation of 7-methyl-thioxanthone-2-carboxylic acid by Calonectria decora, *Appl. Environ. Microbiol.,* 34, 56, 1977.

1161. Cinquini, M. and Cozzi, F., Synthesis of optically active N-alkylidenesulphanamides, *J. Chem. Soc. Chem. Commun.,* 502, 1977.

1162. Lefevre, F., Martin, M. L., and Capmau, M. L., Resolution d'enantiomeres alleniques par resonance magnetique nucleaire, *C. R. Acad. Sci. Paris,* 275, 1387, 1972.

1163. Van den Berg, C. R., Beck, H. C., and Benschop, H. P., Stereochemical analysis of the nerve agents soman, sarin, tabun, and VX by proton NMR spectroscopy with optically active shift reagents, *Bull. Environ. Contam. Toxicol.,* 33, 505, 1984.

1164. Zon, G., Brandt, J. A., and Egan, W., Determination of enantiomeric homogeneity (optical purity) of cyclophosphamide by nuclear magnetic resonance spectroscopy, *J. Natl. Cancer Inst.,* 58, 1117, 1977.

1165. Houalla, D., Sanchez, M., and Wolf, R., Mise en evidence de la chiralite de l'edifice spirophosphoranique, *Org. Magn. Reson.,* 5, 451, 1973.

1166. Epps, L. A., Wiener, K., Stewart, R. C., and Marzilli, L. G., Tris(3-nitroso-2,4-pentanedionato)cobalt(III): strong evidence for a facial geometry analogous to ferroverdin, *Inorg. Chem.,* 16, 2663, 1977.

1167. Kaplan, F. A. and Roberts, B. W., Annelation of tricarbonyliron complexes of ortho-disubstituted[4]annulenes. Synthesis of a dibenzobicyclo[6.2.0]decapentaene via ortho,ortho' cyclobisacylation of a biphenyl, *J. Am. Chem. Soc.,* 99, 518, 1977.

1168. Le Plouzennec, M., Le Moigne, F., and Dabard, R., Premiers exemples d'induction et de synthese asymetriques en serie du cymantrene, *J. Organomet. Chem.,* 111, C38, 1976.

1169. Stephenson, C. R., Transition-metal mediated asymmetric synthesis. IV. Chelation of malonate esters by a chiral lanthanide shift reagent: a general method to measure the enantiomeric excess of tricarbonyl(cyclohexadienyl)iron(1+) salts, *Aust. J. Chem.,* 35, 1939, 1982.

1170. Stephenson, C. R., Transition metal mediated asymmetric synthesis. II. Complete resolution of (2R, 5S)-(−)-tricarbonyl-[1-5-η-(2-methoxy-5-methylcyclohexadienyl)]iron(1+)hexafluorophosphate(1−) by selective crystallization of enantiomers, *Aust. J. Chem.,* 34, 2339, 1981.

1171. Birch, A. J., Kelly, L. F., and Narula, A. S., Organometallic compounds in organic synthesis. 17. Reactions of tricarbonylcyclohexadienyliron salts with O-silylated enolates, allyl silanes and aspects of their synthetic equivalents, *Tetrahedron,* 38, 1813, 1982.

1172. Schurig, V., NMR-spektroskopischer nechweis der chiralitat pi-komplexierter prochiraler olefine, *Tetrahedron Lett.,* p. 1269, 1976.

1173. Hamer, G. and Shaver, A., Cyclopentadienyl platinum(IV) complexes: H-1, C-13 nuclear magnetic resonance and optically active shift reagent study, *Can. J. Chem.,* 58, 2011, 1980.

1174. Klaui, W. and Neukomm, H., Unterscheidung enantiotoper/diastereotoper protonen und regioselektive spinentkopplung in metallorganischen komplexen mit hilfe von shift reagenzien, *Org. Magn. Reson.*, 10, 126, 1977.
1175. De Renzi, A., Morelli, G., Panunzi, A., and Wurzburger, S., Enantiomeric discrimination in n^2-olefin complexes by chiral lanthanide shift reagents, *Inorg. Chem. Acta*, 76, L285, 1983.
1176. Rahm, A. and Pereyre, M., Determination de la purete enantiomerique d'un carbure organostannique au moyen de la RMN: utilisation des complexes d'europium de composes organostanniques fonctionnels, *Tetrahedron Lett.*, p. 1333, 1973.
1177. Meyers, A. I. and Ford, M. E., Oxazolines. XX. Synthesis of achiral and chiral thiiranes and olefins by reaction of carbonyl compounds with 2-(alkylthio)-2-oxazolines, *J. Org. Chem.*, 41, 1735, 1976.
1178. Dambska, A. and Janowski, A., Use of silver trifluoroacetate together with lanthanide shift reagents for simplification of H-1 NMR spectra of aromatic hydrocarbons, *Org. Magn. Reson.*, 13, 122, 1980.
1179. Dambska, A. and Janowski, A., Use of lanthanide shift reagents together with silver trifluoroacetate for quantitative analysis of mixtures of aromatic hydrocarbons, *Chem. Analityczna*, 25, 77, 1980.
1180. Wenzel, T. J. and Sievers, R. E., Binuclear complexes of lanthanide(III) and silver(I) and their function as shift reagents for olefins, aromatics, and halogenated compounds, *Anal. Chem.*, 53, 393, 1981.
1181. Wenzel, T. J. and Sievers, R. E., Nuclear magnetic resonance studies of terpenes with chiral and achiral lanthanide(III)-silver(I) binuclear shift reagents, *J. Am. Chem. Soc.*, 104, 382, 1982.
1182. Wenzel, T. J. and Sievers, R. E., Binuclear shift reagents for nuclear magnetic resonance spectrometry of aromatic and polycyclic aromatic compounds, *Anal. Chem.*, 54, 1602, 1982.
1183. Wenzel, T. J. and Lalonde, D. R., Jr., New binuclear NMR shift reagents for olefins and aromatics, *J. Org. Chem.*, 48, 1951, 1983.
1184. Wenzel, T. J., A better solvent for binuclear lanthanide(III)-silver(I) NMR shift reagent studies, *J. Org. Chem.*, 49, 1834, 1984.
1185. Bennett, M. J., Cotton, F. A., Legzdins, P., and Lippard, S. J., The crystal structure of cesium tetrakis(hexafluoroacetylacetonato)yttrate(III). A novel stereoisomer having dodecahedral eight-coordination, *Inorg. Chem.*, 7, 1770, 1968.
1186. Burns, J. H. and Danford, M. D., The crystal structure of cesium tetrakis(hexafluoroacetylacetonato)europate and -americate. Isomorphism with the yttrate, *Inorg. Chem.*, 8, 1780, 1969.
1187. Allen, G., Lewis, J., Long, R. F., and Oldham, C., A novel form of co-ordination of acetylacetone to platinum(II), *Nature (London)*, 202, 589, 1964.
1188. Watson, W. H., Jr. and Lin, C. T., The crystal structure of trisilver dinitrate tris(acetylacetonato)nickelate(II) monohydrate, *Inorg. Chem.*, 5, 1074, 1966.
1189. Offermann, W. and Fritzsche, U., Silver pi-complexes in chloroform. Stoichiometry, stabilities and proton NMR bound shifts, *Inorg. Chim. Acta*, 73, 113, 1983.
1190. Horrocks, W. D., Jr., Lanthanide shift reagents. A model which accounts for the apparent axial symmetry of shift reagent adducts in solution, *J. Am. Chem. Soc.*, 96, 3022, 1974.
1191. de Boer, J. W. M., Sakkers, P. J. D., Hilbers, C. W., and de Boer, E., Lanthanide shift reagents. II. Shift mechanisms, *J. Magn. Reson.*, 25, 455, 1977.
1192. Audit, M., Demerseman, P., Goasdoue, N., and Platzer, N., H-1 and C-13 NMR studies of binuclear lanthanide(III)-silver(I) shift or relaxation reagents: competition between several bonding sites in olefinic, aromatic and heteroaromatic compounds, *Org. Magn. Reson.*, 21, 698, 1983.
1193. Dambska, A. and Janowski, A., The origin of shifts induced by complex silver-lanthanide shift reagents in H-1 NMR spectra of unsaturated hydrocarbons, *J. Magn. Reson.*, 59, 13, 1984.
1194. Reilley, C. N., Good, B. W., and Allendoerfer, R. D., Separation of contact and dipolar lanthanide induced nuclear magnetic resonance shifts: evaluation and application of some structure independent models, *Anal. Chem.*, 48, 1446, 1976.
1195. Smith, W. B., The C-13 NMR of olefins complexed with a binuclear Ag(I)-Yb(III) chelate, *Org. Magn. Reson.*, 17, 124, 1981.
1196. McKenna, M., Wright, L. L., Miller, D. J., Tanner, L., Haltiwanger, R. C., and DuBois, M. R., Synthesis of inequivalently bridged cyclopentadienyl dimers of molybdenum and a comparison of their reactivities with unsaturated molecules and with hydrogen, *J. Am. Chem. Soc.*, 105, 5329, 1983.
1197. Abravanel, M., Demerseman, P., Goasdoue, N., and Platzer, N., Carbon-13 NMR study of weakly nucleophilic compounds with a binuclear relaxation reagent: $Ag(O_2CCF_3)Gd(fod)_3$ (fod = 6,6,7,7,8,8,8-heptafluoro-2,2-dimethyl-3,5-octanedione), *C. R. Seances Acad. Sci. Ser. 2*, 294, 513, 1982.
1198. Hepner, F. R., Trueblood, K. N., and Lucas, H. J., Coordination of silver ion with unsaturated compounds. IV. The butenes, *J. Am. Chem. Soc.*, 74, 1333, 1952.
1199. Wenzel, T. J., Metal Chelate Complexes as Nuclear Magnetic Resonance Shift Reagents, Paramagnetic Relaxation Reagents, and Fuel Additives, Ph.D. thesis, University of Colorado, Boulder, 1981.

1200. Wenzel, T. J., Secondary deuterium isotope effects with lanthanide(III)-silver(I) NMR shift reagents, *Spectrosc. Lett.*, 17, 77, 1984.

1201. Rackham, D. M., Crutchley, F. M., Tupper, D. E., and Boddy, A. C., Lanthanide shift reagents. Paper 21. Interaction of arenes with silver/lanthanide shift reagent and comparison with chromium tricarbonyl complexes, *Spectrosc. Lett.*, 14, 379, 1981.

1202. Beverwijk, C. D. M., van der Kerk, G. J. M., Leusink, A. J., and Noltes, J. G., Organosilver chemistry, *Organomet. Chem. Rev. Sect. A*, 5, 215, 1970.

1203. Parker, R. G. and Roberts, J. D., Nuclear magnetic resonance spectroscopy. Carbon-13 spectra of the silver(I) complexes of cyclopentene and cyclohexene, *J. Am. Chem. Soc.*, 92, 743, 1970.

1204. Beverwijk, C. D. M. and van Dongen, J. P. C. M., C-13 NMR study of pi-complex formation. I. Characteristic shifts of the carbon resonances of alkenes induced by silver(I), *Tetrahedron Lett.*, 4291, 1972.

1205. Partenheimer, W. and Johnson, E. H., Heats of reaction of triphenylphosphine with compounds of the type hexafluoroacetylacetonato(olefin)silver(I), *Inorg. Chem.*, 12, 1274, 1973.

1206. Muhs, M. A. and Weiss, F. T., Determination of equilibrium constants of silver-olefin complexes using gas chromatography, *J. Am. Chem. Soc.*, 84, 4697, 1962.

1207. Gil-Av, E. and Herling, J., Determination of the stability constants of complexes by gas chromatography, *J. Phys. Chem.*, 66, 1208, 1962.

1208. Rackham, D. M., Lanthanide shift reagents. Paper 22. The influence of mixed silver/praseodymium shift reagents on the proton NMR spectra of heteroaromatics, *Spectrosc. Lett.*, 14, 639, 1981.

1209. Krasutskii, P. A., Yurchenko, A. G., Rodionov, V. N., and Kulik, N. I., Structure of pi-complexes of norbornadiene and 7-tert-butoxynorbornadiene with silver europium fod(AgEu(fod)$_4$), *Teor. Eksp. Khim.*, 20, 54, 1984.

1210. Krasutsky, P. A., Yurchenko, A. G., Rodionov, V. N., and Jones, M., Jr., Lanthanide shift reagents as a structural probe for alkenes, *Tetrahedron Lett.*, 23, 3719, 1982.

1211. Khachaturov, A. S. and Ivanova, V. P., Use of two-center shift reagents for increasing the resolution of NMR spectra of unsaturated elastomers, *Zh. Obshch. Khim.*, 54, 1335, 1984.

1212. Shapiro, Y. Y., Dozorova, N. P., Turov, B. S., and Efimov, V. A., Use of lanthanide shift reagents for analysis of the microstructure of polydienes and polyalkenamers by proton NMR spectroscopy, *Vysokomol. Soedin. Ser. A*, 25, 955, 1983; *Chem. Abstr.*, 99, 6287q.

1213. Aoyagi, F., Maeno, S., Okuno, T., Matsumoto, H., Ikura, M., Hikichi, K., and Matsumoto, T., Gymnopilins, bitter principles of the big-laugher mushroom Gymnopilus spectabilis, *Tetrahedron Lett.*, 24, 1991, 1983.

1214. Audit, M., Davoust, D., Goasdoue, N., and Platzer, N., N-15, H-1, and C-13 NMR. A study of the complexation of binuclear lanthanide(III)-silver(I) shift or relaxation reagents with substrates containing multiply bonded nitrogen, *Magn. Reson. Chem.*, 23, 33, 1985.

1215. Akins, D. L., Resonance-enhanced Raman scattering by aggregated 2,2'-cyanine on colloidal silver, *J. Colloid. Interface Sci.*, 90, 373, 1982.

1216. Wenzel, T. J. and Zaia, J., Organic-soluble lanthanide nuclear magnetic shift reagents for sulfonium and isothis-uronium salts, *Anal. Chem.*, in press.

1217. Burkert, P. K., Fritz, H. P., Gretner, W., Keller, H. J., and Schwarzhans, K. E., PMR-spektren von komplexen der seltenen erden, *Tetrahedron Lett.*, p. 31, 1968.

1218. Corey, E. J. and Chaykovsky, M., Dimethyloxosulfonium methylide ((CH_3)$_2$SOCH$_2$) and dimethylsulfonium methylide ((CH_3)$_2$CH$_2$). Formation and application to organic synthesis, *J. Am. Chem. Soc.*, 87, 1353, 1965.

1219. Bost, R. W. and Everett, J. E., Sulfur studies. XVI. The synthesis of certain higher alkyl sulfonium salts and related compounds, *J. Am. Chem. Soc.*, 62, 1752, 1940.

1220. Trost, B. M. and Melvin, L. S., Jr., Synthesis of sulfonium salts, in *Sulfur Ylides*, Academic Press, New York, 1975, 6.

1221. Trost, B. M. and Bogdanowicz, M. J., Preparation of cyclopropyldiphenylsulfonium and 2-methylcyclopropyldiphenylsulfonium fluoroborate and their ylides. Stereochemistry of sulfur ylides, *J. Am. Chem. Soc.*, 95, 5298, 1973.

1222. LaRochelle, R. W., Trost, B. M., and Krepski, L., Preparation and chemistry of vinyl sulfonium ylides. New synthetic intermediates, *J. Org. Chem.*, 36, 1126, 1971.

1223. Tang, C. S. F. and Rapoport, H., Reaction of sulfonium ylides with diene esters, *J. Org. Chem.*, 38, 2806, 1973.

1224. Acheson, R. M. and Harrison, D. R., The synthesis, spectra, and reactions of some S-alkylthiophenium salts, *J. Chem. Soc. (C)*, p. 1764, 1970.

1225. Acheson, R. M. and Harrison, D. R., S-Alkylthiophenium salts, *J. Chem. Soc. Chem. Commun.*, p. 724, 1969.

1226. Brumlik, G. C., Kosak, A. I., and Pitcher, R., The synthesis of the S-methylthiophenium ion, *J. Am. Chem. Soc.*, 86, 5360, 1964.

1227. Anderson, K. K., Caret, R. L., and Karup-Neilsen, I., Nucleophilic substitution at tricoordinate sulfur(IV). Stereochemistry of dialkylarylsulfonium salt formation from alkyl aryl sulfoxides, *J. Am. Chem. Soc.*, 96, 8026, 1974.
1228. Urquhart, G. G., Gates, J. W., Jr., and Connor, R., n-Dodecyl (lauryl) mercaptan, in *Organic Synthesis: Collection,* Vol. 3, John Wiley & Sons, New York, 1955, 363.
1229. Speziale, A. J., Ethanedithiol, in *Org. Syn. Coll.,* Vol. 4, John Wiley & Sons, New York, 1963, 401.
1230. Pasto, D. J. and Johnson, C. R., *Organic Structure Determination,* Prentice-Hall, Englewood Cliffs, N.J., 1969, 350.
1231. Krausutskii, P. A., Rodionov, V. N., and Yurchenko, A. G., Use of lanthanide shift reagents for studying 3,7-dimethylenebicyclo[3.3.1]nonane derivatives, *Teor. Eksp. Khim.*, 19, 126, 1983.
1232. Krasutskii, P. A., Rodionov, V. N., Tikhonov, V. P., and Yurchenko, A. G., Silver d-camphor-10-sulfonate as a new chiral reagent for enantiomeric resolution of olefins, *Teor. Eksp. Khim.*, 20, 58, 1984.
1233. Peterson, P. E. and Jensen, B. L., The preparation of 1,3-bis(trimethylsilyl)allene and the observation of the enantiomers using olefin complexing chiral shift reagents, *Tetrahedron Lett.*, 25, 5711, 1984.
1234. Offermann, W. and Mannschreck, A., Application of NMR spectroscopy of chiral association complexes 14. Chiral recognition of terpene and cyclohexene hydrocarbons by H-1 and C-13 NMR, *Org. Magn. Reson.*, 22, 355, 1984.
1235. Offermann, W. and Mannschreck, A., Chiral recognition of alkene and arene hydrocarbons by H-1 and C-13 NMR. Determination of enantiomeric purity, *Tetrahedron Lett.*, 22, 3227, 1981.
1236. Reger, D. L., Cyanide, isocyanide, and nitrile derivatives of cyclopentadienyliron. Interaction of chiral metal complexes with an optically active shift reagent, *Inorg. Chem.*, 14, 660, 1975.
1237. Bleaney, B., Dobson, C. M., Levine, B. A., Martin, R. B., Williams, R. J. P., and Xavier, A. V., Origin of lanthanide nuclear magnetic resonance shifts and their uses, *J. Chem. Soc. Chem. Commun.*, p. 791, 1972.
1238. Bleaney, B., Nuclear magnetic resonance shifts in solution due to lanthanide ions, *J. Magn. Reson.*, 8, 91, 1972.
1239. Horrocks, W. D., Jr., The temperature dependencies of lanthanide-induced NMR shifts: evaluation of theoretical approaches and experimental evidence, *J. Magn. Reson.*, 26, 333, 1977.
1240. McGarvey, B. R., Temperature dependence of the pseudocontact shift in lanthanide shift reagents, *J. Magn. Reson.*, 33, 445, 1979.
1241. Golding, R. M. and Pyykko, P., On the theory of pseudocontact NMR shifts due to lanthanide complexes, *Molec. Phys.*, 26, 1389, 1973.
1242. Bovee, W. M. M. J., Alberts, J. H., Peters, J. A., and Smidt, J., Temperature dependence of the lanthanide-induced shifts, structure, and dynamics of adducts of quinuclidine and Ln(fod)$_3$ chelates as studied by variable-temperature NMR shift and relaxation measurements, *J. Am. Chem. Soc.*, 104, 1632, 1982.
1243. Stout, E. W., Jr. and Gutowsky, H. S., On the temperature dependence of lanthanide-induced NMR shifts, *J. Magn. Reson.*, 24, 389, 1976.
1244. Hill, H. A. O., Williams, D., and Zarb-Adami, N., Origin of isotropic shifts in lanthanide complexes: a study of the temperature dependence of the H-1 NMR spectra of the tetrakis-N,N-diethyldithiocarbamatolanthanate(III) anions, *J. Chem. Soc. Faraday Trans. 2*, 72, 1494, 1976.
1245. Beaute, C., Cornuel, S., Lelandais, D., Thoai, N., and Wolkowski, Z. W., Temperature dependence of NMR shifts induced by tris(dipivalomethanato)ytterbium, *Tetrahedron Lett.*, p. 1099, 1972.
1246. Lee, L. and Sykes, B. D., The temperature dependence of lanthanide-induced H-1 NMR shifts observed for the interaction of Yb(III) with the calcium-binding protein parvalbumin, *J. Magn. Reson.*, 41, 512, 1980.
1247. Cheng, H. N. and Gutowsky, H. S., Temperature dependence of lanthanide induced chemical shifts, *J. Phys. Chem.*, 82, 914, 1978.
1248. Stiles, P. J., Dipolar contributions to nuclear magnetic shielding by anisotropic molecules, *Chem. Phys. Lett.*, 30, 259, 1975.
1249. van Zijl, P. C. M., van Wezel, R. P., MacLean, C., and Bothner-By, A. A., Evaluation of the dipolar contribution to lanthanide shift reagent induced isotropic shifts, *J. Phys. Chem.*, 89, 204, 1985.
1250. Voronov, V. K., Nature of paramagnetic shifts induced by lanthanide shift reagents, *Izv. Akad. Nauk SSSR Ser. Khim.*, p. 423, 1977.
1251. Rafalski, A. J., Barciszewski, J., and Wiewiorowski, M., Calculation of lanthanide NMR shift dependence on proton distance. An analytical approach, *Tetrahedron Lett.*, p. 2829, 1971.
1252. Goodisman, J. and Matthews, R. S., The interpretation of lanthanide-induced shifts in H-1 nuclear magnetic resonance spectra, *J. Chem. Soc. Chem. Commun.*, p. 127, 1972.
1253. Horrocks, W. D., Jr., Evaluation of dipolar nuclear magnetic resonance shifts, *Inorg. Chem.*, 9, 690, 1970.

1254. Kainosho, M., Ajisaka, K., and Tori, K., Evidence for the presence of contact term contribution of lanthanide-induced isotropic shifts in C-13 and F-19 NMR spectra of aliphatic compounds: caution for applications of lanthanide shift reagents, *Chem. Lett.*, p. 1061, 1972.
1255. Sanders, J. K. M. and Williams, D. H., Evidence for contact and pseudo-contact contributions in lanthanide-induced P-31 NMR shifts, *Tetrahedron Lett.*, p. 2813, 1971.
1256. Tori, K., Yoshimura, Y., Kainosho, M., and Ajisaka, K., Importance of complex formation and contact shifts in the application of lanthanide shift reagents to H-1 and C-13 NMR spectra of aromatic compounds, *Tetrahedron Lett.*, p. 3127, 1973.
1257. Reuben, J., Structural information from chemical shifts in lanthanide complexes, *J. Magn. Reson.*, 50, 233, 1982.
1258. Horrocks, W. D., Jr., Taylor, R. C., and La Mar, G. N., Isotropic proton magnetic resonance shifts in pi-bonding ligands coordinated to paramagnetic nickel(II) and cobalt(II) acetylacetonates, *J. Am. Chem. Soc.*, 86, 3031, 1964.
1259. Walker, I. M., Rosenthal, L., and Quereshi, S., Fermi contact and dipolar nuclear magnetic resonance shifts in paramagnetic ion-paired systems. Studies on some anionic lanthanide complexes, *Inorg. Chem.*, 10, 2463, 1971.
1260. Desreux, J. F. and Reilley, C. N., Evaluation of contact and dipolar contributions to H-1 and C-13 paramagnetic NMR shifts in axially symmetric lanthanide chelates, *J. Am. Chem. Soc.*, 98, 2105, 1976.
1261. Gansow, O. A., Loeffler, P. A., Davis, R. E., Willcott, M. R., III, and Lenkinski, R. E., Evaluation of lanthanide-induced carbon-13 contact vs. pseudocontact nuclear magnetic resonance shifts, *J. Am. Chem. Soc.*, 95, 3389, 1973.
1262. Gansow, O. A., Loeffler, P. A., Davis, R. E., Lenkinski, R. E., and Willcott, M. R., III, Contact vs. pseudocontact contributions to lanthanide-induced shifts in the nuclear magnetic resonance spectra of isoquinoline and of endo-norbornenol, *J. Am. Chem. Soc.*, 98, 4250, 1976.
1263. Ajisaka, K. and Kainosho, M., Gd(fod)$_3$-induced contact shifts. A versatile new method to estimate contact and pseudocontact shift contributions to observed lanthanide-induced shifts, *J. Am. Chem. Soc.*, 97, 330, 1975.
1264. Hirayama, M., C-13 contact shifts of aromatic amines induced by Gd(fod)$_3$ and spin delocalization mechanism: evidence for the occurrence of direct spin transfer, *Chem. Lett.*, p. 1497, 1980.
1265. Hirayama, M., Akutsu, K., and Fukuzawa, K., The analysis of lanthanoid-induced C-13 shifts for naphthylamines. Contact shifts induced by Gd-chelates and the spin-delocalization mechanisms, *Bull. Chem. Soc. Jpn.*, 55, 704, 1982.
1266. Hirayama, M., Carbon-13 contact shifts and spin delocalization mechanism for substituted aniline complexes with gadolinium(III) chelates, *J. Chem. Soc. Perkin Trans.*, 2, p. 443, 1982.
1267. Quaegebeur, J. P. and Yasukawa, T., Application of lanthanide nuclear magnetic resonance shift reagents in conformational analysis of flexible chain molecules. A pseudocontact-shift and relaxation investigation, *J. Phys. Chem.*, 86, 204, 1982.
1268. Elgavish, G. A., Lenkinski, R. E., and Reuben, J., Analysis of the shift mechanisms and relaxation processes in lanthanide shift reagent systems, *Proc. 11th Rare Earth Res. Conf.*, 2, 885, 1974.
1269. Bergen, H. A. and Golding, R. M., An analysis of NMR shifts in lanthanide complexes, *Aust. J. Chem.*, 30, 2361, 1977.
1270. Golding, R. M. and Halton, M. P., A theoretical study of the N-14 and O-17 NMR shifts in lanthanide complexes, *Aust. J. Chem.*, 25, 2577, 1972.
1271. Reuben, J., On the origin of chemical shifts in lanthanide complexes and some implications thereof, *J. Magn. Reson.*, 11, 103, 1973.
1272. Reilley, C. N., Good, B. W., and Desreux, J. F., Structure independent method for dissecting contact and dipolar nuclear magnetic resonance shifts in lanthanide complexes and its use in structure determination, *Anal. Chem.*, 47, 2110, 1975.
1273. Lee, M. H. and Reilley, C. N., Mixed lanthanide shift reagents, *Taehan Hwahakhoe Chi*, 26, 24, 1982.
1274. Reuben, J., Effects of chemical equilibrium and adduct stoichiometry in shift reagent studies, in *Nuclear Magnetic Resonance Shift Reagents*, Sievers, R. E., Ed., Academic Press, New York, 1973, 341.
1275. de Boer, J. W. M., Sakkers, P. J. D., Hilbers, C. W., and de Boer, E., Lanthanide shift reagents. III. Chemical exchange of substrates and fod ligands, *J. Magn. Reson.*, 26, 253, 1977.
1276. Erasmus, C. S. and Boeyens, J. C. A., Crystal structure of the praseodymium beta-diketonate of 2,2,6,6-tetramethyl-3,5-heptanedione, Pr$_2$(thd)$_6$, *Acta Crystallogr.*, B26, 1843, 1970.
1277. De Villiers, J. P. R. and Boeyens, J. C. A., The crystal and molecular structure of the hydrated praseodymium chelate of 1,1,1,2,2,3,3-heptafluoro-7.7-dimethyl-4,6-octanedione, Pr$_2$(fod)$_6$ · 2H$_2$O, *Acta Crystallogr.*, B27, 692, 1971.

1278. Onuma, S., Inoue, H., and Shibata, S., The crystal and molecular structure of tris-(2,2,6,6-tetramethylheptane-3,5-dionato)lutetium(III), *Bull. Chem. Soc. Jpn.*, 49, 644, 1976.
1279. Cramer, R. E. and Seff, K., Crystal and molecular structure of a nuclear magnetic resonance shift reagent, the dipyridine adduct of tris-[(2,2,6,6-tetramethylheptane-3,5-dionato)europium(III), Eu(dpm)$_3$(py)$_2$], *J. Chem. Soc. Chem. Commun.*, p. 400, 1972.
1280. Cramer, R. E. and Seff, K., The crystal and molecular structure of a europium shift reagent complex, the dipyridine adduct of tris(2,2,6,6-tetramethylheptane-3,5-dionato)europium(III), Eu(dpm)$_3$(py)$_2$, *Acta Crystallogr. Sect. B*, 28, 3281, 1972.
1281. Bye, E., The crystal structure of the adduct between quinuclidine and the NMR shift reagent Eu(dpm)$_3$, *Acta Chem. Scand.*, A28, 731, 1974.
1282. Cunningham, J. A. and Sievers, R. E., Structures of europium complexes and implications in lanthanide nuclear magnetic resonance shift reagent chemistry, *Inorg. Chem.*, 19, 595, 1980.
1283. Uebel, J. J. and Wing, R. M., Lanthanide shift reagents. X-ray structure of the seven-coordinate 3,3-dimethylthietane 1-oxide complex with tris(dipivalomethanato)europium(III), Eu(dpm)$_3$, and its implication on pseudocontact shift calculations, *J. Am. Chem. Soc.*, 94, 8910, 1972.
1284. Horrocks, W. D., Jr., Sipe, J. P., III, and Luber, J. R., Lanthanide shift reagents. The X-ray structure of the eight-coordinate bis(4-picoline) adduct of 2,2,6,6-tetramethylheptane-3,5-dionatoholmium, Ho(dpm)$_3$(4-pic)$_2$, *J. Am. Chem. Soc.*, 93, 5258, 1971.
1285. Wasson, S. J. S., Sands, D. E., and Wagner, W. F., Crystal and molecular structure of 3-methylpyridine-tris(2,2,6,6-tetramethyl-3,5-heptanedionato)lutetium(III), *Inorg. Chem.*, 12, 187, 1973.
1286. Cunningham, J. A. and Sievers, R. E., The structure of an adduct of a chiral lanthanide nuclear magnetic resonance shift reagent, *J. Am. Chem. Soc.*, 97, 1586, 1975.
1287. Kepert, D. L., Structures of monoadducts of tris(beta-diketonato) lanthanoid shift reagents, *J. Chem. Soc. Dalton Trans.*, 617, 1974.
1288. Evans, D. F. and Wyatt, M., Direct observation of free and complexed substrate in a lanthanide shift reagent system, *J. Chem. Soc. Chem. Commun.*, p. 312, 1972.
1289. Evans, D. F. and Wyatt, M., Solvation numbers for lanthanide shift reagent systems, *J. Chem. Soc. Chem. Commun.*, p. 339, 1973.
1290. Cramer, R. E. and Dubois, R., Implications of the low-temperature nuclear magnetic resonance spectrum of the 3-picoline diadduct of tris[2,2,6,6-tetramethylheptane-3,5-dionato]europium, *J. Am. Chem. Soc.*, 95, 3801, 1973.
1291. Cramer, R. E., Dubois, R., and Seff, K., Calculation of lanthanide induced shifts from molecular structure, *J. Am. Chem. Soc.*, 96, 4125, 1974.
1292. Cramer, R. E., Dubois, R., and Furuike, C. K., Calculations of lanthanide-induced shifts from molecular structure. II, *Inorg. Chem.*, 14, 1005, 1975.
1293. Feibush, B., Richardson, M. F., Sievers, R. E., and Springer, C. S., Jr., Complexes of nucleophiles with rare earth chelates. I. Gas chromatographic studies of lanthanide nuclear magnetic resonance shift reagents, *J. Am. Chem. Soc.*, 94, 6717, 1972.
1294. Denning, R. G., Rossotti, F. J. C., and Sellars, P. J., Chiral lanthanide nuclear magnetic resonance shift reagents, *J. Chem. Soc. Chem. Commun.*, p. 381, 1973.
1295. Hirota, M. and Otsuka, S., Infrared spectroscopic studies on the shift reagent-substrate complexes, *Chem. Lett.*, p. 667, 1975.
1296. Kojima, H., Nonaka, H., and Hirota, M., Thermodynamic properties for the adduct formation of some shift reagents with aniline, *Bull. Chem. Soc. Jpn.*, 55, 2988, 1982.
1297. Catton, G. A., Hart, F. A., and Moss, G. P., Studies of the conformations of the NMR shift reagent tris(2,2,6,6-tetramethylheptane-3,5-dionato)europium(III) and its adducts by means of fluorescence spectra, *J. Chem. Soc. Dalton Trans.*, 221, 1975.
1298. Babushkina, T. A., Zolin, V. F., and Koreneva, L. G., Interpretation of lanthanide-induced shifts in the NMR spectra. The case of nonaxial symmetry, *J. Magn. Reson.*, 52, 169, 1983.
1299. Brittain, H. G. and Richardson, F. S., Emission-titration studies of the formation of adducts of tris(6,6,7,7,8,8,8-heptafluoro-2,2-dimethyloctane-3,5-dionato)europium(III) with substrates in solution, *J. Chem. Soc. Dalton Trans.*, 2253, 1976.
1300. Zolin, V. F., Koreneva, L. G., Obukhov, A. E., Zvolinskii, V. P., and Kordova, I. R., Use of luminescence spectra of adducts of europium beta-diketonates with methyl-substituted pyridines for interpretation of NMR data produced using lanthanide shift reagents, *Teor. Eksp. Khim.*, 18, 193, 1982.
1301. Brittain, H. G. and Richardson, F. S., Circularly polarized emission studies on the chiral nuclear magnetic resonance lanthanide shift reagent tris(3-trifluoroacetyl-d-camphorato)europium(III), *J. Am. Chem. Soc.*, 98, 5858, 1976.
1302. Brittain, H. G. and Chan, C. K., Circularly polarized luminescence studies of the adducts formed with mixed-ligand Eu(III) β-diketone complexes and dimethyl sulfoxide or N,N-dimethyl formamide, *Polyhedron*, 4, 39, 1985.

1303. Andersen, N. H., Bottino, B. J., and Smith, S. E., Evidence for octa-co-ordination in alcohol-lanthanide-shift-reagent complexes and its implications in the search for the geometric factor governing observed shifts, *J. Chem. Soc. Chem. Commun.*, p. 1193, 1972.

1304. Peters, J. A., Schuyl, P. J. W., and Knol-Kalkman, A. H., Mass spectrometry of the paramagnetic NMR shift reagents Eu(fod)$_3$ and Yb(fod)$_3$ and their adducts with propylamine, *Tetrahedron Lett.*, 23, 4497, 1982.

1305. Deranleau, D. A., Theory of the measurement of weak molecular complexes. I. General considerations, *J. Am. Chem. Soc.*, 91, 4044, 1969.

1306. Mackie, R. K. and Shepherd, T. M., Lanthanide induced paramagnetic shifts — derivation of equilibrium constants and absolute shifts in Eu(dpm)$_3$ systems, *Org. Magn. Reson.*, 4, 557, 1972.

1307. Armitage, I., Dunsmore, G., Hall, L. D., and Marshall, A. G., Evaluation of the binding constants, bound chemical shifts, and stoichiometry of lanthanide-substrate complexes, *Can. J. Chem.*, 50, 2119, 1972.

1308. ApSimon, J. W., Beierbeck, H., and Fruchier, A., On the stoichiometry of lanthanide shift reagent-substrate complexes, *J. Am. Chem. Soc.*, 95, 939, 1973.

1309. Kelsey, D. R., Nuclear magnetic resonance paramagnetic shift reagents. The use of internal protons as standards for structural determinations. A method for determination of complexation equilibrium constants, *J. Am. Chem. Soc.*, 94, 1764, 1972.

1310. Armitage, I., Dunsmore, G., Hall, L. D., and Marshall, A. G., Determination of stoichiometry for binding of organic substrates to lanthanide shift reagents, *Chem. Ind.*, p. 79, 1972.

1311. Bouquant, J. and Chuche, J., Reactifs lanthanidiques de deplacement chimique I — methode simple de determination de K and Δ, *Tetrahedron Lett.*, p. 2337, 1972.

1312. Roth, K., Grosse, M., and Rewicki, D., Stoichiometrie und bestandigkeit von komlexen mit NMR-shift-reagenzien, *Tetrahedron Lett.*, p. 435, 1972.

1313. Goldberg, L. and Ritchey, W. M., A quantitative study of effects of rare earth chelates on NMR chemical shifts. I. Treatment of 1:1 complexes and chemical shifts of the pure complexes, *Spectrosc. Lett.*, 210, 1972.

1314. Ejchart, A. and Jurczak, J., NMR studies of dynamic systems. Relationship between chemical shift and thermodynamic parameters, *Bull. Acad. Pol. Sci.*, 19, 725, 1971.

1315. Armitage, I. M., Hall, L. D., Marshall, A. G., and Werbelow, L. G., Determination of molecular configuration from lanthanide-induced proton NMR chemical shifts, in *Nuclear Magnetic Resonance Shift Reagents*, Sievers, R. E., Ed., Academic Press, New York, 1973, 313.

1316. Bouquant, J. and Chuche, J., NMR study of the interaction between lanthanide reagents and Lewis bases: scope of the method and comparison with other methods, *Bull. Soc. Chim. Fr.*, p. 959, 1977.

1317. Kawaki, H., Okazaki, Y., Fujiwara, H., and Sasaki, Y., NMR study on the convenient determination of association constant and bound chemical shift between Lewis base and tris(dipivaloylmethanato)europium, *Yakugaku Zasshi*, 102, 521, 1982.

1318. Shapiro, B. L., Johnston, M. D., Jr., Towns, R. L. R., Godwin, A. D., Pearce, H. L., Proulx, T. W., and Shapiro, M. J., Some aspects of the use of lanthanide-induced shifts in organic chemistry, in *Nuclear Magnetic Resonance Shift Reagents*, Sievers, R. E., Ed., Academic Press, New York, 1973, 227.

1319. Reuben, J., Complex formation between Eu(fod)$_3$, a lanthanide shift reagent, and organic molecules, *J. Am. Chem. Soc.*, 95, 3534, 1973.

1320. Lenkinski, R. E., Elgavish, G. A., and Reuben, J., Criteria and algorithms for the characterization of weak molecular complexes of 2:1 stoichiometry from nuclear magnetic resonance data. Applications to a shift reagent system, *J. Magn. Reson.*, 32, 367, 1978.

1321. Caccamese, S., Finocchiaro, P., Librando, V., Maravigna, P., Montaudo, G., and Recca, A., A computer program for the simulation of NMR lanthanide induced shifts and its analytical applications, *Ann. Chim.*, 68, 303, 1978.

1322. Stilbs, P., LISRIT, and iterative computer program for simultaneous simulation of NMR shift and relaxation reagent data, *Chem. Scripta*, 7, 59, 1975.

1323. Rafalski, A. J., Barciszewski, J., and Karonski, M., Application of digital computers in NMR lanthanide shift analysis, *J. Mol. Struct.*, 19, 223, 1973.

1324. Hajek, M., Suchanek, M., and Vodicka, L., Application of computer for the evaluation of parameters of some nonlinear models solution of the pseudocontact model in NMR spectroscopy, *Anal. Chem. (Prague)*, 11, 83, 1976.

1325. ApSimon, J. W. and Beierbeck, H., Lanthanide shift reagents. A novel method for fitting the pseudocontact shielding equation to experimental induced shifts, *Tetrahedron Lett.*, p. 581, 1973.

1326. Panyushkin, V. T. and Buikliskii, V. D., Possible use of a computer for an evaluation of the correlation between lanthanide-induced paramagnetic shifts in the NMR spectra and structural parameters of a lanthanide-substrate system, *Zh. Strukt. Khim.*, 22, 169, 1981.

1327. Velez, H. and Francisco, G., Induced chemical shifts. Adjustment criteria for determination of the optimum position, *Rev. Cienc. Quim.*, 14, 313, 1983.
1328. Wing, R. M., Early, T. A., and Uebel, J. J., A simple graphical method for analysis of lanthanide induced shifts, *Tetrahedron Lett.*, p. 4153, 1972.
1329. Stiles, P. J. and Wing, R. M., Multipolar expansions for nuclear magnetic shielding and their application to shift reagents, *J. Magn. Reson.*, 15, 510, 1974.
1330. Uebel, J. J., Pacheco, C., and Wing, R. M., Diamagnetic corrections to paramagnetic lanthanide induced NMR shifts, *Tetrahedron Lett.*, p. 4383, 1973.
1331. Bikeev, S. S. and Samitov, Y. Y., Frequency dependence and conditions for the neutralization of shifts in nuclear magnetic resonance spectra induced by a lanthanide shift reagent, *Zh. Fiz. Khim.*, 54, 2124, 1980.
1332. Voronov, V. K., Effect of resonance frequency of the spectrometer on shifts induced by paramagnetic shift reagents, *Izv. Akad. Nauk SSSR Ser. Khim.*, p. 2795, 1975.
1333. ApSimon, J. W., Beierbeck, H., and Fruchier, A., Shift reagent spectra. The automatic determination of relative bound shifts and the automatic assignment of signals in the parent spectrum. Effects of concentration, temperature and solvent on relative bound shifts, *Org. Magn. Reson.*, 8, 483, 1976.
1334. ApSimon, J. W., Beierbeck, H., and Fruchier, A., Conversion of nonlinear shift reagent plots to linear plots, *Can. J. Chem.*, 50, 2725, 1972.
1335. ApSimon, J. W. and Beierbeck, H., A method for compensating for experimental errors in shift reagent work, *J. Chem. Soc. Chem. Commun.*, p. 172, 1972.
1336. Raber, D. J., Johnston, M. D., Jr., Campbell, C. M., Janks, C. M., and Sutton, P., Structure elucidation with lanthanide-induced shifts 4. Bound shifts vs. relative shifts, *Org. Magn. Reson.*, 11, 323, 1978.
1337. Raber, D. J. and Hardee, L. E., Structure elucidation with lanthanide-induced shifts 13. A critical evaluation of methods for obtaining bound shifts and association constants, *Org. Magn. Reson.*, 20, 125, 1982.
1338. Armitage, I., Dunsmore, G., Hall, L. D., and Marshall, A. G., Calculation of binding constants and bound chemical shifts for the association of lanthanide shift reagents with organic substrates, *J. Chem. Soc. Chem. Commun.*, p. 1281, 1971.
1339. Williams, D. E., NMR shift reagents: intrinsic LIS parameters in mixtures, *Tetrahedron Lett.*, p. 1345, 1972.
1340. Cramer, R. E. and Maynard, R. B., Calculation of lanthanide-induced shifts from molecular structure. III. Other lanthanides, *J. Magn. Reson.*, 31, 295, 1978.
1341. Cramer, R. E., Furuike, C. K., and Dubois, R., Comments on the sensitivity of lanthanide-induced shifts to the metal substrate distance, *J. Magn. Reson.*, 19, 382, 1975.
1342. Hawkes, G. E., Leibfritz, D., Roberts, D. W., and Roberts, J. D., Nuclear magnetic resonance shift reagents. The question of the orientation of the magnetic axis in lanthanide-substrate complexes, *J. Am. Chem. Soc.*, 95, 1659, 1973.
1343. Honeybourne, C. L., Concerning the location of the principal magnetic axis in lanthanide shift-reagent adducts, *Tetrahedron Lett.*, p. 1095, 1972.
1344. Hawkes, G. E., Marzin, C., Liebfritz, D., Johns, S. R., Herwig, K., Cooper, R. A., Roberts, D. W., and Roberts, J. D., Lanthanide-induced C-13 NMR shifts, in *Nuclear Magnetic Resonance Shift Reagents*, Sievers, R. E., Ed., Academic Press, New York, 1973, 129.
1345. Huber, H., Location of the magnetic main axis in cylindrically symmetric complexes of lanthanide shift reagents, *Tetrahedron Lett.*, p. 3559, 1972.
1346. Hinckley, C. C. and Brumley, W. C., Effects of random coordinate error in analyses of lanthanide-induced pseudocontact shifts. Axially symmetric case, *J. Am. Chem. Soc.*, 98, 1331, 1976.
1347. Raber, D. J., Janks, C. M., Johnston, M. D., Jr., Schwalke, M. A., Shapiro, B. L., and Behelfer, G. L., Structure elucidation with lanthanide-induced shifts. 10. Generation of atomic coordinates: empirical force field calculations vs. other methods, *J. Org. Chem.*, 46, 2529, 1981.
1348. Davis, R. E. and Willcott, M. R., III, Interpretation of the pseudocontact model for nuclear magnetic resonance shift reagents. II. Significance testing on the agreement factor R, *J. Am. Chem. Soc.*, 94, 1744, 1972.
1349. Willcott, M. R., III, Lenkinski, R. E., and Davis, R. E., Interpretation of the pseudocontact model for nuclear magnetic resonance shift reagents. I. The agreement factor, R, *J. Am. Chem. Soc.*, 94, 1742, 1972.
1350. Richardson, M. F., Rothstein, S. M., and Li, W. K., Significance testing of lanthanide shift reagent data, *J. Magn. Reson.*, 36, 69, 1979.
1351. Rothstein, S. M., Bell, W. D., and Richardson, M. F., Significance testing of lanthanide shift reagent data. II, *J. Magn. Reson.*, 41, 310, 1980.
1352. Li, W. K. and Lee, S. Y., Application of rank correlation to lanthanide induced shift data, *Org. Magn. Reson.*, 13, 97, 1980.

1353. Sullivan, G. R., Limitations of lanthanide shift reagents for determination of conformation of nonrigid molecules, *J. Am. Chem. Soc.*, 98, 7162, 1976.
1354. Roberts, J. D., Hawkes, G. E., Husar, J., Roberts, A. W., and Roberts, D. W., Nuclear magnetic resonance spectroscopy. Lanthanide shift reagents, useful tools for study of conformational equilibria in solution?, *Tetrahedron*, 30, 1833, 1974.
1355. Johnston, M. D., Jr., Raber, D. J., DeGennaro, N. K., D'Angelo, A., and Perry, J. W., Structure elucidation with lanthanide-induced shifts. The use of bound shifts and high-symmetry substrates, *J. Am. Chem. Soc.*, 98, 6042, 1976.
1356. Davis, R. E. and Willcott, M. R., III, Assessment of the pseudocontact model via agreement factors, in *Nuclear Magnetic Resonance Shift Reagents*, Sievers, R. E., Ed., Academic Press, New York, 1973, 143.
1357. Armitage, I. M., Hall, L. D., Marshall, A. G., and Werbelow, L. G., Use of lanthanide nuclear magnetic resonance shift reagents in determination of molecular configuration, *J. Am. Chem. Soc.*, 95, 1437, 1973.
1358. Lenkinski, R. E. and Reuben, J., A multisite model for lanthanide shift reagent coordination to monofunctional substrates. Effects of rotational and site averaging on shifts and relaxation rates, *J. Am. Chem. Soc.*, 98, 4065, 1976.
1359. Kornilov, M. Y. and Turov, A. V., Use of lanthanide shift reagents for the structural analysis of compounds containing methyl and other conformationally mobile groups, *Ukr. Khim. Zh.*, 45, 346, 1979.
1360. Karhan, J., Hajek, M., Ksandr, Z., and Vodicka, L., Application of the relaxation reagent Gd(fod)$_3$ (1,1,1,2,2,3,3-heptafluoro-7,7-dimethyl-4,6-octanedionate)-gadolinium(III) in H-1 NMR spectroscopy of adamantoid compounds, *Coll. Czech. Chem. Commun.*, 44, 533, 1979.
1361. La Mar, G. N. and Metz, E. A., Dipolar relaxation in shift reagents as a solution structural probe, *J. Am. Chem. Soc.*, 96, 5611, 1974.
1362. La Mar, G. N. and Faller, J. W., Strategies for the study of structure using lanthanide reagents, *J. Am. Chem. Soc.*, 95, 3817, 1973.
1363. Faller, J. W., Adams, M. A., and La Mar, G. N., The application of lanthanide relaxation reagents in CMR, *Tetrahedron Lett.*, p. 699, 1974.
1364. Hajek, M., Mohyla, I., Ksandr, Z., and Vodicka, L., Application of the relaxation reagent (2,2,6,6-tetramethyl-3,5-heptanedionato)-gadolinium(III) in H-1 NMR studies of adamantane derivatives, *Coll. Czech. Chem. Commun.*, 41, 3591, 1976.
1365. Peters, J. A., van Bekkum, H., and Bovee, W. M. M. J., The use of H-1 and C-13 spin-lattice relaxation rates in the structure determination of adducts of adamantane-1-carbonitrile and lanthanide chelates in solution; comparison with the results of measurements of lanthanide induced shifts, *Tetrahedron*, 38, 331, 1982.
1366. Welti, D. H., Linder, M., and Ernst, R. R., Comparison of molecular geometries determined by paramagnetic nuclear magnetic resonance relaxation and shift reagents in solution, *J. Am. Chem. Soc.*, 100, 403, 1978.
1367. Faller, J. W. and La Mar, G. N., Chemical spin-decoupling and lanthanide reagents, *Tetrahedron Lett.*, p. 1381, 1973.
1368. Johnston, M. D., Jr., Caines, G. H., and Zektzer, A. S., The lanthanide-induced-shift-assisted determination of proton spin-lattice relaxation times, *J. Magn. Reson.*, 60, 415, 1984.
1369. Blackmer, G. L. and Roberts, R. L., Suppression of nuclear Overhauser enhancements in C-13 NMR by chemical shift reagents, *J. Magn. Reson.*, 10, 380, 1973.
1370. Mersh, J. D. and Sanders, J. K. M., Diamagnetic relaxation reagents can increase the growth rates and ultimate sizes of nuclear Overhauser enhancements, *J. Magn. Reson.*, 51, 345, 1983.
1371. Reuben, J. and Fiat, D., Non-Curie behavior of isotropic nuclear resonance shifts, *J. Chem. Phys.*, 47, 5440, 1967.
1372. Reuben, J. and Fiat, D., Proton chemical shifts in aqueous solutions of rare-earth ions as an indicator of complex formation, *J. Chem. Soc. Chem. Commun.*, p. 729, 1967.
1373. Lewis, W. B., Jackson, J. A., Lemons, J. F., and Taube, H., Oxygen-17 NMR shifts in aqueous solutions of rare-earth ions, *J. Chem. Phys.*, 36, 694, 1962.
1374. Phillips, W. D., Looney, C. E., and Ikeda, C. K., Influence of some paramagnetic ions on the magnetic resonance absorption of alcohols, *J. Chem. Phys.*, 27, 1435, 1957.
1375. Reuben, J., Bioinorganic chemistry: lanthanides as probes in systems of biological interest, in *Handbook on the Physics and Chemistry of Rare Earths*, Gschneidner, K. A., Jr. and Eyring, L., Eds., North-Holland, Amsterdam, 1979, 515.
1376. Reuben, J., Aqueous lanthanide shift reagents 4. Interaction of Pr(III), Nd(III), and Eu(III) with xylitol. Origin of induced shifts in polyols, *J. Am. Chem. Soc.*, 99, 1765, 1977.

1377. Sherry, A. D., Stark, C. A., Ascenso, J. R., and Geraldes, C. F. G. C., Lanthanide ethylenediaminetetra-acetate chelates as aqueous shift reagents: evidence for effective axial symmetry in bidentate cytidine 5′-monophosphate and alanine complexes, *J. Chem. Soc. Dalton Trans.*, 2078, 1981.

1378. Bryden, C. C. and Reilley, C. N., Europium luminescence lifetimes and spectra for evaluation of 11 europium complexes as aqueous shift reagents for nuclear magnetic resonance spectroscopy, *Anal. Chem.*, 54, 610, 1982.

1379. Bryden, C. C., Reilley, C. N., and Desreux, J. F., Multinuclear nuclear magnetic resonance study of three aqueous lanthanide shift reagents: complexes with EDTA and axially symmetric macrocylic polyamino polyacetate ligands, *Anal. Chem.*, 53, 1418, 1981.

1380. Bryden, C. C., Reilley, C. N., and Desreux, J. F., Multinuclear NMR study of three aqueous lanthanide shift reagents: complexes with EDTA and two macrocyclic ligands, *Rare Earths Mod. Sci. Technol.*, 3, 53, 1982.

1381. Zolin, V. F. and Koreneva, L. G., Use of the rare earth elements as label for investigating biologically active compounds. II. Luminescence spectra and structure of complexes of europium used as shift reagents for investigations by the NMR method, *Biofizika*, 20, 198, 1975.

1382. Dushkin, A. V., Zolin, V. F., Koreneva, L. G., Molin, Y. N., Obynochnyi, A. A., and Sagdeev, R. Z., New shift-reagents for NMR investigations of molecular structure in polar solvents, *Zh. Strukt. Khim.*, 15, 410, 1974.

1383. Wong, C. P., Venteicher, R. F., and Horrocks, W. D., Jr., Lanthanide porphyrin complexes. A potential new class of nuclear magnetic resonance dipolar probe, *J. Am. Chem. Soc.*, 96, 7149, 1974.

1384. Horrocks, W. D., Jr. and Hove, E. G., Water-soluble lanthanide porphyrins: shift reagents for aqueous solutions, *J. Am. Chem. Soc.*, 100, 4386, 1978.

1385. Sinha, S. P. and Green, R. D., NMR studies of lanthanide(III) complexes. High field shifts in complexes of 1,10-phenanthroline, *Spectrosc. Lett.*, 4, 399, 1971.

1386. Donato, H., Jr. and Martin, R. B., Dipolar shifts and structure in aqueous solutions of 3:1 lanthanide complexes of 2,6-dipicolinate, *J. Am. Chem. Soc.*, 94, 4129, 1972.

1387. Kieboom, A. P. G., Sinnema, A., van der Toorn, J. M., and van Bekkum, H., C-13 NMR study of the complex formation of sorbitol(glucitol) with multivalent cations in aqueous solution using lanthanide(III)nitrates as shift reagents, *Rec. J. Roy. Neth. Chem. Soc.*, 96, 35, 1977.

1388. Hart, F. A., Moss, G. P., and Staniforth, M. L., Lanthanide-induced differential shifts in NMR spectra of aqueous solutions, *Tetrahedron Lett.*, p. 3389, 1971.

1389. Geraldes, C. F. G. C. and Williams, R. J. P., Conformational study of a dinucleoside monophosphate in aqueous solution using the lanthanide probe method, *J. Chem. Soc. Perkin Trans. 2*, p. 1279, 1982.

1390. Pennington, B. T. and Cavanaugh, J. R., Lanthanide ion nuclear magnetic resonance probe studies of benzoic acids. The agreement with an axially symmetric model, *J. Magn. Reson.*, 31, 11, 1978.

1391. Sherry, A. D., Yang, P. P., and Morgan, L. O., Separation of contact and pseudocontact contributions to C-13 lanthanide induced shifts in non-axially-symmetric lanthanide ethylenediaminetetraacetate chelates, *J. Am. Chem. Soc.*, 102, 5755, 1980.

1392. Sherry, A. D. and Pascual, E., Proton and carbon lanthanide-induced shifts in aqueous alanine. Evidence for structural changes along the lanthanide series, *J. Am. Chem. Soc.*, 99, 5871, 1977.

1393. Levine, B. A. and Williams, R. J. P., The determination of the conformations of small molecules in solution by means of paramagnetic shift and relaxation pertubations of NMR spectra, *Proc. R. Soc. London Ser. A*, 345, 5, 1975.

1394. Elgavish, G. A. and Reuben, J., Aqueous lanthanide shift reagents. 10. Proton and carbon-13 studies of the interaction of the aquoions with amino acids, *J. Magn. Reson.*, 42, 242, 1981.

1395. Birnbaum, E. R., Yoshida, C., Gomez, J. E., and Darnall, D. W., Investigations of amino acid complexes of Nd(III), *Proc. 9th Rare Earth Res. Conf.*, 1, 264, 1971.

1396. Sherry, A. D., Yoshida, C., Birnbaum, E. R., and Darnall, D. W., Nuclear magnetic resonance study of the interaction of neodymium(III) with amino acids and carboxylic acids. An aqueous shift reagent, *J. Am. Chem. Soc.*, 95, 3011, 1973.

1397. Howland, D. L. and Flurry, R. L., Jr., A PMR study of the Pr(III)Cl$_3$ and Eu(III)Cl$_3$ complexes of L-histidine, *J. Inorg. Nucl. Chem.*, 38, 1568, 1976.

1398. Singh, M., Reynolds, J. J., and Sherry, A. D., Lanthanide induced shift and relaxation rate studies of aqueous L-proline solutions, *J. Am. Chem. Soc.*, 105, 4172, 1983.

1399. Mossoyan, J., Asso, M., and Benlian, D., Ligand conformation in lanthanide complexes by NMR paramagnetic shifts: L-proline and L-valine, *Org. Magn. Reson.*, 13, 287, 1980.

1400. Inagaki, F., Tasumi, M., and Miyazawa, T., Conformation of hydroxy-L-proline in aqueous solution. Comparison of results from lanthanide ion probe measurements and proton coupling constants, *J. Chem. Soc. Perkin Trans. 2*, p. 167, 1976.

1401. Pennington, B. T. and Cavanaugh, J. R., Proton magnetic resonance study of the equilibria between lanthanide(3+) ions and amino acids and the nature of the induced shifts, *J. Magn. Reson.*, 29, 483, 1978.
1402. Elgavish, G. A. and Reuben, J., Lanthanide effects on the proton and carbon-13 relaxation rates of sarcosine. Evidence for isostructural amino acid complexes along the lanthanide series, *J. Am. Chem. Soc.*, 100, 3617, 1978.
1403. Shelling, J. G., Bjornson, M. E., Hodges, R. S., Taneja, A. K., and Sykes, B. D., Contact and dipolar contributions to lanthanide-induced NMR shifts of amino acid and peptide models for calcium binding sites in proteins, *J. Magn. Reson.*, 57, 99, 1984.
1404. Marinetti, T. D., Snyder, G. H., and Sykes, B. D., Lanthanide interactions with nitrotyrosine. A specific binding site for nuclear magnetic resonance shift probes in proteins, *J. Am. Chem. Soc.*, 97, 6562, 1975.
1405. Reuben, J. and Elgavish, G. A., Aqueous lanthanide shift reagents. 9. Evaluation of a model for the separation of contact and dipolar shift contributions: effects of magnetic asymmetry, *J. Magn. Reson.*, 39, 421, 1980.
1406. Inagaki, F., Takahashi, S., Tasumi, M., and Miyazawa, T., Lanthanide-induced nuclear magnetic resonance shifts and molecular structure of L-azetidine-2-carboxylic acid. II. Determination of molecular conformation, *Bull. Chem. Soc. Jpn.*, 48, 1590, 1975.
1407. Inagaki, F., Takahashi, S., Tasumi, M., and Miyazawa, T., Lanthanide-induced nuclear magnetic resonance shifts and molecular structure of L-azetidine-2-carboxylic acid. I. Evaluation of the shifts intrinsic to the lanthanide-substrate 1:1 complex, *Bull. Chem. Soc. Jpn.*, 48, 853, 1975.
1408. Mossoyan, J., Asso, M., and Benlian, D., Peptide-lanthanide cation equilibria in aqueous phases. I. Bound shifts for L-carnosine-praseodymium complexes, *J. Magn. Reson.*, 46, 289, 1982.
1409. Mossoyan, J., Asso, M., and Benlian, D., Peptide-lanthanide cation equilibria in aqueous solution. II. Isomorphous replacement and separation of the contact and pseudo-contact contributions to NMR shifts induced by a paramagnetic ion, *J. Magn. Reson.*, 55, 188,1983.
1410. Levine, B. A. and Williams, R. J. P., Conformation of peptides in water, in *Environmental Effects on Molecular Structure and Properties,* Pullman, B., Ed., D. Reidel, Dordrecht, Holland, 1976, 95.
1411. Bayer, E. and Beyer, K., Signalzuordnung in C-13 NMR spektren von oligopeptiden und seuqenzbestimmung mit hilfe von Pr(ClO$_4$)$_3$ als verschiebungsreagens, *Tetrahedron Lett.*, p. 1209, 1973.
1412. Lee, L., Corson, D. C., and Sykes, B. D., Structural studies of calcium-binding proteins using nuclear magnetic resonance, *Biophys. J.*, 47, 139, 1985.
1413. Lee, L., Sykes, B. D., and Birnbaum, E. R., A determination of the relative compactness of the Ca(II) binding sites of a Ca(II) fragment of troponin-C and parvalbumin using lanthanide-induced H-1 NMR shifts, *FEBS Lett.*, 98, 169, 1979.
1414. Lee, L. and Sykes, B. D., Strategies for the use of lanthanide NMR shift probes in the determination of protein structure in solution, application to the EF calcium binding site of carp parvalbumin, *Biophys. J.*, 32, 193, 1980.
1415. Lee, L. and Sykes, B. D., Use of lanthanide-induced nuclear magnetic resonance shifts for determination of protein structure in solution: EF calcium binding site of carp parvalbumin, *Biochemistry*, 22, 4366, 1983.
1416. Lee, L. and Sykes, B. D., Nuclear magnetic resonance determination of metal-proton distances in the EF site of carp parvalbumin using the susceptibility contribution to the line broadening of lanthanide-shifted resonances, *Biochemistry*, 19, 3208, 1980.
1417. Lee, L. and Sykes, B. D., The use of lanthanide NMR shift probes in the determination of the structure of calcium binding proteins in solution; application to the EF calcium bonding site of carp parvalbumin, *Dev. Biochem.*, 14, 323, 1980.
1418. Lee, L. and Sykes, B. D., The elucidation of the principal axis system of the magnetic susceptibility tensor of ytterbium(+3) in the EF site of carp parvalbumin as the central element required in the use of lanthanide-induced NMR shifts for the determination of protein structure in solution, *Biochem. Struct. Determ. NMR*, p. 169, 1982.
1419. Morallee, K. G., Nieboer, E., Rossotti, F. J. C., Williams, R. J. P., Xavier, A. V., and Dwek, R. A., The lanthanide cations as nuclear magnetic resonance probes of biological systems, *J. Chem. Soc. Chem. Commun.*, p. 1132, 1970.
1420. Dwek, R. A., Morallee, K. G., Nieboer, E., Richards, R. E., Williams, R. J. P., and Xavier, A. V., Molecular conformation determinations of inhibitor/enzyme complexes with respect to the Gd(III) reported site, *Proc. 9th Rare Earth Res. Conf.*, p. 518, 1971.
1421. Dwek, R. A., Richards, R. E., Morallee, K. G., Nieboer, E., Williams, R. J. P., and Xavier, A. V., The lanthanide cations as probes in biological systems. Proton relaxation enhancement studies for model systems and lysozyme, *Eur. J. Biochem.*, 21, 204, 1971.

1422. Campbell, I. D., Dobson, C. M., Williams, R. J. P., and Xavier, A. V., The use of lanthanide ions for the determination of the 3D structure of proteins in solution, *Proc. 10th Rare Earth Res. Conf.* 2, 791, 1973.
1423. Campbell, I. D., Dobson, C. M., and Williams, R. J. P., Nuclear magnetic resonance studies on the structure of lysozyme in solution, *Proc. R. Soc. London Ser. A*, 345, 41, 1975.
1424. Campbell, I. D., Dobson, C. M., and Williams, R. J. P., Assignment of the proton NMR spectra of proteins, *Proc. R. Soc. London Ser. A*, 345, 23, 1975.
1425. Campbell, I. D., Dobson, C. M., and Williams, R. J. P., Studies of exchangeable hydrogens in lysozyme by means of Fourier transform proton magnetic resonance, *Proc. R. Soc. London Ser. B*, 189, 485, 1975.
1426. Darnall, D. W., Birnbaum, E. R., Gomez, J. E., and Smolka, G. E., Rare earth metal ions as probes of calcium ion binding sites in proteins, *Proc. 9th Rare Earth Res. Conf.*, 1, 278, 1971.
1427. Tanswell, P., Westhead, E. W., and Williams, R. J. P., The active site of phosphoglycerate kinase studied by nuclear magnetic resonance, *Biochem. Soc. Trans.*, 2, 79, 1974.
1428. Di Bello, C., Acampora, M., Filira, F., and Tondello, E., β-Dicarbonylamides as chelating agents. NMR determination of the equilibria formation with lanthanide metal ions, *Inorg. Chim. Acta*, 11, 165, 1974.
1429. Marinetti, T. D., Snyder, G. H., and Sykes, B. D., Nitrotyrosine chelation of nuclear magnetic resonance shift probes in proteins: application to bovine pancreatic trypsin inhibitor, *Biochemistry*, 16, 647, 1977.
1430. Marinetti, T. D., Snyder, G. H., and Sykes, B. D., Nuclear magnetic resonance determination of intramolecular distances in bovine pancreatic trypsin inhibitor using nitrotyrosine chelation of lanthanides, *Biochemistry*, 15, 4600, 1976.
1431. Geraldes, C. F. G. C. and Ascenso, J. R., Solvent effects on the conformation of nucleotides. II. Nuclear magnetic shift and relaxation effects induced by lanthanide ions on adenosine 5'-monophosphate in water-dimethyl sulphoxide, *J. Chem. Soc. Dalton Trans.*, p. 267, 1984.
1432. Geraldes, C. F. G. C., A case of non-axial symmetry of lanthanide NMR dipolar shifts as related to the conformational uses of the lanthanide probe method, *J. Mol. Struct.*, 60, 7, 1980.
1433. Geraldes, C. F. G. C. and Santos, H., Solvent effects on the conformation of nucleotides. I. The conformation of 5'-adenosine monophosphate in water-dimethyl sulphoxide using nuclear Overhauser effects and lanthanide relaxation probes, *J. Chem. Soc. Perkin Trans. 2*, p. 1693, 1983.
1434. Barry, C. D., Glasel, J. A., Williams, R. J. P., and Xavier, A. V., Quantitative determination of conformations of flexible molecules in solution using lanthanide ions as nuclear magnetic resonance probes: application to adenosine-5'-monophosphate, *J. Mol. Biol.*, 84, 471, 1974.
1435. Barry, C. D., Glasel, J. A., North, A. C. T., Williams, R. J. P., and Xavier, A. V., The conformations of adenosine mononucleotide in water and dimethylsulfoxide, *Biochem. Biophys. Res. Commun.*, 47, 166, 1972.
1436. Babushkina, T. A., Buikliskii, V. D., Zolin, V. F., Koreneva, L. G., and Sheveleva, I. S., Study of nucleotide conformation in solutions by the method of lanthanide shift of the NMR spectrum, *Biofizika*, 26, 187, 1981.
1437. Fazakerley, G. V. and Wolfe, M. A., Determination of the solution conformation of adenosine 2':3'-monophosphate by nuclear magnetic resonance with lanthanide probes, *Eur. J. Biochem.*, 74, 337, 1977.
1438. Barry, C. D., North, A. C. T., Glasel, J. A., Williams, R. J. P., and Xavier, A. V., Quantitative determination of mononucleotide conformations in solution using lanthanide ion shift and broadening NMR probes, *Nature (London)*, 232, 236, 1971.
1439. Lavallee, D. K. and Zeltmann, A. H., Conformation of cyclic b-adenosine 3',5'-phosphate in solution using the lanthanide shift technique, *J. Am. Chem. Soc.*, 96, 5552, 1974.
1440. Hayashi, F., Akasaka, K., and Hatano, H., Conformation of adenosine-3':5'-cyclic monophosphate in solution. Application of the NMR-DESERT method with a lanthanide ion-induced shift, *J. Magn. Reson.*, 27, 419, 1977.
1441. Barry, C. D., Martin, D. R., Williams, R. J. P., and Xavier, A. V., Quantitative determination of the conformation of cyclic 3',5'-adenosine monophosphate in solution using lanthanide ions as nuclear magnetic resonance probes, *J. Mol. Biol.*, 84, 491, 1974.
1442. Tanswell, P., Thornton, J. M., Korda, A. V., and Williams, R. J. P., Quantitative determination of the conformation of ATP in aqueous solution using the lanthanide cations as nuclear-magnetic-resonance probes, *Eur. J. Biochem.*, 57, 135, 1975.
1443. Birdsall, B., Birdsall, N. J. M., Feeney, J., and Thornton, J., A nuclear magnetic resonance investigation of the conformation of nicotinamide mononucleotide in aqueous solution, *J. Am. Chem. Soc.*, 97, 2845, 1975.

1444. Oida, T., Yokoyama, S., Inagaki, F., Tasumi, M., and Miyazawa, T., Molecular structure of NMN and UpU in aqueous solution as studied by the lanthanide-ion probe method, *Nucleic Acids Chem.*, 4th, 2, 47, 1976.
1445. Birdsall, B., Feeney, J., Glasel, J. A., Williams, R. J. P., and Xavier, A. V., A method of assigning C-13 nuclear magnetic resonance spectra using europium(III) ion-induced pseudocontact shifts and C-H heteronuclear spin decoupling techniques, *J. Chem. Soc. Chem. Commun.*, p. 1473, 1971.
1446. Birdsall, B. and Feeney, J., The C-13 and H-1 nuclear magnetic resonance spectra and methods of their assignment for nucleotides related to dihydronicotinamide adenine dinucleotide phosphate (NADPH), *J. Chem. Soc. Perkin Trans. 2*, p. 1643, 1972.
1447. Zolin, V. F. and Koreneva, L. G., Study of the conformation of some europium-nucleotide complexes by means of PMR and luminescence spectroscopy, *Magn. Reson. Relat. Phenom. Proc. 20th Congr. AMPERE*, p. 561, 1978.
1448. Zolin, V. F., Koreneva, L. G., and Lagodzinskaya, G. V., Use of rare earth elements as probe for studying biologically active compounds. VIII. Study of the conformation of some nucleotides using a lanthanide shifting reagent and luminescence spectroscopy, *Biofizika*, 24, 945, 1979.
1449. Dobson, C. M., Williams, R. J. P., and Xavier, A. V., Separation of contact and pseudo-contact contributions to shifts induced by lanthanide(III) ions in nuclear magnetic resonance spectra, *J. Chem. Soc. Dalton Trans.*, p. 2662, 1973.
1450. Yokoyama, S., Inagaki, F., and Miyazawa, T., Advanced nuclear magnetic resonance lanthanide probe analyses of short-range conformational interrelations controlling ribonucleic acid structures, *Biochemistry*, 20, 2981, 1981.
1451. Furie, B., Griffin, J. H., Feldman, R. J., Sokoloski, E. A., and Schechter, A. N., The active site of staphylococcal nuclease: paramagnetic relaxation of bound nucleotide inhibitor nuclei by lanthanide ions, *Proc. Natl. Acad. Sci. U.S.A.*, 71, 2833, 1974.
1452. Barry, C. D., Glasel, J. A., North, A. C. T., Williams, R. J. P., and Xavier, A. V., The quantitative conformations of some dinucleoside phosphates in solution, *Biochim. Biophys. Acta*, 262, 101, 1972.
1453. Jones, C. R. and Kearns, D. R., Paramagnetic rare earth ion probes of transfer ribonucleic acid structure, *J. Am. Chem. Soc.*, 96, 3651, 1974.
1454. Taga, T., Kuroda, Y., and Ohashi, M., Structures of lanthanoid complexes of glyceric acid, gluconic acid, and lactobionic acid from the lanthanoid-induced H-1 NMR shifts: pH dependence of the lanthanoid-substrate equilibria, *Bull. Chem. Soc. Jpn.*, 51, 2278, 1978.
1455. Spoormaker, T., Kieboom, A. P. G., Sinnema, A., van der Toorn, J. M., and van Bekkum, H., PMR study of the complex formation of alditols with multivalent cations in aqueous solutions using praseodymium(III) nitrate as shift reagent, *Tetrahedron Lett.*, p. 3713, 1974.
1456. Kieboom, A. P. G., Spoormaker, T., Sinnema, A., van der Toorn, J. M., and van Bekkum, H., H-1 NMR study of the complex formation of alditols with multivalent cations in aqueous solution using praseodymium(III)nitrate as shift reagent, *Rec. J. R. Neth. Chem. Soc.*, 94, 53, 1975.
1457. Grasdalen, H., Anthonsen, T., Harbitz, O., Larsen, B., and Smidsrod, O., NMR-studies of the interaction of metal ions with poly-(1,4-hexuronates). V. Quantitative separation of contact and dipolar contributions to lanthanide induced shifts in H-1 NMR spectra of methyl α-D-gulo- and methyl β-D-hamamelopyranosides, 1,6-anhydro-β-D-manno-, 1,6-anhydro-β-D-talo- and 1,6-anhydro-β-D-allo-pyranoses and epi-inositol, *Acta Chem. Scand.*, A32, 31, 1978.
1458. Grasdalen, H., Anthonsen, T., Larsen, B., and Smidsrod, O., NMR-studies of the interaction of metal ions with poly-(1,4-hexuronates). III. Proton magnetic resonance study of the binding of lanthanides to methyl α-D-gulopyranoside in aqueous solution, *Acta Chem. Scand.*, B29, 17, 1975.
1459. Angyal, S. J., Greeves, D., and Pickles, V. A., Stereospecific contact interactions in the nuclear magnetic resonance spectra of polyol-lanthanide complexes, *J. Chem. Soc. Chem. Commun.*, p. 589, 1974.
1460. Angyal, S. J., Shifts induced by lanthanide ions in the NMR spectra of carbohydrates in aqueous solution, *Carbohydr. Res.*, 26, 271, 1973.
1461. Angyal, S. J. and Greeves, D., Complexes of carbohydrates with metal cations. VII. Lanthanide-induced shifts in the PMR spectra of cyclitols, *Aust. J. Chem.*, 29, 1223, 1976.
1462. Angyal, S. J., Greeves, D., Littlemore, L., and Pickles, V. A., Complexes of carbohydrates with metal cations. VIII. Lanthanide-induced shifts in the PMR spectra of some methyl glycosides and 1,6-anhydrohexoses, *Aust. J. Chem.*, 29, 1231, 1976.
1463. Anthonsen, T., Larsen, B., and Smidsrod, O., NMR studies of the interaction of metal ions with poly(1,4-hexuronates). II. The binding of europium ions to sodium methyl α-D-galactopyranosiduronate, *Acta Chem. Scand.*, 27, 2671, 1973.
1464. Grasdalen, H., Anthonsen, T., Larsen, B., and Smidsrod, O., NMR studies of the interaction of metal ions with poly(1,4-hexuronates). IV. Proton magnetic resonance study of lanthanide binding to sodium methyl α-D-galactopyranosiduronate in aqueous solutions, *Acta Chem. Scand.*, B29, 99, 1975.

1465. Gorin, P. A. J. and Mazurek, M., Structure of a phosphonomannan, as determined by the effect of lanthanide ions on its carbon-13 magnetic resonance spectrum, *Can. J. Chem.*, 52, 3070, 1974.

1466. Lawaczeck, R., Kainosho, M., Girardet, J. L., and Chan, S. I., Effects of structural defects in sonicated phospholipid vesicles on fusion and ion permeability, *Nature (London)*, 256, 584, 1975.

1467. Shapiro, Y. E., Viktorov, A. V., Volkova, V. I., Barsukov, L. I., Bystrov, V. F., and Bergelson, L. D., C-13 NMR investigation of phospholipid membranes with the aid of shift reagents, *Chem. Phys. Lipids*, 14, 227, 1975.

1468. Sears, B., Hutton, W. C., and Thompson, T. E., Effects of paramagnetic shift reagents on the C-13 nuclear magnetic resonance spectra of egg phosphatidylcholine enriched with C-13 in the N-methyl carbons, *Biochemistry*, 15, 1635, 1976.

1469. Michaelis, L. and Schlieper, P., 500 MHz H-1 NMR of phospholipid liposomes. Lanthanide shift on glycerol-gamma and acyl-chain C2 resonances, *FEBS Lett.*, 147, 40, 1982.

1470. Hunt, G. R. A. and Tipping, L. R. H., Trans-bilayer pseudocontact shifts produced by lanthanide ions in the H-1 and P-31 NMR spectra of phospholipid vesicular membranes and their temperature variation, *J. Inorg. Biochem.*, 12, 17, 1980.

1471. Barsukov, L. I., Victorov, A. V., Vasilenko, I. A., Evstigneeva, R. P., and Bergelson, L. D., Investigation of the inside-outside distribution intermembrane exchange and transbilayer movement of phospholipids in sonicated vesicles by shift reagent NMR, *Biochim. Biophys. Acta*, 598, 153, 1980.

1472. Bergelson, L. D. and Bystrov, V. F., Use of shift and broadening reagents in the NMR investigation of membranes, *Fed. Eur. Biochem. Soc. Meet. (Proc.)*, 35, 33, 1975.

1473. Andrews, S. B., Faller, J. W., Gilliam, J. M., and Barrnett, R. J., Lanthanide ion-induced isotropic shifts and broadening for nuclear magnetic resonance structural analysis of model membranes, *Proc. Natl. Acad. Sci. U.S.A.*, 70, 1814, 1973.

1474. Michaelson, D. M., Horwitz, A. F., and Klein, M. P., Transbilayer asymmetry and surface homogeneity of mixed phospholipids in cosonicated vesicles, *Biochemistry*, 12, 2637, 1973.

1475. Viktorov, A. V., Vasilenko, I. A., Barsukov, L. I., Evstigneeva, R. P., and Bergel'son, L. D., Formation of asymmetrical vesicles as a result of spontaneous intermembrane phospholipid exchange, *Dokl. Akad. Nauk SSSR*, 245, 1243, 1979.

1476. Viktorov, A. V., Vasilenko, I. A., Barsukov, L. I., Evstigneeva, R. P., and Bergel'son, L. D., New approach to studying the transmembrane asymmetry of phospholipids using NMR with shift reagents, *Dokl. Akad. Nauk SSSR*, 234, 207, 1977.

1477. Hauser, H., Phillips, M. C., Levine, B. A., and Williams, R. J. P., Ion binding to phospholipids. Interaction of calcium and lanthanide ions with phosphatidylcholine(lecithin), *Eur. J. Biochem.*, 58, 133, 1975.

1478. Assmann, G., Sokoloski, E. A., and Brewer, H. B., Jr., Phosphorus-31 nuclear magnetic resonance spectroscopy of native and recombined lipoproteins, *Proc. Natl. Acad. Sci. U.S.A.*, 71, 549, 1974.

1479. Kostelnik, R. J. and Castellano, S. M., 250 MHz proton magnetic resonance spectrum of a sonicated lecithin dispersion in water. The effect of ferricyanide, manganese(II), europium(III), and gadolinium(III) ions on the choline methyl resonance, *Exp. NMR Conf.*, 13th, 7, 1972.

1480. Shapiro, Y. E., Viktorov, A. V., Barsukov, L. I., and Bergel'son, L. D., Paramagnetic hydrophilic probing of phospholipid membranes by NMR, *Biofiz. Membr.*, p. 22, 1981.

1481. Levine, Y. K., Lee, A. G., Birdsall, N. J. M., Metcalfe, J. C., and Robinson, J. D., Interaction of paramagnetic ions and spin labels with lecithin bilayers, *Biochem. Biophys. Acta*, 291, 592, 1973.

1482. Hunt, G. R. A., Kinetics of ionophore-mediated transport of praseodymium(3+) ions through phospholipid membranes using proton NMR spectroscopy, *FEBS Lett.*, 58, 194, 1975.

1483. Fernandez, M. S., Celis, H., and Montal, M., Proton magnetic resonance detection of ionophor mediated transport of praseodymium ions across phospholipid membranes, *Biochem. Biophys. Acta*, 323, 600, 1973.

1484. Fernandez, M. S. and Cerbon, J., Importance of the hydrophobic interactions of local anesthetics in the displacement of polyvalent cations from artificial lipid membranes, *Biochem. Biophys. Acta*, 298, 8, 1973.

1485. Hayer, M. K. and Riddell, F. G., Shift reagents for K-39 NMR, *Inorg. Chim. Acta*, 92, L37, 1984.

1486. Pike, M. M., Yarmush, D. M., Balschi, J. A., Lenkinski, R. E., and Springer, C. S., Jr., Aqueous shift reagents for high-resolution cationic nuclear magnetic resonance. 2. Mg-25, K-39, and Na-23 resonances shifted by chelidamate complexes of dysprosium(III) and thulium(III), *Inorg. Chem.*, 22, 2388, 1983.

1487. Pike, M. M. and Springer, C. S., Jr., Aqueous shift reagents for high-resolution cationic nuclear magnetic resonance, *J. Magn. Reson.*, 46, 348, 1982.

1488. Chu, S. C., Pike, M. M., Fossel, E. T., Smith, T. W., Balschi, J. A., and Springer, C. S., Jr., Aqueous shift reagents for high-resolution cationic nuclear magnetic resonance. III. Dy(TTHA)$^{3-}$, Tm(TTHA)$^{3-}$, and Tm(PPP)$_2^{7-}$, *J. Magn. Reson.*, 56, 33, 1984.

1489. Nieuwenhuizen, M. S., Peters, J. A., Sinnema, A., Kieboom, A. P. G., and van Bekkum, H., Multinuclear NMR study of the complexation of lanthanide(III) cations with sodium triphosphate: induced shifts and relaxation rate enhancements, *J. Am. Chem. Soc.*, 107, 12, 1985.

1490. Gupta, R. K. and Gupta, P., Direct observation of resolved resonances from intra- and extracellular sodium-23 ions in NMR studies of intact cells and tissues using dysprosium(III)tripolyphosphate as paramagnetic shift reagent, *J. Magn. Reson.*, 47, 344, 1982.

1491. Perrin, C. L., Anionic lanthanide shift reagents. Application to the assignment of NH peaks of amidinium ions, *Org. Magn. Reson.*, 16, 11, 1981.

1492. Elgavish, G. A. and Reuben, J., Aqueous lanthanide shift reagents. 3. Interaction of the ethylenediaminetetraacetate chelates with substituted ammonium cations, *J. Am. Chem. Soc.*, 99, 1762, 1977.

1493. Pike, M. M., Fossel, E. T., Smith, T. W., and Springer, C. S., Jr., High-resolution Na-23 NMR studies of human erythrocytes: use of aqueous shift reagents, *Am. J. Physiol.*, 246, C528, 1984.

1494. Balschi, J. A., Cirillo, V. P., LeNoble, W. J., Pike, M. M., Schreiber, E. C., Jr., Simon, S. R., and Springer, C. S., Jr., Magnetic resonance studies of metal cation transport across biological membranes: use of paramagnetic lanthanide ions, *Rare Earths Mod. Sci. Technol.*, 3, 15, 1982.

1495. Pike, M. M., Simon, S. R., Balschi, J. A., and Springer, C. S., Jr., High-resolution NMR studies of transmembrane cation transport: use of an aqueous shift reagent for Na-23, *Proc. Natl. Acad. Sci. U.S.A.*, 79, 810, 1982.

1496. Rosenthal, L., Walker, I. M., and Wiggins, D., Effect of paramagnetic anions on cations possessing stereochemically non-equivalent methylene groups, *Org. Magn. Reson.*, 11, 333, 1978.

1497. Quereshi, M. S., Rosenthal, L., and Walker, I. M., Contact versus dipolar NMR shifts in ion-paired systems: a comparison of lanthanide and first row metal complexes of organic cations, *J. Coord. Chem.*, 5, 77, 1976.

1498. Vogel, H. J., Andersson, T., and Braunlin, W. H., Trifluoperazine binding to calmodulin: a shift reagent Ca-43 NMR study, *Biochem. Biophys. Res. Commun.*, 122, 1350, 1984.

1499. Civan, M. M., Degani, H., Margalit, Y., and Shporer, M., Observations of Na-23 in frog skin by NMR, *Am. J. Physiol.*, 245, C213, 1983.

1500. Brophy, P. J., Hayer, M. K., and Riddell, F. G., Measurement of intracellular potassium ion concentration by NMR, *Biochem. J.*, 210, 961, 1983.

1501. Vijverberg, C. A. M., Peters, J. A., Kieboom, A. P. G., and van Bekkum, H., A study of lanthanide(III) (hydroxy)carboxylate complexes in aqueous medium using lanthanide induced oxygen-17 NMR shifts, *Rec. J. R. Neth. Chem. Soc.*, 99, 403, 1980.

1502. Reuben, J. and Fiat, D., Nuclear magnetic resonance studies of solutions of the rare-earth ions and their complexes. III. Oxygen-17 and proton shifts in aqueous solutions and the nature of aquo and mixed complexes, *J. Chem. Phys.*, 51, 4909, 1969.

1503. Elgavish, G. A. and Reuben, J., Aqueous lanthanide shift reagents. I. Interaction of the ethylenediaminetetraacetate chelates with carboxylates. pH Dependence, ionic medium effects, and chelate structure, *J. Am. Chem. Soc.*, 98, 4755, 1976.

1504. Elgavish, G. A. and Reuben, J., Aqueous lanthanide shift reagents. 5. Interaction of the aquoions with methoxyacetate. Additional pathways for lanthanide-induced shifts, *J. Am. Chem. Soc.*, 99, 5590, 1977.

1505. Kostromina, N. A. and Ternovaya, T. V., NMR study of ethylenediaminetetraacetato-compounds of lanthanum, yttrium, and lutetium, *Russ. J. Inorg. Chem.*, 17, 825, 1972.

1506. Choppin, G. R., NMR Study of Structure of Lanthanide Complexes in Solution, ERDA Energy Res. Abstr. (Abstr. No. 26213), Washington, D.C., 1976, 1.

1507. Kostromina, N. A. and Novikova, L. B., Proton magnetic resonance investigation of complexes of ethylenediaminediacetic acid with lanthanum and lutetium, *Zh. Neorg. Khim.*, 20, 1793, 1975.

1508. Desreux, J. F., Nuclear magnetic resonance spectroscopy of lanthanide complexes with a tetraacetic tetraaza macrocycle. Unusual conformation properties, *Inorg. Chem.*, 19, 1319, 1980.

1509. Reyes-Zamora, C. and Tsai, C. S., Nuclear magnetic resonance isotropic shifts of europium oxaloacetates, *J. Chem. Soc. Chem. Commun.*, p. 1047, 1971.

1510. Kullberg, L. and Choppin, G. R., Nuclear magnetic resonance study of the structure in solution of lanthanide complexes with benzene-1,2-dioxydiacetate, *Inorg. Chem.*, 16, 2926, 1977.

1511. Krishnan, C. V., Friedman, H. L., and Springer, C. S., Jr., Lifetimes of complexes of the antibiotic X-537-A(lasalocid A) with physiological cations in methanol. Determination by NMR using a dissociative shift reagent, *Biophys. Chem.*, 9, 23, 1978.

1512. Davoust, D., Rebuffat, S., Giraud, M., and Molho, D., Utilisation des sels de lanthanides en RMN: determination de la configuration d'acides carboxyliques α,β-ethyleniques, *C. R. Acad. Sci. Paris*, 280, 815, 1975.

1513. Lerman, C. L. and Whitacre, E. B., Specific deuteration of gamma-aminolevulinic acid by pyridoxal-catalyzed exchange and analysis of products using an aqueous lanthanide nuclear magnetic resonance shift reagent, *J. Org. Chem.*, 46, 468, 1981.

1514. Levine, B. A., Thornton, J. M., and Williams, R. J. P., Conformational studies of lanthanide complexes with carboxylate ligands, *J. Chem. Soc. Chem. Commun.*, p. 669, 1974.
1515. Delepierre, M., Dobson, C. M., and Menear, S. L., Nuclear magnetic shift and relaxation effects resulting from complexation of lanthanide ions with endo-cis-bicyclo[2.2.1]hept-5-ene-2,3-dicarboxylic acid, *J. Chem. Soc. Dalton Trans.*, 678, 1981.
1516. Davoust, D., Molho, D., and Platzer, N., H-1 NMR. Interaction of lanthanide salts with phenols in basic solution, *Org. Magn. Reson.*, 14, 49, 1980.
1517. Davoust, D., Massias, M., Molho, D., Platzer, N., and Basselier, J. J., Etude en RMN di l'association des sels de lanthanides avec des phenols; application a la determination de structure de polyphenols naturels, *Org. Magn. Reson.*, 11, 547, 1978.
1518. Davoust, D., Rebuffat, S., Molho, D., Platzer, N., and Basselier, J. J., Etude en RMN de l'association des sels de lanthanides avec des monophenols, *Tetrahedron Lett.*, p. 2825, 1976.
1519. Koppikar, D. K. and Soundararajan, S., Preparation, IR and nuclear magnetic resonance studies on the rare-earth perchlorate complexes of 3-methylpyridine-1-oxide, *J. Inorg. Nucl. Chem.*, 38, 1875, 1976.
1520. Smith, K., Turner, J. R., Davies, J. S., Byrn, S. R., and McKenzie, A. T., Novel heterocyclic systems. V. Differentiation of isomeric pyridinium betaines by use of water-soluble europium shift reagents, and conformation by X-ray crystallography, *J. Chem. Res. (S)*, p. 107, 1981.
1521. Celotti, M. and Fazakerley, G. V., Conformation of various tetracycline species determined with the aid of a nuclear magnetic resonance relaxation probe, *J. Chem. Soc. Perkin Trans. 2*, p. 1319, 1977.
1522. Birnbaum, E. R. and Stratton, S., Studies of ethylenediamine complexes of the lanthanide(III) perchlorates using nuclear magnetic resonance spectroscopy, *Inorg. Chem.*, 12, 379, 1973.
1523. Birnbaum, E. R. and Moeller, T., Observations on the rare earths. LXXXII. Nuclear magnetic resonance and calorimetric studies of complexes of the tripositive ions with substituted pyridine molecules, *J. Am. Chem. Soc.*, 91, 7274, 1969.
1524. Davidenko, N. K. and Zinich, N. N., Lanthanide-ion-induced paramagnetic shifts in NMR spectra in aqueous solutions, *Teor. Eksp. Khim.*, 12, 704, 1976.
1525. Chang, S. and le Noble, W. J., Study of ion-pair return in 2-norbornyl brosylate by means of O-17 NMR, *J. Am. Chem. Soc.*, 105, 3708, 1983.
1526. Zaev, E. E., Study of the shifting effect of rare earth salts on proton NMR spectra of sodium dodecyl sulfate and solubilizate in micellar aqueous solutions, *Kolloidn. Zh.*, 42, 752, 1980.
1527. Zaev, E. E. and Khalilov, L. M., Study of pseudocontact shifts in the proton NMR signal of acetone in the second coordination sphere of rare earth element aqua ions, *Zh. Strukt. Khim.*, 21, 53, 1980.
1528. Reuben, J., Aqueous lanthanide shift reagents. 8. Chiral interactions and stereochemical assignments of chemically and isotopically chiral ligands, *J. Am. Chem. Soc.*, 102, 2232, 1980.
1529. Reuben, J., Chiral interactions in aqueous solution mediated by lanthanoid ions. NMR spectral resolution of enantiomeric nuclei, *J. Chem. Soc. Chem. Commun.*, p. 68, 1979.
1530. Peters, J. A., Vijverberg, C. A. M., Kieboom, A. P. G., and van Bekkum, H., Lanthanide(III) salts of (S)-carboxymethyloxysuccinic acid: chiral lanthanide shift reagents for aqueous solution, *Tetrahedron Lett.*, 24, 3141, 1983.
1531. Kabuto, K. and Sasaki, Y., The europium(III)-(R)-propylenediaminetetra-acetate ion: a promising chiral shift reagent for H-1 NMR spectroscopy in aqueous solution, *J. Chem. Soc. Chem. Commun.*, p. 316, 1984.
1532. Faller, J. W., Blankenship, C., and Sena, S., Paramagnetic phosphine shift reagents: new probes for the study of structures of transition-metal complexes in solution, *J. Am. Chem. Soc.*, 106, 793, 1984.
1533. Faller, J. W., Blankenship, C., Whitmore, B., and Sena, S., Paramagnetic phosphine shift reagents. 2. Study of structures of (substituted-allyl)palladium complexes in solution, *Inorg. Chem.*, 24, 4483, 1985.
1534. Lanzilotta, R. P., Bradley, D. G., McDonald, K. M., and Tokes, L., Microbiological hydroxylation of carbons 18 and 19 in 9-oxa-13(cis and trans)-prostenoic acids, *Appl. Environ. Microbiol.*, 32, 726, 1976.

INDEX

A

Acenaphthene, 161
Acetamidinium ion, 197
Acetanilide, 48
Acetate, 197
2-Acetidone, 49
Acetone, 25, 157, 176, 186
Acetonitrile, 25, 46, 157
4-Acetonylidene-2,6-dimethyl[4H]pyran, 81
Acetophenone, 25
1-Acetoxycyclopentadiene, 33
2-Acetoxycyclopentadiene, 33
1-Acetyladamantane, 24
1-Acetyl-2-alkyl-3-ethylidenecyclopropane, 26
N-Acetyl-L-aspartic acid, 186, 188
N-Acetyl-L-aspartyl-L-glycyl-L-aspartylamide, 188
Acetylbenzofuran, 81
Acetylcyclohexane, 26
Acetylcyclopentane, 26
1-Acetylimidazole, 109
N-Acetyl-L-3-nitrotyrosine ethyl ester, 186, 189
1-Acetylpyrazole, 109
Acetyl steroids, 35
1-Acetyltetrazole, 109
1-Acetyl-1,2,4-triazole, 109
Achiral lanthanide chelates, 5—125
    applications, 14—123
        metal complexes and substrates with a silicon atom, 66—71
        monofunctional substrates, 14—64
        organic salts, 64—66
        polyfunctional substrates, 71—121
        polymers, 121—123
    chemically induced dynamic nuclear polarization, 125
    coupling constants and, 123—124
    experimental techniques for, 9—14
    ligand in, 7—9
    metal in, 5—6
    preparation of, 10—11
    purification of, 10—11
    secondary deuterium isotope effects, 124—125
    selection of, 5—9
    separation of diastereotopic protons, 124
    solvents for, 9—10
Acid chlorides, 95, 134
Acridine, 44—45
Acrolein, 154
Acrylonitrile, 46, 154
Acyclic alcohols, 124
Acyclic amines, 42—43
Acyclic carbinols, 15
Acyclic ketones, 25
Acyclic organonitriles, 46
Adamantane, 31, 77
Adamantane carbonitrile, 47, 181
Adamantane-1-carboxylic acid, 37
2,4-Adamantanediol, 77
Adamantanedione, 86, 91
2-Adamantanethiol, 58
1-Adamantanol, 19—20, 175, 178
Adamantanone, 23—24, 31, 86
1-Adamantyl-4-(1-pentenyl)sulfone, 151—152
Adenosine diphosphate, 190, 195
Adenosine 2′,3′-monophosphate, 190
Adenosine 5′-monophosphate, 190
Adenosine pyrophosphate, 195
Adenosine tetrapolyphosphate, 195
Adenosine triphosphate, 190
Adenosyl-(3′,5′)-adenylic acid, 191
Adenosyl-(3′,5′)-cytidylic acid, 191
Alanine, 143, 186—187, 203
L-Alanine-L-alanine, 188
Alanine-glycine, 188
Alcohols, 182
    achiral substrates
        monofunctional, 14—23
        polyfunctional, 72—79
    chiral substrates, 136—140
Aldehydes, 36—37, 94, 145
Alditol, 191
Aldohexosylaldohexose, 98
Aliphatic acids, 32
Aliphatic alcohols, 14—15
Aliphatic amines, 42
1,3-Alkadienyl phosphonate, 60
Alkenes, 158—159
3-Alkenoic acid, 38
Alkenols, 15
Alkoxysilane, 71
2-Alkoxytetrahydropyran, 96
3-Alkyl-1-adamantane carbonitrile, 182
2-Alkyl-2-adamantanol, 20
Alkylamines, 42, 135
N-Alkylaniline, 42
2-Alkyl-4-tert-butylcyclohexanone, 27
1-Alkylcyclohexanol, 18
2-Alkylcyclohexanone, 27
9-Alkyl-2-decalone, 30
10-Alkyl-2-decalone, 30
Alkylhalides, 166
Alkylidene furanone, 81
2-Alkylidenehydrazono-3-methyl-2,3-dihydrobenzothiazole, 117
N-Alkylidenesulphinamide, 151
2-Alkyl-1-indanone, 24
3-Alkyl-1-indanone, 24
1-Alkyl-4-methyl-4-nitrocyclohexa-2,5-dienol, 34
Alkyl phenols, 23
2-Alkylpiperidine, 43
3-Alkylpropanol, 134
N-Alkylpyridinium ion, 196
Alkynes, 160
Allenes, 152
Allenic alcohols, 138

Allenic diesters, 143
Allenic esters, 143
α-Allethrin, 81
Allofuranoside, 78, 92
Allopyranoside, 78
Allyl cyanide, 46
N-[2-(Allylsulfinyl)benzyl] N-methylmethanamine, 151
Aluminum shift reagent, Al(acac)$_3$, 67
Aluminum-27 spectrum, 67
Amides, 47—49, 109—111, 149
  preparation of, 134
Amidinium ion, 196
5-Amine-6-phenylnorborn-2-ene, 44
Amines, 41—44, 99—102, 135, 147—148, 201
Amino acid methyl esters, 133, 143
Amino acids, 38, 111, 131, 186—189, 203
1-Aminoadamantane, 44
β-Amino alcohols, 148
3-Aminobutyric acid, 186—187
4-Aminobutyric acid, 186—187
bis(4-Aminocyclohexyl)methane, 100
Aminoethyldiphenyl phosphine, 99
α-Aminoketones, 131
γ-Aminolevulinic acid, 199
3-Amino methyl benzoate, 99
4-Amino methyl benzoate, 99
1-Aminooctane, 13
α-Aminophosphonic acid ester, 121
2-Aminopropane, 147
2-Aminopyridine, 105
3-Aminopyridine, 105
4-Aminopyridine, 105
Ammonium ion, 196
Ammonium salts, 64, 66, 165—166
Amphetamine, 43, 100, 124, 135, 147
α-Amylase, 189
β-Amyrin, 23
Androsta-4-ene-3-one, 86
Androstan-2β-ol, 22
Anemonin, 93
β-Angelicalactone, 35
Anhydrides, 38
1,6-Anhydro-β-D-allopyranoside, 192
3,6-Anhydrogalactopyranoside, 98
1,6-Anhydro-β-D-glucopyranoside, 192
3,6-Anhydroglucopyranoside, 98
1,6-Anhydrohexose, 192
1,6-Anhydro-β-D-mannopyranoside, 192
1,6-Anhydro-β-D-talopyranoside, 192
Aniline, 42—43, 164, 175
Anisole, 38
Anthracene, 161
Anthraquinone, 82, 102
Antipyrine, 112
Aqueous shift reagents, see also Water-soluble lanthanide shift reagents, 2—3
  applications of, 186—202
  chiral, 202—203
  selection of, 185—186
Arabinofuranoside, 92

Arabinopyranoside, 92
Arabitol, 191
Aristoteline, 101
Aromatic amines, 43, 99
Aromatic compounds, 155, 167
Aromatic hydrocarbons, 160—161
Aromatic ketones, 25
Aromatic organonitriles, 46
Arsine oxide, 185
3-(Aryl)-3,3,5-trimethylcyclohexanone, 27
Asparagusic acid methyl ester, 119
Association constant, 26
Atomic coordinates, 180—181
Axial symmetry, effective, 172
N-(4-Aza-9-fluorenylidene)amine, 107
N-(4-Aza-9-fluorenylidene)arylamine, 118
L-Azetidine-2-carboxylic acid, 187—188
Azides, 53—54, 117—118
Azines, 117—118
Azobenzene, 53—54, 164
Azo compounds, 53—54, 117—118
Azoisobutyramidinium ion, 197
Azo-2-methylpropane, 54
Azoxyalkane, 52
Azoxy compounds, 52, 115—116

B

Barbituric acid, 149
Benzaldehyde, 37
Benzamide, 149
Benzamidinium ion, 197
Benzene, 157, 160—161, 174
Benzene-1,2-dioxydiacetate, 198—199
Benzhydrol, 137
Benzo[9]annulenone, 141
Benzoate, 198
Benzocycloalkenol, 16
Benzocyclohexanone, 30
1,4-Benzodiazepine-2-one, 113
Benzodiazine, 106
5,6-Benzo-3,4-dihydro-2-oxo-1,2,3-oxathiazan, 58
1,4-Benzodioxan, 114
Benzofuran, 164
Benzofuranoid neolignin, 84
Benzoic acid, 37
Benzonitrile, 164
Benzophenone, 25
Benzo[a]pyrene, 161
Benzo(f)quinoline, 45
Benzo(f)quinoline N-oxide, 50
Benzo[c]quinolizine-1,2,3,4-tetracarboxylate, 144
Benzothiazole, 164
1-(x-Benzo[b]thienyl)ethyl acetate, 90
Benzothiophene, 165
Benzoxazole, 164
Benzoyl propionyl peroxide, 125
Benzyl alcohol, 14—15, 140
α-Benzyl cyclohexanol, 18
Benzyldimethylsulfonium tetrafluoroborate, 166

Benzylidene, 164
Benzylidene acetal, 73
Benzylideneaniline, 54, 118
3-Benzylidene-2-piperidone, 49
3-Benzylidene-2-pyrollidone, 49
Benzylidine aniline, 95
Benzyl isothiocyanate, 59
Benzylisothiouronium chloride, 166
Benzyl methyl sulfoxide, 151
Benzylnitrile, 46
N-Benzylpyridinium chloride, 65
Benzyl toluene thiolsulfinate, 58
Biacetyl-bis-N,N-dimethyl hydrazone, 154
Bianthrone, 141
Bianthryl, 133
Biaryls, 131, 133
Bicyclic lactam, 113—114
Bicyclic organonitrile, 46
Bicyclic phosphine oxide, 63
Bicyclobenzo[b]-1,4-diazabicyclo[3.2.1]octane, 100, 102
Bicyclo[2.2.1]heptadiene, 90
Bicyclo[4.1.0]heptadiene, 144
Bicyclo[3.2.0]heptane, 75, 139
Bicyclo[4.1.0]heptane, 34
Bicycloheptanone, 28
Bicyclo[3.2.0]heptene, 75
Bicyclo[4.1.0]heptene, 34
Bicyclo[2.2.1]hept-5-ene-2,3-dicarboxylic acid, 199—200
Bicyclo[3.1.0]hexane, 144
Bicyclo[3.1.0]hexan-3-one, 24, 28
Bicyclo[3.1.0]hexene, 20
Bicyclo[3.1.0]hexylmorpholine, 97, 108
Bicyclo[3.3.1]nonane, 75, 167
Bicyclo[3.3.1]nonan-3-ol, 21
Bicyclo[3.3.1]nonan-3-one, 29
Bicyclo[3.3.1]nonan-9-one, 29
Bicyclo[6.1.0]nona-2,4,6-triene, 21, 35
Bicyclononene, 21
Bicyclo[3.3.1]nonene, 35
Bicyclo[2.2.2]octane, 110
Bicyclo[3.3.1]octane, 90
Bicyclo[3.3.0]octane-3,7-dione, 84
Bicyclooctene, 21
Bicyclo[3.2.1]octene, 144
Bicyclo[4.2.0]oct-7-en-3-one, 29
Bicycloundecanol, 21
Binaphthyl, 133
Binuclear lanthanide(III)-silver(I) shift reagents, 2, 66, 155, 168
Biphenyls, 74, 89, 133, 139—140, 144—145, 148
Bis(4-aminocyclohexyl)methane, 100
Bis(phenylsulfinyl)methane, 120, 151
Bleaney's theorem, 169—170
Blocking groups
  for hydroxyl groups, 78—79
  for keto groups, 87
Bornene, 168
Borneol, 19, 175, 182—183
Boron-containing compounds, 150

Bound shifts, 176, 178—180
Bovine serum albumin, 189
Broadening, 5, 13—14
Bromides, 161
1-Bromoacetyladamantane, 24
Bromobenzene, 161
1-Bromobutane, 161
1-Bromopentane, 161
1-(p-Bromophenyl)ethyl-tert-butyl sulfoxide, 55
3-(4-Bromophenyl)-3-pyridylallylamine, 105
3-(4-Bromophenyl)-3-pyridylallylmethylamine, 101, 103
4-(5-Bromo-2-thienyl)-3-buten-2-one, 80
Bulk diamagnetic susceptibility, 13
Butadiene, 140
Butadiene-nitrile rubber, 122
2-Butanethiol, 131
tert-Butanol, 186
2-Butanone, 25
1-Butene-2,4-dicarboxylic acid, 89
tert-Butoxycarbonyl α-amino acid ester, 111
7-tert-Butoxynorbornadiene, 165
Butylamine, 43
sec-Butylamine, 134
tert-Butylamine, 42
n-Butylammonium cation, 194
n-Butyl ammonium iodide, 166
Butylcyclohexanol, 17
3-tert-Butylcyclohexanol, 132
4-tert-Butylcyclohexanol, 17, 132, 178
4-tert-Butylcyclohexanone, 24—27
4-tert-Butylcyclohexyl methyl ether, 41
4-tert-Butylcyclopentane-1,2-epoxide, 39
tert-Butyldimethylsilyl ether, 79
n-Butyldimethylsulfonium iodide, 166
n-Butyldimethylsulfonium tetrafluoroborate, 166
3-(4-tert-Butyl-2,6-dinitroanilino)-2-butanol, 73
2,3-Butylene oxide, 146
tert-Butylhydroperoxide, 41
3-(tert-Butylhydroxymethylene)-d-camphor, 127
n-Butylisocyanide, 46
sec-Butylisothiouronium bromide, 166
n-Butylisothiouronium chloride, 166
sec-Butylisothiouronium chloride, 166—167
n-Butylisothiouronium iodide, 166
4-tert-Butyl-1-methyl-1-cyclohexanol, 17
3-tert-Butyl-6-methylenecyclohexanol, 138
N-Butyl-N(4-methyloxazol-2-yl)-2-methylpropionamide, 109
tert-Butylnitrile, 46
4-tert-Butyl-1-phenylcyclohexanol, 17
2-sec-Butylpyridine, 148
4-sec-Butylpyridine, 148
4-tert-Butylpyridine, 201
4-tert-Butyl-1,2,2-trimethylcyclohexanol, 17
n-Butyltriphenylphosphonium ion, 196
Butyrate, 197
Butyric acid, 37
γ-Butyrolactone, 35

C

Cadmium, 188
Cadmium-113 NMR spectrum, 188
Calciferol, 22
Calcium, 196
Calcium-43, 194—197
Calcium-containing compounds, 188—189
Calmodulin, 196
Camphanate esters, 133—134
Camphanic acid chloride, 133
Camphene, 160, 167—168
Camphor, 24, 27—28, 127—128
Camphorimide, 49
Camphoroxime, 53
Canellin B, 84
Cannivonine b, 76, 139
Carbamates, 50
Carbinols, 15, 131—133, 142
Carbohydrates
    achiral shift reagents and, 73, 78, 88, 92, 98, 111, 117
    aqueous shift reagents and, 185, 191—193
    mixtures of, 92
Carbon-13 NMR spectrum, 171, 183—188
    of alcohols, 14, 139
    of membranes, 194
    of nitrogen heterocycles, 104
    of olefins, 159—160
Carbon disulfide, 157
Carbon tetrachloride, 157
Carboxylates, 185, 197—200
Carboxylic acids, 37—38, 134, 145—146, 203
(Carboxymethoxy)succinate, 198
(S)-Carboxymethyloxysuccinic acid, 203
delta-3-Carene, 160
Carnosine, 188
Carotene, 80
Carotenoid, 73, 94, 96
Carveol, 18, 132
Carvone, 128
Cations, 194—197
Chelate bonding, 165
    of alcohols, 72
    of amines, 100
    of esters, 87
    of ethers, 95
    of ketones, 79
    of nitrogen heterocycles, 103
Chemical exchange spin decoupling, 123
Chemically induced dynamic nuclear polarization (CIDNP), 125
Chiral substrates, 1—2, 127—154, 167—168
    absolute configuration, 131—134
    aqueous, 202—203
    experimental techniques using, 129—131
    ligand in, 127—129
    metal in, 127
    optical purity, 135—154
    preparation of, 129
    purification of, 129
    selection of, 127—129
    solvents for, 129
Chlorazepine-2,5-dione, 82
Chlorides, 161
Chloroazepine-2,3-dione, 113
Chloro-3-benzaldehyde, 37
Chlorobenzene, 161
1-Chlorobutane, 161
3-Chlorocyclobenzaprine, 43
Chlorocyclohexane, 161
2-Chlorocyclohexanone, 27
m-Chloroethyl benzoate, 125
Chloroform, 157, 174
Chloromethyl dimethyl methoxysilane, 71
p-Chloronitrobenzene, 52
1-Chloropentane, 161
7-Chloro-5-phenyl-2,3-dihydro-1H-1,4-benzodiazepine, 118
3-(p-Chlorophenyl)-3,5,5-trimethylcyclohexanone, 176
Chlorophyll a, 87
Chloroprene, 122
4-Chloropyridine, 45
Chloroquine, 107
4-Chloro-3,3,4-trifluoro-2-methyl-2-butanol, 15
β-Chlorovinyltrimethoxysilane, 71
Cholestane, 78
5α-Cholestan-2α-episulfoxide, 56
5α-Cholestan-3β-ol, 22
Cholesterol, 22, 193—194
Cholic acid, 78
Chrysanthemyl alcohol, 16
CIDNP, see Chemically induced dynamic nuclear polarization
Cinnamic acid, 81, 95—96, 108
Circular dichroism, 131, 175
Circularly polarized luminescence, 175
Citramalate, 202—203
Citrate, 198
Cobalt complexes, 68—70, 153—154
Cobalt shift reagents, 67, 173
Cobalt-59 NMR spectrum, 67
Complexation shifts, 22, 94, 169, 178
Conferol, 77
Configurational analysis
    absolute, 127, 131—134, 202
    limitations of, 181—183
Conformational analysis, 26, 32, 181—183
Conformational preference, 94
Contact effects, 61
Contact mechanism, 1
Contact shifts, 159, 169—172
    of amines, 42
    of amino acids, 186—187
    of carbohydrates, 192
    of esters, 31
    of ketones, 25
    of metal complexes, 67
    of nitriles, 46
    of nitrogen oxides, 50

of phosphates, 60
Costunolide, 36
Cotton effect, 131
Coumarin, 77, 93, 107
Coupling constant, 14, 19, 123—124
Cresol, 23, 33
Crotonic acid, 37
Crown ether, 95—96
Cubenol, 21
Cumene hydroperoxide, 41
Cyanine dye, 65, 164
2-Cyanobicyclo[2.2.1]heptane, 47
9-Cyanobicyclo[6.1.0]nona-2,4,6-triene, 47
2-Cyanobutane, 131
9-Cyano-4,9-dimethylbicyclo[6.1.0]nona-2,4,6-triene, 47
Cyano group, 46
Cyanoisoquinoline, 106
2-Cyanopyridine, 105
3-Cyanopyridine, 105
4-Cyanopyridine, 105
α-Cyanosulfoxide, 151
1-Cyano-2-vinylcyclobutane, 46
Cyclic-3′,5′-adenosine monophosphate, 190
Cyclic ethers, 38, 95
Cyclic phosphine oxide, 62
Cyclic sulfite, 56—57
Cycloalkanone, 79
Cyclobutane, 16
Cyclobutanone, 81
Cyclobutenone, 81
Cyclodipeptide, 115
Cycloheptane, 16
Cycloheptanol, 18
Cycloheptanone, 28, 80
4-Cycloheptene carbinol, 18
Cyclohexadienone, 27
Cyclohexane, 16, 34, 110, 157
Cyclohexane carbonitrile, 47
Cyclohexanediaminetetraacetic acid, 185
1,4-Cyclohexanedione, 82
Cyclohexane-1,2,3,4/5-pentol, 192
Cyclohexane-1,2,3,5/4-pentol, 192
Cyclohexane-1,2,3,4,5/O-pentol, 192
Cyclohexane 1,3,5-triol, 192
Cyclohexanol, 14, 17, 132
Cyclohexanone, 24—27, 29, 176
Cyclohexene oxide, 38
Cyclohexenone, 25—26
Cyclohexyl amines, 43
1-Cyclohexyl-2-phenylazetidin-3-ol, 74
Cyclooctanone, 80
Cyclopentadieneone, 68
Cyclopentane, 16, 110
Cyclopentanol, 14, 17
Cyclopentene, 143—144
Cyclopentenone, 26
Cyclophosphoramide, 152—153
Cyclopropane, 16
Cyclopropanol, 74
Cyclopropyl ketone, 25—26

Cyclopropylmethanol, 16
Cyclotriphosphazene, 107
Cytidine-5′-monophosphate, 191
Cytidyl-(3′,5′)-adenylic acid, 191
Cytidyl-(3′,5′)-guanylic acid, 191

D

Deactylphytuberin, 76
1,1,1,5,5,6,6,7,7,7-Decafluoro-2,4-heptanedione, 8
2-Decalone, 26, 29
Decoupling, 183
Dehydronorbornanol, 19
1-β-D-Deoxyribofuranosylthymine-5′-monophosphate, 190
Deuterium, 25, 124—125, 161
Diacetoxysilane, 71
1,2-Diacetyl-3-stearoyl-sn-glycerol, 142
2,3-Diacetyl-1-stearoyl-sn-glycerol, 142
2,4-Dialkoxy[3.3.1]nonan-9-one, 84
Dialkoxysilane, 71
Dialkyladipate, 122
Dialkyl amines, 42
N,N-Dialkylaniline, 42
Dialkylarylsulfonium salt, 66
Dialkyl nitrosamine, 51
2,6-Dialkylpiperidine, 43
2,6-Dialkyl-3-piperidinol, 74
Diamides, 109—110
2,3-Diaminobutane, 153
1,2-Diaminopropane, 153
6,6-Diarylbicyclo[3.1.0]hexan-3-ol, 20
Diastereotopic resolution, 55
1,5-Diazabicyclo[3.1.0]hexan-2-one, 115
2,3-Diaza-1,3,5,7-cyclooctatetraene, 54
Diazacyclophane, 101—102
1,2-Diazido-1,2-di-$tert$-butylethane, 118
Diazoketones, 80, 117
1,2:5,6-Dibenzanthracene, 161
5H-Dibenzo[a,d]cycloheptene-5-ol, 18
5H-Dibenzo[a,d]cyclohepten-5-one, 28
Dibenzsulfone, 151—152
Di-n-butyl ether, 38
Di-n-butylmethylsulfonium iodide, 166
Di-n-butylmethylsulfonium tetrafluoroborate, 166
Dibutyl peroxide, 41
Di-n-butylsulfoxide, 55, 124
d,d-Dicampholylmethane, 129, 167
α,α-Dichlorocyclobutanone, 28
Dicumene peroxide, 41
Dieldrin, 39, 146
Dieneone, 86
Dienones, 26
Diethylacetamide, 48
Diethylacetic acid, 38
Diethylamine, 200
Diethylamine hydrobromide, 166
1,5-Diethyl-1,6-dihydro-2,4-dimethoxy-6-oxopyrimidine, 113
Diethylenetriaminepentaacetic acid, 185, 197

Diethylphenyl phosphonate, 60
Diethyl thioformamide, 59
1,1-Difencholylmethane, 167
2-2'-Difluoro-6,6'-dimethoxybiphenyl, 147
Diglycerides, 138, 140, 142
Diglycine, 188
Dihydroasparagusic acid, 119
Dihydrocostunolide, 36
Dihydrodicyclopentadiene-9,10-diol, 76
1,4-Dihydronaphthalene-1,4-oxide, 39
Dihydronicotinamide adenine dinucleotide phosphate, 190
Dihydropinifolic acid diol, 76
5,6-Dihydroquinoline, 45
7,8-Dihydroquinoline, 45
Dihydrotachysterol, 22
4,5-Dihydroxy-4,5-dihydroaldrin, 77, 139—140
4,5-Dihydroxy-4,5-dihydroisodrin, 77
4,5-Dihydroxyoctanedioic acid *bis* lactone, 94
N,N-Diisopropylamide, 48
Di-isopropyldisulfide, 58
Diketene, 189
2,5-Dimethoxyamphetamine, 147
o-Dimethoxy aromatic compounds, 95
1,2-Dimethoxybenzene, 163
6,7-Dimethoxy-2-benzonorbornenone, 146
3,3-Dimethoxycholestane, 98
2,2'-Dimethoxy-6,6'-dimethoxymethyl, 147
1,2-Dimethoxyethane, 95
5,6-Dimethoxyisatylidene, 113
1,5-Dimethoxynaphthalene, 96
2,6-Dimethoxynaphthalene, 96
Dimethylacetamide, 48, 170
Dimethylacetic acid, 38
N,N-Dimethylamide, 48
Dimethylamine hydrochloride, 166
p-Dimethylaminobenzaldehyde, 94
2[{5-(Dimethylamino)-methyl-2-furanyl}thio}ethyl]amino-2-methylamino-1-nitroethene, 100
[3-(Dimethylamino)propyl]trimethylammonium iodide, 65, 166
2-Dimethylaminotetrahydropyran, 102
2,2'-Dimethylbianthrone, 141
3,3'-Dimethylbianthrone, 141
Dimethyl-1-1'-biphenyl-2,2'-dicarboxylate, 69
1-[2-(1,3-Dimethyl-2-butenylidiene)hydrazino]phthalazine, 107
1,2-Dimethyl-1,2-cyclobutanedicarboxylate, 143
3,4-Dimethylcyclohexanone, 27
1,3-Dimethylcyclohexylamine, 43
1,3-Dimethylcyclopentanol, 17
1,4-Dimethyl-2,3-diazabicyclo[2.2.2]oct-2-ene, 54
3,7-Dimethyl-3,7-diphenyl-4,6-nonanedione, 128
N,N-Dimethyldodecylamine, 65
3,7-Dimethylenebicyclo[3.3.1]nonane, 163
Dimethylformamide, 47—49, 173, 175
Dimethylfumarate iron tetracarbonyl complex, 153
Dimethyl-1,7-heptanedioate, 88
3,4-Dimethylhexa-2,5-dione, 140

3,3-Dimethyl-4-(3-hydroxybutyl)-5-hydroxymethyl-cyclohexan-1-one, 75, 132
4,4-Dimethyl-3-keto steroids, 31
Dimethylmaleate iron tetracarbonyl complex, 153
3,7-Dimethyl-7-methoxymethylcycloheptatriene, 147
1,4-Dimethyl-4-nitrocyclohexa-2,5-dienol, 18, 40
3,7-Dimethyl-4,6-nonanedione, 128
5,5-Dimethyl-2-norbornanone, 28
6,6-Dimethyl-2-norbornanone, 28
3,7-Dimethyloctanoic acid, 142
3,7-Dimethyloct-6-enoic acid, 142
N,N-Dimethyl-2-phenylcyclopropane-1-carboxamide, 49
N,N-Dimethyl-1-phenylethylamine, 147
Dimethylphenylphosphonite, 63
3-(2',4'-Dimethylphenyl)-1H,3H-quinazoline-2,4-dione, 113
Dimethylpropionamide, 48
2,6-Dimethylpyridine, 44, 201
3,5-Dimethylpyridine, 201
2,4-Dimethylquinoline, 44
2,2-Dimethyl-2-silapentane-5-sulfonate, 186
Dimethyl sulfone, 56
Dimethylsulfoxide, 55, 136, 150, 157, 173—176, 201
9,9-Dimethyl-1,2,3,4-tetrahydroanthrone, 31
Dimethyl-3,3,4,4-tetramethyl-1,2-cyclobutane dicarboxylate, 143
3,3-Dimethylthietane-1-oxide, 55, 173
N,N-Dimethylthiocarbamate, 59
5,6-Dimethyl-3-thio-3-phenyl-3-phosphatricyclo[5.3.2.0]-dodeca-4,8,11-trien-10-one, 86
3,5-Dimethylvalerolactone, 36
L-α-Dimyristoylphosphatidylcholine, 193
Dinucleoside phosphates, 191
α-Diols, 131
6,8-Dioxabicyclo[3.2.1]octane, 96—97
Dioxane, 186
1-(1,3-Dioxan-5-yl)uracil, 113
2,2-Dioxo-1,3,2-dioxathiolane, 57
Dipalmitoyllecithin, 194
L-α-Dipalmitoylphosphatidylcholine, 193
L-α-Dipalmitoylphosphatidylethanolamine, 193
Dipeptides, 111, 185—186, 188
Di(perfluoro-2-propoxypropionyl)methane, 128
Diphenylamine hydrochloride, 166
2,3-Diphenyl-5-carboxymethylisoxazolidine, 89
3,8-Diphenyl-1,2-diaza-1-cyclooctane, 54
7,11-Diphenyl-2,4-diazospiro[5.5]undecan-1,3,5,9-tetranone, 85
2,2-Diphenylethanol, 15
2,2-Diphenylethylamine, 42
1,6-Diphenylhexatriene, 162
2,3-Diphenyl-3-hydroxypropionic acid, 73, 87
3,6-Diphenylimidazol[1,2-b]-as-triazine, 107
r-1,C-2-Diphenyl-*tert*-3-methoxycyclobutane-1,2-diolcarbonate, 93
7,7-Diphenyl-2-oxabicyclo[4.2.0]octane, 40
7,8-Diphenyl-2-oxabicyclo[4.2.0]octane, 40
Diphenyls, 96
Diphenylsulfide, 58

Diphenylsulfine, 57, 120
Diphenylsulfone, 56
2,6-Dipicolinate, 197
Dipicolinic acid, 194
Dipolar shift, 171—172
Dipolar shift equation, 6, 159, 170, 177
Dipropylacetamide, 48
2,6-Di-2-propylacetanilide, 48
Dipropylacetic acid, 38
Disaccharide, 95
Di-spiro(bicyclo[4.2.1]non-3-ene-9,2'-oxirane-3',9"-bicyclo[4.2.1]non-3"-ene, 39
L-α-Distearoylphosphatidylcholine, 193
3,5-Disubstituted nortricyclene, 90
Dithiacyclophane, 119
1,2-Dithiolane-4-methylcarboxylate-1-oxide, 89, 119
1,2-Dithiole-3-thione, 119
n-Dodecyloctaethyleneglycol monoether, 73
Downfield shift reagents, 6—7
Dysprosium, 5—7, 13, 174

E

Effective axial symmetry, 176—177
11-Eicosenol, 14
Eldrin, 39
Electronic effects, 5
 of alcohols, 15
 of amines, 41—42
 of esters, 31
 of ketones, 24
 of nitriles, 46
β-Enaminosulfoxide, 151
Enantiomeric excess, 127
Enantiomeric resolution, 127, 130, 135—136, 167, 203
Enantiomers, quantification of mixture of, 131
Ephedrine, 136—137, 147
Epicubenol, 21
Epi-inositol, 192
Epi-β-santalene, 168
Epoxides, 38, 95, 98
Epoxy-2,3-bicyclo[4.3.0]nonane, 39
Epoxy-2,3,3-bicyclo[3.3.0]octane, 39
1,2-Epoxydodecane, 38
1,2-Epoxyoctane, 38
9,10-Epoxy stearate, 98
Equilibrium constant, 176
Erbium, 5—6, 186
Erythritol, 191
Erythrocyte, 196
Esters, 31—35, 87—92, 142—145
Ethanethiol, 58
Ethers, 38—41, 95—99, 146—147
p-Ethoxybenzylidene-p-n-butyl aniline, 15
1-[(Ethoxycarbonyl)methyl]cyclopentadiene, 33
2-[(Ethoxycarbonyl)methyl]cyclopentadiene, 33
2-Ethoxyethanol, 72
3-Ethoxypropionitrile, 108
Ethyl acetate, 31

Ethylbenzene, 159—160
Ethyl benzoate, 125
2-Ethylbenzothiophene, 165
2-Ethylcyclohexanone, 124
1-Ethyl-5-(1'-cyclohexenyl)-5-methylbarbital, 149
1-Ethyl-1,6-dihydro-2,4-dimethoxy-5-methyl-6-oxo-pyrimidine, 113
Ethyl-2-diisopropylaminoethylmethylphosphono-thioate, 152
Ethyldimethylphosphoramidocyanidate, 152
Ethyl diphenyl phosphine, 64
Ethyldiphenylphosphonite, 63
Ethyldiphenylsulfonium tetrafluoroborate, 166
Ethylene, 122
Ethylenediamine, 201
Ethylenediaminetetraacetic acid, 185, 194, 197
Ethylene glycol bis(aminoethyl)tetraacetic acid, 197
Ethylene oxide, 121
Ethylene thioketal, 87
Ethyl ethane thiolsulfinate, 58
Ethylfluorogermane, 64
Ethyl-2-hydroxy-1-cycloheptane carboxylate, 110
Ethyl-2-hydroxy-1-cyclohexane carboxylate, 110
Ethyl-2-hydroxy-1-cyclooctane carboxylate, 110
Ethyl-2-hydroxy-1-cyclopentane carboxylate, 89, 110
L-γ-Ethylideneglutamic acid, 88
Ethylindole-2-carboxylate, 90
Ethylnitrile, 46
Ethyl 2-phenylcyclopropane carboxylate, 33
3-Ethyl-5-phenyl-2,4-imidazolidinedione, 149
3-Ethyl-3-phenyl-2,6-piperidinedione, 149
4-Ethylpyridine, 201
N-Ethylquinolinium chloride, 65
N-Ethylquinolinium iodide, 166
O-Ethylthioacetate, 59
Europium shift reagents, 5—7, 127, 155, 174—175
 Eu(dcm)$_3$, 129
 Eu(dfhd)$_3$, 64
 Eu(dpm)$_3$, 9, 173—175
 Eu(dpm)$_3$py$_2$, 1
 Eu(fod)$_3$, 10—11, 67, 173—176, 183
 Eu(NO$_3$)$_3$, 185
 Eu(tfn)$_3$, 58
Eusiderin B, 95—96
Experimental techniques
 for achiral shift reagents, 9—14
 for binuclear shift reagents, 157—158
 for chiral shift reagents, 129—131

F

Fatty acids, 31—32, 73, 88
Fenchol, 19
Flavone, 83, 140—141, 200
Fluorenone, 31, 82
Fluorinated alcohols, 15
Fluorine atom, 70
Fluorine-19 NMR spectrum, 15, 32, 40, 132, 188
4-Fluoroaniline, 42

4-Fluorophenol-parvalbumin, 188
2-Fluorophenylisocyanate, 52—53
Formaldehyde, 121
Formamidine sulfinic acid, 197
Formylmethylenthiopyran, 94
Friedelan-3β-ol, 23
Fructopyranoside, 78
Fumarodinitrile, 154
Furan, 38
Furo[2.3-e]-1,4-diazepin-5-one, 112—113

G

Gadolinium shift reagents, 171, 183, 186
   Gd(dcm)$_3$, 147
   Gd(dpm)$_3$, 61, 183
   Gd(fod)$_3$, 66, 183
Galactitol, 191
Galactopyranoside, 78, 92
Gasoline, 160
Geraniol, 15—16
Geranyl acetate, 33
Glaucine, 148
Glucofuranoside, 78, 92
Gluconic acid, 191
Glucopyranoside, 78, 92
α-Glucoseamine pentaacetate, 111
β-Glucoseamine pentaacetate, 111
Glutamic acid, 186—187
Glutaric acid, 191
Glyceric acid, 191
Glycerol, 191
Glycolate, 198, 202
Glycols, 122
Glycyl-L-alanine, 188
Glyme, 95—96
Graphical procedures, 176
Griseofulvin, 85
Guanyl-(3′,5′)-cytidylic acid, 191

H

H(acac), 156
Halides, 64, 165—166
Halogenated compounds, 155, 161
Hamilton R-factor, 181
H(t-cam), 127
H(dcm), 129, 167
H(dfhd), 8, 11
H(dfm), 167
H(dpm), 7—10, 156
Helical chirality, 140
1,10-Henanthroline, 103
3-(Heptafluorobutyryl)-d-camphor, 167
6,6,7,7,8,8,8-Heptafluoro-2,2-dimethyl-3,5-octanedione, 8, 156
1-(2-Heptafluorofuran)-1,1-difluoro-4-(2-thienyl)-2,4-butanedione, 8

4,4,5,5,6,6,6-Heptafluoro-1-(2-furyl)-1,3-butanedione, 156
4,4,5,5,6,6,6-Heptafluoro-1-(2-naphthyl)-1,3-hexanedione, 156
4,4,5,5,6,6,6-Heptafluoro-1-phenyl-1,3-hexanedione, 156
4,4,5,5,6,6,6-Heptafluoro-1-(2-thienyl)-1,3-hexanedione, 156
Heptanenitrile, 46
1,1,1,3,3,3-Hexaalkylsiloxane, 71
6-Hexadecyl-1,2-oxathiane-2,2-dioxide, 57
Hexa-2,4-diene, 162
2,4-Hexadienol, 15
1,1,1,5,5,5-Hexafluoro-2,4-pentanedione, 156
32aB,4,5,6,7,7,-a8-Hexahydro-2-benzofuranone, 35
Hexamethyl-3-hydroxybicyclo[2.2.1]hept-2-ene, 19
Hexamethylphorphoramide, 174
Hexamethylphosphoramide, 64
n-Hexane, 174
n-Hexanoic acid, 37
1-Hexene, 157, 159
2-Hexene, 159
1-Hexene-2,4,6-tricarboxylic acid, 89
Hexobarbital, 149
2-n-Hexyl-5-n-propylresorcinol, 134
H(facam), 128, 167
H(fod), 1, 7—8, 10—11, 156
H(hfa), 156
H(hfbc), 128, 167
H(hfh), 156
H(hfth), 156
Hinokiic acid, 35
Histidine, 186—187
Holmium shift reagents, 5—6, 173, 174
Horrock's equation, 169
H(phd), 156
H(ptb), 156
H(tfa), 156
H(tfb), 156
H(tfn), 8
H(thd), 7
H(tnb), 156
H(tta), 156
Hydantoin, 112
Hydrogen-1 NMR spectrum, 14, 152, 186, 188, 191
Hydrogen-2 NMR spectrum, 138
2-Hydroxyacetophenone, 200
Hydroxyalkyl adamantane, 20
α-Hydroxyamines, 131
Hydroxybenzaldehyde, 74, 94, 200
7-Hydroxybicyclo[4.3.1]decatriene, 21
6-Hydroxybicyclo[3.2.0]hept-2-ene, 20
9-Hydroxybicyclo[3.3.1]nonane, 21
9-Hydroxybicyclo[3.3.1]nonane-3-carboxylate, 75
9-Hydroxy brendan-4-one, 90
3-Hydroxybutyrate, 197
10-Hydroxycamphor, 75
Hydroxy-12β-cananine, 78
α-Hydroxycarboxylates, 202
Hydroxycarboxylic acid, 203

Hydroxycarboxylic acid esters, 131
2-Hydroxycarboxylic acid methyl ester, 133
3-Hydroxycarboxylic acid methyl ester, 133
2-Hydroxy-1-cycloakyl carboxylate, 75
4-Hydroxy-4,5-dihydroaldrin, 21
β-Hydroxy esters, 137
N-Hydroxyethylethylenediaminetriacetic acid, 185, 197
3β-Hydroxy-β-homocholestan-7,8a-lactone, 78, 93
2-Hydroxy-1-(2-hydroxyethyl)adamantane, 77
10-Hydroxyisoborneol, 75
α-Hydroxyisobutyrate, 202
4-Hydroxy-1-isopropyl-2-oxaadamantane, 77
β-Hydroxyketones, 137
Hydroxyl groups, blocking of, 78—79
1-Hydroxymethyl-2-alky-3-ethylidenecyclopropane, 16
2-Hydroxy-2-methylbutanoic acid, 137
2-Hydroxy-2-methylcyclobutanone, 76
1-Hydroxymethyl-3,7-dimethylenebicyclo[3.3.1]nonane, 163
6-Hydroxymethyl-2-methoxy-5,6-dihydro-2H-pyran, 133—134
Hydroxymethyl-5-norbornene, 19
α-Hydroxymonoacetates, 131
Hydroxynaphthaldehyde, 74, 94
7-Hydroxynorbornadiene, 19
7-Hydroxynorbornene, 19
9-Hydroxy-7-oxatetracyclo[6.3.0.0.0]undecane, 77
4-Hydroxypentanoate, 197
3-Hydroxphenylalanine, 203
2-(Hydroxyphenylmethyl)cyclohexanol, 74
Hydroxy-L-proline, 186—187
4-Hydroxypyridine-2,6-dicarboxylic acid, 194
Hydroxy steroids, 22

I

Iditol, 191
Illudin M-acetate, 76
Imidazole, 194
Imidazolium salts, 154
Imides, 49, 112—115
Imines, 54, 118
Iminium salts, 64—66, 154
Impurities, 129, 158
Incremental dilution method, 12—13
Indanol, 17
Indanone, 31
Indol-3-acetate, 199
Indole, 164
Infrared spectroscopy, 174—175
Inokosterone-3,22,26-triacetate, 139
Inositol, 192
Internal reference standard, 10, 15, 186
Iodides, 161
Iodobenzene, 161
1-Iodobutane, 161
1-Iodohexane, 161
Iron complexes, 68—70, 153—154

Isoanhydrovitamin A, 40
Isoborneol, 19
Isobutanol, 15
Isobutylamine, 42
Isobutyramide, 48
Isobutyric acid, 203
3-Isochromanone, 35
Isocyanates, 52—53
Isofenchol, 19
Isoflavone, 83
Isooxazole, 164
Isoprene-nitrile rubber, 122
Isopropylamine, 135
2-Isopropylcyclohexanone, 124
Isopropyl isopropane, 58
Isopropyl isopropane thiolsulfinate, 58
3-Isopropyl-6-methylenecyclohexanol, 138
Isopropylmethylphosphonofluoridate, 152
Isopropylnitrile, 46
Isoquinoline, 44—45
Isoquinoline alkaloids, 148
Isothiocyanates, 59
Isothiouronium salts, 165—166
Iterative procedures, 176

J

Jack-knife test, 181
Job's method, 175
Juvenile hormone, 88, 98

K

Kendall tau statistic, 181
3-Keto androstane, 31
4-Keto androstane, 31
15-Keto androstane, 31
16-Keto androstane, 31
17-Keto androstane, 31
δ-Ketoendrin, 30
Keto groups, blocking of, 87
Ketones, 23—31, 79—87, 140—142
Ketosteroids, 25, 31
β-Ketosulfoxide, 151
Ketoterpenes, 27

L

Lactams, 49, 112—118, 124, 149—150
Lactate, 198, 202
Lactobionic acid, 191
Lactones, 35—36, 92—94, 135—138, 145
Landanosine, 148
Lanthanide anion species, 65
Lanthanide shift reagents
  achiral, see Achiral lanthanide chelates
  binuclear, see Binuclear lanthanide(III)-silver(I) shift reagents

chiral, see Chiral substrates
theory of, 169—184
  applications of pseudocontact shift equation, 177—183
  relaxation phenomena, 183—184
  separation of dipolar and contact shifts, 171—172
  shift mechanism, 169—171
  stoichiometry and symmetry, 172—177
water-soluble, see Water-soluble lanthanide shift reagents
Lanthanide tris β-diketones, 1
Lanthanum shift reagents, 67, 157, 184
Lasalocid A, 199
Lecithin, 193
Lewis acid, shift reagents for, 70
Licarin B, 96
Ligand
  in achiral shift reagent, 7—9
  in binuclear shift reagent, 156—157
  in chiral shift reagent, 127—129
Ligand exchange, 172
Limonene, 162, 168
Linalool, 138, 163
Linkage isomerism, 70
Lithium-7 NMR spectrum, 195
L-S bond length, 180—181
Luminescence spectroscopy, 175, 185, 191
Lupanine, 114
Lutetium shift reagents, 157, 173
Lysine, 185—186
Lysozyme, 189

M

Magnesium-25, 194—197
Malate, 198
Malonate, 198
D-Mandelate, 202
Manganese complexes, 68—69, 153
Mannitol, 191
Mannopyranoside, 78, 92
α-D-Mannose-1-phosphate, 193
α-D-Mannose-6-phosphate, 193
Mass spectrometry, 175—176
McConnell-Robertson equation, 169
Meliacin, 92
Membranes, 185, 193—194, 196
p-Mentha-2,5-dien-7-oic acid, 34
p-Menthan-7-oic acid, 34
Menthene, 168
p-Menth-2-en-7-oic acid, 34
Menthol, 18, 175
Menthone, 27
Mephobarbital, 149
Metal
  in achiral shift reagent, 5—6
  in binuclear shift reagent, 155—156
  in chiral shift reagent, 127
Metal complexes, 66—71, 153
Metal β-diketonates, 66—68
Metal β-ketoimines, 66
Methadone, 147
Methane, 129
Methanol, 15
Methionine, 143
Methohexital, 149
1-Methone, 128
Methorphan, 148
Methoxyacetate, 197—198
Methoxybenzaldehyde, 94
Methoxybenzene, 38, 96, 164
7-Methoxybicyclo[4.3.1]decatriene, 41
1-Methoxy-4-bromo-2-butyne, 40
Methoxy-n-butane, 38, 176
2-Methoxycarbonylspiro(cyclopropane-1,1'-idene), 34
3-Methoxycycloheptanone, 82
3-Methoxycyclohexanone, 82
3-Methoxycyclooctanone, 82
3β-(1'-Methoxyethoxy)-5α-cholestane, 98
3β-(1'-Methoxyethoxy)cholest-5-ene, 98
Methoxyfluorene, 38
Methoxy indanol, 75
Methoxyindanone, 82
o-Methoxymethylbenzoate, 87
Methoxymethylphenylethylalcohol, 73
1-Methoxynaphthalene, 38
2-Methoxynaphthalene, 38
1-Methoxy-2-n-octyloxyethane, 95
6-Methoxyphenalone, 82
Methoxyphenol, 72, 74, 200
p-Methoxyphenyltrimethylsilane, 71
3-Methoxypropionitrile, 108
Methoxy styrene, 41
α-Methoxy-α-trifluoromethylphenylacetic acid (MTPA), 131
α-Methoxy-α-trifluoromethylphenylacetic acid (MTPA) esters, 131—133
N-Methylacetamide, 47—48
Methyl acrylate, 154
Methyl acrylonitrile, 46
1-Methyl-2-adamantanol, 20
Methyladamantanone, 23—24
Methyl alanate, 135
Methyl 2-alkyl-3-ethylidene-1-carboxylatecyclopropane, 33
Methyl α-D-allopyranoside, 192
1-(Methylamino)anthraquinone, 82
2-Methyl-3-aminopinane, 44
Methyl-2-benzaldehyde, 37
Methyl-3-benzaldehyde, 37
Methyl benzene, 160
N-Methylbenzimidazole, 103—104
Methyl benzoate, 33, 89
Methylbenzofuran, 164
Methylbenzylalcohol, 15
p-Methylbenzylamine, 43
Methyl betulinate, 35
2-Methyl-2-butanol, 140
3-Methyl-3-butanone, 25

2-Methyl-2-butene, 159
2-Methyl-5-*tert*-butyl-1,3-dioxane, 96
2-Methyl-5-*tert*-butyl-2-oxo-1,3,2-dioxaphosphorinane, 61
1-Methyl-5-butyl-5-phenylbarbital, 149
Methyl-*tert*-butylsulfoxide, 136
Methyl butyrate, 32
Methyl-*p*-chlorophenylsulfoxide, 136
Methylchrysanthemate, 33
Methyl cinnamate, 32
1-Methylcyclohexanol, 18
3-Methylcyclohexanol, 132
4-Methylcyclohexanone, 27
Methylcyclohexene, 159, 168
3-Methylcyclopentanol, 17
Methyl cyclopropyl ketone, 26
1-Methyl-1,2-diaminoethane, 67
Methyldiazaphenanthrene, 107
2-Methyl-5,6-dihydro-α-pyran-6,6-dicarboxylic acid, 145—146
Methyl-9,10-dihydroxystearate, 73
Methyl-3,11-dimethyl-10,11-epoxy-7-ethyl-2-farnesoate, 88
N-Methyl-2,4-diphenylimidazole, 103
Methyldiphenylsilanol, 71
Methyl dodecanoate, 32
Methyl elaidate, 32
3-Methylene-7-benzylidenebicyclo[3.3.1]nonane, 163, 167
4-4′-Methylenedicinnamic acid, 94
Methylene(2-methyl)cyclohexane, 155, 167
5-Methylenenorborn-2-ene, 162—163
Methyl ephedrine, 136—137
Methyl-10,11-epoxy-2,6-farnesoate, 88
Methyl ethers, 40
2-Methyl-1-(ethoxycarbonyl)cyclohexane, 34
4-Methyl-1-(ethoxycarbonyl)cyclohexane, 34
Methyl ethyl ammonium ion, 196
N-Methylethylenediaminetriacetic acid, 197
Methylfenchol, 19
N-Methylformamide, 48
Methyl α-D-galactopyranosiduronate, 192—193
Methyl α-D-glucopyranoside, 192
Methyl glycoside, 192
Methyl groups, 182
  enantiotopic, 150, 202
Methyl β-D-hamamelopyranoside, 192
Methyl heptanoate, 32
Methyl hexanoate, 32
Methylhydrogen polysiloxane, 123
Methyl-12-hydroxystearate, 73
Methyl-13-hydroxy-9-*cis*-11-transoctadecadienoate, 133
1-Methylimidazole, 164
N-Methylimidazole, 103—104
1-Methyl-2-imidazole carboxaldehyde, 94
2-Methylindanol, 17
7-Methylindanol, 17
N-Methylindole, 164
N-Methylisobutyramide, 47
Methyl isopropyl ketone, 124

Methyl isothiocyanate, 59
Methyl linoleate, 99
O-Methylmandelic acid, 134
Methyl β-D-mannofuranoside, 192
3-Methyl-2-methylaminotetrahydropyran, 101
4-Methyl-2-methylaminotetrahydropyran, 101
2-Methyl-2-methylbutanol, 138
Methyl-3-methylcyclohexene-4-carboxylate, 144
Methyl 2-methyloctanoate, 32
Methyl-2-(2′-methyl-1′-propenyl)tetrahydropyran, 40
1-Methylnaphthalene, 161
Methyl-1,5-naphthyridine, 106
N-Methylnicotinium iodide, 65, 166
Methylnitrile, 46
Methyl nonanoate, 32
Methylnopinol, 20
Methyl oleanolate, 35
Methyl oleate, 31—32
N-Methylpavine, 148
3-Methyl-1-pentene, 167
Methyl petroselinate, 31
1-Methylphenanthrene, 161
9-Methylphenanthrene, 161
Methyl-2-phenylcyclopropane carboxylate, 33, 143
N-Methyl-1-phenylethylamine, 135
α-Methyl-β-phenylmercaptopropiophenone, 140
Methylphenyl polysiloxane, 123
Methylphenylsulfone, 56
Methyl phenyl sulfoxide, 55, 135, 151
N-Methylpiperidone, 49
2-Methyl-2-propanol, 175
N-Methylpropionamide, 47
2-Methyl-2-*n*-propylcyclopentane-1,3-diol, 74
Methylpyrazine, 105
Methylpyridine, 125
2-Methylpyridine, 44
3-Methylpyridine, 173
4-Methylpyridine, 45, 201
3-Methylpyridine-1-oxide, 200—201
N-Methylpyridinium cation, 194
1-Methylpyrrolidine-2-one, 47
N-Methylpyrrolidone, 49
Methyl ricinoleate, 73
1-Methylsilatrane, 71
4-Methylstyrene, 162
Methylsulfide, 58
3-Methyl sulfonylcyclobenzaprine, 102
N-Methylsydnome, 94, 117
2-Methyltetrahydropyran, 40
S-Methyltetrahydrothiophenium iodide, 166
Methyltetrahydrothiophenium ion, 166
S-Methyltetrahydrothiophenium tetrafluoroborate, 166
S-Methylthianaphthenium tetrafluoroborate, 166
Methyl-7-(2-thienylacetamide)-3-methylenecepham-4-carboxylate, 111
2-Methyl-1-*p*-tolylsulfonyl-1,2-epoxybutane, 146
2-Methyl-1-*p*-tolylsulfonyl-1,2-epoxypropane, 146
Methyl-*p*-tolylsulfoxide, 136, 151
1-Methyl-1,2,3-triazole, 104

1-Methyl-1,2,4-triazole, 104
Methyl tricyclo[3.3.0.0]octane-2-carboxylate, 144
Methylverbanol, 20
Methylverbenol, 20
Micelles, 202
Monoalkyl amines, 42
Monoamines, 131
Monocyclic amines, 43
Monocyclic aromatic hydrocarbons, 160—161
Monocyclic ketones, 25
Monocyclic organonitriles, 46
Monocyclic terpenes, 18
Monoglycerides, 138, 142
Monoheptylglycol ether, 73, 96
Monoterpene, 34
Mosher's reagent, 131
MTPA, see α-Methoxy-α-trifluoromethylphenyl acetic acid
Multicyclic amines, 44
Multicyclic carbinols, 139
Multicyclic hydroxyl-containing compounds, 18—19
Multicyclic ketones, 25
Multicyclic olefins, 159—160
Multicyclic polyfunctional compounds, 75
Multicyclic ring compounds, 20, 34
Myo-inositol, 192
Myrtanol, 20
Myrtenol, 20

N

Naphthalene, 161
Naphtho[9]annulenone, 141
Naphthoflavone, 83
Naphthomycin, 77, 115
Naphthoquinone, 82
3-(1-Naphthyl)-5,5-dimethylcyclohexanone, 27
3-(1-Naphthyl)-1,3,5,5-tetramethylcyclohexanol, 17
4[β(1-Naphthyl)vinyl]pyridine, 45
1,8-Naphthyridine, 103, 105—106
Nematic phase, 45
Neodymium, 5—6
Neo-inositol, 192
Neopentanol, 175
Neothiobinupharidine, 119
Nerol, 15—16
Neryl acetate, 33
Nickel complexes, 153—154
Nicotinamide, 109
Nicotinamide mononucleotides, 190
Nicotine, 105
Nitrated lactone, 136
Nitriles, 46—47, 95, 108—109, 164, 182
Nitrilotriacetate, 198, 203
Nitrilotriacetic acid, 185, 194
Nitrobenzene, 52
2-α-(p-Nitrobenzoyl)methylenepyrrolidine, 81
2-Nitrobutane, 131
Nitro compounds, 52, 116
Nitrogen-14, 195

Nitrogen-14 NMR spectrum, 49, 54, 104
Nitrogen-15 NMR spectrum, 111
Nitrogen heterocycles, 44—45, 50, 102—108, 148, 201
Nitrogen oxides, 50—51, 115—116
Nitroisoquinoline, 106
Nitromethane, 15, 52
Nitrones, 115—116, 185
o-Nitrophenol, 200
p-Nitrophenol, 200
Nitrosamines, 150
Nitroso-4-tert-butylpiperidine, 51
N-Nitroso camphidine, 51
Nitroso compounds, 51
N-Nitroso-6,7-dihydro-1,11-dimethyl-5H-dibenz[c,e]azepine, 51
p-Nitrotoluene, 52
3-Nitrotyrosine, 189
4-Nonene, 159
n-Nonylisothiouronium bromide, 166
Nootkatone, 29
Nopinol, 20
d-Nopinone, 128
Norbornadiene, 162, 168
Norbornanes, 18—19
Norbornanol, 34
2-Norbornanol, 18—19
Norbornene, 160
5-Norbornen-2-ol, 19
Norborn-5-en-2-ol, 182
Norbornylamine, 44
2-Norbornylbrosylate, 201—202
Norbornyltrifluoroacetate, 19
Norcamphor, 24
Norephedrine, 147
Norleucine, 143
3-Nortricyclanol, 19
Nortropine, 72, 101
Nortropinone, 101
Norvaline, 143
Nuclear magnetic resonance
 high field, 13—14
 low-temperature, 174
Nuclear Overhauser effect, 36, 100, 184, 190
Nucleic acids, 185, 190—191
Nucleotides, 185, 190—191

O

1-Octanol, 13—14, 134
2-Octene, 159
1-Octylfluoride, 64
n-Octylfluoride, 64
sec-Octylfluoride, 64
2-Octylphenylsulfone, 151—152
2-Octyne, 160
Olefinic ketones, 24
Olefins, 2, 14—15, 155, 159—162, 167
Optical purity, 127, 135—154
Optical spectroscopy, 175

Organic salts, 64—66, 154, 165—167
Organoborons, 150
Organobromides, 64
Organochlorides, 64
Organofluorides, 64
Organohalides, 64, 161
Organoiodides, 64
Organometallics, 68—70, 153—154, 202
Organonitriles, 46
3-Oxabicyclo[3.3.1]nonane, 40
1-Oxadethiapenicillin, 111
Oxadiazine, 135
Oxazaphospholine-2-oxide, 61
1,3-Oxazolidin-2-one, 84, 114
Oxazophospholines, 60—61
Oxides, 45, 200—201
6-Oxidodipyrido[2,1-b:2′,3′-d]thiazolium, 201
9-Oxidodipyrido[2,1-b:3′,2′-d]thiazolium, 201
Oximes, 53, 116—117
β-Oximinosulfoxide, 151
4-Oxo-4,5-dihydroaldrin, 30
2-[4-(1-Oxo-2-isoindolinyl)-phenyl]propionic acid, 134
Oxonium salts, 64—66
2-Oxo-1,2,3-oxathiazan, 58
9-Oxo-13-prostenoic acid, 137—138
Oxycarboxylic acid, 203
Oxydiacetate, 198, 203
Oxydilactate, 203
β,β′-Oxydipropionitrile, 108
Oxygen, basicity of, 67
Oxygen-17 NMR spectrum
  of alcohols, 15
  of cations, 195—198
  of esters, 31
  of lactams, 49
  of lactones, 35, 94
  of sulfur-containing compounds, 57
Oxyloacetate, 198
Oxynitrile, 108

P

Papaverine, 148
Parvalbumin, 188—189
Penicillamine, 147—148
Penicillin, 112
Penicillin benzylester, 111
Penicillin-6-methyl ester, 111
Pentacyclodecane, 97
Pentacyclo[5.3.0.0.0.0]decan-6-ol, 21
Pentacyclo[5.3.0.0.0.0]decan-6-one, 29
Pentafluorobutylamine, 42
2,2,3,4,4-Pentamethyl phosphetan oxide, 62
Pentane, 157
2,4-Pentanedione, 153, 156
n-Pentylisothiouronium bromide, 166
Peptides, 186—189
Perhydrophenalenol, 21
Peridinin, 73

Peroxides, 41, 99
Phenacetin, 48
Phenalenone, 31
Phenanthene, 161
Phenanthridine, 45
1,10-Phenanthroline, 173
Phenolate ion, 200
Phenol-formaldehyde resin, 122
Phenols, 14—23, 33, 89, 200
Phenoxybutyronitrile, 108
Phenyl acetate, 33
Phenylalanine, 186
2-Phenyl-3,1-benzoxathian-4-one, 145
2-Phenyl-2-butanol, 130
Phenyl-*tert*-butyl carbinol, 15
3-Phenylbutyric acid, 32
4-Phenylcyclohexanone, 26
2-Phenylcyclopropane carboxaldehyde, 145
2-Phenylcyclopropane carboxylic acid, 146
2-Phenylcyclopropylamine, 43
α-Phenyl-α-*N*-dimethylnitrone, 51
1-Phenyl-3,4-dimethylphospholene oxide, 62
2-Phenylethanol, 134, 137
1-Phenylethylamine, 128, 134—135, 147, 153
α-Phenylethylamine, 135
1-Phenylethyl sulfoxide, 55
N-Phenylmaleimide, 85
2-Phenylmercaptoethyl methyl ketone, 140
2-Phenylmercaptopropyl phenyl ketone, 140
Phenyl-1-(3-methyl-2-butenyl)sulfone, 152
3-Phenyl-7-methylcoumarin, 35—36
N-(2-Phenyl-1-methyl)propyl-*N*-phenylamine, 43
Phenylnitrile, 46
bis[3-Phenyl-5-oxoisoxazol-4-yl]pentamethine-oxonol, 73
8-Phenyl-8-oxo-8-phosphabicyclo[3.2.1]-octan-3-one, 120
1-Phenylphospholene oxide, 62
2-Phenyl-1-propanol, 15
2-Phenylpropionaldehyde, 37, 145
α-Phenylpropionaldehyde, 145
2-Phenylpropylamine, 42
2-Phenylpropyl methyl ether, 40
N-Phenylpyridinium perchlorate, 65
bis(Phenylsulfinyl)methane, 120, 151
1-Phosphabicyclo[2.2.1]heptane 1-oxide, 62
1-Phosphabicyclo[2.2.2]octane 1-oxide, 62
Phosphates, 60, 185
Phosphatidyl choline, 193—194
Phosphatidyl glycerol, 193
Phosphatidyl inositol, 193—194
Phosphatidyl serine, 193—194
Phosphetan oxide, 62
Phosphine oxide, 61—63, 185
Phosphines, 64, 99, 155, 162
Phosphites, 63
Phosphoglycerate kinase, 189
Phospholane 1-oxide, 61
Phospholipid membranes, 193—194
Phosphonates, 60—61, 120, 185
Phosphonium bromide, 66

Phosphonium chloride, 66
Phosphonium ion, 196
Phosphonium salts, 64—66
Phosphonomannan, 193
Phosphoramides, 64
Phosphoranes, 63
Phosphorinanes, 60—61
Phosphorus-31 NMR spectrum, 60—64, 152, 162
  of alcohols, 139
  of carbohydrates, 191—193
  of membranes, 194
  of nucleotides, 191
Phosphorus-containing compounds, 60—64, 120—121, 152—153, 162
Phosphoryl group, 120
Photodieldrin, 39
Phthalazino[2.3-b]phthalazine-5,12-(14H,7H)-dione, 114
3-Picoline, 174
4-Picoline, 6, 173
β-Picoline, 176
8-Picoline N-oxide, 50
Pimarol, 23
Pinacolone, 182
α-Pinene, 159—160, 168
β-Pinene, 159—160
Pinitol, 89
Pinocarveol, 20, 182
Piperidine, 42—43, 51, 81, 89, 200
Piperidinium ion, 196
Piperidinium salts, 165
Piperine, 110
Pivalophenone, 142
N-Pivaloylamide, 47
Platinum-195 NMR spectrum, 68
Platinum complexes, 154
Polyamides, 110, 122
Polyazanaphthalene, 106
Polybutadiene, 163
Poly(*tert*-butylmethacrylate), 122
Poly[β-(2-cyanoethyl)β-propiolactone], 122
Polycyclic aromatic hydrocarbons, 160—161
Polycyclic nitrogen heterocycles, 103
Polyethylene glycol, 121
Poly(ethyleneisophthalate), 122
Poly(ethyleneterphthalate), 122
Polyfunctional steroids, 77—78, 86, 92
Polyfunctional compounds, 5, 74, 90, 163—165
Polyglycoldimethyl ether, 95
D-Poly-β(−)-hydroxybutyrate, 122
Polymers, 89, 121—123
Poly(methylmethacrylate), 122
Poly(methylpolysiloxane), 70
Poly(β-methyl-β-propiolactone), 122
Polyoctenylene, 163
Polyoxadioxo[n]paracyclophane, 91—92
Polypentenylene, 163
Polyphenols, 200
Poly(propylene oxide), 121
Poly(vinylacetate), 122
Poly(vinylethylether), 122

Poly(vinylmethylether), 122
Porphyrin, 77, 186
Potassium, 196
Potassium-39, 194—197
Praseodymium shift reagents, 5—7, 127, 174
  Pr(dpm)$_3$, 14, 175
  Pr(dpm)$_6$, 173
  Pr(facam)$_3$, 173
  Pr(fod)$_3$, 156, 174, 176, 183
  Pr$_2$(fod)$_6$, 173
  Pr(hfbc)$_3$, 167
  Pr(NO$_3$)$_3$, 185
Pregnadienone, 78
Preparation
  of achiral shift reagents, 10—11
  of binuclear shift reagents, 157—158
  of chiral shift reagents, 129
Presqualene alcohols, 16
Primary amines, 99—100
Principal magnetic axis, location of, 180
Proline, 186—187
Propanoate, 197
2-Propanol, 140, 176
Propanthiol, 58
1-Propylamine, 175—176
n-Propylamine, 42
Propylenediaminetetraacetic acid, 203
Propylene oxide, 146
1-Propylfluoride, 64
2-Propyl phenyl sulfoxide, 55
Prostaglandins, 73
Proteins, 185—189
Protons
  diastereotopic, 15, 27, 38, 43, 166
    separation of, 124
  enantiotopic by internal comparison, 146, 151, 153
Pseudocontact shift, 1, 169
Pseudocontact shift equation, 2—3, 11, 169—171
  application of, 177—183
  simplified form of, 170
Pseudoephedrine, 147
Pseudotropine, 72
Psicopyranoside, 78
Pulegols, 18
Pulegone, 128
Purification
  of achiral shift reagents, 10—11
  of binuclear shift reagents, 157—158
  of chiral shift reagents, 129
3-H-Pyrano[3,4-f]-quinoline-3-one, 106
Pyranosides, 78
Pyrazine, 105
Pyrazolotropone, 85
Pyrene, 161
Pyridazine, 105—106
Pyridine, 44—45, 105, 173—175, 178, 183—184
Pyridine-2,6-dicarboxylate, 185
Pyridine N-oxide, 50
Pyridinium ion, 196
Pyridoxylalanine, 186

Pyridoxylasparagine, 186
Pyridoxylaspartic acid, 185—186
1-Pyridylalkanol, 137
1-(2-Pyridyl)benzotriazole, 104
1-(4-Pyridyl)-1-phenylethane, 148
1-(2-Pyridyl)-1-phenylpropane, 148
Pyrimidine, 105
Pyrimidine 1-oxide, 51
Pyrimidine 3-oxide, 51
Pyrimidine N-oxide, 115
Pyrrole, 42, 164
Pyrrolidine, 42—43, 89
Pyrrolidinium ion, 196
Pyrroline, 43
Pyrromethene, 89
Pyruvate, 198

Q

Quinine, 75
Quinoline, 44—45, 123
Quinoline N-oxide, 50
Quinone, 79—80
Quinuclidine, 44, 170, 173

R

Ratio method, 171
Reagent selection
   achiral shift, 5—9
   aqueous shift, 185—186
   binuclear shift, 155—157
   chiral shift, 127—129
Relaxation data, 22, 61, 66, 159, 186—188, 198
Relaxation phenomena, 183—184
Reserpine, 91
Retinol, 16, 33
Rhodopin, 16
Ribitol, 191
Ribofuranoside, 78
9-β-D-Ribofuranosyladenine-5′-monophosphate, 190
Ribopyranoside, 78
Ribose-5-phosphate, 190
Rotation, 47, 50
   barriers to, 18, 149
   free, 182
Rubidium-87, 195

S

Salsolidine, 148
Samarium, 5—6, 174
Sarcosine, 186
Scatchard plot, 175
Scavengers, 11—12, 129
Secondary alcohols, 72
Secondary amines, 72, 100—101
Serine, 186

Sesquinorbornene, 160
Shift mechanism, 169—171
Shift reagents
   achiral, see Achiral shift reagents
   aqueous, see Aqueous shift reagents
   binuclear, see Binuclear shift reagents
   chiral, see Chiral shift reagents
Significance testing, 181
1-Silacyclopentadienyl ring, 69
Silanes, 71
Silanols, 70
Silatranes, 71
Silicon-containing compounds, 70—71
Silver heptafluorobutyrate, 155
Silver shift reagents, 156—157, 167
Silver trifluoroacetate, 155
Sodium, 196
Sodium-23, 194—197
Sodium dodecylsulfate, 202
Solvents
   for achiral shift reagents, 9—10
   for binuclear shift reagents, 157
   for chiral shift reagents, 129
   nematic, 15
   purity of, 10
Sorbic acid, 32
Sorbitol, 191—192
Spearman's r-test, 181
Sphingomyelin, 193—194
Spiro[3,4]octan-1-one, 29
Spirophosphorane, 63
Steric effects
   in alcohols, 15
   in amines, 41—42
   in ketones, 23—24
   in metal complexes, 67
   in nitriles, 46
   in olefins, 159—160
Steric hindrance, 5, 105
Steroids, 132
   esters, 35
   hydroxy, 14, 22, 77—78
   polyfunctional, 86, 92
Stoichiometry, 172—177
Structure, of binuclear shift reagents, 158—159
Styrene, 122, 159, 162
Substrate, recovery of, 13
Sucrose, 92
Sulfanate esters, 56
Sulfate diesters, 56—57
Sulfides, 58, 95, 118—120
Sulfilimines, 150
Sulfines, 57—58, 150—151
Sulfites, 56—57
Sulfones, 56, 118—120, 145, 150
Sulfonium salts, 64, 66, 165—166
β-Sulfonylsulfoxide, 151
Sulfoxides, 55, 118—120, 135—136, 150—151
Sulfoximes, 150
Sulfur-containing compounds, 54—59, 118—120, 150—152, 165, 201—202

Sulphinamates, 58
Sulphinamides, 150
Sultones, 56—57
Symmetry, 172—177

T

Talopyranoside, 78
Temperature, 131, 143, 169—171
Terbium, 5—6
Terpenes, 18
p-Terphenyl, 161
Terpin-1-en-4-ol, 138
Tertiary amines, 72, 101—102
Testosterone, 31
1,4,7,10-Tetraazacyclododecane-N,N',N'',N'''-tetraacetic acid, 185, 194, 197
1,4,8,11-Tetraazacyclotetradecane-N,N',N'',N'''-tetraacetic acid, 185
1,3,5,8-Tetraazanaphthalene, 103
Tetra-n-butylammonium iodide, 166
Tetracycline, 201
1,1,1,2,2,3,3,7,7,8,8,9,9,9-Tetradecafluoro-4,6-nonanedione, 8
Tetraethylammonium N,N-diethyldithiocarbamatolanthanate, 170, 172
Tetrafluoroborate salts, 165
2,2,3,3-Tetrafluoropropanol, 15
Tetrahydroanemonin, 93
Tetrahydrodicyclopentadiene-9,10-diol, 76
Tetrahydrofuran, 38, 70
Tetrahydrofuranyl-5-fluorouracil, 149
1,2,3,4-Tetrahydronaphthalene-oxide, 39
5,6,7,8-Tetrahydro-α-naphthoic acid, 33
Tetrahydropalmatine, 148
Tetrahydropavine, 148
3β-Tetrahydropyranyloxy-5α-cholestane, 98
3β-Tetrahydropyranyloxycholest-5-ene, 98
Tetrahydropyrimidine, 149
1,2,7,11b-Tetrahydropyrrolol[1.2-d][1,4]benzodiazepine-3,6(5H)dione, 76
1,2,4,5-Tetramethylbenzene, 160
1,1,3,3-Tetramethylbutylphenoxyethanol, 73
4,4,7,7-Tetramethylcyclononanone, 28
3,3-Tetramethylenedioxydipropionitrile, 108
Tetramethylenesulfide, 56
Tetramethylenesulfone, 56
Tetramethylenesulfoxide, 56
2,2,6,6-Tetramethyl-3,5-heptanedione, 8, 156
3,4,5,6-Tetramethylphenanthrene, 168
Tetramethylsilane, 10
Tetramethylurea, 174
Tetramisole, 148
2,3,4,5-Tetraphenylcyclopentadienone, 85
Tetra-p-sulfonatophenylporphin, 194
Tetrethylammonium chloride, 65
Theory, of binuclear shift reagents, 158—159
Thiacarbamates, 58
3-Thiadecalone, 85
4-Thiadecalone, 85

Thiamylal[5-allyl]-5-(1-methylbutyl)-2-thiobarbituric acid, 149
Thiazole, 164
α-(2-Thienyl)ethylamine, 135
Thietane-1-oxide, 55
Thioacetates, 59
2-Thioadamantanone, 59
Thioaldehydes, 69
Thioamides, 59
Thiocarbamates, 59
Thiocarbamic acid, 59
Thiocarbonyls, 150
Thiocyanates, 59, 70
Thioketones, 59
2-Thiolene 1,1-dioxide, 56
3-Thiolene 1,1-dioxide, 56
Thiols, 58
Thiosulfinates, 58
Thionyl group, 87, 119
Thiophene, 165
Thiourea, 59
Threitol, 191
Threonine, 186
Thujanebicyclo[3.1.0]hexane, 20
Thulium, 5—7, 174, 186
3',5'-Thymidine diphosphate, 191
Tin complexes, 68, 70
Tocopherol, 33
Tocopherol acetate, 90
Toluamide, 49
o-Toluate, 198
p-Toluate, 198
Toluene, 160
Toluic esters, 33
o-Toluidine, 42
1-$para$-Tolyl-2-phenyl-1-propane, 25—26
Trachyloban-19-ol, 22
Transfer RNA, 191
Transition metal chelates, 171
Trehalose, 92
Trialkoxysilane, 71
Trialkyl amines, 42
Trialylsulfonium salt, 66
1,4,7-Triazacyclononane-N,N',N''-triacetic acid, 185, 194
1,4,5-Triazanaphthalene, 103
Triazole, 104
Triazolobenzodiazepine, 104
bis-Tricarbonylchromium complexes, 69
Tricarbonyl(3,5-heptadien-2-ol)iron, 69
Tricarbonyl[1,2,3,4,-n-2-methoxy-1,3-cyclohexadiene]iron, 153
Trichloroacetylcarboxamide, 89
Tricyclic compounds, 90
Tricyclic dodecatriene, 21
Tricyclo[4.4.1.1]dodecan-11-ol, 21
Tricycloheptane, 34—35
Tricyclononadiene, 21
Trieneone, 86
Triethylamine, 174
Triethylenetetraaminehexaacetic acid, 194

Triethylphosphate, 60
Triethyl phosphine, 64
Triethyl phosphite, 63
Triethyl phosphonate, 60
Triethylthiophosphate, 60
Trifluoperazine, 196
Trifluoroacetate ester, 79
N-Trifluoroacetyl-d-alanine, 38
3-(Trifluoroacetyl)-d-camphor, 167
Trifluoroaniline, 42
Trifluoroethyl ester, 32
Trifluorethyl ether, 40
4,4,4-Trifluoro-1-(2-furyl)-1,3-butanedione, 156
Trifluoroindanol, 17
4,4,4-Trifluoro-1-(2-naphthyl)-1,3-butanedione, 156
1,1,1-Trifluoro-2,4-pentanedione, 156
4,4,4-Trifluoro-1-phenyl-1,3-butanedione, 156
2,2,2-Trifluorophenylethanol, 135
4,4,4-Trifluoro-1-(2-thienyl)-1,3-butanedione, 156
Triglycerides, 88, 142—143
Triglycine, 188
3,7,12-Trihydroxycholanate, 78
Trimethoxy averufin, 82
2,4,6-Trimethylacetophenone, 25
N,N,N-Trimethylanilinium ion, 196
1,2,3-Trimethylbenzene, 160
1,2,4-Trimethylbenzene, 160
1,3,5-Trimethylbenzene, 160—161
4,4,6-Trimethylbicyclo[4.1.0]heptan-2-ol, 20
Trimethylcarbamate, 50
3,5,5-Trimethyl-3-(p-chlorophenyl)cyclohexanone, 27
3,3,5-Trimethyl-2-cyclohexen-1-ol, 138
3,7,11-Trimethyldodecanoic acid, 142
N,N,N-Trimethyldodecylammonium chloride, 65
1,5,5-Trimethyl-3-(1-naphthyl)cyclohexanol, 17
Trimethylphosphate, 60
Trimethyl phosphonate, 60
1,2,2-Trimethylpropylmethylphosphonofluoridate, 152
2,4,6-Trimethylpyridine, 201
Trimethylsilylated sugar, 71
Trimethylsilyl ether, 79
3-Trimethylsilylpropionate, 186
Trimethylsulfonium iodide, 166
4,7,7-Trimethyltricyclo[2.2.1.0]heptan-3-ol, 20
Tripeptides, 111, 188
Triphenyl phosphine, 64, 162
Triphenyl phosphine oxide, 200
Triphenylphosphite, 63, 162
1,2,5-Triphenylphosphole oxide, 62
Tripolyphosphoric acid, 194
Tris(3-nitroso-2,4-pentanedionato)cobalt, 153
Triterpenoid, 23
Tri(m-tolyl)arsine oxide, 200
Tropinone, 102
Troponin C, 189
Trypsin, 189
Trypsin inhibitor, 189
Trypsinogen, 189
Tryptophan, 186

Tungsten complexes, 70

U

Undecylic acid, 37
Upfield shift reagents, 6—7
Urethane, 47
Uridine-3'-monophosphate, 191
12-Ursen-28-oate, 35
12-Ursen-11-one, 31

V

Valeranone, 29
n-Valerate, 197
Valine, 186—187
Valinomycin, 115
Vapor phase osmometry, 174
Verbanol, 20, 124
Verbenol, 20
Vinyl acetate, 122
Vinyl aldehyde, 36—37
Vinyl chloride, 122
4-Vinylcyclohexene, 162, 168
8-Vinyl-8-hydroxytricyclo[5.3.0.0]deca-3,5-diene, 21
Vinyl phosphate, 60
Vinly phosphonate, 60
N-Vinylpyrrolidone, 49
Vitamin A, 33
Vitamin D, 22, 35

W

Water-soluble lanthanide shift reagents, 185—203
  applications
    to biological systems, 186—197
    other, 197—202
  chiral, 202—203
  selection of, 185—186
Wrong way shift, 6, 171
  in amides, 47
  in azoxy compounds, 115
  in carbohydrates, 192
  in ketones, 25

X

Xanthone, 82
Xanthose, 200
X-ray crystallography, 173—174
Xylene, 155
m-Xylene, 160
o-Xylene, 160—161
p-Xylene, 160—161
2,6-Xylidine, 42
Xylitol, 191—192

Xylopyranoside, 78, 92

Y

Ytterbium shift reagents, 5—7, 127, 155, 174, 186

Yb(dpm)$_3$, 173
Yb(facam)$_3$, 167
Yb(fod)$_3$, 13, 156, 175
Yb(hfbc)$_3$, 167
Yb(tfn)$_3$, 31

PROPERTY OF
SIGMA-ALDRICH COMPANY LTD